# High-Temperature Superconductivity

# High-Temperature Superconductivity

Edited by
V. L. Ginzburg
and
D. A. Kirzhnits
P. N. Lebedev Physical Institute
of the Academy of Sciences of the USSR

*Translated from Russian by*
**A. K. Agyei**
University of Ghana
Legon – Accra, Ghana

*Translation edited by*
**Joseph L. Birman**
City College of New York
New York, New York

SPRINGER SCIENCE+BUSINESS MEDIA, LLC

Library of Congress Cataloging in Publication Data

Problema vysoko-temperaturnoi sverkhprovodimosti. English. High-temperature superconductivity.

Translation of a rev. and updated version of: Problema vysoko-temperaturnoi sverkhprovodimosti.
Bibliography: p.
1. Superconductivity. 2. Materials at high temperatures. I. Ginzburg, V. L. (Vitalii Lazarevich), 1926– . II. Kirzhnits, D. A. (David Abramovich). III. Agyei, A. K. IV. Birman, Joseph Leon, 1927– . V. Title.
QC612.S8P7613 537.6′23 82-5295
AACR2
ISBN 978-1-4613-3917-5 ISBN 978-1-4613-3917-5 (eBook)
DOI 10.1007/978-1-4613-3917-5

Contributing Authors:

L. N. Bulaevskii, V. L. Ginzburg, D. I. Khomskii,
D. A. Kirzhnits, Yu. V. Kopaev, E. G. Maksimov, and G. F. Zharkov

The original Russian text, published by *Nauka* in Moscow in 1977, has been expanded and revised by the authors for the present edition. This translation is published under an agreement with VAAP, the Copyright Agency of the USSR.

PROBLEMA VYSOKOTEMPERATURNOI SVERKHPROVODIMOSTI
V. L. Ginzburg and D. A. Kirzhnits

ПРОБЛЕМА ВЫСОКОТЕМПЕРАТУРНОЙ СВЕРХПРОВОДИМОСТИ


Originally published by Consultants Bureau, New York in 1982

# English Editor's Foreword

I first learned of the existence of this book on high-temperature superconductivity when I received a copy in the office of one of the co-editors, Prof. V. L. Ginzburg, shortly after publication. I had known of the work on problems and prospects of achieving high-temperature superconductors by the members of the I. E. Tamm Department of Theoretical Physics of the P. N. Lebedev Physical Institute. I was naturally anxious to read and study this volume, which integrates the work of more than a decade.

Lest one think that the contributions contained here are of the nature of a reflective looking backward, two important considerations should be kept in mind. First, achievement of high-$T_c$ superconductivity is very much a current and future goal. Elsewhere, one of the authors has described it in these words: "Yes, high-temperature superconductivity is a dream, but a sufficiently realistic one." Second, the current physics literature contains reports of new and astounding findings—perhaps some of these will later be recognized as precursors to achieving the "dream." As I write this, the latest *Physical Review Letters* (11 August 1980) carries a report suggesting that the well-known insulator CdS, which has been investigated for decades for photoluminescence, photoconductivity, and Raman and Brillouin light scattering, may be a high-temperature superconductor or at least a "superdiamagnet!"

Many basic scientific problems remain before we reach a firm understanding of the *sina qua non* for high-$T_c$ superconductors, or of the diamagnetic anomalies in CdS and similar anomalies reported a few years ago for CuCl. I commend this book warmly to the English-speaking and -reading audience interested in these problems.

Joseph L. Birman
*Physics Department*
*City College of the City University of New York*

# Preface to the English Edition

The Russian edition of this book reflected the state of the problem of high-temperature superconductivity at the beginning of 1977. It is natural, therefore, that in preparing the book for translation we should endeavor to make additions and changes that reflect the results published over the last few years. The greatest progress during this time has occurred in the understanding of the problems connected with the electron–phonon interaction in metals and with lattice stability. Therefore, the contents of Chapters 3 and 4 of the book required the greatest changes and additions.

However, the schedule requirements and limitations imposed upon the process of publishing a translated work precluded a fundamental revision of the text, and we had to limit ourselves to relatively short additions to the text and to the expansion of the references.

We hope, nevertheless, that the translation of the book will be of benefit to readers and will help stimulate efforts in the solution of the important and interesting problem of high-temperature superconductivity.

L. N. Bulaevskii
V. L. Ginzburg
D. I. Khomskii
D. A. Kirzhnits
Yu. V. Kopaev
E. G. Maksimov
G. F. Zharkov

# Preface

The problem of high-temperature superconductivity is one of the most interesting physical problems, and one of the problems that is potentially important from the point of view of technical applications. Although research in this field has been going on for about 10 years now, it has been growing relatively slowly and has not yet led to any spectacular results; there nevertheless has been progress and the volume of published work is increasing all the time. It is quite probable that this trend will continue in the next few years, even if there are no major successes, such as the production of superconductors with a critical temperature tens of degrees higher than the presently known highest critical temperature.

In such a situation, it is hardly necessary to justify the publication of this book, which attempts to present the current state of the problem. The authors form the nucleus of a group engaged in research on the theory of superconductivity at the I. E. Tamm Department of Theoretical Physics of the P. N. Lebedev Physical Institute, Academy of Sciences of the USSR. Although we do not limit ourselves to the investigation of the problem of the critical temperature of superconductors and the means of raising it, this theme is the focus of our attention. Therefore, writing this book was for us an organic part of our work in progress.

The problem of high-temperature superconductivity is now beginning to interest an ever-increasing number of people, not all of whom are highly qualified physicists. Therefore, Chapter 1 (written by V. L. Ginzburg) is devoted to the formulation of the problem and an elementary introduction to it. This chapter also present the general characteristics of the problem in light of the results obtained (the results are described in greater detail in the other chapters).

Chapter 2 (by D. A. Kirzhnits) is devoted to the exposition of the general theory of superconductivity in the form widely used by us (the dielectric formalism). It is only when such a formalism is used that the critical temperature turns out to be especially clearly connected with the parameters of the metal.

In Chapter 3 (by E. G. Maksimov and D. I. Khomskii) the problem of describing the electron–phonon system in metals and the question of crystal-lattice stability are discussed.

Then Chapter 4 (by E. G. Maksimov and D. I. Khomskii) gives the present state of the problem of the computation of the critical temperature, $T_c$, for three-dimensional quasiisotropic metals as a function of the parameters characterizing the metal in the normal state.

The problem of lattice stability is also touched upon in Chapter 5 (by Yu. V. Kopaev), where the primary topics of discussion are the electronic phase transitions and their connection with the critical temperature. Chapter 6 (by L. N. Bulaevskii) is devoted to quasi-two-dimensional and layered compounds; Chapter 7 (by L. N. Bulaevskii), to one-dimensional and quasi-one-dimensional systems; Chapter 8 (by G. F. Zharkov), to dielectric–metal–dielectric "sandwiches." The problems touched upon in Chapters 6 and 8 are especially closely connected with each other. In Chapter 9 (by Yu. V. Kopaev) nonequilibrium superconductivity, which can set in, for example, during laser irradiation, is considered.

Considerable assistance was rendered in the writing of Chapters 5 and 8 by A. I. Rusinov and Yu. A. Uspenskii, respectively.

The general editing of the text was by V. L. Ginzburg and D. A. Kirzhnits, but all the authors participated in it. A good deal of work was done by L. N. Bulaevskii in preparing the manuscript for publication.

Thus, the book is indeed an integrated work, although it undoubtedly does retain in many respects the characteristics of a collection of papers by different authors. To the extent that these characteristics do not hinder understanding the book, we did not try to eliminate them, and, in particular, we had no misgivings about the presence of repetitions that facilitate the reading.

Let us emphasize that the above-noted structure of the book is to a large extent a reflection of the state of the problem of high-temperature superconductivity itself. The range of superconducting structures in question is unusually broad, and it is still impossible to single out any one of the known directions of search for high-temperature superconductors with any degree of confidence.

The authors are aware of a number of shortcomings of the book and the possibility that it may quickly become obsolete. But we are convinced of the importance of the problem of high-temperature superconductivity and of the appropriateness of the presentation of the investigations in this field. Therefore, we hope that our efforts have not been in vain and that the publication of the book is justified.

In conclusion, the authors take this opportunity to thank all those (and especially the reviewer of the book, I. O. Kulik) who by their comments contributed to the improvement of the manuscript.

*September 1976* *The Authors*

# Contents

# Notation

| | |
|---|---|
| $T_c$ | Critical temperature† |
| $T_p$ | Structural transition temperature |
| $v_F$ | Fermi velocity |
| $p_F$ | Fermi momentum |
| $q_F, k_F$ | Fermi wave number |
| $\Delta$ | Energy gap |
| $E_F$ | Fermi energy |
| $\varepsilon(\omega, q)$ | Permittivity (or dielectric function, or permeability) |
| $\Theta_D$ | Debye temperature |
| $\omega, \Omega$ | Frequency |
| $N(0)$ | Density of states |
| $V$ | Electron interaction |
| $\Pi$ | Polarization operator |
| $\varkappa$ | Inverse Debye screening radius |
| $G$ | Electron Green function |
| $D$ | Phonon Green function |
| $\Sigma$ | Electron self-energy |
| $\lambda$ | Dimensionless electron–phonon interaction constant |
| $f(E)$ | Electron distribution function |
| $N(\omega)$ | Phonon distribution function |
| $E$ | Energy |
| $N$ | Electron density |
| $F$ | Free energy |
| $\sigma$ | Conductivity |

†In some chapters it is assumed that $\hbar = 1$ and $k_B = 1$, i.e., that the temperature varies in energy units; $k_B$ is the Boltzmann constant.

# 1

# The Problem of High-Temperature Superconductivity (General Review)†

## 1. Introduction

Superconductivity was discovered in 1911, and the history of the study of this phenomenon can be divided (arbitrarily) into three periods. The first period (from 1911 through 1945) is characterized by a modest scale of investigations and the absence of practical applications. The study of superconductivity was difficult, especially because of the necessity of working with liquid helium. Before World War II, liquid helium could be produced in only about 10 laboratories in the world (there were two such laboratories in the USSR). Further, the parameters of superconductors–the critical temperature $T_c$, the critical magnetic field $H_c$, and the critical current $j_c$–for the then known materials did not allow the construction of powerful magnets or other practical devices. Finally, superconductivity remained a puzzling phenomenon since no consistent macroscopic, let alone, microscopic, theory of superconductivity has as yet been developed.

The situation completely changed in the second period, which began after World War II and ended about 10 years ago. The production of liquid helium and, hence, the feasibility of the study and use of superconductors ceased to be a problem, except for the cost and scarcity of helium. The nature of superconductivity was understood, and the corresponding theory was developed in a number of directions even better than was the case for many other fields of solid state physics. Superconducting alloys having critical temperatures $T_c$ as high as 18 K, and remaining superconducting in fields $H_c$ up to 100-200 kG (and in even stronger fields) and for quite high critical currents $j_c$ were discovered. As a

†In this chapter the author used in part the text of his earlier review [1].

TABLE 1.1

| Material | $T_c$, K | Year of discovery of superconductivity |
|---|---|---|
| Hg | 4.1 | 1911 |
| Pb | 7.2 | 1913 |
| Nb | 9.2 | 1930 |
| $Nb_3Sn$ | 18.1 | 1954 |
| $Nb_3(Al_{0.75}Ge_{0.25})$ | 20–21 | 1966 |
| $Nb_3Ga$ | 20.3 | 1971 |
| $Nb_3Ge$ | 23.2 | 1973 |

result, superconducting magnets for fields as high as 100 kG, as well as a number of other superconducting instruments and devices, have been constructed.

The third period in the study of superconductivity, which began about 10 years ago is characterized, first, by the production of materials with still higher $T_c$ values [of the presently known materials the highest $T_c$ ($\approx$ 23 K) is possessed by $Nb_3Ge$] and critical fields $H_c \gtrsim 600$ kG [2]. The second, and especially important characteristic, is that problems pertaining equally well to physics and materials science have now become the major and decisive problems. We are referring to the feasibility of controlling and varying the superconducting parameters and the possibility of understanding the factors governing them. The problem of the critical temperature $T_c$ is an especially acute one.

The point is that the values of $H_c$ and $j_c$ for superconducting materials have, over the history of the investigations of superconductivity, increased hundreds of times. For example, for the first superconductor discovered–mercury–even at absolute zero, $H_c \approx 400$ G. Now, however, many alloys are known for which the values of $H_c \approx$ 100–200 kG, and, thus, the "magnetic-current barrier," which prevented the wide application of superconductors, has been overcome. At the same time, the known $T_c$ values have increased over the past 60–65 years by only a factor of 5 to 6 (Table 1.1) and, above all, the use of superconductors still requires cooling by liquid helium. More precisely (as can be seen from Tables 1.1 and 1.2), in principle, liquid hydrogen can also be used now for the cooling, but this is still impracticable, with the exception, perhaps, of the case when we are working with the $Nb_3Ge$ alloy. The point here is that liquid hydrogen can be used only under pumping (the boiling point, $T_b$, of liquid hydrogen at atmospheric pressure is equal to 20.3 K, while the melting point $T_m$ = 14 K). Moreover, it is not advantageous to work with superconductors in the region of

TABLE 1.2

| Substance: | He | $H_2$ | Ne | $N_2$ | $O_2$ | $H_2O$ |
|---|---|---|---|---|---|---|
| $T_b$, K: | 4.2 | 20.3 | 27 | 77.4 | 90 | 373.46 |

temperatures close to $T_c$, and sometimes it is even impossible because of the lowering of the values of $H_c$ and $j_c$ (let us recall that as $T \longrightarrow T_c$ the parameters $H_c \longrightarrow 0$ and $j_c \longrightarrow 0$).

At present the main obstacle standing in the way of a still wider use of superconductivity is the existence of the "temperature barrier"–the fact that for the known materials $T_c \lesssim 23$ K. It is natural, therefore, that the elucidation of the causes impeding the production of materials with $T_c > 20$–$23$ K and the attempts to overcome the corresponding difficulties are at the focus of attention. Here, of course, the problem is on the whole a broader one and consists of the general development of the "materials science of superconductors." Below, however, we shall consider, at least explicitly, only the problem of the critical temperature.†

Here we can distinguish two problems. First, we can try to achieve temperatures $T_c \approx 25$ K (perhaps even 30 K) with the aid of the traditional methods of production and adaptation of new alloys. If we are indeed able to produce such superconductors and utilize them fairly successfully in technology, then there will arise simultaneously the possibility of the wide use of liquid hydrogen for their cooling. The stage of "hydrogen" or "medium-temperature" superconductivity will thereby be opened. The object of this stage will consist, in particular, in the construction of liquid-hydrogen-cooled superconducting magnets for the production of fields in the hundred-kilogauss range. Such success could be of decisive importance, say, for the construction of magnetic systems in facilities for controlled thermonuclear fusion.

Second, it is necessary to investigate the possibility of radically raising the $T_c$ with new types of materials or systems. Specifically, we have in mind the possibility of producing "high-temperature" superconductors with $T_c \gtrsim 90$ K, which can be cooled by liquid air (nitrogen) or even superconductors with critical temperatures of the order of room temperature, i.e., with $T_c \approx 300$ K.

The "dream" of high-temperature superconductivity probably arose long ago, but the first extensive, realistic discussion of it occurred, as far as we know, in 1964. We have in mind the suggestion to investigate the the superconductivity of long organic molecules [3] or superconductivity in the vicinity of metal-dielectric interfaces [4] using, above all, the electronic (exciton) mechanism instead of the conventional phonon mechanism, the dominant mechanism in superconductors, to provide attraction between the conduction electrons near the Fermi surface. (For greater details about this, see [5] and below.)

The advances made in the study of the problem during the ten years follow-

†Both the theoretical and empirical data indicate that the critical field, $H_{c,\,\max}$, for the disappearance of superconductivity at $T = 0$ generally speaking increases with increasing $T_c$ (in the simplest approximation, for example, we can use the formula $H_{c,\,\max} = 1.76\sqrt{2}\,k_B T_c / g\mu_B \approx 18500\,T_c$ kG, where $T_c$ is measured in K). In this connection, there is no special reason to fear that the production of superconductors with high $T_c$ values will be accompanied by a decrease in the parameters $H_c$ and $j_c$.

ing the appearance of [3] cannot be described as impressive. True, we have gained a significantly better insight into a number of theoretical questions, but in the crucial area–in the experimental field–we have gained only limited experience and have taken only the first steps. In our opinion, this situation is not at all discouraging, for considering the difficulty of the problem, we cannot but admit that only very insignificant efforts have been made to solve it. Besides, as follows from a number of examples, the solution of major problems under conditions when success cannot be guaranteed usually takes many years. In any case, the problem of high-temperature superconductivity unquestionably exists, and it is one of the most important and interesting problems of both contemporary physics and of materials science.

The present chapter is devoted to a general introduction to the problem of high-temperature superconductivity and to a discussion of the general features of its present state in light of the material that will be dealt with in the following chapters. In this chapter the material is introduced with brief remarks of an elementary nature dealing with the nature of and the mechanisms underlying superconductivity. Undoubtedly, these remarks are totally inadequate for a real familiarization with the theory of superconductivity. Besides, they are often insufficiently exact and all embracing. Nevertheless, we are convinced that this mode of exposition, which is intended for a great many readers, is justified. Those readers who are sufficiently well conversant with the theory of superconductivity can, without difficulty, choose, in the present chapter as well as in the subsequent ones, the material that is of interest to them.

## 2. The Nature and Mechanisms of Superconductivity. High-Temperature Superconductivity (Formulation of the Problem)

As a result of the success of the electron-gas model in its application to metals in the normal (not superconducting) state, the phenomenon of superconductivity for a long time seemed especially puzzling. In fact, the applicability of the electron-gas model to the electrons in a metal, despite the fact that the kinetic energy of the electrons in a metal is of the order of their interaction potential (Coulomb) energy, on account of which they form a liquid rather than a gas, was only slightly less surprising.

The theory of a Fermi liquid and the concept of pseudopotentials and investigations related to them have made it possible for us to understand the behavior of metals in the normal state. In particular, it has been ascertained that the presence of repulsion between the electrons does not impair the applicability of the gas model in the qualitative respect. On the other hand, if the conduction electrons in the vicinity of the Fermi surface attract each other, then the Fermi particle (quasiparticle) distribution that is characteristic of the normal state turns out to be unstable.

The point is that in the presence of attraction it is more advantageous for the quasiparticles in the vicinity of the Fermi surface to form pairs. These pairs have integral spin and undergo something like a Bose–Einstein condensation, although because of the large size of the pairs compared to the distance between pairs, the connection with the Bose–Einstein condensation should not be taken literally. As a result of the formation of the pairs and their "condensation," there arises in the energy spectrum of the system a gap whose width is temperature dependent and is greatest at $T = 0$ K (here and below we do not discuss the so-called gapless superconductors). To break up a pair into two quasiparticles, it is necessary to expend an amount of energy not less than $2\Delta(T)$, where $\Delta(T)$ is the minimum energy of one quasiparticle. The critical temperature $T_c$ is determined from the condition $\Delta(T_c) = 0$ and is, in the cases under discussion, proportional to the maximum gap width $\Delta(0)$.

Thus, the occurrence of superconductivity in metals is the result of attraction between the electrons in the vicinity of the Fermi level. In the simplest case such an attraction (interaction) is characterized by some dimensionless "interaction function," $U(\xi, \xi')$, which in a number of cases can be assumed to depend on the difference $\xi - \xi'$ [clearly, under such conditions $U = U(\xi - \xi')$]. The arguments $\xi$ and $\xi'$ are the energies of the interacting electrons, as measured from the Fermi surface, which is assumed here to be spherical (the isotropic case). As is well known, $\xi = v_F(p - p_F)$, where $v_F$ and $p_F = \hbar q_F = mv_F$ are the velocity and momentum at the Fermi level $E_F$ (it is assumed that $\xi << E_F = p_F^2/2m$).

If the interelectron interaction energy does not vanish at the Fermi level in the first-order perturbation theory, then attraction corresponds to the condition $U(\omega) < 0$ (we have set $\hbar\omega = \xi - \xi'$). The theoretical problem in its simplest variant under discussion consists of the determination, given the function $U(\omega)$, of the gap $\Delta(T)$ and, thus, the critical temperature, $T_c$, at which $\Delta(T_c) = 0$.

Another aspect of the problem consists of the determination of the mechanism underlying superconductivity, or, formally, in the computation or estima-

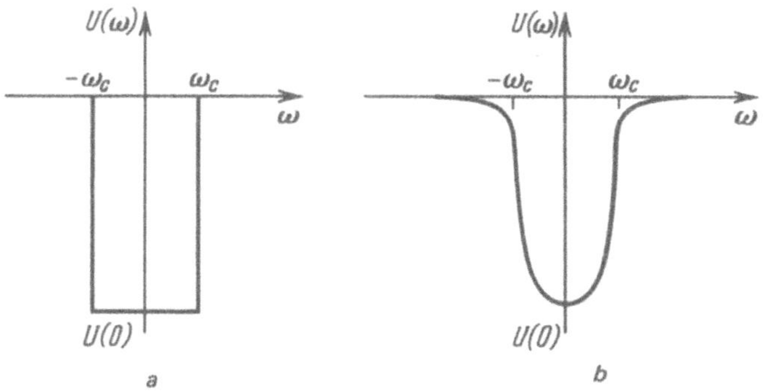


**Figure 1.1. (a) Interaction energies in the form of a square well (as assumed in the BCS theory). (b) A smoothed out well for the interaction energy.**

tion of the function $U(\omega)$ itself that gives rise to superconductivity. Below we shall briefly discuss both of these questions, but this and the following sections of this chapter can in no way be a substitute for the reviews of the microscopic theory of superconductivity (see Chapter 2 and, for example, [6–8]).

In the first successful microscopic theory of superconductivity, constructed in 1957 by Bardeen, Cooper, and Schrieffer (BCS), it was assumed that the function $U(\omega)$ has the form of a square well [see also Fig. 1.1a and the slight refinement made below in going from formula (1.22) to (1.23)]:

$$U(\omega) = -N(0)V = \text{const}, \qquad |\omega| < \omega_c,$$
$$U(\omega) = 0, \qquad |\omega| > \omega_c. \tag{1.1}$$

In this case when $N(0)V \ll 1$ [the merit in setting $U(0) = -N(0)V$ will become apparent below

$$T_c = \left(\frac{\gamma}{\pi k_B}\right)\Delta(0) = \left(\frac{2\gamma}{\pi}\right)\left(\frac{\hbar\omega_c}{k_B}\right)\exp\left(-\frac{1}{N(0)V}\right) =$$
$$= 1.14\frac{\hbar\omega_c}{k_B}\exp\left(-\frac{1}{N(0)V}\right), \tag{1.2}$$

where $\gamma = e^C = 1.781\ldots$ ($C = 0.577\ldots$ is the Euler constant) and $k_B = 1.38 \times 10^{-16}$ erg/deg is the Boltzmann constant.

The approximation of the function $U(\omega)$ by a square well can have some physical meaning only if a similar result is obtainable for a well with smooth edges (Fig. 1.1b), which, in fact, is what happens. For a well of the type shown in Fig. 1.1b, we obtain the BCS formula in the form

$$T_c = \Theta e^{-1/g}, \tag{1.3}$$

where

$$g \ll 1, \tag{1.4}$$

i.e., the weak-coupling approximation is used here. In the formula (1.3) the parameter $\theta$ characterizes the width of the well–the region of attraction for the particles located in the vicinity of the Fermi level; the parameter $g$ is a measure of the attraction and corresponds to the well depth. As is evident from (1.2) and (1.3), in the model (1.1), $\theta = 1.14(\hbar\omega_c/k_B)$ and $g = N(0)V$.

Attraction between the conduction electrons in a metal can arise only when their interaction with the ions (the lattice) and the other electrons (i.e., generally speaking, the electrons at lower energy levels) is taken into account. In some approximation the effect of all the other particles (electrons, ions) on the interaction energy, $V$, for some two considered electrons located at a distance $r$ from each other is taken into account through the introduction of a dielectric constant, $\varepsilon$, so that $V = e^2/\varepsilon r$. When $\varepsilon < 0$, the energy $V < 0$, i.e., we have attraction.

The quantity $\varepsilon$ is, in the general case, not a constant, as was assumed above, but depends on the frequency $\omega$ and the wave vector $\mathbf{q}$. The corresponding quantity $\varepsilon(\omega, \mathbf{q})$ relates, in the case of an isotropic medium, the electric induction $\mathbf{D}$ with the electrostatic (longitudinal) electric field $\mathbf{E}$:

$$\mathbf{D}(\omega, \mathbf{q}) = \varepsilon(\omega, \mathbf{q})\,\mathbf{E}(\omega, \mathbf{q});$$

here

$$\mathbf{E}(\omega, \mathbf{q}) = \frac{1}{(2\pi)^4}\int \mathbf{E}(\mathbf{r}, t)\, e^{-i(\mathbf{qr}-\omega t)}\, d\mathbf{r}\, dt.$$

A similar Fourier expansion is implied for $\mathbf{D}$ (for greater details, see [9, 10]).

The Fourier transform of the interaction energy has the form

$$V(\omega, \mathbf{q}) = \frac{4\pi e^2}{\varepsilon(\omega, q)\, q^2}, \tag{1.5}$$

and it is precisely this quantity that should be used to characterize the interaction when account is taken of the dependence of $\varepsilon$ on $\omega$ and $\mathbf{q}$.

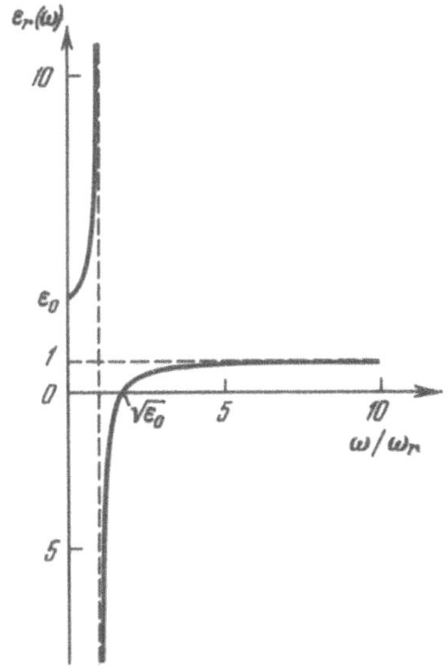


**Figure 1.2. The permittivity as a function of the frequency for the simple model that treats a medium as consisting of noninteracting harmonic oscillators with eigenfrequency $\omega_r$.**

In the simple model that treats a medium as consisting of noninteracting harmonic oscillators with eigenfrequency $\omega_r$ (a charge $e$ oscillates in each of such oscillators),

$$\varepsilon(\omega, q) \equiv \varepsilon_r(\omega) = 1 + \frac{\omega_r^2(\varepsilon_0 - 1)}{\omega_r^2 - \omega^2}, \tag{1.6}$$

where, by definition of the quantity $\varepsilon_0$, the numerator $\omega_r^2(\varepsilon_0 - 1) = (1/m)4\pi e^2 n_r$, $n_r$ being the oscillator concentration. The function $\varepsilon_r(\omega)$ is well known (Fig. 1.2); it is negative in the frequency range $\omega_r < \omega < \varepsilon_0^{1/2}\omega_r$. Clearly, in the same frequency range $V(\omega, q) < 0$.

Another example is the fairly widely used "jellium" model, which describes in some approximation a structureless electron-ion plasma (see [7, 10] and Chapter 2). In this case if $\omega \ll qv_F$ and damping is neglected, then

$$\varepsilon(\omega, q) = 1 + \frac{\varkappa^2}{q^2} - \frac{\omega_i^2}{\omega^2}, \tag{1.7}$$

where

$$\omega_i = \sqrt{\frac{4\pi e^2 Z^2 n_i}{M}} = \sqrt{\frac{4\pi e^2 Z n}{M}} -$$

is the plasma frequency for the ions ($eZ$ is the ion charge, $M$ the ion mass, $n_i$ the ion concentration, $n = n_i Z$ the electron concentration, and $1/\varkappa$ the screening radius) and

$$\varkappa^2 = \frac{6\pi e^2 n}{E_F} = \frac{4e^2 m q_F}{\pi\hbar^2}.$$

In the jellium model (1.7)

$$V(\omega, q) = \frac{4\pi e^2}{\varepsilon(\omega, q)\, q^2} = \frac{4\pi e^2}{q^2 + \varkappa^2}\left(1 + \frac{\omega_q^2}{\omega^2 - \omega_q^2}\right), \tag{1.8}$$

where the $\omega_q$ are the frequencies of the longitudinal waves that can propagate in a medium with the permittivity (1.7).

As is well known [9, 10], the frequencies of such longitudinal waves are determined from the condition

$$\varepsilon(\omega_q, q) = 0, \tag{1.9}$$

and in the case (1.7)

$$\omega_q^2 = \frac{\omega_i^2 q^2}{q^2 + \varkappa^2}. \tag{1.10}$$

In (1.8) the term $4\pi e^2/(q^2 + \varkappa^2)$ corresponds to the Coulomb interaction between the two electrons in question, as screened by the other conduction electrons, while the term proportional to $[\omega_q^2/(\omega^2 - \omega_q^2)]$ describes the influence of the ions, i.e., in application to a crystal, the influence of its lattice. Clearly, in the range $0 < \omega < \omega_q$ the interaction $V < 0$.

The above-introduced function $U(\omega)$ is obtained from $V(\omega, q)$ in the simplest case by some averaging over $q$. This will be discussed further in the following section and in greater detail in Chapter 2. But it is already clear from the foregoing that the region of attraction is determined by the eigenfrequencies of the medium [the frequencies $\omega_r$ and $\omega_i$, respectively, in (1.6) and (1.7)]. Specifically, in the BCS model (1.1)-(1.3) the frequency $\omega_c \sim k_B\theta/\hbar$ corresponds to a characteristic frequency range in which $V < 0$ [in (1.1) the quantity $|V|$ is, in accord with tradition, denoted by $V$].

For the known superconductors, at least in the majority of cases, the attraction between the electrons results from their interaction with the lattice. In the language of quantum mechanics this means that the interaction between the electrons occurs as a result of the exchange of phonons—the energy quanta of the lattice vibrations. Therefore, in such a case we speak of a phonon mechanism of superconductivity.

For the phonon mechanism the influence of the lattice on the electrons can be considerable and, in particular, there can be attraction between them only in the region of frequencies $\omega$ not exceeding the highest phonon frequencies $\omega_{ph,\,max}$ [see (1.8); as is clear from (1.10), in the jellium model $\omega_{ph,\,max} \sim \omega_i$ when $q^2 \gg \varkappa^2$]. As is well known, $\omega_{ph,\,max} \sim (u/a) \sim 10^{13}$ sec$^{-1}$ where the sound speed $u \sim 10^5$ cm/sec and $a \sim 10^{-8}$ cm is the lattice constant. At the same time, $\omega_{ph,\,max} \sim (k_B\theta_D/\hbar)$, where $\theta_D$ is the Debye lattice temperature, which determines the thermal capacity, etc., of the lattice. On account of the foregoing, it is clear that for the phonon mechanism of superconductivity, in the BCS formula (1.3)

$$\Theta \sim \Theta_D \quad \text{(phonon mechanism of superconductivity).} \tag{1.11}$$

The Debye temperature, $\theta_D$, for metals usually does not exceed several hundred degrees (e.g., $\theta_D \sim (\hbar\omega_{ph,\,max}/k_B) \sim 100$ K for $\omega_{ph,\,max} \sim 10^{13}$ sec$^{-1}$). Hence, on account of the presence of the exponential factor in formula (1.3), it is already understandable why for the phonon mechanism, especially in the weak-coupling approximation (1.4), the values of $T_c$ cannot exceed a few tens of degrees. To obtain more exact estimates for $T_c$, it is necessary to go beyond the limits of the weak-coupling approximation (1.4), to take the specific form of the phonon spectrum into account, etc. All this has, to a large extent, already been done (see Section 3 and Chapter 4), and confirms the conclusion that

$$T_c \lesssim 30 - 40 \text{ K} \quad \text{(phonon mechanism of superconductivity).} \quad (1.12)$$

It should be noted, though, that the estimate (1.12) is based on the consideration of metals with more or less known parameters and cannot be directly applied to all hypothetical cases. Thus, for metallic hydrogen, estimates lead to values of $T_c \sim 10$–$200$ K (see Chapter 4) since in this case $\theta_D \sim 3 \times 10^3$ K and, moreover, the parameter $g$ may not be very small. (Since, as before, $T_c \ll \theta_D$, strictly speaking, this example in no way contradicts what was asserted above in the discussion of the BCS formula with $\theta \sim \theta_D$.) Other possible exceptions are the "organic" metals (see below and Chapter 7) and systems with "interfering" electronic structural and superconducting transitions (see below and Chapter 5).

On the whole, however, we can assert that for materials of the "ordinary" type with the phonon mechanism of superconductivity we can be guided by the limitation (1.12). Besides, metallic hydrogen and its alloys (i.e., hydrogen with an admixture of other elements) are found can exist in a stable state only under very high pressures, and the possibility of their existence in a sufficiently stable metastable state when the pressure is relieved has not yet been established. As to organic metals with the requisite properties, the feasibility of their production is not sufficiently apparent and will be considered separately.

Comparing the estimate (1.12) with the attained highest $T_c$ value ($\approx 23$ K), we see that the feasibility, noted in Section 1, of the production of "medium-temperature" superconductors still exists in principle within the framework of the phonon mechanism of superconductivity. But this is, generally speaking, impossible in the case of high-temperature superconductors, i.e., superconductors with $T_c \gtrsim 90$ K (with the specified exception of metallic hydrogen), and it is necessary to turn to another mechanism. Such a mechanism is known, and can, in its general form, be called the electronic mechanism—we have in mind attraction between the conduction electrons as a result of their interaction with other electrons entering into the composition of the solid. We shall call this mechanism the exciton mechanism.†

The point is that different waves—excitons of the electronic-type, or simply excitons—can propagate in the electron system even without allowance for the influence of the lattice (to say nothing, of course, of its role in guaranteeing the quasineutrality of the whole system). We are particularly interested in longitudinal waves (longitudinal excitons) of this kind, the most well-known example of which is plasma waves (plasmons). If such excitons exist in the requisite region of frequencies $\omega$ and wave vectors $\mathbf{q}$ (we have in mind the region $q \sim q_F$), then

†The use in the present case of the term "electronic mechanism of superconductivity" would have led to confusion since the superconductivity of metals is, in all cases, an electronic phenomenon.

they can "substitute" for phonons in the process that ensures attraction between the conduction electrons.

For electronic-type excitons, however, the characteristic frequency, $\omega_{ex} \equiv \Omega_e$, corresponds to an energy of the order of tenths of an electron volt or even several electron volts. For reasons that will become apparent below (see Section 3), the frequency, $\Omega_e$, of the excitons that give rise to excitonic superconductivity should nevertheless, at least in simple models, satisfy the condition $\Omega_e \ll \omega_F$, where $\omega_F = E_F/\hbar$ ($E_F$ is the energy at the Fermi level). Since normally $E_F \lesssim$ 1-10 eV, the characteristic temperature, $\theta_e$, for the exciton mechanism of superconductivity should be of the order of

$$\Theta_e \sim \frac{\hbar\Omega_e}{k_B} \sim 10^3 - 10^4 \text{K}. \tag{1.13}$$

At the same time, within the framework of the BCS model (1.1)-(1.3), for the exciton mechanism of superconductivity

$$\Theta \sim \Theta_e \quad \text{(exciton mechanism of superconductivity).} \tag{1.14}$$

Under such conditions, high $T_c$ values are attainable even in the case of the weak coupling (1.4). For example, for $\theta = \theta_e = 10^4$ K and $g = \frac{1}{4}$, the temperature $T_c = 180$ K. Thus, in principle the problem is quite simple: we need to substitute excitons for phonons, and then limitations of the type (1.12) disappear. But in reality, of course, the problem has thus far only been formulated.

In the first place, a question arises as to how reliable and convincing are the arguments, which are based on the greatly simplified BCS model. This will be discussed in the following section and in Chapter 2, and the permissibility of using the BCS approximation for making estimates will be confirmed in certain respects. The above-expounded approach will thereby be justified.

In the second place, we must find out whether the choice of parameter values,

$$\begin{gathered} \Theta_e \sim \frac{\hbar\Omega_e}{k_B} \sim 10^3 - 10^4 \text{K}, \\ g = g_e \gtrsim \frac{1}{5} - \frac{1}{3}, \end{gathered} \tag{1.15}$$

necessary for obtaining high-temperature superconductivity is realistic.

No reliable answer to this question has as yet been found, and, consequently, the possibility of the existence of high-temperature superconductivity based on the exciton mechanism has neither been proved nor refuted. Sections 4 and 5 of the present chapter are devoted to a discussion of this aspect of the matter and, in essence, of ways of using and "activating" the exciton mechanism. But it is expedient even now, in developing the introductory remarks made above and summing them up, to enumerate the directions in which research in the field of superconductivity is proceeding and to whose description the present book is devoted.

Our primary interest is the problem of the computation or at least estimation of the critical temperature $T_c$. This problem is, of course, important even within the framework of the phonon mechanism, to say nothing of the fact that the corresponding investigation is important also in the analysis of the exciton mechanism of superconductivity. General formulas are derived for $T_c$ in the case of quasiisotropic metals in Chapter 2-4 (a brief discussion is presented in Section 3 below). These formulas and the results of the corresponding estimates indicate how "medium-temperature" superconductors (15 K $\lesssim T_c \lesssim$ 40 K; such superconductors are, it is true, often called high-temperature superconductors) can be produced on the basis of the phonon mechanism. For metallic hydrogen (and, perhaps, also for hydrogen-containing and organic metals) the phonon mechanism can, in principle, also lead to high-temperature superconductivity ($T_c >$ 40–90 K).

The exciton mechanism in its simplest form amounts to the replacement of phonons by excitons with $\theta \sim \theta_e \sim 10^3$–$10^4$ K [see (1.13)]. It is, generally speaking, quite difficult to secure the existence of the requisite excitons in a metal, especially in a three-dimensional metal of the more or less ordinary type (but with, say, polarizable impurities). Therefore, different possibilities are discussed of enhancing the exciton mechanism through the use of one-dimensional (quasi-one-dimensional) metallic compounds (Chapter 7), or two-dimensional (quasi-two-dimensional, layered) compounds or "sandwiches" (Chapters 6 and 8).

Another important direction is connected with the use of materials with metallic conductivity (more often semimetals or doped semiconductors), which are, roughly speaking, found in states close to the region of electronic (excitonic) phase transitions (Chapter 5). Under such conditions, as a result of the reconstruction (phase transition) of the system, the density of the electronic states near the Fermi surface can change drastically. And although under such conditions the interaction that gives rise to superconductivity has basically a phonon character, the BCS formula (1.3) cannot be applied directly. This possibility deserves as much attention as the others, but since it falls somewhat outside their class, we shall virtually not discuss it in the present chapter. The same thing should be said of superconductivity under nonequilibrium conditions (Chapter 9).

## 3. Some Remarks on the Computation of the Critical Temperature

Let us discuss briefly (a more detailed discussion is presented in Chapter 2) the computation of the critical temperature $T_c$. It will be shown in the process that the simple BCS model [in which the interaction has the form of some potential well, (1.1), and the critical temperature is given by the formulas (1.2) and (1.3)] is fairly well applicable for a qualitative and, under certain conditions, for a quantitative description of superconductivity. Therefore, the general conclusions drawn in Section 2 turn out to be valid.†

The computation of the critical temperature $T_c$ and the other parameters of a metal is, generally speaking, possible only when a number of assumptions and approximations is made. Thus, the interaction between two electrons located in a metal and having momenta $\mathbf{p}_1$ and $\mathbf{p}_2$ depends, in the general case, on both of these momenta and the corresponding energies. Furthermore, in a periodic lattice the role of momentum is played by the quasimomentum.

Meanwhile, in the majority of papers the interaction is assumed to be isotropic, the difference between momentum and quasimomentum is neglected (i.e., the so-called Umklapp processes are neglected), the energy and momentum are assumed to be connected by the relation $E = p^2/2m$, and, finally, only the dependence of the corresponding matrix element of the interaction energy $V(\mathbf{p}_1, \mathbf{p}_2, \mathbf{p}_3)$ on the transferred energy $h\omega$ and momentum $h\mathbf{q}$ is taken into consideration. In other words, it is assumed that (see Chapter 2)

$$V(\mathbf{p}_1, \mathbf{p}_2, \mathbf{p}_3) = V(\omega, q), \quad \hbar\omega = \xi - \xi', \qquad \xi = v_F \cdot (p - p_F), \quad \hbar\mathbf{q} = \mathbf{p} - \mathbf{p}', \tag{1.16}$$

where $\mathbf{p} - \mathbf{p}' = \mathbf{p}_1 - \mathbf{p}_4 = \mathbf{p}_3 - \mathbf{p}_2$, and $\mathbf{p}_3$ and $\mathbf{p}_4$ are the momenta of the electrons after their elastic scattering [on account of the momentum conservation law, which allows us to express $\mathbf{p}_4$ in terms of $\mathbf{p}_1$, $\mathbf{p}_2$, and $\mathbf{p}_3$, the momentum $\mathbf{p}_4$ does not figure in the above written-out element $V(\mathbf{p}_1, \mathbf{p}_2, \mathbf{p}_3)$]. Under the conditions (1.16), it is convenient to write

$$V(\omega, q) = \frac{4\pi e^2}{\varepsilon_{\text{eff}}(\omega, q)\, q^2}. \tag{1.17}$$

If $\varepsilon_{\text{eff}}(\omega, q) = \varepsilon(\omega, q)$, then (1.17) goes over into the expression (1.5), whose meaning was explained above. Generally speaking, however, the quantities $\varepsilon_{\text{eff}}$ and $\varepsilon$ are different. Still, these quantities are related and in some cases are even close to each other. (Thus, in the weak-coupling approximation we can assume that $\varepsilon_{\text{eff}} = \varepsilon$; see Chapter 2.) Furthermore, a consistent computation of $\varepsilon_{\text{eff}}(\omega, q)$ is normally not carried out, and below we generally only indicate some simplest $\varepsilon(\omega, q)$ dependences as examples. Under such conditions it does not make any sense to distinguish between $\varepsilon_{\text{eff}}$ and $\varepsilon$, and this is taken into account below.

†Readers who are prepared to take this assertion on trust can limit themselves to essentially the last part of this section.

Thus, the very use of the basic interaction between the conduction electrons in the form (1.5) is tied in with a number of approximations. But this is not all since the relation between $V(\omega, q)$ and $T_c$ is also not a simple one. First, it is necessary to go over from $V(\omega, q)$ to the interaction function $U(\omega)$. Second, it is necessary to solve an integral equation for the gap $\Delta(T)$ with a kernel that depends on $U(\omega) = U(\xi - \xi')$. Then the temperature $T_c$ is determined from the condition for the existence of a nontrivial solution $\Delta \neq 0$ since

$$\Delta(T_c) = 0, \quad \Delta(T < T_c) \neq 0.$$

The simplest and quite widely used relation between $V$ and $U$ is†

$$U \equiv U_0(\omega) = N(0)\,\overline{V} = \alpha \int_{\omega/v_F}^{2q_F} \frac{dq}{\varepsilon(\omega, q)\,q}. \tag{1.18}$$

Here $N(0)$ is the density of states at the Fermi level for a free-electron gas:

$$N(0) = \frac{1}{2}\left(\frac{dn}{dE}\right)_{E_F} = \frac{mq_F}{2\pi^2\hbar^2}, \qquad n = \frac{q_F^3}{3\pi^2},$$
$$E_F \equiv \hbar\omega_F = \frac{\hbar^2 q_F^2}{2m}, \qquad p_F = mv_F = \hbar q_F, \tag{1.19}$$

where $m$ is the mass and $n$ the concentration of the electrons and $\overline{V} \equiv \overline{V}(\omega, q)$ is the value of $V(\omega, q)$ averaged over the Fermi surface:

$$\overline{V} = \frac{1}{2}\int_{-1}^{1} \frac{4\pi e^2 d\cos\theta}{\varepsilon(\omega, q)\,q^2} \approx \int_{q^2_{\min}}^{q^2_{\max}} \frac{4\pi e^2 dq^2}{4q_F^2 q^2 \varepsilon(\omega, q)} = \frac{2\pi e^2}{q_F^2}\int_{\omega/v_F}^{2q_F} \frac{dq}{q\varepsilon(\omega, q)}$$
$$\hbar^2 q^2 = (\mathbf{p} - \mathbf{p}')^2 = p^2 + p'^2 - 2pp'\cos\theta, \tag{1.20}$$
$$q_{\min} = \frac{p - p'}{\hbar} = \frac{\omega}{v_F}, \quad q_{\max} = \frac{p + p'}{\hbar} = 2q_F.$$

The parameter $\alpha$ in (1.18) can be expressed in terms of the electron concentration

†As will be noted below and shown in Chapter 2, the use of the function $U = U_0$ is, generally speaking, not justified. But since such an approximation is quite graphic and widely used, it is pertinent to discuss it here.

$$\alpha = \frac{e^2}{\pi\hbar v_F} = \frac{e^2 m}{\pi\hbar^2 q_F} = \frac{e^2 m}{(3\pi^2)^{1/3}\,\pi\hbar^2 n^{1/3}}. \tag{1.21}$$

The parameter $\alpha$ has the meaning of the ratio (up to a small numerical factor) of the mean Coulomb interaction between two electrons, $e^2/\bar{r} = e^2 n^{1/3}$, to their kinetic energy at the Fermi level, $E_F = \hbar^2 q_F^2/2m = (3\pi^2)^{2/3}\hbar^2 n^{2/3}/2m$. In normal metals $\alpha \sim 1$, while in a highly compressed material $\alpha \ll 1$.

When the function $U_0(\omega)$, defined according to (1.18), is used, the above-indicated integral equation for the gap $\Delta(\xi, T)$ for $T \longrightarrow T_c$ has the form

$$\Delta(\xi, T) = -\frac{1}{2}\int_{-\infty}^{+\infty} U_0(\xi - \xi')\,\Delta(\xi, T)\,\frac{\tanh(\xi'/2k_B T_c)}{\xi'}\,d\xi'. \tag{1.22}$$

The solution of this equation for any complicated function $U_0(\xi - \xi')$, or, in the more general case, the function $U(\xi, \xi')$, presents familiar problems. Obviously, the BCS approximation (1.1) has been, and is still, used particularly often in this connection; more exactly, it is assumed that $U_0(\xi, \xi') = \text{const} = -NV$ for $|\xi| < h\omega_c$ and $|\xi'| < h\omega_c$. Then the gap $\Delta$ for $\xi, \xi' < h\omega_c$ turns out to be independent of $\xi$, and therefore from (1.22) and (1.1) we immediately obtain an equation for the temperature at which $\Delta(\xi, T_c) \equiv \Delta(T_c) = 0$:

$$N(0)\,V \int_0^{\hbar\omega_c} \frac{d\xi'}{\xi'} \tanh\left(\frac{\xi'}{2k_B T_c}\right) = 1. \tag{1.23}$$

If $NV \ll 1$, then from this we can easily obtain the formula (1.2). To obtain the relation between $T_c$ and $\Delta(0)$, it is necessary to solve in addition an equation, similar to (1.22), for $\Delta(T)$ at $T = 0$.

If we assume some reasonable expressions for the permittivity $\varepsilon(\omega, q)$ and then compute the function $U_0(\omega)$ according to (1.5) and (1.18), we obtain dependences entirely different from the BCS approximation (1.1). For example, for the jellium model (1.7), we obtain

$$U_0(\nu) = \frac{\alpha}{2(1-\nu^{-2})}\ln\left|1 + \frac{1}{\alpha}\left(1 - \frac{1}{\nu^2}\right)\right|, \tag{1.24}$$

$$\nu^2 = \frac{\omega^2}{\omega_i^2}, \quad \alpha = \frac{e^2}{\pi\hbar v_F}.$$

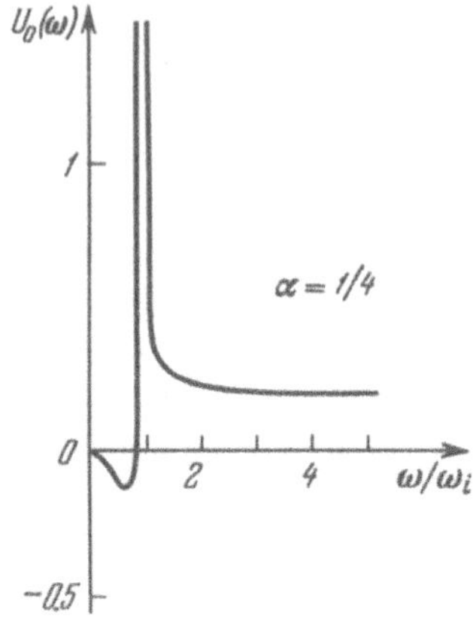


**Figure 1.3. The interaction energy for the jellium model [according to (1.24)].**

The function (1.24) for $\alpha = \frac{1}{4}$ is shown in Fig. 1.3 and, as has been stated, is quite different from the BCS well shown in Fig. 1.1. In this connection, it may be noted that the great success achieved by the original BCS theory [6, 11] on comparing the theory with experiment was surprising in some respect.†

The BCS approximation turned out, on the whole, to be good for two reasons. The first is the "suppression" of the role of the Coulomb repulsion in the theory of superconductivity (for greater details, see Chapter 2 and, for example, [7]). The point is that the interaction $V$ and, consequently, $U_0(\omega)$ usually contain terms with entirely different frequency dependences. One of these terms, $V_c$, corresponds to a screened Coulomb repulsion, while the second term, $V_a$, reflects the interaction (attraction) resulting from the exchange of phonons and/or excitons.

The foregoing is evident even from the expression (1.8) for the jellium model. In this expression, just as in more general cases,

$$V_c(\omega, q) = \frac{4\pi e^2}{q^2 + \varkappa^2}, \quad \omega \lessgtr \omega_F = \frac{E_F}{\hbar}, \quad \omega \ll q v_F. \tag{1.25}$$

The nondependence of $V_c$ on $\omega$ up to frequencies of the order of $\omega_F$ (actually, what we are talking about is a weak dependence as long as $\omega < \omega_F$) results from the instantaneous nature of the Coulomb interaction and also the "noninertial-

†The BCS theory was the first successful attempt at the construction of a consistent microscopic theory of superconductivity for some reasonable model of the superconducting medium. In this respect, the outstanding role of the work [11] by Bardeen *et al.* is beyond doubt. It is only somewhat startling that the rather arbitrary approximation (1.1) led to a large number of quantitative results that correspond with reality.

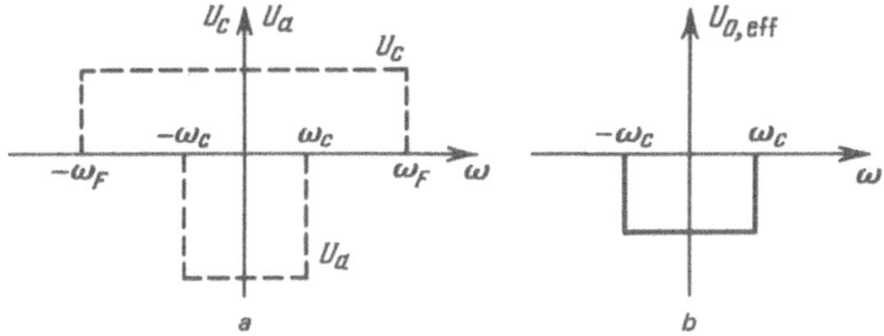


**Figure 1.4. (a) The interaction energy [according to (1.27)]. (b) The effective interaction energy when account is taken of the fact that the lower frequencies make a greater contribution than the higher frequencies [see (1.29)]. The well depth $U_{0\ \text{eff}} = U_a + U_c/[1 + U_c \ln (\omega_F/\omega_c)]$.**

lessness" of the screening for frequencies $\omega << \omega_F$.† For $\omega > \omega_F$ the interaction $V_c \approx 4\pi e^2/q^2$, but this frequency range turns out to be unimportant, and we can set $V_c = 0$ there.

As to the term $V_a(\omega, q)$, it corresponds to attraction (i.e., $V_a < 0$) only in some frequency range where $\omega_c << \omega_F$. Thus, for the jellium model (1.8)

$$V_a(\omega, q) = \frac{4\pi e^2}{q^2 + \varkappa^2} \cdot \frac{\omega_q^2}{\omega^2 - \omega_q^2}. \tag{1.26}$$

Here $V_a < 0$ only for frequencies $\omega < \omega_q$ or, if we are talking about the important values of $q \sim q_F$, for frequencies $\omega < \omega_{q_F} \sim \omega_i$ [see (1.10); the frequency $\omega_i$ in the present case plays the role of the frequency $\omega_c \sim \omega_{\text{ph, max}}$]. In the region where $V_a < 0$, the phonon interaction indeed has the form of a well somehow reminiscent of the BCS approximation (see Fig. 1.1).

On account of the foregoing, the total interaction $V = V_c + V_a$ corresponds in a rough approximation to the function $U_0$ (shown in Fig. 1.4a):

$$\begin{gathered} U_0 = U_c + U_a = N(0)\overline{V}_c + N(0)\overline{V}_a, \\ \overline{V}_c = C_1 > 0, \quad \omega < \omega_F; \quad \overline{V}_c = 0, \quad \omega > \omega_F; \\ \overline{V}_a = C_2 < 0, \quad \omega < \omega_c \ll \omega_F; \quad \overline{V}_a = 0, \quad \omega \gg \omega_c, \end{gathered} \tag{1.27}$$

where $C_{1,2}$ are constants.

†The screening is caused by the electrons that traverse the mean distance between them of $n^{-1/3} \sim q_F^{-1}$ during the time $\tau_F \sim 1/v_F q_F \sim 1/\omega_F$. Further, the screening radius $1/\kappa = \pi^{1/2}\hbar/2e(mq)_F^{1/2} = \frac{1}{2}\alpha^{1/2} q_F \sim 1/q_F$ for $\alpha \sim 1$ and, consequently, the time required to traverse the distance $1/\kappa$ in metals is also of the order of $1/\omega_F$. From this it is clear that for frequencies $\omega << \omega_F$ the screening can take place "instantaneously"–during a period of time significantly shorter than the period $\tau = 2\pi/\omega$.

When $C_1 \sim |C_2|$, the difference between the functions (1.27) and (1.1) is quite radical; for example, when $C_1 > |C_2|$, attraction does not occur at any frequency. But, in fact, in the integral equation (1.22), the contributions of the terms $U_c$ and $U_a$ when $C > |C_2|$ turn out to be entirely different as a result of the above-indicated unidentical frequency dependences of these terms. The corresponding computations can be found in, for example, [7]; their result indicates a greater relative importance of the low frequencies $\omega$ and consists in the fact that the role of the Coulomb term is suppressed: we should not compare $U_a$ and $U_c$, but $U_a$ and $U_c^*$, where

$$U_c^* = \frac{U_c}{1 + U_c \ln (\omega_F/\omega_c)}. \tag{1.28}$$

It is clear that when $U_c \ln (\omega_F/\omega_c) >> 1$ the quantity $U_c^* \approx [1/\ln (\omega_F(\omega_c))] << 1$. The total effective interaction that is equivalent to (1.27) from the point of view of the determination of $T_c$ can therefore be written as follows:

$$\begin{aligned} U_{0,\text{eff}} &= U_a + \frac{U_c}{1 + U_c \ln (\omega_F/\omega_c)}, & \omega &< \omega_c; \\ U_{0,\text{eff}} &= 0, & \omega &> \omega_c. \end{aligned} \tag{1.29}$$

The interaction $U_{0,\text{ eff}}$ can be close to $U_a$ even when $U_c \sim |U_a|$, as illustrated by Fig. 1.4b. This is the first of the indicated causes of the significant broadening of the region of applicability of the BCS approximation (1.1).

The second cause is connected with the fact that the interaction function $U(\omega)$ is far from being always replaceable by the function $U_0(\omega)$, defined according to (1.18). In fact, we have not even specially justified the approximation (1.18), a justification which would have been necessary: the function $U(\omega)$ should not at all reduce *a priori* to the matrix element $V(\omega, q)$ averaged over $q$. The relation between $U(\omega)$ and $V(\omega, q)$ can be reliably derived only on the basis of detailed computations, more or less consistently determining the conditions for the onset and the character of the superconducting state in metals. Such computations have been repeatedly carried out (see [8], Chapter 2 and the literature cited there), but they are, unfortunately, very complicated, and it is difficult to follow them even in monographs, let alone in shorter papers. In our opinion, these computations are generally comprehensible to only a small circle of theoretical physicists working in the field of many-particle theory, remaining unintelligible to many specialists working in other fields. Therefore, we refer the reader to Chapters 2–4 and limit ourselves here to only a few remarks concerning the results of the refined theory.

We shall assume as before that the matrix element of the interaction between the electrons is described by the expression $V(\omega, q)$. Then the role of the interaction function $U(\omega)$ in the theory of superconductivity is played by the quantity (see Chpater 2, Section 3)

$$U(\omega) = N(0)\,\overline{V}_{\text{eff}},$$
$$V_{\text{eff}} = \frac{4\pi e^2}{q^2}\left(1 + \frac{2}{\pi}\int_0^\infty \frac{\operatorname{Im}\left[\frac{1}{\varepsilon(\omega', q)}\right]}{\omega + \omega'}\,d\omega'\right), \tag{1.30}$$

where $\overline{V}_{\text{eff}}$ is related to $V_{\text{eff}}$ in the same way as $\overline{V}$ is related to $V$ [see (1.20)]; notice that on account of the dispersion relations,

$$\left.\begin{aligned} \operatorname{Re}\left[\frac{1}{\varepsilon(\omega, q)}\right] &= 1 + \frac{2}{\pi}\int_0^\infty \frac{\omega' \operatorname{Im}\left[\frac{1}{\varepsilon(\omega', q)}\right]}{(\omega')^2 - \omega^2}\,d\omega', \\ \operatorname{Im}\left[\frac{1}{\varepsilon(\omega, q)}\right] &= -\frac{2\omega}{\pi}\int_0^\infty \frac{\operatorname{Re}\left[\frac{1}{\varepsilon(\omega', q)}\right] - 1}{(\omega')^2 - \omega^2}\,d\omega', \end{aligned}\right\} \tag{1.31}$$

where the principal values of the integrals are to be taken here.

For an equilibrium system the quantity $\operatorname{Im}(1/\varepsilon) \leqslant 0$. Furthermore, there is no resonance denominator in (1.30). As a result, the function $U(\omega)$ turns out to be a real function of $\omega$ and is, in addition, smooth compared to the function $U_0(\omega)$, which can have a resonance (see Fig. 1.3). Such a "smoothing out" together with the already noted Coulomb-interaction-suppression effect, which remains in the more rigorous theory, leads to a still better correspondence of the consistent computations with the BCS theory.

Thus, the use of the BCS formulas (1.2) and (1.3) for obtaining the estimates (1.11), (1.12), (1.14), and (1.15) and on the whole for a general orientation in the problem of high-temperature superconductivity was indeed justified.

At the same time, it is perfectly clear that we should by no means limit ourselves at the present stage to such estimates; it is possible and necessary to go considerably further. Thus, in application to the phonon mechanism of superconductivity, the value of $T_c$ is computed with allowance for details of the phonon spectrum and a number of other factors, though still with the use of a number of approximations. Specifically, McMillan [12] has derived the formula

$$\begin{aligned} T_c &= \frac{\Theta_D}{1.45}\exp\left[-\frac{1.04\,(1+\lambda)}{\lambda - \mu^*\,(1+0.62\lambda)}\right], \\ \lambda &= 2\int_0^\infty \frac{\alpha^2(\omega)\,F(\omega)\,d\omega}{\omega}, \quad \mu \equiv U_c = N(0)\,\overline{V}_c, \\ \mu^* &\equiv U_c^* = \frac{\mu}{1 + \mu \ln (E_F/k_B\Theta_D)}, \end{aligned} \tag{1.32}$$

which is quite widely used in the literature. In this formula, $F(\omega)$ is the density of phonon states and $\alpha^2(\omega)$ is a measure of the electron-phonon coupling. The parameter $\lambda$ can be written in the form

$$\lambda = \frac{N(0)\,\overline{G^2}}{M\overline{\omega^2}},$$

where $\overline{G^2}$ and $\overline{\omega^2}$ are, respectively, the mean values of the square of some electronic matrix element and the square of the phonon frequency [$M$ is the ion mass and $N(0)$ is the density of states]. It is significant that $\lambda$ decreases with increasing $\overline{\omega^2}$ and, generally speaking, with increasing Debye temperature $\theta_D$.

The formula (1.32) is in point of fact a semiempirical formula: It pertains to a number of more general expressions of the type

$$T_c = \Theta \exp\left[-\frac{1+\lambda}{\lambda - \mu^*(1+\lambda)}\right], \tag{1.33}$$

in which the parameters $\theta$, $\lambda$, and $\mu^*$ can be expressed in terms of the parameters of the metal by formulas similar to those given in (1.32).

The most exact and validated of these formulas is the following [13]:

$$T_c = \frac{\hbar\overline{\omega}_{\log}}{1.43k_B} \exp\left\{-\frac{1+\lambda}{\lambda - \mu^*\left(1 + 0.5\frac{\lambda}{1+\lambda}\right)}\right\},$$

$$\ln\overline{\omega}_{\log} = \frac{2}{\lambda}\int_0^\infty \ln\omega\,\alpha^2(\omega)\,F(\omega)\,\frac{d\omega}{\omega}, \tag{1.32a}$$

$$\lambda = 2\int \frac{\alpha^2(\omega)\,F(\omega)\,d\omega}{\omega}, \quad \mu^* = \frac{\mu}{1 + \mu\ln\left(\omega_F/\overline{\omega}_{\log}\right)}.$$

The derivation of this formula is given in Chapter 4, where $F(\omega)$ is denoted by $g(\omega)$ and the constants $\hbar$ and $k_B$ are left out since $\overline{\omega}_{\log}$ is expressed in energy units. Notice that for $\lambda << 1$ (weak coupling) the formula (1.33) goes over essentially into the BCS formula (1.3), where $g = \lambda - \mu^*$:

$$T_c = \Theta e^{-\frac{1}{\lambda-\mu^*}}. \tag{1.3a}$$

The transition form (1.3a) to the expression (1.33) is achieved as a result of a "renormalization" of the interaction constant $\lambda$ by means of the replacement

$\lambda \longrightarrow \lambda/(1+\lambda)$. More exactly [as is evident, for example, from (1.32) and (1.32a) and as is explained in greater detail in Chapter 4], the transition to the strong-coupling case in the formula (1.3a) requires the substitution $\lambda \longrightarrow \lambda/(1+\lambda)$ and the redefinition of the quantities $\theta$, $\lambda$, and $\mu^*$.

For a number of metals $\mu^* \approx 0.1\text{-}0.2$, but the parameter $\lambda$ can be large, or, more exactly, it may not satisfy the weak-coupling condition (e.g., for Pb, according to various estimates, $\lambda \approx 1.3\text{-}1.6$, the value $\lambda = 1.55$ being used in Chapter 4). Analysis based on the use of the formulas (1.32) and (1.33) leads to the result (see [12, 13] and Chapter 4, where references are also given) that, for quasiisotropic metals and the phonon mechanism of superconductivity, the highest value of $T_c$ satisfies, generally speaking, the criterion (1.12).

The formula (1.33), like (1.3), can also be used in the case of the exciton mechanism of superconductivity, only the values of $\theta$, $\lambda$, and $\mu^*$ changing when this is done. In this sense, the assertion made in Section 2 remains valid in the strong-coupling case when the condition (1.4), i.e., $g = (\lambda - \mu^*) \ll 1$, is violated. In view of this, a number of conclusions concerning the exciton mechanism of superconductivity can be drawn on the basis of a general analysis of expressions of the type (1.33). We shall pursue this course, but, first, it seems expedient to explain the essence of the exciton mechanism, as illustrated by a very simple model: the generalized jellium model.

## 4. The Exciton Mechanism of Superconductivity (The Generalized Jellium Model; General Considerations)

In the generalized jellium model, when damping is neglected, the permittivity has the form

$$\varepsilon(\omega, q) = 1 + \frac{\varkappa^2}{q^2} - \frac{\omega_i^2}{\omega^2} + \frac{\Omega^2(\varepsilon_0 - 1)}{\Omega^2 - \omega^2}. \tag{1.34}$$

If we want to take damping into account, then we should replace the denominator $\Omega^2 - \omega^2$ by $\Omega^2 - \omega^2 - i\delta, \delta > 0$. It can be seen from a comparison of (1.34) with (1.6) and (1.7) that what is in consideration here is the conventional jellium model supplemented by a set of oscillators with an eigenfrequency, $\omega_r \equiv \Omega$, that is regarded as being electronic, i.e., as corresponding to an electron transition. Therefore, it is assumed that

$$\Omega \gg \omega_i. \tag{1.35}$$

In a medium with the permittivity (1.34), the propagation of longitudinal waves

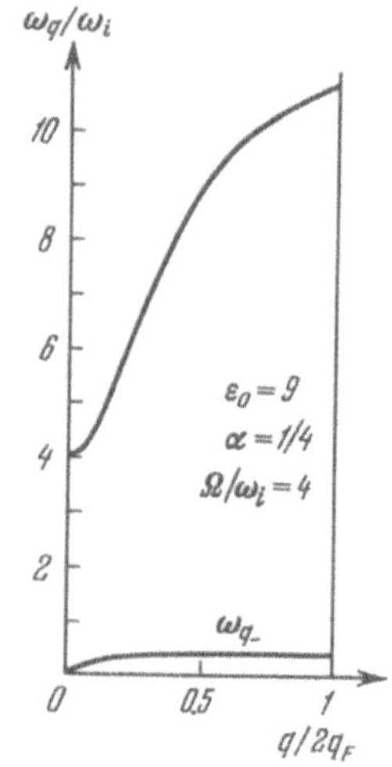


**Figure 1.5. Dispersion curves in the generalized "jellium" model [according to (1.38) and (1.39)].**

with frequencies $\omega(q) \equiv \omega_q$ is possible provided†

$$\varepsilon(\omega_q, q) = 1 + \frac{\varkappa^2}{q^2} - \frac{\omega_i^2}{\omega_q^2} + \frac{\Omega^2(\varepsilon_0 - 1)}{\Omega^2 - \omega_q^2} = 0. \tag{1.36}$$

The solution to the dispersion equation (1.36) is a follows:

$$\omega_q^2 = \frac{(\varepsilon_0 q^2 + \varkappa^2)\,\Omega^2 + \omega_i^2 q^2 \pm A}{2(q^2 + \varkappa^2)}, \tag{1.37}$$

where $A = \left\{\left[(\varepsilon_0 q^2 + \varkappa^2)\,\Omega^2 + \omega_i^2 q^2\right]^2 + 4(\varepsilon_0 - 1)\,\omega_i^2 \Omega^2 q^4\right\}^{1/2}.$

†For greater details see [9, 10]. Let us note that even in an isotropic medium it is, generally speaking, necessary to use the permittivity tensor

$$\varepsilon_{ij}(\omega, \mathbf{q}) = \varepsilon_{tr}(\omega, q)(\delta_{ij} - s_i s_j) + \varepsilon_l(\omega, q) s_i s_j,$$

where $\mathbf{s} = \mathbf{q}/q$ and $\varepsilon_{tr}(\omega, 0) = \varepsilon_l(\omega, 0)$. Since here we are interested in only longitudinal waves, for all intents and purposes $\varepsilon(\omega, q) = \varepsilon_l(\omega, q)$ in (1.36), and the index $l$ has, for simplicity, been dropped.

It should be borne in mind that when $q \sim q_F$, the possibility of using the quantity $\varepsilon_{ij}$ in a crystal is limited. Furthermore, at large $q$ it is not only the longitudinal waves that play a role, to say nothing of the fact that for an aribtrary direction of propagation the waves (phonons, excitons) propagating in the crystalline medium are neither longitudinal nor transverse. All these points are ignored in the approximation under discussion.

If the condition $\Omega^2 >> \omega_i^2$, (1.35), is satisfied, then we obtain two approximate solutions (Fig. 1.5):

$$\omega_{q-}^2 \equiv \omega_{q,\mathrm{ph}}^2 = \frac{\omega_i^2 q^2}{\varepsilon_0 q^2 + \varkappa^2}, \tag{1.38}$$

$$\omega_{q+}^2 \equiv \Omega_e^2(q) = \frac{(\varepsilon_0 q^2 + \varkappa^2)\,\Omega^2}{q^2 + \varkappa^2}. \tag{1.39}$$

Evidently, the solutions (1.38) and (1.39) can be found at once from (1.36) if it is assumed that $\omega_q^2 << \Omega^2$ or $\omega_q^2 >> \omega_i^2$.

The solution (1.38) represents the phonon part of the spectrum, while the solution (1.39) corresponds to the spectrum of the longitudinal electronic excitons. Here $2q_F$ is the highest encountered value of $q$ [see (1.20)]. From (1.39) we obtain

$$\Omega_e(0) = \Omega, \quad \Omega_e(2q_F) = \sqrt{\frac{\varepsilon_0 + \alpha}{1 + \alpha}}\,\Omega \tag{1.40}$$

since

$$\alpha \cdot (2q_F)^2 = \varkappa^2, \ \alpha = \frac{e^2 m}{\pi \hbar^2 q_F}, \ \varkappa^2 = \frac{4e^2 m q_F}{\pi \hbar^2}. \tag{1.41}$$

Under conditions when

$$\varepsilon_0 \gg 1, \quad \alpha \ll 1, \tag{1.42}$$

it is obvious that

$$\Omega_c(2q_F) \approx \sqrt{\varepsilon_0}\,\Omega. \tag{1.43}$$

Thus, the exciton frequency band extends from $\Omega$ to $\varepsilon_0^{1/2}\Omega$, it being precisely the short-wave excitons with $\Omega_e(q) \sim \Omega_e(2q_F) \approx \varepsilon_0^{1/2}\Omega$ that determine the exciton contribution to the interaction between the electrons [see the integral (1.20) and the expression (1.34)].

To estimate the temperature $T_c$, we will write down the interaction for the generalized jellium model with the permittivity (1.34):

$$V = \frac{4\pi e^2}{\varepsilon(\omega, q)\, q^2} \approx$$

$$\approx \frac{4\pi e^2}{q^2 + \varkappa^2}\left\{1 + \frac{\omega_i^2 q^2 (q^2 + \varkappa^2)(\varepsilon_0 q^2 + \varkappa^2)^{-2}}{\omega^2 - \omega_{q,\mathrm{ph}}^2} + \right.$$

$$\left. + \frac{(\varepsilon_0 - 1)\,\Omega^2 q^2 (q^2 + \varkappa^2)^{-1}}{\omega^2 - \Omega_e^2(q)}\right\}. \tag{1.44}$$

Averaging over the angles, we find that in order to magnitude $\overline{V}(\omega) \sim V(\omega, q \sim 2q_F)$. Furthermore, when $\alpha << 1$, we can set $(2q_F)^2 + \varkappa^2 = 2q_F^2$ [see (1.41)]. Therefore, let us substitute [see (1.38) and (1.39)]

$$\omega_{q,\mathrm{ph}} = \frac{\omega_i}{\sqrt{\varepsilon_0}}, \quad \Omega_e = \Omega_e(2q_F) = \sqrt{\varepsilon_0}\,\Omega. \tag{1.45}$$

into $\overline{V}$. As a result, we arrive at the expression

$$\overline{V} = \overline{V}_c\left\{1 + \frac{\omega_i^2/\varepsilon_0^2}{\omega^2 - \omega_i^2/\varepsilon_0} + \frac{(\varepsilon_0 - 1)\,\Omega^2}{\omega^2 - \varepsilon_0\Omega^2}\right\}, \tag{1.46}$$

where $\overline{V}_c$ is the Coulomb interaction $V_c = 4\pi e^2/(q^2 + \varkappa^2)$ averaged over $q$. Evidently, $\overline{V}_c \sim (4\pi e^2/q_F^2)$, but the quantity $\overline{V}_c$ figures in the expression as a common factor, and therefore it is more proper to introduce it as a parameter.

In accordance with what was said in Section 3, a rough computation of $T_c$ can be carried out after making a further simplification, namely, after replacing the function $\overline{V}(\omega)$ by a set of square wells [see (1.27) and Fig. 1.4]. Such a computation shows that for $\varepsilon_0 >> 1$ (i.e., in the case of great "strength of the oscillators" corresponding to the excitons) the exciton contribution predominates, and the interaction with the phonons is unimportant.

Further, if

$$\Omega_e = \sqrt{\varepsilon_0}\,\Omega \ll \omega_F, \tag{1.47}$$

then the Coulomb interaction possesses a broader spectrum than the exciton interaction, on account of which it turns out to be "suppressed." As a result, we return to the result (1.29) with $\omega_c$ replaced by $\Omega_e$, and so we obtain the BCS formula:

$$T_c \sim \Theta_e e^{-\frac{1}{\lambda - \mu^*}},$$

$$\Theta_e = \frac{\hbar\Omega_e}{k_B} = \frac{\hbar}{k_B}\sqrt{\varepsilon_0}\,\Omega, \quad \lambda = N(0)\,|\overline{V}_a|,$$

$$\mu = N(0)\,\overline{V}_e, \quad \mu^* = \frac{\mu}{1+\mu \ln(\omega_F/\Omega_e)}. \tag{1.48}$$

Then for the generalized jellium model specifically, as is clear from (1.46) with $\omega = 0$ and $\varepsilon_0 >> 1$, $\lambda = \mu$ [this applies also in the case of the conventional jellium model (1.7), (1.8); for greater details, see Chapter 2]. Incidentally, it is often asserted in the literature that the stability condition for the system has the form $V(0, q) \geqslant 0$ or $\varepsilon(0, q) \geqslant 0$ when the permittivity $\varepsilon(\omega, q)$ is used. As can easily be verified, this means that $\lambda \leqslant \mu$ and, consequently, that the jellium models with $\lambda = \mu$ correspond from this point of view to the highest possible value of $\lambda$. Meanwhile, as shown in Chapter 2, the requirement that $\varepsilon(0, q) \geqslant 0$ at large values of $q$, which are important in the theory of superconductivity, is not at all necessary and, consequently, the jellium model cannot be considered to be a limiting theory (in the indicated sense).

The formula (1.48) has been written in the more general form (i.e., without the assumption that $\lambda = \mu$, which pertains to the jellium model), allowing for a transition to models more general than the jellium model. Such a transition is so much more justified that we dwelt on the jellium models only for some clearness and concreteness. If, on the other hand, we proceed at once from the general expressions, as is done in Chapters 2-4, then the phonons and excitons emerge with entirely equal roles to play, and correspond to different (low- and high-frequency) peaks of the number density function, $F(\omega)$, for the states in the expression for $\lambda$ [see (1.32)]. From this it is clear that the result (1.48) is in fact equivalent to the BCS formula (1.3a). In the strong-coupling case we obtain a formula similar to the expressions (1.32) and (1.33). Thus, in the case of superconductivity connected with the exciton mechanism, as a result of the substitution in (1.48) of $1/(1 + \lambda)$ for $\lambda$, we obtain

$$T_c \sim \Theta_e \exp\left\{-\frac{1+\lambda}{\lambda - \mu^*(1+\lambda)}\right\},$$
$$\mu^* = \frac{\mu}{1+\mu \ln(\Theta_F/\Theta_e)}, \quad \Theta_F = \frac{\hbar}{k_B}\omega_F. \tag{1.49}$$

From this expression, as, indeed, from the formula (1.48), it is clear that for suitably chosen values of the parameters $\theta_e$, $\lambda$, and $\mu^*$ the temperature $T_c$ can be appreciable. Thus, the center of gravity of the problem undoubtedly lies in the rating or estimation of the various possibilities of obtaining the most advantageous values for the parameters.

Certain limitations on that score can already be seen from the formula (1.49). Indeed, as the temperature $\theta_e$ (i.e., as the exciton frequency $\Omega_e = \varepsilon_0^{1/2}\Omega$ increases, the temperature $T_c$ increases on account of the growth of the pre-exponential factor but decreases as a result of the growth of the parameter $\mu^*$. If, as an example, we assume that the parameters $\lambda$ and $\mu$ are constants and only the temperature $\theta_e$ can vary, then the temperature $T_c$ has its highest value at $\theta_e = \theta_{e,\,max}$ and is equal to

$$T_{c,\max} \sim \Theta_F \exp\left\{-\frac{4(1+\lambda)}{\lambda} + \frac{1}{\mu}\right\} = \Theta_{e,\max} \exp\left\{-\frac{2(1+\lambda)}{\lambda}\right\},$$

$$\Theta_{e,\max} = \Theta_F \exp\left\{-\frac{2(1+\lambda)}{\lambda} + \frac{1}{\mu}\right\}. \tag{1.50}$$

Setting $\theta_F = (E_F/k_B) \sim 10^5$ K ($E_F \sim 10$ eV), and assuming that we can use the values $\lambda \sim 1$ and $\mu \lesssim 0.5$, which are realized for certain known superconductors under conditions when the phonon mechanism operates, we obtain the estimate $T_{c,\,max} \lesssim 300$ K, $\theta_{e,\,max} \sim 10^3$ K.

If we use the isotropic and homogeneous model of a metal, in which case $\varepsilon(\omega, q)$ in (1.5) has the meaning of a dielectric constant and also assume that $\lambda \leqslant \mu \lesssim 0.5$ (as has already been pointed out, for the jellium model $\lambda = \mu$; the value $\mu \lesssim 0.5$ follows from various estimates of the contribution of the Coulomb repulsion), then the temperature $T_{c,\,max} \lesssim 10$ K (this value is obtained when $\theta_F \sim 10^5$, $\lambda = \mu = 0.5$). This fact led to the publication in the literature of assertions of the ineffectiveness of the exciton mechanism.

This point of view is, however, based on a misunderstanding. It is sufficient to say that we could have similarly "proved" that the phonon mechanism is also ineffective. Meanwhile, as has already been emphasized, we know from the experimental data that we are not at all obliged to use the condition $\lambda \leqslant \mu$ in formulas of the type (1.33) and that in a number of cases $\lambda \gtrsim 1$, $\mu \lesssim 0.5$. Even aside from the incorrectness or, more exactly, the dispensability of the stability condition $\lambda \leqslant \mu$, the point here is simply that the exchange of phonons or excitons with momenta $q \sim 2q_F$ is primarily responsible for the superconductivity-inducing attraction between the electrons in the vicinity of the Fermi level.

For such short-wave excitations (phonons or excitons), a metal can no longer be considered to be a continuous medium, and, generally speaking, the umklapp processes become important. Therefore, the use of the permittivity $\varepsilon(\omega, q)$ is literally inadmissible, and the corresponding approach has a limited region of applicability. Nevertheless, a qualitative allowance for the Umklapp processes and, generally, for the inhomogeneity of the metal is attained by carrying out the integration in the formula (1.18) and similar formulas right up to the momentum $2p_F = 2\hbar q_F$. That is how we dealt with the problem. Briefly, in the case of the exciton mechanism, as for the phonon mechanism, we do not see any basis for using in formulas of the type (1.33) and (1.49) the values $\lambda \leqslant \mu \lesssim 0.5$, and it is quite legitimate, at least on the basis of the experimental data, to assume a value of $\lambda \sim 1$. The fact that for the phonon mechanism the foregoing is an already established fact while, for the exciton mechanism the question of the actual value of $\lambda_{max}$ remains open, is another matter. But with this reservation, we can, within the framework of all that is known, give only the following estimate:

$$T_c \lesssim 300\ \mathrm{K}\ \text{(exciton mechanism of superconductivity)}. \qquad (1.51)$$

Naturally, even within the limits of the above-used assumptions, the estimate (1.51) is a very rough one. But it can quite clearly be seen that there are hardly any grounds for the estimate $T_c \gtrsim 10^3$ K, even with optimally chosen parameters for more or less real materials. The possibility of lowering the limit (1.51) is another matter. Perhaps for some as yet unknown reasons, it is impossible in principle to attain a sufficiently strong coupling with $\lambda \sim 1$ for the exciton mechanism and, in general, produce a material with values of the parameters $\lambda$, $\mu^*$, $\theta_F$, and $\theta_e$ that correspond to high $T_c$. When a formula of the type (1.49) is used, the problem of high-temperature superconductivity reduces to the investigation of such a possibility. The difficulty in solving the problem arises out of the fact that we are talking largely about numerical values and then under conditions of a strong (exponential) dependence of $T_c$ on $\lambda$ and $\mu$.

Below we shall proceed to a qualitative discussion of a number of possibilities that open up along the route to raising $T_c$. But now let us note again that, in terms of the permittivity $\varepsilon(\omega, q)$ and for $\varepsilon(0, q) > 0$, the most suitable materials are those with the highest possible value of $\varepsilon(0, q \sim q_F)$. For materials with $\varepsilon(0, q \sim q_F) < 0$, the situation is even more auspicious (these assertions are justified in Chapter 2; see also [14]).

The foregoing pertained, generally speaking, only to systems in thermodynamic equilibrium. In particular, it was precisely under this assumption that we computed the critical temperature in Section 3. The transfer of the system into a nonequilibrium state† with an inverted electron distribution (e.g., under the action of the electromagnetic field in a "pumping" laser wave) opens up new possibilities. One of them is connected with the fact that the coupling constant $\lambda$ and the quasiparticle (normal-electron) distribution function $f$ enter essentially in the combination $\lambda(1 - 2f)$. Therefore, for the inverted distribution, when $(1 - 2f) < 0$, superconductivity is possible in systems with repulsion between the electrons (in this case, in the adopted notation, $\lambda < 0$). The critical temperature of nonequilibrium systems is not longer given by the formulas presented above and can be high even when, for example, $|\lambda| \lesssim 0.5$. Nonequilibrium systems will be exclusively considered in Chapter 9.

†A given substance has at a given temperature only one equilibrium state (or, say, several such states if we also count the metastable, but practically equilibrium, phases). At the same time, there are infinitely many nonequilibrium states, and they are characterized not by a number—a temperature $T$—but by electron, phonon, etc., distribution functions. From this we can clearly see the exceptional wealth of possibilities in the investigation of nonequilibrium superconducting systems. It may be inferred that the study of such systems will be one of the most important directions in further investigations in the field of superconductivity.

## 5. Ways of Producing High-Temperature Superconductors (Some Possibilities and Applications)

Since the question of the feasibility of the production of high-temperature superconductors has not been resolved, a broad approach to the problem—a search for the most diverse ways, a discussion of the various proposals—is especially appropriate here. And such ways and proposals already abound in the literature. Below we briefly discuss their characteristics, dividing quite conditionally all superconducting systems (materials) into three-dimensional, quasi-one-dimensional, and quasi-two-dimensional systems. Such a division is conditional mainly because, on the atomic scales of interest to us, a metal is inhomogeneous, and there is no sharp distinction between, say, a three-dimensional lattice and an ensemble of very closely spaced atomic layers or chains. Nevertheless, the respective differences that manifest themselves in, in particular, the occurrence of macroscopic anisotropy can be fairly marked, especially when different models are used.

**A. Three-Dimensional Systems.** The known superconducting materials, with the exception of a number of layered compounds (extensive investigation of which began only relatively recently, largely in connection with the problem of high-temperature superconductivity), are three-dimensional systems. More precisely, the properties of some of these materials (in particular, $Nb_3Sn$) are sometimes related to the presence in them of a straight-chain structure and the corresponding distinctive features of the electronic spectrum, but in the macroscopic respect these materials are homogeneous and quasiisotropic, as are, in general, the overwhelming majority of metals.

It seems natural that high-temperature superconductors should be sought first and foremost among three-dimensional materials. If we are talking about the phonon mechanism, then this is indeed the case. For example, metallic hydrogen can be classified with three-dimensional materials, although its structure has probably strongly pronounced one-dimensional features. In this case, as has already been noted, the high $T_c$ values, if they are realized, can be connected with the high value of the Debye temperature $\theta_D \sim 3 \times 10^3$ K and the fairly large parameter $\lambda$ (for greater details, see Chapter 4).

Generally speaking, it should be noted that for a number of the already investigated metals the Debye temperature is high ($\theta_D \sim 10^3$ K) and the low critical temperatures $T_c$ are connected with the smallness of $\lambda$ or the difference $g = \lambda - \mu^*$. In view of this, attention is attracted by the various hydrogen-containing substances ($LiH_2F$, PdHHg, etc.) under, or after having been subjected to, pressure and, mainly, the possibility of producing an "organic" metal—a metal made up of quite large organic molecules.

The organic character of the material (i.e., the presence of carbon) is, of course, not fundamental here. It is just that carbon is light, and there actually exist organic molecules with quite high vibrational frequencies.† The combination of such molecules into a solid should ensure a large $\theta_D$ value.

At the same time, there are reasons to believe (see Chapter 7) that the Coulomb repulsion in such a metal, if it were to form (in the overwhelming majority of the known cases organic molecules combine to form semiconductors), would be suppressed. This means that the parameter $\mu$ would be small and that even when $\lambda \sim 0.3$ we could expect values of $T_c \sim \theta_D e^{-1/\lambda} \lesssim 100$ K.

As for the exciton mechanism, three-dimensional systems are, at least at first glance, not suitable. Aside from the extensive empirical data (many three-dimensional systems have been investigated, but the exciton mechanism has nowhere clearly manifested itself, even if it has, in general, turned out to be important), we simply have in mind the absence in metals (at least in normal metals) of suitable exciton bands. In fact, in the description of the properties of a metal with the aid of the permittivity $\varepsilon(\omega, q)$, the condition $\varepsilon(\omega, q) = 0$ corresponds to the longitudinal-exciton band. The corresponding frequencies $\omega_q \equiv \Omega_e(q)$ should also satisfy the condition $\Omega_e << \omega_F$ [see (1.47)].‡

The known exciton branch, called the plasmon branch, does not meet the last condition. Thus, for small $q << q_F$, the characteristic plasmon frequency

$$\omega_p = \sqrt{\frac{4\pi e^2 n}{m}} = 5.64 \cdot 10^4 \sqrt{n}, \tag{1.52}$$

where $n$ is the electron concentration. In a good metal $n \sim 10^{23}$ cm$^{-3}$ and $\omega_p \sim 10^{16}$, i.e., $\hbar\omega_p \sim E_F \sim 10$ eV [for the degenerate electron gas $\omega_p = (4\pi e^2 n/m)^{1/2} = (16\alpha/3)^{1/2}_{F}\omega_F$; see (1.19) and (1.21)].

Furthermore, plasmons are clearly weakly damped only when

$$\frac{q^2 v_F^2}{\omega_p^2} = \frac{3}{4\alpha}\left(\frac{q^2}{q_F^2}\right) = \frac{3q^2}{\varkappa^2} \ll 1. \tag{1.53}$$

Even when $\alpha \sim 1$, not to mention the case $\alpha << 1$, the plasmons with $q \sim 2q_F$ are, in general, very strongly damped and, to put is more accurately, do not exist as a branch of elementary excitations.

Thus, the plasmons in question cannot ensure attraction between the conduction electrons. As to excitons with frequencies $\Omega_{e,\,\max} << \omega_p \sim \omega_F$ (or

†The assertion that the organic nature of the hypothetical metals under discussion is in itself not important is supported by the discovery of the possession of metallic properties by the compound $(SN)_x$. Moreover, this metal (which does not contain a single metallic atom!) turns out to be superconducting ([15] and Chapter 7). The corresponding value of $T_c$ ($\approx$ 0.3 K) is not high, but this cannot in any way dampen interest in the study of metallic materials containing light nonmetallic atoms.

‡Actually, this condition is too rigid since we need the fulfillment of the weaker condition $\overline{\omega}_{\log} << \omega_F$ [see (1.32a) and Chapter 2].

even with frequencies $\Omega_{e,\max} < \omega_p$), as far as we know, they do not exist in a good metal (we have in mind here excitations that are weakly damped over a wavelength, or, formally speaking, frequencies $\Omega_e$ whose real part $\mathrm{Re}\Omega_e$ is greater than the imaginary part $\mathrm{Im}\Omega_e$).

On the whole, this fact is understandable: the bound electrons with energy below the Fermi level $E_F$ should form an aggregate. An exception could be, say, electrons bound to some impurities and separated from the conduction electrons by a sufficiently high barrier. But in the latter case there is no reason to expect a sufficiently strong coupling between the excitons and conduction electrons, i.e., the parameter $\lambda$ will be small.

If the excitations are weakly damped, then the existence of the branch corresponding to them is obvious. As the damping becomes stronger, the excitations as it were lose their individuality but can still make a large contribution to the interaction between the electrons. In terms of the permittivity $\varepsilon(\omega, q)$, the strong damping of the longitudinal excitons implies that the dispersion condition $\varepsilon(\omega_q, q) = 0$ is satisfied when $q$ is real only for complex frequencies, $\omega$, with a large imaginary part. This circumstance by itself does not yet prevent the corresponding frequency region from making a significant contribution to the interaction between the electrons. The same thing can be said about the case in which the dispersion condition $\varepsilon(\omega_q, q) = 0$ does not have solutions at all at large $q$, and there can be no talk of excitations—they cease to exist. On the other hand, the dispersion condition $\varepsilon = 0$ can have solutions at large $q$, but the contribution of this region to the interaction turns out, upon integration, to be small.

In the case of plasmons, even if in some cases they do exist also at $q \sim 2q_F$, their role is minor (the fact that the plasma frequency is too high is already sufficient). It may be inferred that the other electronic excitons in normal metals either are very weakly bound to, or "dissolve" in, the conduction electrons and do not form a separate excitation branch. Therefore, the conclusion drawn in the text that there do not exist in normal metals electronic excitons that secure attraction between the electrons is, in fact, not restricted to only weakly damped excitations.

The foregoing is automatically taken into account in the theory expounded in Chapter 2; that theory is based on the use, for determining $T_c$, of the permittivity $\varepsilon(\omega, q)$. In this case the parameter $\lambda$ is, in the final analysis, determined by the value of $\varepsilon(0, q)$, and we may, if we wish, forget the exciton spectrum and the graphic picture in which the interaction between the electrons is linked with the exchange of excitons.

Undoubtedly, the qualitative considerations presented above, like the analysis presented at the end of Chapter 2, is not yet sufficient for us to "close" the possibility of an effective operation of the exciton mechanism in three-dimensional systems. Moreover, the literature (see [15]) contains a number of suggestions concerning the use of the exciton mechanism in such three-dimensional systems as the transition metals with "light" $s$ and "heavy" $d$ electrons, "alloys" of metals with nonmetals (dielectric admixtures or impurities), heavily doped semiconductors, etc. Among the objections against the possibility of using the exciton mechanism in the transition metals, we can cite the negative influence of impurities and the hybridization of the bands (see Chapter 4).

Needing to be discussed separately are metals, akin to semiconductors (they are often called semimetals), whose state is close to the region of the electronic phase transition—a transition during which the electronic spectrum changes [16]. Under such conditions the interelectron interaction plays an important role and can, in particular, lead to a drastic change in the density of electron states in the vicinity of the Fermi level [we are, thus, talking about a change in the quantity $N(0)$]. As a result, the critical temperature $T_c$ can rise sharply even in the case of the phonon-mediated attraction between the conduction electrons. In a sense, it is as if we are dealing with "interference" between an electronic structural transition and the superconducting transition.

The consideration of this variant in the present chapter based, as it is, on the simpler notions of the phonon and exciton mechanisms, would be difficult. Therefore, we refer the reader to Chapter 5, and confine ourselves to pointing out the great interest attached to the corresponding direction in the search for high-temperature superconductors (this direction is, incidentally, closely tied with the problem of lattice instability and the quasi-one-dimensional nature of the electron spectrum and the corresponding materials).

The study of models of the semiconducting type has led, as a "by-product," to the discovery of the possibility of the existence of materials with spontaneous currents [17, 18]. In such materials, there exists in the ground state below some phase-transition temperature, $T_c$, an electric current $\mathbf{j} \neq 0$. The spontaneous currents should be closed, and the corresponding state analogous, as it were, to the presence of some kind of orbital antiferromagnetism. At the same time, on account of the large size of the current rings, a material with spontaneous currents may behave in an external magnetic field like a superdiamagnet (see [19, 20]; we call a substance for which the diamagnetic susceptibility $\chi$ is, in order of magnitude, comparable or even close to the susceptibility, $\chi = -1/4\pi$, of an ideal diamagnet a superdiamagnet). It is possible that the diamagnetic anomalies observed in cuprous chloride (CuCl) [21, 22] give, if they are not connected with some side effects, precise evidence of a transition into the superdiamagnetic state (it is very important that the conductivity preserve its semiconducting character in such a transition). It is interesting that the hypothesis of spontaneous currents was used in the past to explain superconductivity [23-25]. However, it seemed even then [25], before the elucidation of nature of superconductivity in 1957, that a substance with spontaneous currents would behave not like a superconductor, but rather like an orbital antiferromagnet. We refer the reader to [17-25] and Chapter 5 for details and only note here that at the present stage the investigation of substances with spontaneous currents has turned out to be connected with the problem of high-temperature superconductivity (the cause of this connection is clear in Chapter 5). At the same time, it should be emphasized that a system with spontaneous currents can in no way be regarded as superconducting.

**B. On One- and Two-Dimensional Superconductors. The Role of the Fluctuations. Connection with the Problem of High-Temperature Superconductivity.** Superconductivity was discovered and studied as a volume, three-dimensional effect. True, finite-sized samples, including even very thin (on the macroscopic scale) films and wires have always been used. But until recently, this circumstance was not considered to be significant. And what is more, the question arose whether the existence of two-dimensional superconductivity [e.g., superconductivity for electrons in surface (Tamm) states] was possible [26].

On the basis of the superconductivity theories (both macroscopic and microscopic) in which the fluctuations are neglected, there can be no objections to the consideration of not only two-dimensional, but also one-dimensional, superconductors. But if the fluctuations are not neglected, then, as has been demonstrated [27, 28], a superconducting phase transition, which involves the appearance of "long-range order,"† cannot occur in a two- or one-dimensional system in a state of thermodynamic equilibrium. Here a system is considered to be two-dimensional (one-dimensional) not only when it is an infinitely thin film (filament); it is sufficient for this purpose that its momentum space be two dimensional (one dimensional). For example, for a filament (tube) filled with an ideal gas, the particle energy

$$E = \frac{p_z^2}{2m} + E_n, \quad E_n \sim \frac{\hbar^2 n^2}{2mr^2},$$

where the $z$ axis coincides with the direction of the filament and $r$ is the filament radius. The gas in the filament can be considered to be strictly one dimensional if virtually all the particles are at the lowest level $n = 1$ and, consequently, $E =$ const + $(p_z^2/2m)$.

We note that the cited results concerning the absence of the phase transition and existence of long-range order are, in fact, more powerful than implied above—they are generally applicable to filaments (cylinders) and plates that are infinite in only one or two directions.

The physical cause of this situation is clear, especially in the case of filaments. It is sufficient to rupture the filament at just one place for the long-range

†The superconducting state is characterized by the complex order parameter $\Psi = |\Psi|e^{i\phi}$ (if we do not consider gapless superconductors and highly inhomogeneous systems, then the modulus $|\Psi|$ is proportional to the gap width $\Delta$). The presence of long-range order implies that the values of $\Psi$ for any two points of the superconductor are correlated no matter how far apart the two points are. This means that when we neglect fluctuations (which should be sufficiently weak) both $|\Psi|$ and the phase $\phi$ vary along the superconductor in quite a law-governed fashion, so that, for example, $|\Psi|$ = const and $\phi = 0$ or $\phi = kz$.

order and the superconductivity to disappear. At the same time, to effect such a rupture, it is sufficient to make a segment of the filament with cross section of the order of $r^2$ and length $l$ significantly greater than the atomic dimensions go over into the normal state, i.e., to "close" the energy gap for such a segment. Since the volume, $r^2 l$, of the breaking fluctuation is finite, such a fluctuation can, in principle, arise at $T \neq 0$.

It is clear, however, that in the case of a macroscopic radius $r$ the appearance of the fluctuation in question is highly improbable. But in thermodynamics the question of equilibration time is not considered, and the possibility of the destruction of a long-range order leads, even in the case when the probability for the destruction is extremely small, to the conclusion that the long-range order cannot exist in the equilibrium state. The situation for two-dimensional systems is similar: The probability for the appearance of a strip (formally a line) in the normal state is still high enough for the long-range order to be destroyed over an infinitely long period of time (i.e., in the equilibrium state). It must, however, be borne in mind that, in the quantitative respect, there is a great difference between a filament and a plate (film): The probability for the fluctuation that destroys the superconductivity (in general, the long-range order) is incomparably weaker in the case of a film than for a filament (see also Chapter 6).

Filaments (fine wires) and films (with radius $r$ or thickness $d$ exceeding hundreds, sometimes even tens, of angstroms) can remain in the superconducting state for practically an infinitely long period of time since the probability for the states that destroy the superconducting current is extremely small.

The problem, touched upon here, of the role of fluctuations in superconductivity in one- and two-dimensional systems also has a number of other aspects, which are discussed in the literature (see, for example, [27–28]). We cannot, however, discuss them here in greater detail and confine ourselves to stating that the fact that a system is two- or one-dimensional is, in general, not an impediment from the point of view of the existence of superconducting properties. It is borne in mind here that, in practice, it is systems with dimensions (thickness) significantly (at the least 10 times) greater than the atomic dimension $a \sim 10^{-8}$ cm that are normally discussed. Moreover, the systems that are considered are often not one- or two-dimensional, but quasi-one-dimensional or quasi-two-dimensional systems, which consist of weakly coupled filaments or layers. On account of the presence of couplings, such systems are three-dimensional with respect to long-wave excitations, but for excitations with wavelength shorter than the distance between the filaments or layers, they can, generally speaking, be considered to be one- or two-dimensional.

The foregoing in no way implies that the fluctuations can always be neglected and that we can unconditionally use the "mean-field" approximation and, in particular, a BCS-type theory. On the contrary, the smaller the characteristic dimensions of a system are, the more important are the fluctuations, and it is necessary to keep their actual or possible role in mind and control it. This pertains particularly to quasi-one-dimensional systems since the fluctuations in quasi-two-dimensional and even purely two-dimensional systems are significantly weaker than the fluctuations in one-dimensional systems and can even be regarded as effectively suppressed in two-dimensional samples with finite, but still macroscopic, dimensions. (We are obviously talking about the length and width of arbitrarily thin superconducting films, etc.) But at the same time we cannot *a priori* object to the consideration of one- and two-dimensional systems as potential superconductors only on the basis of the above-presented general results concerning the impossibility of the existence in such systems of equilibrium superconducting long-range order.

All this has a direct bearing on our theme in connection with the above-noted difficulties that have been encountered in the production of three-dimensional high-temperature superconductors. Although these difficulties are, possibly, surmountable, it is quite natural to analyze the corresponding possibilities in application to quasi-one-dimensional and quasi-two-dimensional systems. Moreover, in application to the exciton mechanism of superconductivity, there are direct grounds for such an approach.

Indeed, the main difficulty in the application of the exciton mechanism to three-dimensional systems is connected with the difficulty in producing a metal with the requisite exciton band (see above). But dielectrics (semiconductors) with diverse exciton bands exist. Therefore, the idea has emerged that we could bring together a metal and a dielectric, thus producing metallic filaments surrounded by a dielectric or dielectric–metal–dielectric "sandwich" (Fig. 1.6).

Naturally, close properties should be possessed by the filamentary or layered combinations; the latter, for example, are similar to a pile of sandwiches. In all such systems, the excitons should, according to the idea, exist and propagate largely in the dielectric parts (i.e., in the facings of the sandwich, etc.), while the electrons, concentrated in the metallic filaments or layers, will superconduct.

If superconductivity resulted from the interaction between electrons located

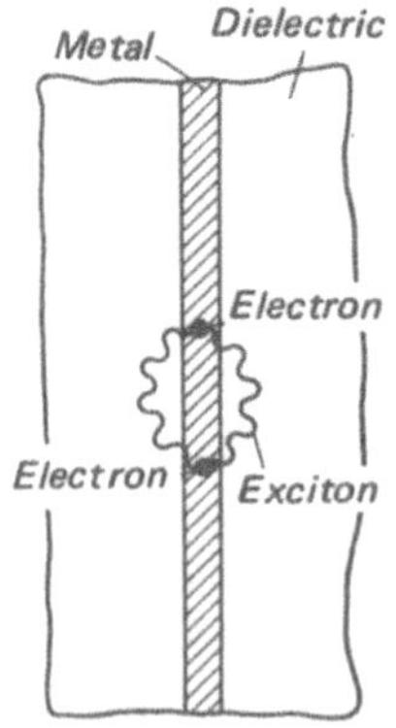


**Figure 1.6. A dielectric–metal–dielectric sandwich.**

at large, macroscopic distances, then the use of the inhomogeneity of the system—its division into a metallic and a dielectric part—perhaps would not have met with objections and would have led to success. But, in fact, for superconductors, small atomic distances, at which ordinary three-dimensional solids are inhomogeneous, play a role. Furthermore, it should be borne in mind that under conditions of relatively short-range forces, the spatial separation of the metallic and dielectric parts does have its own disadvantages: In this case the interaction between the electrons and excitons is weakened (indeed, in order to interact, the excitons must, roughly speaking, penetrate into the metal or the electrons should penetrate into the dielectric).

As a result, the difference between quasi-one-dimensional, quasi-two-dimensional, and three-dimensional metals becomes less sharp and, in a sense, conditional, it being important only in quantitative calculations. It is, however, the last calculations that are of interest to us in the computation of $T_c$, the attainment of high $T_c$ values in quasi-one-dimensional or quasi-two-dimensional systems being possible only if we are able, on account of the inhomogeneity and the existence of excitons in the dielectric part, to gain considerably more than we lose as a result of the weakening of the interaction.

**C. One-Dimensional and Quasi-One-Dimensional Systems.** The property of one-dimensional superconductivity should, according to an idea put forward in [3, 39], be possessed by a long molecular conducting chain with side molecular branches that play the role of a polarizable medium, or, to put it differently, a dielectric shell in which excitons can propagate. The existence of superconductivity with equilibrium long-range order in such a chain is impossible; the fluctuations in the one-dimensional chain will be especially strong even in the nonequilibrium (metastable) state. Specifically, it is still not clear for how long the superconducting current can flow along one filament, and it is possible that it will be difficult to speak of superconductivity.

Difficulties arise, however, even at an earlier stage, in the sense that sufficiently long molecular chains with metallic conductivity have not been successfully synthesized (the polymer chains that have so far been produced are semiconductors). This result is, evidently, not accidental; it is because of the instability of one-dimensional metallic conducting chains (for greater details, see Chapter 7).

A quasi-one-dimensional system that is comprised of an ensemble of parallel conducting filaments coupled among themselves to some extent, but possessing a substantially lower conductivity in the directions perpendicular to them than in their direction, is free from the indicated shortcomings to a certain extent. Materials of this type and, in particular, tetracyanoquinodimethane (TCNQ) based compounds and complex platinum compounds have been intensively investigated in recent years.

These compounds and their properties will be discussed in Chapter 7 (see also [40] for a review). Here we confine ourselves to noting that there are no indications of the possibility of the occurrence of superconductivity in such systems if their anisotropy is very high. This is apparently connected with the existence, for a broad class of quasi-one-dimensional systems, of an instability similar to the one that occurs in one-dimensional metals, namely, the instability with respect to the metal-to-dielectric phase transition. Only when the interaction between the filaments is sufficiently strong will this instability be suppressed. We, however, thus arrive at systems that are nearly three-dimensional but still have one-dimensional features.

Some of these systems—semimetals with quasi-one-dimensional density of states—are promising and are considered in Chapter 5. The same thing can be said about organic metals or $(SN)_x$-type compounds, which have already been mentioned in Section 5A. As to the more strongly pronounced one-dimensional chains and quasi-one-dimensional compounds (the TCNQ-based compounds, for example), as has already been noted, they do not at present seem to be promising candidates in the plan for the realization of high-temperature superconductivity (see, however, [41] and Chapter 7).

**D. Two-Dimensional and Quasi-Two-Dimensional Systems (Dielectric-Metal-Dielectric "Sandwiches" and Layered Compounds).** It would have been possible for us to speak of a two-dimensional superconducting system if, for example, the above-mentioned superconductivity of electrons in surface states [26] has been realized. Since the fluctuations in two-dimensional systems are very strongly suppressed (as compared to those in one-dimensional systems), the absence in the two-dimensional case of strictly equilibrium superconducting long-range order is scarcely of vital importance (see Section 5B).

A more important obstacle, at least at present, is the difficulty encountered in the preparation and study to a sufficient degree of ideal surfaces with a controllable surface-level spectrum. As a result, it is still very difficult to investigate even normal metallic surface conductivity on the surface of a bulk dielectric, to say nothing of metals.

Thus, the study of the problem of surface superconductivity in application to a particular surface of a solid is, perhaps, not a matter for the very near future.†

In any case, attention is now focused on two other cases: dielectric-metal-dielectric sandwiches and layered compounds. Sandwiches with a metallic film of appreciable thickness fall into the quasi-two-dimensional rather than the two-dimensional systems. This applies *a fortiori* to layered compounds when the interaction (specifically, the conduction) between the layers is taken into account. However, in the limit of a very thin metallic film in a sandwich and a very large distance between the highly conducting (metallic) layers in layered

†Let us emphasize at the same time that the problem of surface superconductivity is very interesting (and potentially important from the point of view of possible applications), even in the case when the corresponding critical temperatures are not high.

compounds both of these systems can practically be considered to be two-dimensional (for greater details, see Chapter 6).

It is clear even from very general considerations (see, for example, [5]) that the metallic film in a sandwich should be extremely thin ($d \lesssim 10$-$20$ Å). The requisite exciton band ($\hbar\Omega_e \sim 0.1$-$0.3$ eV) should exist in the dielectric part of the sandwich and must be pronounced at large values of $q \sim q_F$. A more detailed analysis and, strictly speaking, a rough calculation of $T_c$ for sandwiches [42, 43] corroborate these conclusions and at the same time indicate that it will not be easy to attain values of $T_c \gtrsim 40$ K for sandwiches.

This question is elucidated in Chapter 8; here we confine ourselves to citing the expression, discussed there, for the effective constant that describes the coupling between the conduction electrons in the metallic film and the excitons in the dielectric (semiconductor) layers:

$$\lambda_{eff} = -\frac{2\beta}{\pi v_F}\frac{2S_F}{(2\pi)^3}\frac{4\pi e^2}{\bar{q}^2}\int\frac{d\omega}{\omega}\,\mathrm{Im}\left[\frac{1}{\varepsilon(\omega, \bar{q})}\right], \tag{1.54}$$

where $\beta$ is a numerical coefficient that takes account of the fact that the electrons that tunnel into the semiconductor pass only part of the time there, $\hbar\bar{q}$ is some mean value of the momentum transfer, $S_F$ is the area of the Fermi surface in the metallic film ($v_F$ is the electron velocity at this surface), $\varepsilon(\omega, \bar{q}) = \varepsilon' + i\varepsilon''$ is the effective permittivity, and, finally, the integration is performed over the frequency region where the value of

$$\mathrm{Im}\,\frac{1}{\varepsilon(\omega, \bar{q})} = -\frac{\varepsilon''}{(\varepsilon')^2 + (\varepsilon'')^2}$$

is high.

It should be noted that, as is evident from (1.31),

$$-\frac{2}{\pi}\int_0^{\infty}\frac{d\omega}{\omega}\,\mathrm{Im}\left[\frac{1}{\varepsilon(\omega, \bar{q})}\right] = 1 - \frac{1}{\varepsilon(0, \bar{q})}, \tag{1.55}$$

since the quantity $1/\varepsilon(0, \bar{q})$ is real {see the expression (1.31), according to which $\mathrm{Im}[1/\varepsilon(0, \bar{q})] = 0$}. In view of this, if the integration in (1.54) were performed over all frequencies, then the value of $\lambda_{eff}$ would be proportional to $1 - [1/\varepsilon(0, \bar{q})]$.

It should also be remembered that the use of the permittivity $\varepsilon(\omega, \bar{q})$ at the large values of $q \sim q_F$ of interest to us is possible only with important limitations and has, in fact, a conditional character even for an unbounded material. In the presence of boundaries, however, the situation becomes even more significantly complicated, and, above all, becomes potentially rich in new possibilities [44]; therefore, the presently available estimates of $T_c$ for sandwiches are not yet sufficiently general and, consequently, not sufficiently convincing.

Thus, the expression (1.54) has been cited only by way of an illustration and for the purpose of indicating one of the ways of choosing materials for sandwiches (we have in mind the use of materials with the maximum value of

$$-\int \frac{d\omega}{\omega} \operatorname{Im}\left[\frac{1}{\varepsilon(\omega, \bar{q})}\right]\Big).$$

The theoretical and, to an even greater extent, the experimental study of sandwiches has actually only just begun (see [45, 46] and Chapter 8).

The investigation of superconducting layered compounds began comparatively recently; nevertheless, it has already led to a number of interesting results (see [47, 48] and Chapter 6). We note, in particular, the study of a number of dichalcogenides with intercalated molecular layers.

The first example of this type is the pyridine ($C_5H_5N$) intercalated, layered compound $TaS_2$ schematically represented in Fig. 1.7. The intercalation led here to the raising of $T_c$ from 0.7 to 3.5 K. However, the analysis of this specific case, as of a number of other cases (see Chapters 6 and 8), does not give grounds for believing that the exciton mechanism has already been successfully used to raise $T_c$.

The point is that for large momentum transfers (i.e., for $q \sim q_F$) the conduction electrons interact weakly with the rather large pyridine molecules, which are apparently coupled to each other and to the $TaS_2$ layers only by van der Waals forces. Furthermore, the first excited electronic level in pyridine lies too high ($\hbar\Omega \sim 5$ eV), and this impedes the suppression of the Coulomb repulsion [we have in mind the condition $\Omega_e < \omega_F$; see (1.47)].

However, even when $\hbar\Omega \sim 0.1$-$1$ eV, the exciton mechanism will not be sufficiently strong if we are talking about molecular layers: the lower the excited

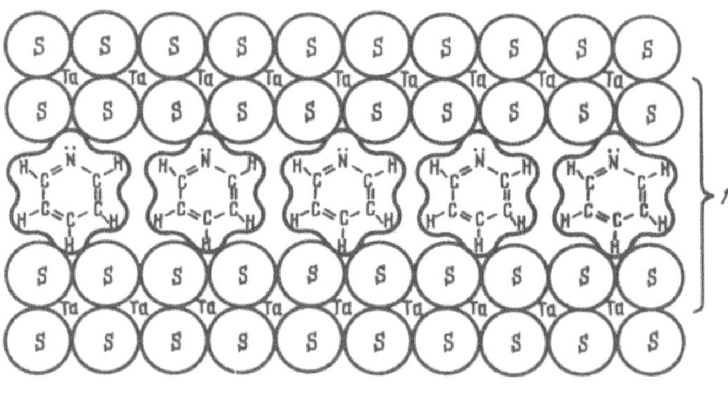


**Figure 1.7. Schematic structure of the Ta, S, and pyridine layers in $TaS_2(Py)_{1/2}$.**

level is, the larger the molecule is, as a result of which the transfer of large momenta $q \sim q_F$ will be impossible (of low probability). In the exciton language, we can say that in these cases there are no excitons, or at most weakly damped excitons with $q \sim q_F$ and a sufficiently large "oscillator strength."

Thus, for layered compounds with molecular layers, on account of the large molecule size ($a \gtrsim$ 10–20 Å), which is necessary for obtaining low excited levels ($\hbar\Omega < 1$ eV), the exciton mechanism turns out to be ineffective. To overcome this difficulty, it is necessary to try to introduce into the dichalcogenide compounds (or other layered compounds) layers of the semiconductor type instead of molecular layers. There can exist in such layers delocalized excitons having propitious values of $\Omega_e$ and at the same time existing at $q \sim q_F$.

Thus, further investigation of layered compounds of the most diverse types is absolutely necessary. In this case, for the elucidation of the mechanism of superconductivity, it would be interesting not only to vary as far as possible the thickness and composition of the conducting and poorly conducting (dielectric) layers, but also to measure the variation of $T_c$ with pressure or deformation (we have in mind the determination of the dependence of $T_c$ on the distance between the conducting layers for a given composition of the compound).

It must be emphasized that the study of layered compounds is very interesting for the physics of superconductivity and its applications, independent of the attempts to produce high-temperature superconductors. It is sufficient to say that the critical fields in layered superconductors are highly anisotropic and that the upper critical field is abnormally high. Furthermore, in the case of "strong" intercalation, when much material (forming several molecular layers) is implanted between the conducting layers, the interaction between the layers is of the Josephson type, and this may prove to be important from the point of view of a number of applications.

In any event, there is no doubt whatsoever that layered superconductors will be extensively studied from different points of view and, in particular, with the object of producing materials with high $T_c$ values. The limitedness of such an approach is connected with the fact that not every molecule or substance can be intercalated into layered compounds. But then we at once obtain, as it were, sandwiches at the atomic level. In the case of artificially produced sandwiches, it would seem much easier to choose different metal–dielectric combinations, but, on the other hand, it is difficult to produce sufficiently thin and at the same time dense and quasihomogeneous metallic films, as well as to monitor and regulate the processes that occur at the metal–dielectric interface. In general, sandwiches and layered compounds supplement each other to a certain extent, and it is difficult to say *a priori* which of these systems will turn out to be the most effective on the road to raising $T_c$.

In connection with the discussion of superconductivity of the "surface" type, let us also mention [49], in which the influence on superconductivity of light metallic atoms adsorbed on the surface is discussed. Even more peculiar are semiconductor systems, which allow pair production from electrons and holes belonging to spatially separated regions (e.g., when the electrons move in one semiconductor, while the holes move in another one, separated from the first by a thin dielectric layer; the motion of the electron-hole pairs under such conditions can correspond to the flow of a superconducting current around the dielectric interlayer [50, 51].

## 6. Conclusion

On the basis of general theoretical considerations, we believe at present that the most reasonable estimate is $T_c \lesssim 300$ K [see (1.51)], this estimate being, of course, for materials and systems under more or less normal conditions (equilibrium or quasiequilibrium metallic systems in the absence of pressure or under relatively low pressures, etc.). In this case, if we exclude from consideration metallic hydrogen and, perhaps, organic metals, as well as semimetals in states near the region of electronic phase transitions (Chapter 5), then it is suggested that we should use the exciton mechanism of attraction between the conduction electrons.

In this scheme, the most promising—from the point of view of the possibility of raising $T_c$—materials are, apparently, layered compounds (Chapter 6) and dielectric-metal-dielectric sandwiches (Chapter 8). However, the state of the theory, let alone of experiment, is still far from being such as to allow us to regard as closed other possible directions, in particular, the use of filamentary compounds. Furthermore, for the present state of the problem of high-temperature superconductivity, the soundest and most fruitful approach will be one that is not preconceived, in which attempts are made to move forward in the most diverse directions.

The investigation of the problem of high-temperature superconductivity is entering into the second decade of its history (if we are talking about the conscious search for materials with $T_c \gtrsim 90$ K with the use of the exciton and other mechanisms). Supposably, there begins at the same time a new phase of these investigations, which is characterized not only by greater scope and diversity, but also by a significantly deeper understanding of the problems that arise. There is still no guarantee whatsoever that the efforts being made will lead to significant success, but a number of new superconducting materials have already been produced and are being investigated. Therefore, it is, in any case, difficult to doubt that further investigations of the problem of high-temperature superconductivity will yield many interesting results for physics and technology, even if materials that will remain superconducting at room (or even liquid-nitrogen) temperatures will not be produced. Besides, as has been emphasized, this ultimate aim does not seem to us to have been discredited in any way. As may be inferred, the next decade will be crucial for the problem of high-temperature superconductivity.

# 2

# The Critical Temperature of a Superconducting System†

## 1. Introduction

In the field of high-temperature superconductivity, the ultimate goal of which is to radically raise the critical temperature $T_c$ of the superconducting transition, it would have been extremely useful if we had for research an expression for $T_c$ as a function of the factors characterizing the material and geometry of the superconducting system. With such an expression at our disposal, we could have determined the system's parameters that are optimal from the point of view of the raising of $T_c$, and the highest $T_c$ value that is attainable for the class of structures in question. This would have signified, first and foremost, the resolution of the principal question of the whole problem of high-temperature superconductivity: Is it possible, in principle, to radically raise the critical temperature of superconductors? And a yes answer to this question would have provided a firm basis for the purposeful search for structures with a high $T_c$ value.

Unfortunately, the problem of the derivation of an expression for $T_c$ in the general case is extremely complex and unamenable to solution. We can solve only certain particular problems about the dependence of $T_c$ on individual parameters of a superconductor and the aid of additional limiting assumptions. The present chapter is devoted to the complete and systematic analysis of one such problem—the problem of determining the $T_c$ of a *homogeneous* and *isotropic* superconductor with an interaction that is arbitrary in form but *weak*. For all its limitedness, the solution to this simplified problem provides interesting information about the main tendencies of the behavior of $T_c$ as the parameters of the superconducting material are varied, and about the main demands—from the point of view of the raising of the critical temperature—on the material of the superconductor.

The assumed limitations are dictated by the desire to simplify the problem with a view to obtaining its solution in such an explicit form as would encom-

†The material of this chapter is based in part on [6] (see also [7]).

pass as wide a class of interelectron interactions as possible. This is especially important for the high-temperature superconductivity problem, whose state is such that it is as yet impossible to give conclusive preference to any one structure from the broad spectrum of structures that promise to have substantially elevated $T_c$. Under such conditions the only reasonable approach is a semi-phenomenological approach within the framework of which the interelectron interaction is not computed from "first principles" but is reduced to directly observable quantities. The assumed limitations also allow us to carry out this program: we are able to express the interelectron interaction and the quantity $T_c$ explicitly in terms of the material's permittivity, which in its turn can be reduced to the characteristics of the inelastic scattering of the electrons by the system in question.

Of course, the limitations under discussion appreciably narrow the region of direct applicability of the obtained results. Thus, the assumption that the system is homogeneous and isotropic prevents the consideration of quasi-one-dimensional and quasi-two-dimensional systems, sandwiches, etc. Also excluded from consideration here are a number of crystal-structure effects—the Umklapp processes, the interaction of the electrons with the transverse vibrational modes, etc. However, the corresponding generalization of the theory does not meet the fundamental difficulties, and a number of results that apply here are cited elsewhere in the book.†

More serious is the assumption that the interaction is weak. It does not allow us to describe quantitatively the direct interelectron Coulomb interaction, which is certainly strong at normal metal densities,‡ as well as other types of interactions (phonon, exciton; see Chapter 1) in those situations when they cannot be considered to be weak. However, the abandonment of the assumption that the interaction is weak will lead, as in any many-body problem without a small parameter, to an extremely difficult situation in the proposed theory.

It should, however, be mentioned that a small parameter that makes the solution of the problem possible is not necessarily connected with the interaction strength. Thus, for a purely phonon interaction of arbitrary strength, the

†The dielectric formalism of the theory of superconductivity expounded in this chapter was recently generalized to the case of inhomogeneous anisotropic systems by Dolgov [1] who considered a number of physical effects concerning the effect on superconductivity of the system's inhomogeneity and anisotropy, in particular, of the presence of the crystal lattice.

‡The interaction of the electrons becomes weak under conditions of highly compressed matter (the interior of "white dwarfs," the crust of pulsars, etc.), but the corresponding $T_c$ value is extremely low [2]. Moreover, under these conditions matter is in a crystalline state of a special kind [3], which makes the assumption that the system in homogeneous and isotropic qualitatively inadmissible. Thus, there are grounds for asserting that the interaction of the electrons of highly compressed matter with the transverse vibrational modes corresponding to the interconversion of energetically close, close-packed lattices sharply raises the quantity $T_c$ [4].

ratio of the width of the characteristic energy region (the Debye energy) to the Fermi energy of the electrons can serve as such a parameter. It is precisely this circumstance that enables us to construct a quantitative superconductivity theory with strong coupling (the Éliashberg theory [5, 6]), which allows us to obtain a numerical solution to the problem for the purely phonon interaction. However, a theory of this sort is unable to describe interactions not containing a small parameter—first and foremost, the Coulomb interaction—without the replacement of this interaction by a constant, a simplification which is often resorted to, but which is valid only in the weak-coupling limit (see Section 4 below). The exciton interaction is also characterized by the absence of a small parameter of the type under discussion. In any event, a strong-coupling theory is incapable of furnishing a solution to the problem formulated in the present chapter, to wit, the derivation for $T_c$ of an explicit expression that encompasses as broad a class of interactions as possible.

Let us add that without the assumption that the interaction is weak, we cannot carry out the above-mentioned semiphenomenological program. The point is that in the general case the interelectron interaction does not, in principle, reduce to observable quantities of the permittivity type (see Section 2 below).

Having accepted the assumption that the interaction is weak and that the system is homogeneous and isotropic, we can easily establish a direct connection between the interelectron interaction and the permittivity $\varepsilon(\omega, \mathbf{k})$ of the material. Accordingly, the interelectron interaction acquires properties possessed by the permittivity. For the problem of high-temperature superconductivity, the most important property of this sort is the nonnegativeness of the static permittivity $\varepsilon(0, \mathbf{k})$ and, accordingly, the static interaction between the electrons. This property, the assertion of whose necessity is prevalent in the literature limits the strength of the attractive forces, which lead to the formation of Cooper pairs, as opposed to the repulsive forces, which impede "pairing." There thus arises an absolute upper limit for the value of the critical temperature of the superconducting transition [7-9].

In Section 2 of this chapter, which contains an analysis of the properties of the permittivity and its connection with the interelectron interaction, the question of the admissible sign of $\varepsilon(0, \mathbf{k})$ is critically reviewed. As a result, we arrive at the conclusion that this property of the permittivity can, in fact, be validated, but only in the long-wave limit. As to the case of large wave vectors (and it is precisely this case that is important for superconductivity), negative values of $\varepsilon(0, \mathbf{k})$ are also entirely admissible here. This result leads to the removal of the corresponding limitations on the value of $T_c$, if, of course, materials with $\varepsilon(0, \mathbf{k}) < 0$ exist or can be artifically produced.

As has now become clear, structures with a negative value of the static permittivity are far from being a rarity. Among them are the classical, highly nonideal plasma, electrolytes (in particular, the melt of sodium chloride [10]), a number of simple metals [11], etc. A detailed review of the problem of the admissible sign of $\epsilon(0, \mathbf{k})$ is contained in [12].

The equations of the theory of superconductivity in which the interelectron interaction is explicitly expressed in terms of the permittivity are considered in Section 3 of this chapter. They have the form of homogeneous linear equations for the energy gap or the wave function of the Cooper pair [13]. The critical temperature is defined as the eigenvalue of these equations. The permittivity that enters into their kernel depends, in the general case, on two variables (the frequency and the wave vector), is highly nonmonotonic in the vicinity of the absorption lines, has a nonzero imaginary part, etc. These and other characteristics make the interaction under consideration radically different from the model forming the basis of the original Bardeen–Cooper–Schrieffer (BCS) theory. Consequently, the equations for the determination of $T_c$ that follow, for example, from the Eliashberg theory turn out to be quite different from the corresponding equation of the BCS theory. Meanwhile, this last theory, being comparatively simple and graphic, describes very satisfactorily a broad range of superconducting phenomena.

The main result of Section 3 is that in the case of the permittivity of the general form, the equation for the determination of $T_c$ can be written in the form of the standard BCS equation. This is achieved by going over in the equations of the theory of superconductivity to a new unknown function, as a result of which the kernel of these equations also changes. Although this new kernel is uniquely connected with the permittivity, it no longer possesses the above-enumerated properties of the latter; on the contrary, it is quite close in form to the kernel of the BCS model. Therefore, the result under discussion can be regarded as a justification of a BCS-type theory, as applied to a broad class of interactions in media with arbitrary spatial and frequency dispersions, and as an explanation of the success of such a theory.

The solution to the obtained equation in the weak-coupling approximation has the form of the standard BCS exponential formula for the critical temperature. Consequently, the object of the analysis performed in Section 4 of this chapter is to represent explicitly the index of the exponential function and the pre-exponential factor in the BCS formula in terms of the permittivity of the material.

Finally, in Section 5 we discuss the obtained results as applied to the problem of high-temperature superconductivity. Special attention is given to the question of the optimization of the critical temperature, i.e., of the determination of those parameters of the material which will lead to the maximum $T_c$.

We take both the Planck constant $\hbar$ and the Bolzmann constant $k_B$ to be equal to unity. The most important formulas, written in the conventional system of units, can be found in Chapter 1.

## 2. The Interelectron Interaction and the Permittivity

**A. The Permittivity.**† The permittivity, in terms of which we shall seek to express the interelectron interaction, is the most important characteristic of a material, and is determined in the following manner (see, for example, [15]).

Let us subject the system under consideration to the action of an external source having a charge density with Fourier transform $\delta\rho_e(\omega, \mathbf{k})$ and a corresponding potential $\delta U_e(\omega, \mathbf{k}) = (4\pi/k^2)\delta\rho_e(\omega, \mathbf{k})$. The electric field of the source produces within the system an induced charge density $\delta\rho(\omega, \mathbf{k})$ and the corresponding potential $\delta U(\omega, \mathbf{k}) = (4\pi/k^2)\delta\rho(\omega, \mathbf{k})$. The longitudinal permittivity $\varepsilon(\omega, \mathbf{k})$ is defined as the coefficient of proportionality between the external, $\delta U_e$, and the total, $\delta U_t = \delta U_e + \delta U$, potentials:

$$\delta U_e(\omega, \mathbf{k}) = \varepsilon(\omega, \mathbf{k})\delta U_t(\omega, \mathbf{k}). \tag{2.1}$$

A similar relation connects the charge densities $\delta\rho_e$ and $\delta\rho_t = \delta\rho_e + \delta\rho$. Introducing the field intensity $\mathbf{E}$, which is determined by the gradient of the total potential, and the induction $\mathbf{D}$, which is connected, according to the equation $\operatorname{div}\mathbf{D} = 4\pi\rho_e$, with the gradient of the external potential, we obtain the well-known relation‡

$$\delta\mathbf{D}(\omega, \mathbf{k}) = \varepsilon(\omega, \mathbf{k})\delta\mathbf{E}(\omega, \mathbf{k}), \tag{2.2}$$

which can also be regarded as a definition of the quantity $\varepsilon$.

The concept of "external source" introduced above is not self-evident and needs to be explained further. The role of the external source is, by definition, played by charges that do not enter into the composition of the system in question. Among them are, first and foremost, the charges located outside the system, e.g., the charges on capacitor plates between which the system itself is located. It is significant that in this way we can only produce an external potential or induction with $k = k_0 \sim 1/L \longrightarrow 0$, where $L$ is the macroscopic dimension of the system. Consequently, external sources of the type under consideration are connected only with the long-wave limit of the permittivity $\varepsilon(\omega, 0)$.

To produce an external field with $k \neq 0$, it is necessary to place the sources inside the system itself (the corresponding charge density $\delta\rho_e$ depends harmonically on the coordinates). We must then assume that the state of the external charges is given and that these charges do not experience the reaction of the system having, for example, a sufficiently large mass. In the opposite case, the

†The material of Sections 2.A–2.C is expounded upon in greater detail in [14].

‡As applied to a dielectric, the foregoing needs to be qualified with regard to the plane geometry of the system [16] or the "shorting" of its ends [15].

external and internal charges of the system would be indistinguishable, and the matter would have essentially been reduced to the consideration of a new, more complex system.

The permittivity has a double physical meaning. First of all, it describes the external-charge screening effects caused by the redistribution of the system's own charges. In the scheme of interest to us, another property of $\varepsilon$–its ability to describe the response of the system to external influences–is more important. The quantity that determines the connection between an action on the system and the result of the action is called the response function:

$$\text{(result of an action)} = \text{(response function)} \times \text{(action)}.$$

This formula reflects the causal relationship between an action on a system and its result. Therefore, the response function obeys the causality principle (cause always precedes effect), and satisfies a dispersion relation of the type of the Kramers-Kronig formula (see, for example, [15]). The question of whether the permittivity is unconditionally regarded as a response function is extremely important since the dispersion relations impose important limitations on $\varepsilon$ and, in the final analysis, on the critical temperature. At the same time, this question is not at all simple, as will be seen from the discussion of it below.

The meaning of the concept of "external action" in the definition of the response function also needs to be made more precise. By this term we mean an influence exerted on the system irrespective of the state of the system, or, in other words, an influence that can be varied, switched on, or switched off at will. It is precisely such a property that must be possessed by the event-cause.† It is significant that the definition given above does not allow us to regard as an external action any influence exerted on the system by external sources.

In any event, under this definition falls an action emanating from the second of the above-considered types of external sources (i.e., a source located within, and free from the reaction of, the system). In this case the external action is connected with the magnitude of the external charge $\rho_e$ or the strength of the induction **D**, while the result of such an action is determined by the magnitude of the total charge $\rho_t$ or of the field intensity **E**. Consequently, the response function will be the quantity $1/\varepsilon(\omega, \mathbf{k})$ [see (2.2)]. It is significant that we cannot in the present situation interchange **D** and **E** and correspondingly take the quantity $\varepsilon$ itself to be the response function.‡ Aside from the violation of the condition that the external action be independent of the state of the sys-

†For a discussion of this and other aspects of the problem of causality, see [17].

‡To do this, it will be necessary to introduce into the system a set of electrodes with harmonically varying $\delta U_t$. It is clear that, in contrast to the introduction of $\delta\rho_e$, this will signify a radical modification of the original system itself.

tem that this would have entailed, we would have had to use as the indicator of a change in the latter the magnitude of the external charge, the state of which is, in fact, given, and does not depend on the state of the system.

As for the external source located outside the system, it exerts an influence only when the capacitor plates are disconnected and we control their charge. In such a situation the response function is the quantity $1/\varepsilon(\omega, 0)$. If, on the other hand, the plates are connected via a battery and we control the corresponding potential difference, then the effected external action is not connected with the external sources (the charge of the plates). The latter, in the situation under consideration, depends on the state of the system, flowing to or from the plates when this state is changed. The external action is now determined by the magnitude of the total potential difference across the plates, i.e., by the magnitude of the field intensity **E**, while the result of such an action is determined by the magnitude of the external charge or of the induction **D**. Consequently, the quantity $\varepsilon(\omega, 0)$ is now the response function. The considered cases exhaust the possible types of influence that can be exerted on the system with the aid of a longitudinal field.

The result of the above-performed analysis (see also [18]) is that the response function is not the quantity $\varepsilon(\omega, \mathbf{k})$, as is often assumed, but its reciprocal, $1/\varepsilon(\omega, \mathbf{k})$ (see [15]). As to the permittivity $\varepsilon(\omega, \mathbf{k})$ itself, it plays the role of a response function only in the long-wave limit $k \longrightarrow 0$.

**B. The Dispersion Relations.** As has already been pointed out, the response function satisfies a dispersion relation of the type of the Kramers-Kronig formula. Such a relation can be written for the quantity $1/\varepsilon(\omega, \mathbf{k})$ for arbitrary values of **k** (see [15]):

$$\frac{1}{\varepsilon(\omega, \mathbf{k})} = 1 - \int_0^\infty \frac{dE^2 \sigma(E, \mathbf{k})}{E^2 - \omega^2 - i\delta}. \tag{2.3}$$

The spectral density

$$\sigma(\omega, \mathbf{k}) = -\frac{1}{\pi} \operatorname{Im}\left[\frac{1}{\varepsilon(\omega, \mathbf{k})}\right]$$

is a nonnegative quantity, being directly connected with the form factor of the inelastic scattering of the fast electrons of the system under consideration, $k^2\sigma(\omega, \mathbf{k})/(4\pi e^2)$, which determines the probability for the loss by an electron of an amount of energy $\omega$ and momentum **k**. Therefore, the relation (2.3) directly relates the permittivity to experimentally measurable quantities.

The dispersion relation for the permittivity itself is valid (according to the results of Section 2.A) only in the long-wave limit

$$\varepsilon(\omega, 0) = 1 + \frac{1}{\pi} \int_0^\infty \frac{dE^2 \operatorname{Im} \varepsilon(E, 0)}{E^2 - \omega^2 - i\delta}, \tag{2.4}$$

where

$$\operatorname{Im} \varepsilon = \pi\sigma |\varepsilon|^2 \geqslant 0.$$

The dispersion relations lead to significant limitations on the magnitude of the static permittivity $\varepsilon(0, \mathbf{k})$. According to (2.3), $1/\varepsilon(0, \mathbf{k}) = 1 - I$, where

$$I = 2 \int_0^\infty \frac{dE}{E} \sigma(E, \mathbf{k}) \geqslant 0,$$

and the quantity $1/\varepsilon(0, \mathbf{k})$ should be less than unity, while the quantity $\varepsilon(0, \mathbf{k})$ itself should satisfy one of the inequalities

$$\varepsilon(0, \mathbf{k}) \geqslant 1, \quad \varepsilon(0, \mathbf{k}) \leqslant 0, \tag{2.5}$$

the first of them corresponding to the values $I < 1$; the second, to the values of $I > 1$. Similarly, from (2.4) it follows directly that

$$\varepsilon(0, 0) = \lim_{k \to 0} \varepsilon(0, \mathbf{k}) \geqslant 1. \tag{2.6}$$

If the relation (2.4) were valid for all $\mathbf{k}$, then from it would have followed the assertion of the positiveness of $\varepsilon(0, \mathbf{k})$ [more precisely, of the validity of the first condition (2.5)].

Of course, as a result of the continuity of $\varepsilon(0, \mathbf{k})$ as a function of $k$, the inequality (2.6) will also be fulfilled in a region of relatively small $k$, the dimension of which will be determined by the parameters of the system.† Therefore, in a real system, either $\varepsilon(0, \mathbf{k}) \geqslant 1$ for all values of $k$ or $\varepsilon(0, \mathbf{k})$ changes its sign, going over from values greater than unity to negative values. More satisfactory is the language based on the quantity $1/\varepsilon$: In the first case this quantity lies between zero and unity for all $k$ (Fig. 2.1a); in the second case it passes through zero, going over from positive to negative values (Fig. 2.1b).

†This region is determined either by the quantity $1/L$, where $L$ is the dimension of the system [metals for which in the infinite-system limit the structure factor contains the contribution, $\sim \delta(\mathbf{k})$, of the static rigid lattice, a nonideal plasma], or by the internal parameters of the system (for semiconductors and dielectrics, by the quantity $E_g/v_F$, where $E_g$ is the dielectric gap and $v_F$ is the velocity at the Fermi level) [12].

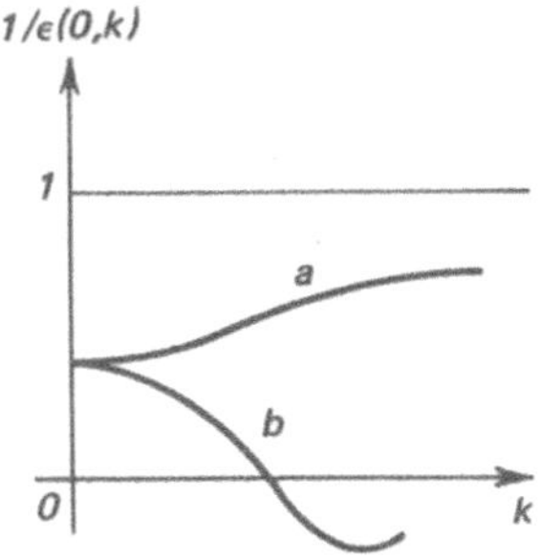


**Figure 2.1. Possible behavior of the inverse permittivity.**

Let us note in this connection that the inequality $\varepsilon(0, \mathbf{k}) \geqslant 0$, given in [16] (see Section 14 in [15] for a general derivation†) in fact admits negative $\varepsilon(0, \mathbf{k})$ values as well, a fact that becomes apparent on going over to the $1/\varepsilon$ language. This result derives from the assertion that when $\varepsilon = 1$ initially, any change in $\varepsilon$ $(1/\varepsilon)$ has a positive (negative) sign. But this does not exclude the possibility that the function $1/\varepsilon$ will, after decreasing to values less than unity ($\varepsilon > 1$), go through zero and thus become negative ($\varepsilon < 0$). In the $\varepsilon$ language, this possibility is masked by the inevitable assumption of an infinite value by $\varepsilon$.

Thus, like the dispersion relation for $\varepsilon$, the generally accepted limitation (2.6) is valid only in the long-wave limit. At large values of $k$, however, we can, without coming into conflict with the dispersion relations, also allow negative values of the static permittivity $\varepsilon(0, \mathbf{k})$ of the system. There then occurs a "overscreening" of the external charge: the field harmonics with large $k$ in the medium and those in free space have opposite signs.

Contrary to the opinion expressed in [15], we see no reason to think that such a situation is physically inadmissible. This is now indisputable because of the discovery of real structures with $\varepsilon(0, \mathbf{k}) < 0$.

The analysis that has been performed in the more general scheme and that has led to the conclusion that no dispersion relations exist for $\varepsilon$ for $k \neq 0$, serves as some caution against the mechanical extension of the formulas of conventional electrodynamics to the case of media with spatial dispersion.

In fact, there exists a natural parameter, $k_0 \sim 1/L$ ($L$ is the dimension of the system), because of which the physical situation for $k \sim k_0$ and the one for $k \gg k_0$ can prove to be entirely different.

**C. Stability of the System.** The nonfulfillment of the dispersion relations (2.3) and (2.4) leads to the instability of the system and to the spontaneous breaking of the system's intrinsic symmetry. This is connected with the appearance of eigenstates having an imaginary frequency. Thus, the nonfulfillment of (2.3) entails the appearance of growing density oscillations that eventually go

†Actually, in this derivation no account was taken of dispersion, and what was in question was just the inequality $\varepsilon \geqslant 1$.

over into static density waves, signifying the destruction of the translational invariance of the system [19]. In view of the above-noted importance of the question of the limitations expressed by the inequalities (2.5) and (2.6), we derive them here by another method, demonstrating in the process that these inequalities are simply criteria for the stability of the system. We shall at the same time refute the arguments, based on stability considerations, that exist in the literature against the second inequality in (2.5).

Let us begin with a general consideration of the stability problem.† Let us consider the stability of the system against fluctuations in some physical quantity—a "field" $\phi$ (we shall assume some "current" $j$ to be conjugate to this quantity). Let the problem consist of the construction of a functional $E\{\phi\}$ that depends on the arbitrary (nonequilibrium) function $\phi$ and has, as a function of $\phi$, a minimum in the equilibrium state. The function $\phi$ that realizes this minimum determines the equilibrium field (which is nonzero in the case of spontaneously broken symmetry), and the corresponding conditions for the minimum coincide with the stability criterion.

In order to make a state (of the system) with an arbitrary $\phi$ an equilibrium state, it is necessary to introduce a specially chosen external current $j$, adding to the Hamiltonian the term $H' = \int j\phi \, d\mathbf{x}$. Limiting ourselves, for the sake of simplicity, to the case of zero temperature, and setting $E\{j\} = \langle H + H' \rangle$, where the angle brackets denote averaging over the state with current, we have

$$\frac{\delta E\{j\}}{\delta j(\mathbf{x})} = \langle \varphi(\mathbf{x}) \rangle \equiv \varphi. \tag{2.7}$$

$E\{j\}$ is an intermediate functional and possesses no minimum properties with respect to $j$ [this is already evident from (2.7) for $\phi \neq 0$]. In order to go over to the sought functional $E\{\phi\}$, it is necessary to subtract from $E\{j\}$ the "extraneous" work connected with the external current and go over from $j$ to the argument $\phi$, using (2.7). The matter reduces to the Legendre transformation $E\{\phi\} = E\{j\} - \int j\phi \, d\mathbf{x} = \langle H \rangle$, where the current involved in the averaging should be expressed in terms of $\phi$. The conditions for the obtained expression, which is the energy of the system for a given value of $\phi$, to have a minimum, assume the form

$$\frac{\delta E\{\varphi\}}{\delta \varphi(\mathbf{x})} = -j(\mathbf{x}) = 0; \quad \frac{\delta^2 E\{\varphi\}}{\delta \varphi(\mathbf{x})\, \delta \varphi(\mathbf{y})} = -\frac{\delta j(\mathbf{x})}{\delta \varphi(\mathbf{y})} > 0. \tag{2.8}$$

†The approach expounded below corresponds to the general "effective-potential" method, which has been developed in recent years in connection with the investigation of spontaneous symmetry breaking in quantum field theory and statistics [20]. In its simplified form, this approach has been used before [16, 21].

In the problem of interest to us, we should set $j = U_e$ and $\phi = \rho$ (see Section 2.A). Then the first condition in (2.8) yields the equality $U_e = 0$, while the second yields the inequality†

$$\frac{\delta^2 E\{\rho\}}{\delta\rho(0, \mathbf{k})\,\delta\rho(0, -\mathbf{k})} = -\frac{\delta U_e(0, \mathbf{k})}{\delta\rho(0, k)} = -\frac{1}{\chi(0, \mathbf{k})} > 0$$

[we are considering the equilibrium (i.e., the static) state of the system]. Here we have introduced the susceptibility

$$\chi(\omega, \mathbf{k}) = \frac{k^2}{4\pi}\left(\frac{1}{\varepsilon(\omega, \mathbf{k})} - 1\right),$$

defined by the relation $\delta\rho(\omega, \mathbf{k}) = \chi(\omega, \mathbf{k})\delta U_e(\omega, \mathbf{k})$. It is easy to see that the obtained stability criterion coincides exactly with the inequalities (2.5). This is not surprising since the spontaneous appearance of density waves should necessarily be accompanied by an unstable state, forbidden by the dispersion relations, in the excitation spectrum of the homogeneous system.

The arguments, based on stability considerations, against the second inequality in (2.5) rest on the expression for the field energy of the system:

$$\frac{1}{8\pi}\int d^3k \frac{|\mathbf{D}(0, \mathbf{k})|^2}{\varepsilon(0, \mathbf{k})} = \frac{1}{8\pi}\int d^3k\, \varepsilon(0, k)\,|\mathbf{E}(0, \mathbf{k})|^2, \tag{2.9}$$

which does not have a minimum when $\varepsilon(0, \mathbf{k}) < 0$. In [15] Pines and Nozières consider the left expression in (2.9), which is identified with the energy of a system of infinitely heavy ions producing the external—with respect to the electrons—field **D**. It is clear, however, that this expression has the meaning of a functional $E\{j\}$ with $\phi = \mathbf{E}$ and $j = \mathbf{D}/4\pi$, and should therefore not have a minimum (see above). The instability of the system should, in the opinion of Pines and Nozières [15], manifest itself in the reconstruction of the ion subsystem. This is, however, impossible because of the infinitely large mass of the ions.

The correct formulation of the problem of stability against the spontaneous appearance of induction corresponds to the placing of the system between the plates of a capacitor that are connected via a battery (see Section 2.A). As has already been noted, we can then consider only the long-wave case; this follows also from the fact that, on account of the equation div $\mathbf{D} = 4\pi\rho_e$, only a spatially uniform induction can arise spontaneously (without the involvement of external sources). The sought functional, which corresponds to $\phi = \mathbf{D}$ and $j = \mathbf{E}/4\pi$ is

†For such a choice, we arrive essentially at the well-known variational principle of the quantum theory of many-body systems (in recent years this principle has more often been called the density-functional method; see, for example, [22] for a review).

obtained from the expression on the left-hand side of (2.9) by subtracting that part of it, $(1/8\pi)\int \mathbf{E}^2\, d^3k$, which pertains to the external sources. The resulting expression,

$$\frac{1}{8\pi}\,\frac{\varepsilon(0,\,0)-1}{\varepsilon^2(0,\,0)}\,|\,\mathbf{D}(0,\,0)\,|^2 \qquad (2.9')$$

leads directly to the condition (2.6).†

Another argument against the second inequality in (2.5) is based on the fact that, as a function of **E**, the expression on the right-hand side of (2.9) does not have a minimum (see [9]). And this expression in fact should not have a minimum. Setting $\phi = \mathbf{E}$ and $j = \mathbf{D}/4\pi$ and performing the Legendre transformation, we arrive at the well-known result that the potential corresponding to **E** is the quantity differing from (2.9) in sign [16]. Retaining in this quantity only the part pertaining to the system itself, as was done above, we arrive at the functional

$$\frac{1}{8\pi}\int d^3k\varepsilon(0,\,\mathbf{k})\,[\varepsilon(0,\,\mathbf{k})-1]\,|\,\mathbf{E}(0,\,\mathbf{k})\,|^2, \qquad (2.9'')$$

whose minimum requires the fulfillment of the conditions (2.5).

Notice that the last two functionals (2.9′) and (2.9″) have a direct physical meaning since each of the states that are found in the computation of the minimum to be rivals can be realized by appropriately choosing the external current $j$. In this, they differ from the functionals introduced in [16] (see Section 18), which, moreover, do not always have a minimum in the equilibrium state.

The question may arise whether the above-performed investigation of the stability of the idealized homogeneous system makes sense when we are fully aware of the fact that it is unstable against the formation of the normal crystal lattice. The answer to this question should be affirmative since upon the violation of the above-derived conditions there arises not the normal, but a special "coherent," crystal [19]. The latter is a structure that is in a sense the opposite of the normal crystal (the "quantum" crystals that are being extensively studied now occupy an intermediate position). Among the differences between coherent and normal crystals is the absence in a coherent crystal of localization of the particles near the sites of the lattice, whose parameter is determined not, as is usually the case, by the particle concentration, but by the law of interaction between the particles. Therefore, the spontaneous appearance of normal and

†For conducting systems, this condition follows also from the necessity for the compressibility of the system to have a positive sign [15].

coherent crystals is a manifestation of two independent modes of instability of the homogeneous state of the system.

It should be noted, however, that, although a slight violation of the above-obtained conditions (slight modulations of the homogeneous density, a second-order phase transition) undoubtedly results in the formation of coherent-crystal state, with the growth of such a violation the crystal can go over into the normal state by breaking into a first-order phase transition. Apparently, that is precisely how matters stand in the crystallization of a nonideal plasma [10].

**D. Connection between the Interelectron Interaction and the Permittivity.** Having elucidated the properties of the permittivity, let us proceed to the central question of this section, namely, the representation of the interelectron interaction in terms of the quantity $\varepsilon(\omega, \mathbf{k})$, Let us first consider the interaction between two external charges located inside the system. According to what has been said above (in Section 2.A), their state is fixed and cannot change under the influence of the system itself. Let the density of the external charge and the potential produced by this charge inside the system be, respectively, equal to $\delta\rho_e(\omega, \mathbf{k})$ and

$$\delta U_t(\omega, \mathbf{k}) = \frac{4\pi\delta\rho_e(\omega, \mathbf{k})}{k^2\varepsilon(\omega, \mathbf{k})}.$$

The action function corresponding to the self-interaction of such a charge has the form

$$-\frac{1}{2}\int dt\, d\mathbf{x}\, \delta\rho_e(t, \mathbf{x})\, \delta U_t(t, \mathbf{x}) = -\frac{1}{2}\int d\omega\, d^3k\, 4\pi \frac{|\delta\rho_e(\omega,\mathbf{k})|^2}{k^2\varepsilon(\omega, \mathbf{k})}.$$

If $\delta\rho_e$ is the sum of the charge densities of the material's components with elementary charges $Q_i$, then the effective interaction in the medium between particles with charges $Q_i$ and $Q_j$ has, in the momentum representation, the form

$$\frac{4\pi Q_i Q_j}{k^2\varepsilon(\omega, \mathbf{k})}.$$

In particular, the interaction between electrons is given by the expression

$$V(\omega, \mathbf{k}) = \frac{4\pi e^2}{k^2\varepsilon(\omega, \mathbf{k})}, \tag{2.10}$$

which is a generalization of the elementary formula for the Coulomb interaction between charges in a medium with $\varepsilon$ = const to the case when there is spatial and frequency dispersion. The expression (2.10) is illustrated by Fig. 2.2.

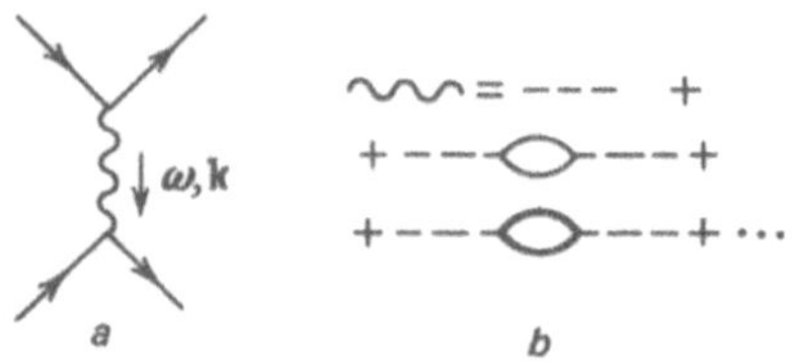


**Figure 2.2. The simplest diagram for the interaction between electrons: the wavy line depicts the interaction (2.10); the solid line, an electron; the heavy line, an ion; the dashed line, the Coulomb interaction.**

Of interest to us, however, is not the interaction between the external electrons, whose state is fixed, but the interaction between the electrons of the system itself, which are capable of feeling the influence of the system in the course of their interaction.† This important circumstance is reflected in the appearance of additional interaction diagrams, among which is the diagram shown in Fig. 2.3. As a result, the effective interaction between the electrons will no longer be described by the simple formula (2.10) but by a complex function of not only the energy and momentum transfers, but also of the total energy and total momentum of the colliding electrons. It should, however, be noted that the diagram in Fig. 2.3 and others like it also can be expressed in terms of the permittivity if we associate the expression (2.10) with each wavy line [see (2.20) below].

The situation changes significantly if we take into consideration the more complex types of interaction diagrams that take account of those effects of the other particles (like the ions) that do not reduce to a simple polarization loop (see Fig. 2.2b). In this case, to which corresponds the diagram shown in Fig. 2.4, the interaction between the electrons is determined not only by the permittivity, but also by the characterisitcs of the ion subsystem that are not reducible to the permittivity (the third-order ion tail shown in Fig. 2.4). Allowance for such diagrams would have required the introduction of new response functions, i.e., response functions different from the quantity $1/\varepsilon(\omega, \mathbf{k})$. The latter quantity, according to the results of Section 2.a and 2.b, describes the response to

†With this effect is connected, in particular, the inequality of the field acting on a particle and the mean field.

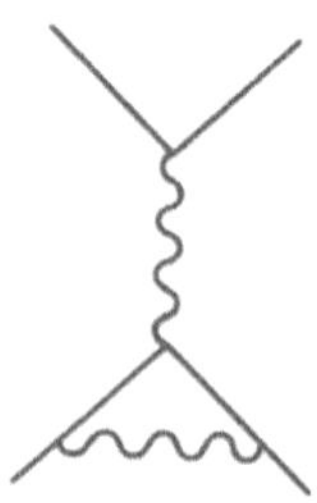

**Figure 2.3. Interaction diagram with the electron vertex part.**

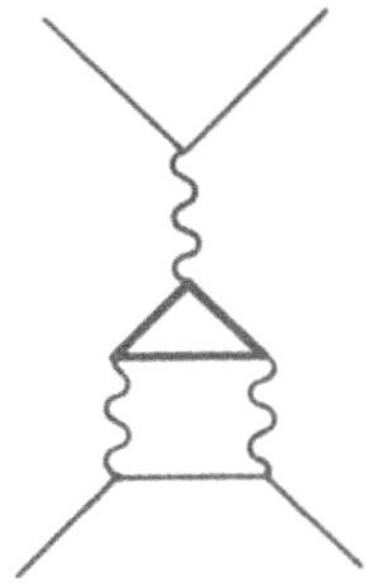

**Figure 2.4. Interaction diagram with a third-order ion tail.**

the external action

$$\int dx\, \psi^{+}(x)\, \delta U_e(x)\, \psi(x),$$

whereas the diagram in Fig. 2.4 corresponds to the action

$$\int dx\, dx'\, dt'\, \psi^{+}(x)\, \delta U_e(x, x')\, \psi(x').$$

Unfortunately, such response functions do not have a direct physical meaning because of the difficulty in realizing actions of this type.†

The assumption, which we adopt, that the interaction is weak, allows us to neglect the diagrams that take account of the reaction of the system on the interelectron-interaction process and, as compared to the diagram in Fig. 2.2a, necessarily contains extraneous interaction lines; in the final analysis such diagrams can be regarded as small corrections. Therefore, in the first approximation, it becomes possible for us to identify the interaction between the electrons of the system and between the external electrons and to use the expression (2.10).

**E. The Simplest Models.** In concluding this section, let us consider some very simple theoretical interelectron-interaction models that will be used below to estimate the critical temperature. Let us introduce the polarization operator $\Pi(\omega, \mathbf{k})$ that describes the polarization loops (see Fig. 2.2b) and is defined as the coefficient of proportionality between the induced charge and the total potential (see Section 2.A):

$$\delta\rho(\omega, \mathbf{k}) = \Pi(\omega, \mathbf{k})\, \delta U_t(\omega, \mathbf{k}). \tag{2.11}$$

†The theory of such response functions and the stability problems connected with them have been considered by Yu. A. Nepomnyashchii, to whom we are grateful for a discussion of his unpublished results.

It is proportional to the polarizability $\alpha$ ($\Pi = -k^2\alpha$) and is connected with the permittivity by the relation

$$\varepsilon(\omega, \mathbf{k}) = 1 - \frac{4\pi}{k^2}\Pi(\omega, \mathbf{k}). \tag{2.12}$$

The left-hand side of (2.11) is a sum of variations of the charge densities of the conduction electrons (subscript "$c$") and the ions (subscript "$i$"), etc.; the polarization operator itself has the form of a similar sum.

An approximate expression for the $\Pi_i$ term that is valid at high frequencies can be found within the framework of the well-known jellium model, in which the ions are regarded as a continuum of free particles. Using the equations of motion and the continuity equation

$$m\delta\dot{\mathbf{v}} = -Ze\nabla\delta U_t, \quad \delta\dot{\rho}_i = \rho_i \operatorname{div}\delta\mathbf{v},$$

we find

$$\Pi_i(\omega, \mathbf{k}) = \frac{k^2\omega_i^2}{4\pi\omega^2},$$

where $\omega_i = (4\pi Ze\rho_i/M)^{1/2}$ is the plasma frequency of the ions and $M$ and $Ze$ are their mass and charge. In the more general model in which it is assumed that the ions are bound and have the bare-frequency spectrum $\omega_0(\mathbf{k})$,

$$\Pi_i(\omega, \mathbf{k}) = \frac{k^2\omega_i^2}{4\pi\left[\omega^2 - \omega_0^2(\mathbf{k}) + i\delta\right]}. \tag{2.13}$$

The form of the $\Pi_c$ term is easily found in the static long-wave limit, where the Thomas–Fermi model is valid, and $\rho_c = \rho_c(E_F + \delta U_t)$ ($E_F$ is the chemical potential or the Fermi energy of the electrons) [22]. With the aid of (2.11) we find

$$\Pi_c(0, \mathbf{k}) = \frac{\partial\rho_c}{\partial E_F} \equiv -\frac{\varkappa^2}{4\pi} \qquad (k \ll \varkappa),$$

where $\varkappa$ is the inverse Debye-screening radius; at zero temperature, in the weak-coupling case, $\varkappa = 2(p_F me^2/\pi)^{1/2}$, where $p_F^3/3\pi^2$ is the electron concentration and $p_F$ is the Fermi momentum. In the opposite limiting case, in which $\omega \gg kv_F$ ($v_F \sim \omega_c/\varkappa$ is a characteristic velocity, equal at $T = 0$ to the velocity, $p_F/m$, at the Fermi level), we have

$$\Pi_c(\omega, \mathbf{k}) = \frac{k^2\omega_c^2}{4\pi\omega^2},$$

where $\omega_c = (4\pi e\rho_c/m)^{1/2}$ is the plasma frequency of the electrons; $m$ is their mass. We shall also give a convenient interpolation formula encompassing both limiting cases (though not describing plasmon damping, etc.; see [7]):

$$\Pi_c(\omega, \mathbf{k}) = \frac{k^2\omega_c^2}{4\pi\left(\omega^2 - \omega_c^2 k^2/\varkappa^2 + i\delta\right)}. \tag{2.14}$$

Combining (2.12)-(2.14), we find†

$$\varepsilon(\omega, \mathbf{k}) = 1 - \frac{\omega_c^2}{\omega^2 - \omega_c^2 k^2/\varkappa^2 + i\delta} - \frac{\omega_i^2}{\omega^2 - \omega_0^2(\mathbf{k}) + i\delta}, \tag{2.15}$$

with

$$\varepsilon(0, \mathbf{k}) = 1 + \frac{\varkappa^2}{k^2} + \frac{\omega_i^2}{\omega_0^2(\mathbf{k})}. \tag{2.15'}$$

The expression (2.15) satisfies the dispersion relations (2.3) and (2.4) corresponding to the spectral density

$$\sigma(\omega, k) = \omega_c^2\delta\left[\omega^2 - \omega_c^2\left(1 + \frac{k^2}{\varkappa^2}\right)\right] + \\ + \omega_i^2\left(\frac{k^2}{k^2 + \varkappa^2}\right)^2\delta\left[\omega^2 - \omega_i^2\frac{k^2}{k^2 + \varkappa^2} - \omega_0^2(\mathbf{k})\right]. \tag{2.16}$$

The excitation spectrum consists of two branches: the plasmon branch

$$\omega = \omega_c\left(1 + \frac{k^2}{\varkappa^2}\right)^{1/2}$$

and the phonon branch

$$\omega = \left[\omega_0^2(\mathbf{k}) + \omega_i^2\frac{k^2}{k^2 + \varkappa^2}\right]^{1/2}.$$

The similarity of the terms in (2.16) allows us to assert, with allowance for (2.10), that the interelectron interaction involves the exchange of not only

†A simpler expression for $\varepsilon$ that is conventionally associated with the jellium model is given in Chapter 1 [see (1.7)].

phonons, but plasmons as well. Either mode of interaction leads to additional attraction, which acts besides the Coulomb repulsion. However, for plasmons this effect is exhausted by the screening of the latter.

The model in question is now capable of describing a system in which $\varepsilon(0, \mathbf{k}) < 0$ (see Sections 2.B and 2.C above). According to (2.15′), this corresponds to $\omega_0^2 < 0$, i.e., to the instability of the "bare" ion subsystem, which exists until its interaction with the conduction electrons is switched on. However, the function $\omega_0^2(\mathbf{k})$ can, at least within the framework of the mathematical model, always be chosen such that the final physical phonon spectrum is stable, such that the condition (2.6) is fulfilled, etc. For other possible ways of physically realizing the inequality $\varepsilon(0, \mathbf{k}) < 0$, see Chapter 3.†

Finally, let us consider the more complex (generalized) jellium model‡ corresponding to the introduction into the system of the uniformly distributed (over the volume) oscillators of frequency $\Omega >> \omega_i$ that would produce outside the system a permittivity $\varepsilon_0 > 1$. Such a model could describe a weakly damped exciton vibrational mode if it existed in a superconductor (see Chapter 1). The corresponding additional term in the polarization operator has the form

$$k^2(\varepsilon_0 - 1)\Omega^2/[4\pi(\omega^2 - \Omega^2 + i\delta)]. \tag{2.17}$$

According to (2.12), this leads to an increase in the initial permittivity $\varepsilon(0, \mathbf{k}) \equiv \varepsilon^{(0)}$ of an amount equal to $\varepsilon_0 - 1$. If prior to the introduction of the oscillators we had $\varepsilon^{(0)} > 1$, then the final value of $\varepsilon$ will all the more be greater than unity. However, for $\varepsilon^{(0)} < 0$, the introduction of the oscillators can lead to a situation in which the quantity $\varepsilon$ falls within the interval, $0 < \varepsilon < 1$, forbidden by the conditions (2.5). There arises in this case, to which corresponds the condition

$$1 - \varepsilon^{(0)} < \varepsilon_0 < 2 - \varepsilon^{(0)}, \tag{2.18}$$

an instability of the above-considered type (a coherent crystal).

We give another expression for the total spectral density in the generalized jellium model for the $k^2 >> \varkappa^2$ case, which is important for the determination of $T_c$:

$$\sigma(\omega, \mathbf{k}) = \omega_c^2\delta\left(\omega^2 - \omega_c^2\frac{k^2}{\varkappa^2}\right) + \frac{\omega_i^2}{\varepsilon_0^2}\delta\left(\omega^2 - \frac{\omega_i^2}{\varepsilon_0} - \omega_0^2\right) + \\ + (\varepsilon_0 - 1)\Omega^2\delta(\omega^2 - \varepsilon_0\Omega^2) \tag{2.19}$$

(see Section 4.E below).

†Let us point out that allowance for the Umklapp processes that occur in the crystal lattice can also lead to an effective expression, similar to (2.15) with $\omega_0^2 < 0$ (see [7, 9]).

‡For another modification of the jellium model, see [23].

## 3. Equation for the Critical Temperature

**A. The Basic Equation.** In this section we derive the basic equation used to determine the critical temperature in the theory of superconductivity, as applied to interelectron interactions of the general type, which correspond to arbitrary spatial and frequency dispersion of the permittivity of the material. The subsequent simplification and transformation of this equation will allow its reduction to the form of the conventional BCS equation.

The dynamical description of the electron subsystem will be based on the following expression for the electron-electron interaction Hamiltonian†:

$$H' = \frac{1}{2} \int dx\, d^4x'\, \psi_\alpha^+(x)\, \psi_\beta^+(x')\, V(x - x')\, \psi_\beta(x')\, \psi_\alpha(x). \tag{2.20}$$

Here $V(x)$ is the Fourier transform of the interaction (2.10), having, as a result of the frequency dispersion of the permittivity, a retarded character; $x = (\mathbf{x}, t)$, $d^4x = d\mathbf{x}dt$. The expression (2.20) describes all the purely electronic effects of higher order (in particular, the diagram shown in Fig. 2.3), leaving out of consideration the ionic effects, such as the effect depicted by the diagram in Fig. 2.4. These latter effects have, however, a minimum of the third order with respect to the interaction and can be neglected in the weak-coupling theory. Notice also that the Hamiltonian (2.20) already includes the electronic polarization effects (see Fig. 2.2b), and therefore the diagrams with electronic loops should be discarded in the subsequent computations.

We shall proceed from the equation, corresponding to (2.20), for the field operator

$$\left[O_x - \int d^4y V(x-y)\, \rho(y)\right] \psi_\alpha(x) = 0,$$

where we have introduced the notation

$$O_x \equiv i\frac{\partial}{\partial t} + E_F - \frac{p^2}{2m}, \quad \rho(x) = \sum_\alpha \psi_\alpha^+(x)\, \psi_\alpha(x),$$

$p$ being the momentum operator and have chosen the chemical potential $E_F$ as the reference point for the energy. Following Gor'kov [5, 13], let us introduce the Cooper-pair wave function

$$F_{\alpha\beta}(x - x') \equiv \sigma_{\alpha\beta} F(x - x') = i\langle T(\psi_\beta(x')\, \psi_\alpha(x))\rangle,$$

where

†Strictly speaking, because of a dependence of (2.20) from time difference, our starting point must not be (2.20) but the corresponding interaction Lagrangean i.e. interaction Hamiltonian (2.20) with opposite sign.

$$\sigma = \begin{pmatrix} 0 & 1 \\ -1 & 0 \end{pmatrix}.$$

A direct computation yields

$$O_x O_{x'} F_{\alpha\beta}(x - x') = iV(x - x') F_{\alpha\beta}(x - x') + \\ + \int d^4y\, d^4z\, V(x - y) V(x' - z) \langle T(\rho(y)\rho(z)\psi_\beta(x')\psi_\alpha(x))\rangle. \quad (2.21)$$

In the lowest weak-coupling approximation, we can retain only the first term on the right-hand side of (2.21). Introducing the free electron Green function

$$G_{\alpha\beta}(x - x') \equiv \delta_{\alpha\beta} G(x - x') = -i\langle T(\psi_\beta^+(x')\psi_\alpha(x))\rangle,$$

which satisfies the equation

$$O_x G(x - x') = \delta(x - x'),$$

and the "energy gap"

$$\Delta(x) = iV(x) F(x), \quad (2.22)$$

we obtain

$$F(x - x') = \int d^4y\, d^4z G(x - y) G(x' - z) \Delta(y - z). \quad (2.23)$$

Let us rewrite (2.22) and (2.23) in the momentum representation and go over to the temperature functions for $T = T_c$ with the aid of the standard substitutions:

$$p \rightarrow (i\omega_n, \mathbf{p}), \quad \int d^4p \rightarrow iT_c \sum_n \int d^3p, \qquad \omega_n = (2n + 1)\pi T_c.$$

Using the expression

$$G(i\omega_n, \mathbf{p}) = [i\omega_n - \xi(\mathbf{p})]^{-1}, \text{ где } \xi(\mathbf{p}) \equiv (p^2/2m) - E_F,$$

we arrive at the equations

$$\Delta(i\omega_n, \mathbf{p}) = (\omega_n^2 + \xi^2(\mathbf{p})) F(i\omega_n, \mathbf{p}), \quad (2.24)$$

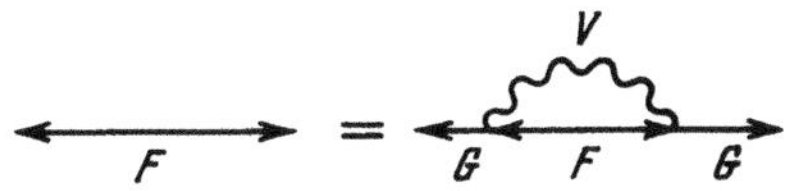


**Figure 2.5. Diagrammatic representation of (2.25) for the determination of $T_c$.**

$$F(i\omega_n, \mathbf{p}) = -\frac{T_c}{\omega_n^2 + \xi^2(\mathbf{p})} \sum_m \int d^3k\, V(i(\omega_n - \omega_m), \mathbf{p} - \mathbf{k})\, F(i\omega_m, \mathbf{k}). \tag{2.25}$$

It is this last equation, which is illustrated in Fig. 2.5, that will be used to determine $T_c$. We also give a schematic form of the analogous equation for the energy gap:

$$\Delta = -T_c \sum_m \int \frac{d^3k}{\omega_m^2 + \xi^2(\mathbf{k})} V\Delta. \tag{2.25'}$$

A disadvantage of (2.25) is that it has the form not of an integral but of an integrosummation equation.† To eliminate this shortcoming, it is convenient to use the spectral representations for the quantities entering into (2.25). Such representations, an example of which is the dispersion relation (2.3), make the frequency dependence of these quantities concrete, which allows us to carry out explicitly the summation over the frequencies. The spectral representation for the interaction is the temperature analog of (2.3), (2.10):

$$V(i\omega_n, \mathbf{p}) = \frac{4\pi e^2}{p^2}\left(1 - \int_0^\infty \frac{dE^2 \sigma(E, \mathbf{p})}{E^2 + \omega_n^2}\right). \tag{2.26}$$

The analogous spectral representation

$$F(i\omega_n, \mathbf{p}) = \int_0^\infty \frac{dE^2 f(E, \mathbf{p})}{E^2 + \omega_n^2} \tag{2.27}$$

can be derived from general principles; it is, however, easier to discern its validity directly from (2.25). Finally, from (2.24), we have

$$\Delta(i\omega_n, \mathbf{p}) = \alpha(\mathbf{p}) - \int_0^\infty \frac{dE^2 \beta(E, \mathbf{p})}{E^2 + \omega_n^2}. \tag{2.28}$$

†For the use of such equations in the theory of superconductivity, see [24].

All the spectral densities, $\sigma, f, \alpha$, and $\beta$, are real functions.

After substituting (2.26) and (2.27) into (2.25), we can perform the summation over the frequencies with the aid of the well-known formula

$$T_c \sum_n f(\omega_n) = \frac{1}{2} \sum_i \tan\left(\frac{\omega_i}{2T_c}\right) \mathrm{Res}_i\, f(\omega),$$

where the $\omega_i$'s are the poles of the analytic function $f(\omega)$, which falls off sufficiently rapidly on a large circle, and the $\mathrm{Res}_i$ are the corresponding residues. A simple calculation leads to an integral equation for the spectral density:

$$f(x, \mathbf{p}) = -\int d^3k \int_0^\infty dy K(x, y, \mathbf{p}, \mathbf{k}) f(y, \mathbf{k}). \tag{2.29}$$

Here the kernel $K$ is an odd function of $y$: $K(x,y,\mathbf{p},\mathbf{k}) = \frac{1}{2}[u(x,y,\mathbf{p},\mathbf{k}) - (y \longrightarrow -y)]$, where

$$u(x, y, \mathbf{p}, \mathbf{k}) = \frac{4\pi e^2}{|\mathbf{p}-\mathbf{k}|^2}\Big[\delta(x^2 - \xi^2(\mathbf{p}))\, M(y, \mathbf{p}, \mathbf{k}) + \\ + \mathrm{P}\frac{1}{x^2 - \xi^2(\mathbf{p})} N(x, y, \mathbf{p}, \mathbf{k})\Big], \tag{2.30}$$

$$M(y, \mathbf{p}, \mathbf{k}) = \tanh\left(\frac{y}{2T_c}\right) - \\ - 2\int_0^\infty \frac{dE\sigma(E, \mathbf{p}-\mathbf{k})(y+E)}{(y+E)^2 - \xi^2(\mathbf{p})}\left[\tanh\left(\frac{y}{2T_c}\right) + \coth\left(\frac{E}{2T_c}\right)\right],$$

$$N(x, y, \mathbf{p}, \mathbf{k}) = \sigma(x - y, \mathbf{p}-\mathbf{k})\left[\tanh\left(\frac{y}{2T_c}\right) + \coth\left(\frac{x-y}{2T_c}\right)\right].$$

Equation (2.29) plays the role of a basic equation for the determination of the critical temperature. It has a complex kernel and is moreover a multidimensional integral equation. It turns out, however, that within the framework of the approximation being used the kernel can be significantly simplified and the equation itself reduced to a one-dimensional integral equation of the type of the BCS equation.

**B. The Simplified Equations.** Underlying the simplification of (2.29) are the following simple considerations. Its kernel is linear in the interaction and is, in the weak-coupling approximation, small. Nevertheless, the equation can be satisfied since its right-hand side depends on $T_c$ in a singular fashion and $T_c$ itself is small. To wit, there appears on the right-hand side the factor $\ln(1/T_c)$, which compensates for the smallness of the interaction, leading at the same time to the exponential dependence of $T_c$ on the parameters of the problem.

To the indicated singularity leads the discontinuity of the density $f(x, \mathbf{p})$ at the point $x = \xi(\mathbf{p})$, a discontinuity that can be explicitly separated out in the form

$$f(x, \mathbf{p}) = \delta(x^2 - \xi^2(\mathbf{p})) A(\mathbf{p}) + \mathrm{P}\frac{1}{x^2 - \xi^2(\mathbf{p})} B(x, \mathbf{p}); \qquad (2.31)$$

this is suggested by the form of the kernel (2.30). If in the latter we directly set $T_c = 0$, replacing the hyperbolic functions by unity, then the right-hand side of (2.29) will diverge logarithmically at the point $y = \xi(\mathbf{k}) = 0$. Therefore, in separating out a singularity of the type $\ln(1/T_c)$, we should retain the quantity $\mathrm{th}(y/2T_c)$, which "cuts off" the integral at the indicated point. The remaining hyperbolic functions could be replaced by unity without prejudice to the convergence of the integral. However, we can more conveniently, and without going outside the limits of the assumed accuracy, make the substitutions†

$$\coth\left(\frac{E}{2T_c}\right) \to \tanh\left(\frac{y}{2T_c}\right), \quad \coth\left(\frac{x-y}{2T_c}\right) \to \varepsilon(x-y)\tanh\left(\frac{y}{2T_c}\right),$$

which are justified for low $T_c$; the accuracy of these substitutions is exponential in $T_c$, i.e., appreciably higher than that of the other simplifications made here. Again using the smallness of $T_c$, we can refer the kernel $K$ in (2.29) to the point $y = \xi(\mathbf{k})$, discarding the term $K(\ldots, y) - K(\ldots, \xi(\mathbf{k}))$, which is regular at this point. Below in Section 3.D we give the exact equations, which allow us to take account of the effects connected with the discarded terms.

The subsequent simplification of the equation that leads to an equation of the BCS type involves a transition from $f(x, \mathbf{p})$ to a new unknown function

$$\Phi(\mathbf{p}) = 2|\xi(\mathbf{p})| \int_0^\infty dx f(x, \mathbf{p}). \qquad (2.32)$$

With allowance for the simplifications made above, we have

$$\left.\begin{aligned} A(\mathbf{p}) &= -\int d^3k \frac{4\pi e^2}{|\mathbf{p}-\mathbf{k}|^2} \frac{\tanh(\xi(\mathbf{k})/2T_c)}{2\xi(\mathbf{k})} \times \\ &\quad \times \left(1 - \int_0^\infty \frac{dE^2 \sigma(E - |\xi(\mathbf{k})|, \mathbf{p}-\mathbf{k})}{E^2 - \xi^2(\mathbf{p})}\right) \Phi(\mathbf{k}), \\ B(x, \mathbf{p}) &= -\int d^3k \frac{4\pi e^2}{|\mathbf{p}-\mathbf{k}|^2} \frac{\tanh(\xi(\mathbf{k})/2T_c)}{2\xi(\mathbf{k})} \times \\ &\quad \times \sigma(x - |\xi(\mathbf{k})|, \mathbf{p}-\mathbf{k}) \Phi(\mathbf{k}) \end{aligned}\right\} \qquad (2.33)$$

†Here $\varepsilon(x)$ is equal to 1 for $x > 0$ and $-1$ for $x < 0$.

Next, combining (2.29)-(2.32) with these relations, we arrive at a simple equation for the function $\Phi$:

$$\left.\begin{aligned}\Phi(\mathbf{p}) &= -\int d^3k \frac{\tanh(\xi(\mathbf{k})/2T_c)}{2\xi(\mathbf{k})} R(\mathbf{p},\mathbf{k})\,\Phi(\mathbf{k}),\\ R(\mathbf{p},\mathbf{k}) &= \frac{4\pi e^2}{|\mathbf{p}-\mathbf{k}|^2}\left(1-2\int_0^\infty \frac{dE\sigma(E,\mathbf{p}-\mathbf{k})}{E+|\xi(\mathbf{p})|+|\xi(\mathbf{k})|}\right).\end{aligned}\right\} \qquad (2.34)$$

We shall discuss the properties of the quantities $\Phi$ and $R$ as well as their connection with $\Delta$ and $V$, respectively, in the following section.

To obtain a one-dimensional integral equation, we must take another step: Using the homogeneity and isotropy of the system, we should go over from the momentum to the new argument $\xi = p^2/2m - E_F$. Let us introduce the notation

$$\Phi(\xi) = \Phi(p)\,|_{p\to p(\xi)}, \quad R(\xi,\xi') = N(\xi')\,\overline{R(\mathbf{p},\mathbf{k})}\,|_{p,k\to p(\xi),p(\xi')},$$

where $p(\xi) = [2m(E_F+\xi)]^{1/2}$ is the inverse of the function $\xi(p)$;

$$N(\xi) = \frac{d^3p}{d\xi} = \frac{mp(\xi)}{2\pi^2}$$

is the level density; the bar denotes averaging over the angle between **p** and **k**. Then the sought equation assumes the form

$$\Phi(\xi) = -\int_{-E_F}^{\infty} d\xi' \frac{\mathrm{th}(\xi'/2T_c)}{2\xi'} R(\xi,\xi')\,\Phi(\xi'), \qquad (2.35)$$

and its kernel can be written in the form

$$R(\xi,\xi') = \frac{me^2}{\pi p(\xi)} \int_{|p(\xi)-p(\xi')|}^{p(\xi)+p(\xi')} \frac{dq}{q}\left[1-2\int_0^\infty \frac{dE\,\sigma(E,q)}{E+|\xi|+|\xi'|}\right]. \qquad (2.36)$$

**C. Connection with the BCS and Éliashberg Equations.** The simplification achieved above of the equation for the determination of the critical temperature is due to the transition to the new unknown function $\Phi$ and to the new kernel $R$. Let us elucidate the meaning of these quantities. To begin with, we give the relation

$$\alpha(\mathbf{p}) = A(\mathbf{p}) + \int_0^\infty \frac{dE\,B(E,\mathbf{p})}{E^2-\xi^2(\mathbf{p})}, \quad \beta(E,\mathbf{p}) = B(E,\mathbf{p}),$$

which follows from a comparison of (2.24), (2.28), and (2.31). With allowance for this relation, the expression obtained for the function $\Phi$ when (2.31) is substituted into (2.32) assumes the form

$$\Phi(\mathbf{p}) = A(\mathbf{p}) + 2|\xi(\mathbf{p})| \int_0^\infty \frac{dE\, B(E, \mathbf{p})}{E^2 - \xi^2(\mathbf{p})} = \alpha(\mathbf{p}) - 2 \int_0^\infty \frac{dE \beta(E, \mathbf{p})}{E + |\xi(\mathbf{p})|}. \tag{2.37}$$

A comparison of the last expression with the time analog of (2.28)

$$\Delta(\omega, \mathbf{p}) = \alpha(\mathbf{p}) - \int_0^\infty \frac{dE^2 \beta(E, \mathbf{p})}{E^2 - \omega^2 - i\delta} \tag{2.38}$$

reveals the connection between $\Phi$ and $\Delta$: The function $\Phi$ is obtained from $\Delta$ by making the substitution

$$(E^2 - \omega^2 - i\delta)^{-1} = \frac{1}{2E}[(E + |\omega|)^{-1} + (E - |\omega| - i\delta)^{-1}] \rightarrow \rightarrow \frac{1}{E}(E + |\omega|)^{-1} \tag{2.39}$$

and moving to the "mass surface" $\omega = \xi(\mathbf{p})$, i.e., introducing the same connection between $\omega$ and $\mathbf{p}$ that obtains for the free electron. The interaction $V$ [see (2.3), (2.10)] undergoes a similar modification, going over into the kernel $R$ (2.34).

The meaning of the substitutions (2.39) can be explained by referring to the condition for the cancellation of the "dangerous" diagrams, a condition that also leads to an equation for the critical temperature [25].† These diagrams correspond to the *creation* from vacuum of a phonon and a pair of electrons, whereas the interaction $V$ corresponds to the exchange of phonons, i.e., the *creation* of a phonon and an electron and the *annihilation* of another electron (and vice versa). As to the referral to the mass surface, it gives a direct answer to the often-asked question: In terms of which variable (the frequency or wave vector) should superconductivity be described?

Let us now compare (2.34) with the corresponding BCS equation (see, for example, [6] and Chapter 1 above):

†The necessity of the substitutions (2.39) as applied to the purely phonon interaction has already been noted in [25].

$$\Delta(\mathbf{p}) = -\int d^3k \frac{\tanh(\xi(\mathbf{k})/2T_c)}{2\xi(\mathbf{k})} W(\mathbf{p}, \mathbf{k})\, \Delta(\mathbf{k}). \tag{2.40}$$

Here the kernel $W$ is a smooth real function of $\xi(\mathbf{p})$ and $\xi(\mathbf{k})$ that decreases rapidly as each of its arguments increases, while the unknown function depends on only one argument, $\xi(\mathbf{p})$. In this $W(\mathbf{p}, \mathbf{k})$ and $\Delta(\mathbf{p})$ differ from the quantities $V(\omega, \mathbf{p})$ and $\Delta(\omega, \mathbf{p})$ entering into (2.25′), which follows from the exact theory. However, the above-described modification, by eliminating these differences, draws the equations closer in form [cf. (2.34) and (2.40)] : the referral to the mass surface eliminates the "superfluous" variables, while the substitutions (2.39) free the kernel from resonance singularities and imaginary parts.

The only remaining qualitative difference between (2.34) and (2.40) is that the kernel of (2.40) is localized near the Fermi level, while the quantity $R$ has a distant Coulomb "tail." The analogy with the BCS equation can be made complete by changing $R$ in such a way that the integration in (2.34) is performed near the Fermi level without changing the result. It is well-known that this leads to the suppression of the direct Coulomb interaction (see [6, 7, 25] and Section 4 below). We shall not repeat the corresponding calculations, but we shall derive the same result by a somewhat different method (see Section 4).

It remains to compare (2.34) with the Éliashberg equation [5, 6] for the purely phonon interaction with

$$\sigma(\omega, \mathbf{p}) = \frac{p^2 \alpha_p^2}{4\pi e^2}\, \delta(\omega - \omega_p),$$

where $\alpha_p$ is the electron–phonon coupling constant and $\omega_p$ is the phonon frequency, which is bounded from above by the Debye frequency $\omega_D$. The smallness of the latter in comparison with $E_F$ leads to the dependence of $\Delta$ on only the frequency. In the time representation [see (2.38)] the Éliashberg equation at $T = T_c$ has the form

$$\operatorname{Re} \Delta(\omega) = \frac{m}{4\pi^2 p_F} \int_0^\infty \frac{d\omega'}{\omega'} \tanh\left(\frac{\omega'}{2T_c}\right) \int_0^{q_D} dq\, q \alpha_q^2 \times \frac{\omega' + \omega_q}{(\omega' + \omega_q)^2 - \omega^2} \operatorname{Re} \Delta(\omega'), \tag{2.41}$$

Here $q_D = m\omega_D/p_F$; the weak-coupling approximation is used. As has already been emphasized in Section 1, the kernel of this equation is singular.

It is not difficult to show that the above-expounded method would have led directly to (2.41) if we had not made the transition to the new variable $\Phi$. Indeed, let us consider the first equality in (2.33), on the right-hand side of which we can, in the weak-coupling approximation, replace $\Phi$ by the quantity

$A$, which, according to (2.37), behaves in a similar fashion at the singular point $\xi(\mathbf{p}) = 0$ (see Section 3.B). Discarding the direct Coulomb term, we have

$$A(\mathbf{p}) = \int d^3k \frac{4\pi e^2}{|\mathbf{p}-\mathbf{k}|^2} \times$$
$$\times \frac{\tanh(\xi(\mathbf{k})/2T_c)}{\xi(\mathbf{k})} \int_0^\infty \frac{dE\,(E+|\xi(\mathbf{k})|)\,\sigma(E,\mathbf{p}-\mathbf{k})}{(E+|\xi(\mathbf{k})|)^2-\xi^2(\mathbf{p})} A(\mathbf{k}).$$

This equation can now easily be reduced to the Éliashberg equation, taking account of the fact that, according to (2.37), (2.38), $A(\mathbf{p}) = \mathrm{Re}\,\Delta(|\xi(\mathbf{p})|, \mathbf{p})$.

**D. Second-Order Perturbation Theory.** The calculations leading to (2.34) were carried out in lowest-order perturbation theory in terms of the interaction $V$. As applied to the computation of $T_c$, this corresponds to the correct description of the exponent in the corresponding formula for $T_c$ [see (1.3) above], the order of magnitude of this exponent being $1/V$. But to describe the pre-exponential factor, which is of zeroth order in $V$, we must go a step further by adding to the kernel $R$ in (2.34) the terms of second order in $V$.

Let us, to begin with, show that the transition to a BCS-type equation is possible not only within the framework of lowest-order perturbation theory, but also in the strong-coupling theory. In the general case, in place of Eq. (2.25) we can write:

$$F(i\omega_n, \mathbf{p}) = -\frac{T_c}{\omega_n^2+\xi^2(\mathbf{p})} \sum_m \int d^3k U(i\omega_n, \mathbf{p}; i\omega_m, \mathbf{k}) F(i\omega_m, \mathbf{k}), \quad (2.42)$$

where the kernel $U$ no longer coincides with the interaction $V$ but contains terms of all orders in $V$, $T_c$ being no longer considered small.

Let us introduce the notation

$$\underset{n,p}{S}(\ldots) \equiv 2\xi(\mathbf{p}) \coth\left(\frac{\xi(\mathbf{p})}{2T_c}\right) T_c \sum_n \frac{(\ldots)}{(\omega_n^2+\xi^2(\mathbf{p}))} \quad \left(\underset{n,p}{S}(1) = 1\right)$$

and define the function $\Phi$ according to the relation

$$\Phi(\mathbf{p}) = \underset{n,p}{S} F(i\omega_n, \mathbf{p})\left(\omega_n^2+\xi^2(\mathbf{p})\right) =$$
$$= 2\xi(\mathbf{p}) \coth\left(\frac{\xi(\mathbf{p})}{2T_c}\right) \int_0^\infty dx \tanh\left(\frac{x}{2T_c}\right) f(x, \mathbf{p}). \quad (2.43)$$

Coupling $F$ and $\Phi$ with the aid of a new kernel $Q$:

$$F(i\omega_n, \mathbf{p}) = -\int d^3k \frac{\tanh(\xi(\mathbf{k})/2T_c)\, Q(i\omega_n, \mathbf{p}, \mathbf{k})}{2\xi(\mathbf{k})\left(\omega_n^2 + \xi^2(\mathbf{p})\right)} \Phi(\mathbf{k}),$$

and using (2.43), we easily arrive at (2.34) with

$$R(\mathbf{p}, \mathbf{k}) = \underset{n,p}{S}\, Q(i\omega_n, \mathbf{p}, \mathbf{k}), \tag{2.44}$$

the new kernel $Q$ satisfying the equation

$$\begin{aligned} Q(i\omega_n, \mathbf{p}, \mathbf{k}) &= \underset{m,k}{S}\, U(i\omega_n, \mathbf{p}; i\omega_m, \mathbf{k}) - \\ &- \int d^3q \frac{\mathrm{th}(\xi(\mathbf{q})/2T_c)}{2\xi(\mathbf{q})} \underset{m,q}{S}\, \underset{r,q}{S} [U(i\omega_n, \mathbf{p}; i\omega_m, \mathbf{q}) - \\ &\qquad - U(i\omega_n, \mathbf{p}; i\omega_r, \mathbf{q})]\, Q(i\omega_m, \mathbf{q}, \mathbf{k}). \end{aligned} \tag{2.45}$$

The kernel $U$ in (2.42) does not possess resonances and is real, having been defined in the temperature representation; the same properties are possessed by the kernel $R$.

Let us proceed to the determination of the second-order corrections to the latter. They contain two terms. The first, $\delta_1 R$, corresponds to a correction to the interaction (2.10) and is contained in the kernel $U$ in (2.42). Physically, this term describes the difference between the interaction of the particles of the system and that of the external particles (see Section 2); to it corresponds, in particular, the diagram in Fig. 2.3. The second term, $\delta_2 R$, corresponds to the nonsingular term $K(\ldots, y) - K(\ldots, \xi(\mathbf{k}))$, which was discarded in Section 3.B. It is contained in the second term of the right member of (2.45) and arises as a result of the first iteration of this equation.

In computing $\delta_{1,2}R$, we can immediately set $T_c = 0$. Furthermore, as will be seen from the results of Section 4, the answer will be for $\xi(\mathbf{p}), \xi(\mathbf{k}) = 0$, and we denote it by $\delta_{1,2}R(0, 0)$. We can derive the simple formula $\lim\limits_{\xi(p) \to 0} \underset{n,p}{S} f(p) = f(0, p_F)$.† As a result we obtain

$$\delta_1 R(0, 0) = \delta U(0, \mathbf{p}; 0, \mathbf{k})\,|_{p,k=p_F}, \tag{2.46}$$

†This formula is valid for functions that are regular at the points $\omega_n = 0$ and $\xi(p) = 0$. If, on the other hand, $f = \dfrac{A(p) - A(0, p_F)}{(i\omega_n - \xi(p))}$, then the answer reads

$$\frac{1}{2}\left(\frac{\partial A}{\partial(i\omega_n)} - \frac{\partial A}{\partial \xi(p)}\right)\Bigg|_{0,p_F}.$$

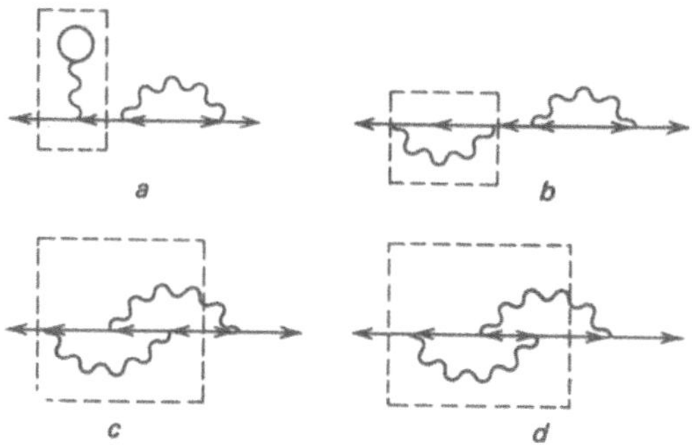


**Figure 2.6. Second-order corrections to the right-hand side of the diagram in Fig. 2.5.**

$$\delta_2 R(0, 0) = -2N(0) \int d^3q \frac{(4\pi e^2)^2}{|\mathbf{p}-\mathbf{q}|^2 |\mathbf{k}-\mathbf{q}|^2} \times$$
$$\times \int_0^\infty dE\, dE' \frac{\sigma(E, \mathbf{p}-\mathbf{q})\, \sigma(E', \mathbf{k}-\mathbf{q})}{(E+E')(E+|\xi(\mathbf{q})|)(E'+|\xi(\mathbf{q})|)}. \qquad (2.47)$$

To determine the quantity $\delta U$ in (2.46), which is a second-order correction to the interaction, let us turn to the formula (2.21) and compute, using Wick's theorem, the second term on the right-hand side of this formula. The result has the form

$$\delta U(p, k) = V(p-k)\,[2G(0)\,V(0)\,G(p) - \Sigma(p)\,G(p) -$$
$$- \int d^4q V(q)\, G(k-q)\, G(p-q) + \begin{pmatrix} p \to -p \\ k \to -k \end{pmatrix}\Big] - \qquad (2.48)$$
$$- \int d^4q\, V(q)\, V(q-p+k)\, G(q+k)\, G(q-p),$$

where we have introduced the electron self-energy part

$$\Sigma(x) = iV(x)\, G(x), \quad \Sigma(p) = -\int d^4k V(p-k)\, G(k). \qquad (2.49)$$

The above-given formulas have been written in the temperature representation:

$$p = (i\omega_n, \mathbf{p}), \int d^4k = T_c \sum_n \int d^3k.$$

The expression (2.48) is illustrated in Fig. 2.6, whose diagrams correspond to the sequence of the terms in (2.48): Diagram (a) is the Hartree self-energy correction to the diagram on the right-hand side of Fig. 2.5; diagram (b), the exchange self-energy correction; diagram (c) the correction to the vertex part (see Fig. 2.3); diagram (d) the anomalous vertex, which includes the function $F$.

Let us begin the computation of (2.46) with the self-energy corrections and point out that the chemical potential in the free Green function also acquires a correction determined by the condition for this function to become infinite at $\omega_n = 0$ and $\xi(\mathbf{p}) = 0$ (or $p = p_F$). This leads to the vanishing of the first term in (2.48) and to the substitution $\Sigma(p) \longrightarrow \Sigma(p) - \Sigma(0, p_F)$ in the second. With allowance for the assertion made in the preceding footnote, we find that

$$\delta_{SE} R(0, 0) = \tilde{\Sigma}(0) R(0, 0), \quad \tilde{\Sigma}(0) = -\frac{1}{2}\left(\frac{\partial \Sigma}{\partial (i\omega_n)} - \frac{\partial \Sigma}{\partial \xi(p)}\right). \tag{2.50}$$

The computation of the self-energy part (2.49) yields

$$\Sigma(p) = -\frac{1}{2}\int d^3k \frac{4\pi e^2}{|\mathbf{p}-\mathbf{k}|^2}\left[\varepsilon(\xi(\mathbf{k})) - 2\int_0^\infty \frac{dE\, \sigma(E, \mathbf{p}-\mathbf{k})}{E\varepsilon(\xi(\mathbf{k})) + \xi(\mathbf{k}) - i\omega_n}\right].$$

As a result, the self-energy correction to $R$ has the form of (2.50) with

$$\tilde{\Sigma}(0) = -\frac{1}{2}\frac{\partial}{\partial \xi(\mathbf{p})}\int d^3k \frac{4\pi e^2}{|\mathbf{p}-\mathbf{k}|^2}\varepsilon(\xi(\mathbf{k})) \times$$
$$\times \left(1 - 2\int_0^\infty \frac{dE\, \sigma(E, \mathbf{p}-\mathbf{k})}{E + |\xi(\mathbf{k})| + \varepsilon(\xi(\mathbf{k}))\,\xi(\mathbf{p})}\right)\Bigg|_{p=p_F}. \tag{2.51}$$

Let us make an observation apropos of the structure of the expression (2.50). The self-energy correction results from the factor $[\omega_n^2 + \xi^2(p)]^{-1} = G(p)G(-p)$ in (2.25), which should now be replaced by $\mathrm{G}(p)\mathrm{G}(-p)$, where $\mathrm{G}(p) = [G^{-1}(p) - 2G(0)V(0) + \Sigma(p)]^{-1}$ is the exact electron Green function. Near the Fermi level $\mathrm{G} = Z/[i\omega_n - v_F(p - p_F)]$, where $Z = 1 + [\partial\Sigma/\partial(i\omega_n)]$ is the quasiparticle "charge" renormalization factor,

$$v_F = \frac{p_F}{m}\left(1 + \frac{\partial \Sigma}{\partial (i\omega_n)} + \frac{\partial \Sigma}{\partial \xi(p)}\right)$$

is the renormalized velocity at the Fermi level. For $\omega_n = 0$ and $p = p_F$, we have

$$\frac{\mathrm{G}(p)\,\mathrm{G}(-p)}{G(p)\,G(-p)} = \frac{Z^2 p_F}{m v_F},$$

from which emerges the combination of the derivatives of the quantity $\Sigma$ figuring in (2.50).

We have yet to compute the components of (2.46) that are connected with the vertex parts. They have a structure characteristic of perturbation theory:

$$\delta R\,(0,\,0) = N\,(0) \int d^3q \,\frac{\varepsilon\,(\xi\,(\mathbf{k}-\mathbf{q}))\,B\,(\mathbf{q},\mathbf{k},\mathbf{p}) - (\mathbf{k} \rightleftarrows \mathbf{p})}{\xi\,(\mathbf{k}-\mathbf{q}) - \xi\,(\mathbf{p}-\mathbf{q})}. \tag{2.52}$$

Introducing the notation

$$V\,(\mathbf{p},\,\xi,\,\xi') = \frac{4\pi e^2}{p^2}\left(1 - 2\int\limits_0^\infty \frac{dE\sigma\,(E,\,\mathbf{p})}{E + |\xi| + |\xi'|}\right),$$

we obtain for the term corresponding to the simple vertex

$$B_v(\mathbf{q},\,\mathbf{k},\,\mathbf{p}) = V(\mathbf{p}-\mathbf{k},\,0,\,0)\,V(\mathbf{q},\,0,\,\xi(\mathbf{k}-\mathbf{q})), \tag{2.53}$$

and for the term corresponding to the anomalous vertex

$$\begin{aligned} B_{\mathrm{av}}\,(\mathbf{q},\,\mathbf{k},\,\mathbf{p}) = {}& {}^1/_2 V\,(\mathbf{q},\,0,\,\xi\,(\mathbf{k}-\mathbf{q}))\,V\,(\mathbf{p}-\mathbf{k}+ \\ & + \mathbf{q},\,0,\,\xi\,(\mathbf{k}-\mathbf{q})) + 2\,\frac{(4\pi e^2)^2\,|\,\xi\,(\mathbf{k}-\mathbf{q})\,|}{q^2\,|\,\mathbf{q}+\mathbf{p}-\mathbf{k}\,|^2}\int\limits_0^\infty \times \\ & \times \frac{dE\,dE'\sigma\,(E,\,\mathbf{q})\,\sigma\,(E',\,\mathbf{q}+\mathbf{p}-\mathbf{k})}{(E+E')(E+|\,\xi\,(\mathbf{k}-\mathbf{q})\,|)(E'+|\,\xi\,(\mathbf{k}-\mathbf{q})\,|)}. \end{aligned} \tag{2.54}$$

Below, in Section 4 we shall explicitly compute all the above-considered corrections to the kernel $R$. Let us note in conclusion that averaging over the directions of the vector $\mathbf{k}$ and the equation of $p$ and $k$ to the quantity $p_F$ are implied in the formulas (2.46) and (2.47). The very meaning of the quantity $\delta R(0, 0)$ [see (2.35), (2.36)] makes these operations necessary.

## 4. The Critical Temperature

**A. The General Expression for $T_c$.** In this section we derive an expression for the critical temperature as an explicit functional of the permittivity of the superconducting material. The quantity $T_c$ is defined as the eigenvalue of the basic equation, which we take in the form (2.35), of the theory of superconductivity. Its kernel in lowest-order perturbation theory is given by the formula (2.36), and the corrections of the next order are given in Section 3.D.

For low values of $T_c$, the right-hand side of (2.35) contains two leading terms: one proportional to $\ln(1/T_c)$ and the other not depending on $T_c$. To separate them out, let us divide the domain of the integration on the right-hand side of (2.35) into three parts: from $-E_F$ to $-\omega_0$, from $-\omega_0$ to $\omega_0$, and from

$\omega_0$ to $\infty$; the quantity $\omega_0$ is absolutely arbitrary and drops out of the final formulas.

In the integrals for the first and third subdomains, we can immediately set $T_c = 0$, while the integral for the second subdomain should be divided into two terms, into one which will enter the quantity $R(\xi, 0)\Phi(0)$; into the second, $R(\xi, \xi')\Phi(\xi') - R(\xi, 0)\Phi(0)$.

Performing the integration by parts, and using the asymptotic estimate†

$$\int_0^{\omega_0} \frac{d\xi}{\xi} \tanh\left(\frac{\xi}{2T_c}\right) = \ln\left(\frac{2\gamma\omega_0}{\pi T_c}\right) + O(T_c),$$

we can easily reduce (2.35) to the form

$$\Phi(\xi) = -\ln\left(\frac{2\gamma\omega_0}{\pi T_c}\right)\Phi(0)\,R(\xi, 0) + A_{\xi'}(R(\xi, \xi')\,\Phi(\xi')),$$

where we have introduced the notation

$$A_{\xi'}(\ldots) = \frac{1}{2}\int_{-E_F}^{\infty} d\xi' \ln\left(\frac{|\xi'|}{\omega_0}\right)\frac{d}{d|\xi'|}(\ldots).$$

Let us now introduce the new variable

$$\chi(\xi) = \Phi(\xi) \times \left[\Phi(0)\ln\left(\frac{\pi T_c}{2\gamma\omega_0}\right)\right]^{-1},$$

which allows us to write down the expression for the critical temperature in the form

$$T_c = \frac{2\gamma\omega_0}{\pi}\exp\left(\frac{1}{\chi(0)}\right), \tag{2.55}$$

where the function $\chi(\xi)$ satisfies the equation

$$\chi(\xi) = R(\xi,\ 0) + A_{\xi}'(R(\xi,\ \xi')\chi(\xi')). \tag{2.56}$$

This equation could have been solved by the method of iterations if the characteristic frequencies of the system were contained in a relatively narrow range. Of real interest, however, is the opposite case of strong incommensurability of

†Here $\gamma = e^C$, where $C$ is the Euler constant.

the frequencies (e.g., of the characteristic phonon frequency and the Fermi energy). Therefore, we shall assume that the quantity $R$ is small but that $R \ln(\xi_2/\xi_1) \sim 1$ ($\xi_1$ and $\xi_2$ are quantities of the order of the lowest and highest of the characteristic frequencies).

Consequently, let us set $R = R_1 + R_2$, where the quantity $R_i$ decreases when $\xi \sim \xi_i$ ($i = 1, 2$), with $\xi_1 \ll \xi_2$. Similar properties are possessed by the terms of the function $\chi = \chi_1 + \chi_2$, which are determined by the equations

$$\chi_i(\xi) = R_i(\xi, 0) + A_{\xi}'(R_i(\xi, \xi')\chi(\xi)).$$

By referring these equations to the point $\xi = 0$, we can give the quantities $\xi_1$ and $\xi_2$ exact meaning with the aid of the relations

$$\left.\begin{aligned} A_{\xi'}(R_1(0, \xi')\chi(\xi')) &= -\ln\left(\frac{\xi_1}{\omega_0}\right) R_1(0, 0)\chi(0), \\ A_{\xi'}(R_2(0, \xi')\chi(\xi')) &= -\left[\ln\left(\frac{\xi_1}{\omega_0}\right)\chi_1(0) + \right. \\ &\quad \left. + \ln\left(\frac{\xi_2}{\omega_0}\right)\chi_2(0)\right] R_2(0, 0). \end{aligned}\right\} \tag{2.57}$$

It is precisely to such a structure that the functions $R_i, \chi_i$, which have the form of a rectangular "step" cut off at $\xi = \xi_i$, would have led. Introducing the notation

$$R_1(0, 0) = -\lambda, \quad R_2(0, 0) = \mu, \tag{2.58}$$

we find

$$\chi_1(0) = -\lambda\left(1 + \mu \ln\left(\frac{\xi_2}{\xi_1}\right)\right)\frac{1}{A}, \quad \chi_2(0) = \mu\frac{1}{A},$$
$$\chi(0) = \left[\mu - \lambda\left(1 + \mu \ln\left(\frac{\xi_2}{\xi_1}\right)\right)\right]\frac{1}{A},$$

where

$$A = \left(1 - \lambda \ln\left(\frac{\xi_1}{\omega_0}\right)\right)\left(1 + \mu \ln\left(\frac{\xi_2}{\omega_0}\right)\right) + \lambda\mu \ln^2\left(\frac{\xi_1}{\omega_0}\right).$$

The final general formula for $T_c$ assumes the form [see (2.55)]:

$$T_c = \frac{2\gamma\xi_1}{\pi}\exp\left(\frac{1}{g}\right), \quad g = -\lambda + \mu\frac{1}{1+\mu\ln(\xi_2/\xi_1)}{}^{*)}. \tag{2.59}$$

The computations just carried out were based essentially on the assumption that the quantities $R_i(\xi, \xi')$ decrease in the same fashion with respect to both variables. Below we shall have to deal with the case when the laws of decrease with respect to the two variables are different. Setting

$$R(\xi, \xi') = \sum_{i,j=1}^{2} R_{ij}(\xi, \xi'),$$

where the $R_{ij}$'s decrease when $\xi \sim \xi_i$, $\xi' \sim \xi_j$, we easily arrive at the conclusion that we should replace $R_i$ by $\widetilde{R}_i \equiv \sum_j R_{ji}$ in (2.57), the expression for the quantity $g$ in (2.59) then assuming the form†

$$g = \widetilde{R}_1(0, 0) + \widetilde{R}_2(0, 0)\frac{1}{1 + R_{22}\ln(\xi_2/\xi_1)}. \tag{2.60}$$

Now, in determining the frequencies $\xi_1$ and $\xi_2$ entering into (2.59), we can use the method of iterations, replacing in (2.57) $\chi_i(\xi)$ by $\chi_i(0)R_i(\xi, 0)/R_i(0, 0)$ [see (2.56)]. This leads to the relation $\ln\xi_1 = \ln\bar{\xi} - \lambda Q/g$, where

$$\ln\bar{\xi} = \frac{1}{2\lambda}\int_{-E_F}^{\infty} d\xi \ln|\xi| \frac{d}{d|\xi|} R_1(0, \xi), \tag{2.61}$$

$$Q = \frac{1}{2\lambda^2}\int_{-E_F}^{\infty} \frac{d\xi}{|\xi|} R_1(0, \xi)\,[R_1(\xi, 0) - R_1(0, 0)]. \tag{2.62}$$

The frequency $\xi_2$ enters only into the second term in the expression for $g$, which term assumes the form

$$\frac{\mu}{1 + \mu\ln(\xi_2/\xi_1)} = \frac{\mu(1 - \lambda Q)}{1 + \mu\ln(\xi_2/\bar{\xi})},$$

where

†In contrast to the notation adopted in the other chapters, here $g$ is a dimensionless quantity.

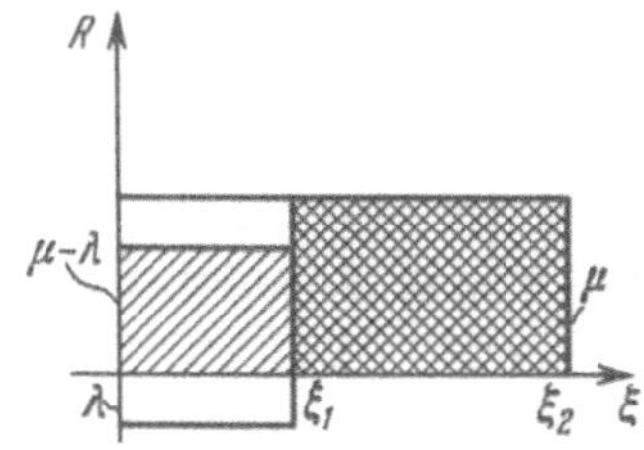


**Figure 2.7. The kernel of (2.35) for the determination of $T_C$. The diagonally hatched area represents the step localized near the Fermi level; the cross-hatched area, the shifted step.**

$$\ln\left(\frac{\xi_2}{\bar{\xi}}\right) = \frac{1}{2}\int_{-E_F}^{\infty}\frac{d\xi}{|\xi|}\left[\frac{R_2(0,\xi)\,R_2(\xi,0)}{R_2^2(0,0)} - \frac{R_1(0,\xi)}{R_1(0,0)}\right]. \tag{2.63}$$

Combining the obtained results within the adopted accuracy, we find in place of the first formula in (2.59) the expression

$$T_c = \frac{2\gamma\bar{\xi}}{\pi}\exp\left(\frac{1}{g} + \frac{\lambda^2}{g^2}Q\right). \tag{2.64}$$

It is important to emphasize that the above-performed analysis is based essentially on the sole assumption that the mean value $\bar{\xi}$ is small compared to $\xi_2$. Therefore, the formulas given are also valid in the case when the function $R_1(0, \xi)$ has components that decrease even when $\xi > \xi_2$, provided their weight is sufficiently small.

Let us return to the expression for the quantity $g$ [see (2.59)], from which it can be seen that part of the interaction that has a wide range of action–in energy terms–turns out to be modified: it weakens in the case of repulsion ($R_2 > 0$), while it intensifies in the case of attraction ($R_2 > 0$). What happens in both cases is that additional attractive forces emerge.

This fact is quite important. It is sufficient to say that for structures with positive static permittivities ($\mu - \lambda > 0$) it is precisely this effect that leads to the possibility of effective attraction between the electrons, i.e., to the phenomenon of superconductivity itself.

The effect under discussion has been repeatedly considered in the literature (for references, see the end of Section 3.C). Following [26], let us give another physical interpretation of the effect. Let us split the quantity $R$ into two terms: the first term represents a step of width $\xi_1$ and height $\mu - \lambda$, "pressed" against the Fermi level, and the second term is a step of width $\xi_2 - \xi_1$ and height $\mu$, shifted with respect to the Fermi level (Fig. 2.7). The second term, which corresponds to the absence of interaction at the Fermi level itself, leads to effective excess attraction irrespective of whether this term by itself corresponds to attraction or repulsion.†

**B. Asymptotic Formula for $T_c$.** Let us first investigate the asymptotic (from the standpoint of interaction weakness) expression for $T_c$, corresponding to the neglect of factors and terms of the order of unity in, respectively, the pre-exponential factor and the logarithms entering into the formulas for $\lambda$ and $g$. This expression corresponds to (2.64) with the second term in the exponent dropped and the quantities $\bar{\xi}$ and $g$ computed with the aid of (2.36).

The above-performed division of the interaction into parts characterized by the scales $\xi_1$ and $\xi_2$ physically corresponds to the fact that the spectral density $\sigma$ consists of two terms of different natures. The first term (with the scale $\xi_1 \sim \bar{\xi}$) corresponds to the phonon, exciton, and the other interactions localized—in the sense that the mean value $\bar{\xi}$ is small (see Section 4.A)—near the Fermi level; to this part of the interaction corresponds the spectral density $\sigma_1$. The second term, $\sigma_2 = \sigma - \sigma_1$, has the scale $\xi_2$, and corresponds to the Coulomb forces.

Let us first determine the quantities $\lambda$ and $\mu$ [see (2.58)]. Introducing the purely Coulomb permittivity $\varepsilon_2$

$$\frac{1}{\varepsilon_2(\omega, \mathbf{k})} = 1 - \int_0^\infty \frac{dE^2 \sigma_2(E, \mathbf{k})}{E^2 - \omega^2 - i\delta}$$

and the coupling constant $\alpha = me^2/\pi p_F$ (which is small in the weak-coupling approximation), we have from (2.36) the expressions

$$\lambda = \alpha \int_0^{2p_F} \frac{dq}{q} \left( \frac{1}{\varepsilon_2(0, \mathbf{q})} - \frac{1}{\varepsilon(0, \mathbf{q})} \right), \quad \mu = \alpha \int_0^{2p_F} \frac{dq}{q} \frac{1}{\varepsilon_2(0, \mathbf{q})}.$$

The integrals over $q$ converge at the lower limit because of the Debye screening, i.e., to the singularities of $\varepsilon$ and $\varepsilon_2$ of the type $\varkappa^2/q^2$ at small $q$ (see Section 2.E).

Setting $\varepsilon(0, q) = [(\varkappa^2/q^2) + 1] f(q)$ [and similarly for $\varepsilon_2(0, q)$] and considering the expressions for $\lambda$ and $\mu$ in the limit as $\alpha \to 0$, we find the logarithmic factor $\ln(2p_F/\varkappa) = \frac{1}{2} \ln(1/\alpha)$ to be attached to the corresponding combinations of the quantities $f(0)$ and $f_2(0)$. What happens is that the functions $\varepsilon(0, \mathbf{q})$ and $\varepsilon_2(0, \mathbf{q})$ become referred to the argument $\bar{q}$, which is determined by the inequalities $\varkappa \ll \bar{q} \ll p_F$. This is not surprising since the dominant contribution to

†In the situation under consideration the interaction of the electrons at the Fermi level constitutes a second-order effect and, as is usual in quantum mechanics, corresponds to effective attraction. This is connected with the fact that before entering into direct interaction, the electrons should virtually leave the Fermi level and get into the interaction region.

the angle-averaged effective interaction (2.36) should be made by distances less than the Debye-screening radius but at the same time greater than the mean particle separation.

It is significant that, owing to the inequality $\bar{q} \gg \varkappa$, the Coulomb permittivity $\varepsilon_2(0, \bar{q})$ can be set equal to unity. This is because of the vanishing of all the screening effects at the momentum $\bar{q}$. Let us note that, for the indicated reason, the spectral density $\sigma_2$ completely drops out of all the expressions that we shall consider below. Summing up all the foregoing, we obtain

$$\mu = \frac{\alpha}{2} \ln \frac{1}{\alpha}, \quad \lambda = \mu \left(1 - \frac{1}{\varepsilon(0, \bar{q})}\right). \tag{2.65}$$

Let us proceed to the determination of the characteristic frequencies. Using (2.61) and the smallness of $\xi_1$ in comparison with $E_F$, we find

$$\ln \bar{\xi} = \int_0^{2p_F} \frac{dq}{q} \int_0^{\infty} \frac{dE}{E} \ln E\sigma_1(E, \mathbf{q}) \Bigg/ \int_0^{2p_F} \frac{dq}{q} \int_0^{\infty} \frac{dE}{E} \sigma_1(E, \mathbf{q}) =$$
$$= \int_0^{\infty} \frac{dE}{E} \ln E\sigma(E, \bar{q}) \Bigg| \int_0^{\infty} \frac{dE}{E} \sigma(E, \bar{q}), \tag{2.66}$$

where $\sigma = -(1/\pi) \operatorname{Im}(1/\varepsilon)$ (see Section 2.B).

Thus, the frequency $\bar{\xi}$ (it can be called the mean-logarithmic frequency) is the result of the averaging of the logarithm of the frequency over the real spectrum of the system. It has a direct physical meaning (and can, in principle, be found directly from experiment), being connected with the Grüneisen lattice constant $\gamma = -\partial \ln \bar{\xi} / \partial \ln v$, where $v$ is the atomic volume; to wit,

$$\bar{\xi} = C\omega_i \exp\left[-\int_0^{v} \frac{dv}{v} \left(\gamma(v) - \frac{1}{2}\right)\right],$$

where $C \approx 0.54$. The quantity $\gamma$ is itself connected through the relation $\gamma = \alpha v K / c_v$ with the volume expansion coefficient $\alpha$, the bulk modulus $K$, and the specific heat $c_v$. In the Debye model $\ln \bar{\xi} = \ln \omega_D - \frac{1}{3}$, but in real cases the deviation of the mean-logarithmic frequency from the Debye frequency $\omega_D$ can be appreciable. Details concerning the formulas just presented can be found in [27].

It remains to compute the frequency $\xi_2$ [more precisely, $\ln(\xi_2/\bar{\xi})$; see (2.63)]. By carrying out the computation with logarithmic accuracy, we can verify that the largest contribution to the first term in this formula is made by the interval $\omega << \xi << E_F$, where

$$R_2(0, \xi) \sim \alpha \int_{\xi/v_F}^{2p_F} \frac{dq}{q} = \alpha \ln\left(\frac{E_F}{\xi}\right).$$

Within the limits of the adopted accuracy, we can set

$$\ln\left(\frac{\xi_2}{\bar{\xi}}\right) = \ln\left(\frac{E_F}{\bar{\xi}}\right). \quad (2.67)$$

Combining the obtained results, we arrive at the following asymptotic (in the $\alpha \longrightarrow 0$ limit) formula for the critical temperature

$$T_c \backsim \bar{\xi} \exp\left(-\frac{1}{\lambda - \mu^*}\right), \quad \mu^* = \frac{\mu}{1 + \mu \ln (E_F/\bar{\xi})}, \quad (2.68)$$

where the quantity $\bar{\xi}$ is given by the formula (2.66), while the constants $\lambda$ and $\mu$ are given by the formulas (2.65) (see also [6] and Chapter 4). As we can see, the answer turns out to be explicitly expressed in terms of the permittivity of the superconducting material, or, more exactly, in terms of its static value and its imaginary part (the spectral density).

So far, we have considered the normal $\bar{\xi} << E_F$ case. The question arises whether the inverse relation $\bar{\xi} > E_F$ is possible. Using the formulas (2.61) and (2.36), we can easily verify that this possibility is not realized: no matter how distant and strongly pronounced a peak of the spectral density $\sigma(\omega, q)$ may be, the quantity $R_1(\xi, 0)$ will begin to diminish the maximum when $\xi \sim E_F$. Therefore, the greatest possible value of $\bar{\xi}$ is of the order of $E_F$.

Let us note in conclusion that all the qualitative considerations presented in this section can easily be illustrated using the models considered in Section 2.E.

**C. The Complete Expression for $T_c$ in the Weak-Coupling Approximation.** In the above-obtained asymptotic expression (2.68), we have dropped the pre-exponential factor of the order of unity and, furthermore, the quantities $\lambda$ and $\ln(\xi_2/\bar{\xi})$ are computed with logarithmic accuracy. In this section we shall find the complete expression for the critical temperature with allowance for all the terms essential to the weak-coupling approximation.

Besides the purely quantitative refinement of the formula (2.68), we shall also achieve a number of qualitative objectives. First of all, the complete expression for $T_c$ will allow the graphic demonstration of the fact that the Coulomb interaction cannot be legitimately taken into consideration in the strong-coupling theory by replacing it by a constant. Furthermore, it is precisely the complete expression for $T_c$ that will allow the complete solution of the problem of the optimization of the critical temperature, i.e., the problem of the determination of that form of the spectral density that corresponds to the maximum value of $T_c$.

The correction terms to the expression (2.68) for $T_c$ are of three types: a pre-exponential factor, $K_1$, that depends on the structure of the spectral function; a similar factor, $K_2$, determined only by the integrated characteristic of the spectrum, i.e., by the quantity $\lambda$; the corrections to the logarithms entering into the expressions for $\lambda$ and $\mu^*$. The general structure of the pre-exponential factors is determined by the expression

$$\exp\left\{-\frac{1}{g^2}\left[\delta R_1(0,0)+\delta R_2(0,0)\frac{\mu^{*2}}{\mu^2}\right]\right\}, \tag{2.69}$$

where the $\delta R_i(0, 0)$ are the correction to the kernel $R$. In the case when it is necessary to take into account the fact that the correction to $R$ decreases according to different laws with respect to the different arguments [see (2.60)], the indicated factor assumes the form

$$\exp\left\{-\frac{1}{g^2}\left[\delta\widetilde{R}_1(0,0)+\delta\widetilde{R}_2(0,0)\frac{\mu^*}{\mu}-\delta R_{22}(0,0)\frac{\mu^*(\mu-\mu^*)}{\mu^2}\right]\right\}. \tag{2.70}$$

The pre-exponential factor of the first type originates from the factor $\exp(\lambda^2 Q/g^2)$ in (2.64) and the correction $\delta_2 R(0, 0)$ [see (2.47)]. The latter depends only on $\sigma_1$, and therefore pertains to the quantity $\delta R_1(0, 0)$ [see (2.69)]. Combining the indicated expressions and separating out the cofactor $\exp(-\lambda^2/2g^2)$, which it is more convenient to transfer to the coefficient $K_2$, we have

$$K_1=\exp\left\{-\frac{2\mu^2}{g^2}\int_0^\infty\frac{dE\,dE'}{EE'}\left(\frac{E^2+E'^2}{E^2-E'^2}\ln\frac{E}{E'}-1\right)\times\sigma(E,\bar{q})\sigma(E',\bar{q})\right\}. \tag{2.71}$$

This expression (and only it, not counting the mean-logathmic frequency) depends on the form of the function $\sigma$.

The pre-exponential factor of the second type is comprised of the factor $2\gamma/\pi$, which occurs in the basic formulas for $T_c$, the factor $\exp(-\lambda^2/2g^2)$, which was discarded in the derivation of (2.71), and factors arising from the corrections, $\delta_1 R(0, 0)$, to the interaction [see (2.46)]. Let us begin the consideration of the latter factors with an analysis of the self-energy part [see (2.50) and (2.51)]. Writing $\widetilde{\Sigma}(0)$ in the form $\widetilde{\Sigma}_1(0) + \widetilde{\Sigma}_2(0)$, we can easily see that the product $\widetilde{\Sigma}_i(0)R_j(0, 0)$ can be identified with the quantity $R_{ij}(0, 0)$ [see (2.60)]. Then the formula (2.70) gives the self-energy contribution to $K_2$:

$$\exp\left\{-\frac{1}{g^2}\left[\widetilde{\Sigma}_1(0)\,g+\widetilde{\Sigma}_2(0)\left(\frac{\mu^{*2}}{\mu}-\lambda\right)\right]\right\}.$$

A simple computation with the aid of (2.51) leads to the expression

$$\widetilde{\Sigma}(0)=\alpha\int_0^{2p_F}\frac{dq}{q}\left[1+2\int_0^{\infty}dE\,\sigma(E,\mathbf{q})\left(\frac{1}{E}-\frac{2}{E+qp_F}\right)\right],$$

whence

$$\widetilde{\Sigma}_1(0)=-\lambda,\quad \widetilde{\Sigma}_2(0)=-\mu.$$

In computing the vertex corrections, we can replace in (2.52) the ratio of differences by a derivative. This gives for the simple vertex [see (2.53)]

$$\delta_v R(0,0)=2\alpha\lambda\int_0^{2p_F}\frac{dq}{q}\left(1-2\int_0^{\infty}\frac{dE\,\sigma(E,\mathbf{q})}{E+qp_F}\right)$$

and for the anomalous vertex [see (2.54)]

$$\delta_{av}R(0,0)=\alpha^2\int_0^{2p_F}\frac{dq}{q}\int_0^{2p_F}\frac{dq'\,q'}{q'^2+q^2}\left(1-\right.$$

$$\left.-2\int_0^{\infty}\frac{dE\,\sigma\left(E,\sqrt{q^2+q'^2}\right)}{E+q'p_F}\right)\left(1-2\int_0^{\infty}\frac{dE\,\sigma(E,\mathbf{q})}{E+qp_F}\right).$$

These expressions reduce, respectively, to $2\mu(\mu-\lambda)$ and $\mu^2/2$. Notice that the terms containing $\mu^2$ belong to the correction $\delta R_2(0,0)$, while those containing $\mu\lambda$ belong to the correction $\delta R_1(0,0)$ [see (2.69)]. It is important to emphasize that the phonon density $\sigma_1$ makes no contribution to the expressions obtained for the vertex parts because of the presence of the large–in comparison with $\xi_1$–quantity $qp_F$ in the denominators of the above-presented formulas (see [28]).†

†The term with $\lambda$ in the expression $\delta_v R(0,0)$ for the simple vertex is connected not with the vertex itself, but with an external–with respect to the vertex–phonon line (see Fig. 2.6c).

Collecting the obtained results, we find the following expression for the coefficient $K_2$, which does not depend on the structure of the spectrum of the system:

$$K_2 = \frac{2\gamma}{\pi} \exp\left(-\frac{1}{2} + \frac{(\lambda + \mu^* - \mu)}{g} + \frac{\mu^* (\mu - 2\mu^*)}{g^2}\right). \qquad (2.72)$$

It remains to find the corrections to the logarithms. Carrying out the computations that led to (2.67) more accurately, we obtain

$$\ln\left(\frac{\xi_2}{\xi}\right) = \ln\left(\frac{4\alpha^{1/2} E_F}{\xi}\right). \qquad (2.73)$$

Consequently, it turns out that the quantity $\xi_2$ lies between the energy $E_F$ and the electron plasma oscillation frequency $\omega_c \sim \alpha^{1/2} E_F$, which is due to the slow decrease of the interaction at energies exceeding $\omega_c$. Similarly, carrying out the computations that led to (2.65) more accurately, we find that the quantity $\mu$ does not contain nonlogarithmic terms, while

$$\lambda \to \lambda + \frac{\alpha\mu}{\lambda} \int_0^{2p_F} dq \ln\left(\frac{q}{2p_F}\right) \frac{d}{dq}\left(\frac{1}{\varepsilon(0,q)}\right).$$

In the most important case when $\varepsilon(0, \mathbf{q}) = \text{const}$ for $q \gg \varkappa$, $\lambda$ also does not contain nonlogarithmic terms.

For brevity of notation, in collecting the obtained results, we transfer part of the terms from (2.72) to the index of the exponential function in (2.68). The final formula for the critical temperature in the weak-coupling approximation assumes the form

$$T_c = \frac{2\gamma}{\pi} \xi K_1 \exp\left(-\frac{1}{2} - \frac{1 + \lambda - \mu}{\lambda - \mu^* (1 + \lambda - \mu)}\right), \qquad (2.74)$$

where

$$\mu^* = \frac{\mu}{1 + \mu \ln\left(\frac{4\alpha^{1/2} E_F}{e\xi}\right)} \qquad (2.75)$$

(here $e$ is the base of the natural logarithms).

Formula (2.74) is similar in appearance to the expression corresponding to the strong-coupling theory (see Chapter 1), but in this formula it should be assumed that $\lambda \ll 1$, $\mu \ll 1$, and $\mu^* \ll 1$.

The computation of the $T_c$ with the aid of the Éliashberg equation with the Coulomb interaction replaced by a constant (see Chapter 4) leads to a formula reminiscent of (2.74) but not coincident with it in a comparable approximation. In particular, the term $-\mu$ in the factor $1 + \lambda - \mu$ entering into the exponent in (2.74) is not reproduced; this term originates from the Coulomb vertex and the self-energy. The corresponding differences serve as clear evidence against the feasibility of constructing a complete quantitative strong-coupling theory, although the successes of such a theory as applied to individual problems are well known [6].

**D. The Critical Temperature in the Simplest Models.** Let us apply the above-obtained general expressions to the case when the system is characterized by a discrete set of undamped vibration modes. This case embraces the models considered in Section 2.E.

As formulated by us, the problem conforms to the longitudinal character of the vibrations, and their spectrum $\omega_j(\mathbf{k})$ ($j = 1, 2, \ldots$) is determined by the equation $\varepsilon(\omega_j(\mathbf{k}), \mathbf{k}) = 0$. Let us introduce the force constants

$$a_j(\mathbf{k}) = \left[\omega^2 \frac{\partial \varepsilon}{\partial \omega^2}\right]^{-1}_{\omega^2 = \omega_j^2(\mathbf{k})},$$

in terms of which we can express the spectral density:

$$\sigma_1(\omega, \mathbf{k}) = \sum_j a_j(\mathbf{k})\, \omega_j^2(\mathbf{k})\, \delta\left(\omega^2 - \omega_j^2(\mathbf{k})\right).$$

For $k = \bar{q}$ this quantity coincides with the total spectral density; below we shall drop the argument $\bar{q}$ in $\omega_j(\bar{q})$ and $a_j(\bar{q})$. Using (2.65), we have

$$\lambda = \sum_j a_j \mu, \tag{2.76}$$

while the formula (2.66) yields

$$\bar{\xi} = \prod_j (\omega_j)^{a_j / \sum_j a_j}. \tag{2.77}$$

Notice that the Coulomb interaction can also be included in the logarithmic averaging performed in (2.77). (This averaging means that each frequency acts to the extent of the relative magnitude of the corresponding force constant.) Limiting ourselves to the asymptotic formula (2.68), we set†

†Here the indices $k$ embrace the set of indices $j$ and include an additional term corresponding to the Coulomb interaction.

$$\Omega_k = (\omega_j, E_F), \quad A_k = \left(a_j\left(1 + \mu \ln\left(\frac{E_F}{\xi}\right)\right), -1\right); \tag{2.78}$$

the additional factor attached to $a_j$ is connected with the enhancement of the attraction (see Section 4.A). As a result, (2.68) assumes the simple and compact form

$$T_c \sim \prod_k (\Omega_k)^{A_k/\sum_k A_k} \exp\left(-\frac{1}{\sum_k A_k \mu}\right). \tag{2.79}$$

Returning to the expression (2.77), let us point out that it is quite severely distorted by the coefficient $K_1$ [see (2.71)]:

$$K_1 = \exp\left\{-\frac{\mu^2}{2g^2}\sum_{i,j}{}' a_i a_j \left[\frac{\omega_i^2 + \omega_j^2}{\omega_i^2 - \omega_j^2} \ln\left(\frac{\omega_i}{\omega_j}\right) - 1\right]\right\}.$$

Thus, in the case of two modes with $\omega_2 >> \omega_1$ we have

$$\xi K_1 = \omega_1^{\frac{a_1}{a_1+a_2} + a_1 a_2 \frac{\mu^2}{g^2}} \omega_2^{\frac{a_2}{a_1+a_2} - a_1 a_2 \frac{\mu^2}{g^2}}. \tag{2.80}$$

The exponents in this expression and in (2.77) can differ sharply. The remaining effects of second-order perturbation theory are not of particular importance and are not considered below. We shall limit the specific analysis to two examples, corresponding, respectively, to one- and two-vibration modes. The first example pertains to the "jellium" model in the form in which it was introduced in Section 2. The formula (2.16) yields†

$$\lambda = a\mu, \quad a = 1 - \frac{1}{\varepsilon(0, \bar{q})} = \frac{\omega_i^2}{\omega_c^2 + \omega_0^2(\bar{q})} = \frac{\omega_i^2}{\omega_{ph}^2}$$

and the final expression for $T_c$ assumes the form

$$T_c \sim \frac{\omega_i}{\sqrt{a}} \exp\left(-\frac{1}{a\mu - \mu^*}\right), \quad \mu^* = \frac{\mu}{1 + \mu \ln\left(\frac{E_F\sqrt{a}}{\omega_i}\right)}. \tag{2.81}$$

As has already been noted in Section 2, for $\omega_0^2 < 0$ we have $\varepsilon(0, \bar{q}) < 0$ and $a > 1$.

†The dependence $\lambda \sim \omega_{ph}^{-2}$ has been repeatedly discussed in the literature (see Chapters 3 and 4).

The second example is the generalized jellium model [see (2.17) and (2.19)], in which $\lambda = (a_1 + a_2)\mu$ and

$$\omega_1 = \left(\frac{\omega_i^2}{\varepsilon_0} + \omega_0^2\right)^{1/2}, \quad \omega_2 = \sqrt{\varepsilon_0}\,\Omega, \quad a_1 = \frac{\omega_i^2}{\varepsilon_0^2\omega_1^2},$$

$$a_2 = 1 - \frac{1}{\varepsilon_0}.$$

The relations (2.68), (2.76), (2.77), and (2.80) yield

$$T_c \sim \bar{\xi} K_1 \exp\left(-\frac{1}{g}\right), \tag{2.82}$$

where

$$\bar{\xi} = \omega_1^{\frac{a_1}{a_1+a_2}} \omega_2^{\frac{a_2}{a_1+a_2}}, \quad g = \mu\left\{a_1 + a_2 - \left[1 + \mu \ln\left(\frac{E_F}{\bar{\xi}}\right)\right]^{-1}\right\},$$

$$K_1 = \left(\frac{\omega_1}{\omega_2}\right)^{a_1 a_2 \frac{\mu^2}{g^2}}.$$

We shall return to the expressions (2.81) and (2.82) in the following section.

In conclusion of this section, let us note that the direct replacement in the obtained general formulas of the delta function $\sigma_1$ (see above) by the corresponding Lorentz form shows that the damping of the excitations adversely affects $T_c$, albeit not to a great extent. We shall return to this question in Section 5.

## 5. Applications to the Problem of High-Temperature Superconductivity (General Conclusions)

**A. Optimization of the Critical Temperature.** The expressions obtained above for the critical temperature should not be regarded as a source of quantitative information about the magnitude of $T_c$ for real superconducting structures. As noted in Section 1, it is difficult to indicate physical conditions under which both of the above-used assumptions—that the interelectron interaction is weak and that the system is homogeneous and isotropic—will be fulfilled. The object of the above-performed investigation was quite different: it was to obtain for the critical temperature a general expression that would reveal the main tendencies in the behavior of $T_c$ as the properties of the material are varied and to formulate in a language close to the experimental language the principal demands—from the standpoint of the raising of $T_c$—on the material of the superconductor. In the remaining part of this chapter we attempt to solve this problem at the qualitative level.

We shall begin with the simplest models considered in Section 4.D, and only afterward shall we proceed to the general case, successively increasing the number of free parameters characterizing the material of the superconductor. In the jellium model with one vibration mode [see (2.81)], there is, for a fixed density and a given composition of the material, only one free parameter: the force constant, $a$, of the phonon interaction. As this force constant increases, the characteristic frequency [the pre-exponential factor in (2.81)] decreases, while the quantity $T_c$ first increases and then decreases, attaining the maximum

$$T_{c,\max} \sim \mu^{1/2}\omega_l \tag{2.83}$$

at $a_{\text{opt}} = 2/\mu >> 1$. Thus, in the model under consideration, the optimization of $T_c$ requires a large negative value of the quantity $1/\varepsilon(0, \bar{q})$ (see Section 4).

Let us proceed to the generalized jellium model [see (2.82)] containing three free parameters: the force constant, $a_1$, of the phonon mode, a quantity $\varepsilon_0$ determined by the density of the oscillators added to the system, and the frequency, $\Omega$, of these oscillators. In this model the characteristic frequency and the coupling constant are no longer connected by a relation as rigid as the one found above. Let us first consider the asymptotic expression for $T_c$, dropping the cofactor $K_1$ in (2.82). This expression depends on only two parameters—$\bar{\xi}$ and $\lambda = (a_1 + a_2)\mu$—its behavior as a function of these parameters being as follows: It increases monotonically as the latter parameter increases, while as $\bar{\xi}$ increases, it first increases (because of the growth of the pre-exponential function) and then decreases (because of the decrease of the logarithm in the expression for $g$). Therefore, the critical temperature as a function of $\bar{\xi}$ has a maximum [7, 9],

$$T_{c,\max} \sim E_F \exp\left(-\frac{1}{\mu}\,\frac{3\varepsilon(0,\bar{q})+1}{\varepsilon(0,\bar{q})-1}\right), \tag{2.84}$$

attained by

$$\bar{\xi}_{\text{opt}} \sim E_F \exp\left(-\frac{1}{\mu}\,\frac{\varepsilon(0,\bar{q})+1}{\varepsilon(0,\bar{q})-1}\right) \tag{2.85}$$

[see also (1.50)]. For (2.85) to be realized, the quantities $\varepsilon_0$ and $\Omega$ should be connected by the relation

$$\frac{1}{\varepsilon_0}\mu\ln\left(\frac{\Omega}{\omega_l}\right) = \frac{1}{\varepsilon(0,\bar{q})}\left[1+\mu\ln\left(\frac{E_F}{\omega_l}\right)\right] + \left[1-\mu\ln\left(\frac{E_F}{\Omega}\right)\right]. \tag{2.86}$$

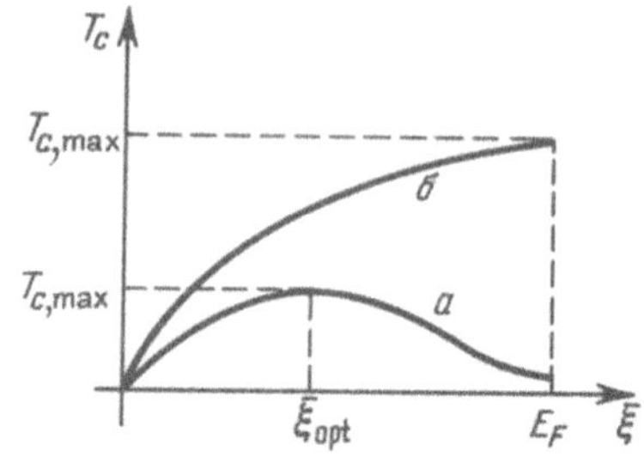


**Figure 2.8. Dependence of $T_c$ on the mean phonon frequency: (a) $\lambda < 2\mu$; (b) $\lambda > 2\mu$.**

It should be noted that the above-presented relations are valid when $\xi < E_F$, i.e., when $1/\varepsilon(0, \bar{q}) > -1$, or, equivalently, when $\lambda < 2\mu$. When $\lambda > 2\mu$, there arises an even more propitious—from the standpoint of the raising of $T_c$—situation. In this case the optimal value of $\bar{\xi}$ coincides with this quantity's limiting value, which is of the order of $E_F$ (see Section 4.B). Consequently, the maximum value of the critical temperature is

$$T_{c,\max} \sim E_F \exp\left(-\frac{1}{\lambda-\mu}\right) \sim E_F \exp\left(\frac{\varepsilon(0,\bar{q})}{\mu}\right) \tag{2.84a}$$

and attains a value of the order of $E_F$ for $\lambda \longrightarrow \infty$ (see Fig. 2.8).

We can consider the following problem within the framework of the generalized jellium model. Let the initial state of the superconductor be characterized by the critical temperature $T_c^{(0)}$. Let us introduce oscillators with frequency $\Omega$ and intrinsic permittivity $\varepsilon_0$ into the material. Let us assume that the quantities $\Omega$ and $\varepsilon_0$, which taken separately are arbitrary [see, however, (2.18)], satisfy the condition (2.86). Then the critical temperature becomes equal to the quantity (2.84), i.e., greater than the initial value $T_c^{(0)}$. It can be shown that the ratio $T_c/T_c^{(0)}$ is in any case greater than the quantity

$$\exp\left(\frac{4}{\mu}\,\frac{1-1/\varepsilon_0}{(1/\varepsilon_0-1/\varepsilon)(1-1/\varepsilon)}\right) \gg 1.$$

Allowance for the coefficient $K_1$ in (2.82) changes the situation in two respects. First, it decreases the absolute value of $T_c$ because of the inequality $\omega_1 << \omega_2$. Thus, under the above-found optimal conditions

$$K_1 \sim \left(\frac{\omega_l}{\sqrt{\varepsilon_0}\,\Omega}\right)^{\frac{4(1/\varepsilon_0-1/\varepsilon)(1-1/\varepsilon_0)}{(1-1/\varepsilon)^2}} \ll 1. \tag{2.87}$$

Second, the expression for $T_c$ now depends on all three of the free parameters of the model under consideration (see above). Therefore, the optimization of the complete expression (2.82) could, in principle, secure the optimum values of the parameters $\Omega$ and $\varepsilon_0$ separately. The corresponding value of $K_1$ would then differ from (2.87) but would nevertheless be much smaller than unity. As will become clear below, this is because of the insufficient number of free parameters in the generalized jellium model.

In the general case, when the number of parameters of the spectral function $\sigma$ is infinite, the optimum value of $T_c$ is realized under the condition that $\sigma$ has an infinitely narrow peak at the optimum frequency $\bar{\xi}_{opt}$. The $T_c$ value itself then coincides with (2.84), while the coefficient $K_1$ is equal to unity.

The proof of the assertions made above is based on the analysis of the general expression for $T_c$, regarded as a functional of the quantity $\sigma(E) \equiv \sigma(E, \bar{q})$ [see (2.65), (2.66), (2.68), and (2.71)]:

$$T_c \sim \bar{\xi} K_1 \exp(-1/g),$$

where

$$\ln \bar{\xi} = \int_0^\infty \frac{dE}{E} \ln E \sigma(E) \Big/ \int_0^\infty \frac{dE}{E} \sigma(E),$$

$$g = \mu \left\{ \int_0^\infty \frac{dE}{E} \sigma(E) - \left[ 1 + \mu \ln \left( \frac{E_F}{\bar{\xi}} \right) \right]^{-1} \right\},$$

$$\ln K_1 = -\frac{2\mu^2}{g^2} \int_0^\infty \frac{dE\, dE'}{EE'} \left[ \frac{E^2 + E'^2}{E^2 - E'^2} \ln \left( \frac{E}{E'} \right) - 1 \right] \sigma(E) \sigma(E').$$

The functional under consideration reduces to a simple function of three integrated arguments:

$$\ln \bar{\xi}, \quad \frac{\lambda}{\mu} = \int_0^\infty \frac{dE}{E} \sigma(E)$$

and the integral in the expression for $\ln K_1$, it being possible to carry out the optimization independently with respect to each of these arguments. The last of them is always nonnegative, vanishing in the class of functions $\sigma(E) = A\delta(E - E_0)$, on which $K_1 = 1$. It is precisely in this class that the optimum of the entire expression for $T_c$ is realized. The optimization with respect to the parameters $A$ and $E_0$ clearly takes us back to the above-found expressions (2.84) and (2.85). In this case all that has been said about the gain in the value of the critical temperature remains valid.

Let us note in conclusion that the remaining second-order perturbation theory effects, except $K_1$, have little influence on the results; the most important effect consists of the replacement of $E_F$ by $\alpha^{1/3}E_F$ in the formulas of this section [see (2.73)].

**B. General Conclusions.** Summing up the above-performed investigation as applied to the problem of high-temperature superconductivity, we can arrive at the following conclusions (see also Section 5.A).

1. The decisive influence on the critical temperature is exerted by the value of the static permittivity, $\varepsilon(0, \bar{q})$, of the superconducting material at large wave vector values, $\bar{q}$, lying between Debye momentum $\varkappa$ and the Fermi momentum $p_F$. The smaller the value of $1/\varepsilon(0, \bar{q})$ is, the higher the critical temperature is. Consequently, a positive permittivity $\varepsilon(0, \bar{q})$ should have the greatest possible value; it is even better when the quantity $\varepsilon(0, \bar{q})$ is negative. These requirements correspond to a material that scatters fast electrons through large angles with the largest possible cross section.

The foregoing follows from the form of the dominant (exponential) factor in the expression for $T_c$ [see (2.65), (2.68), and (2.82)]:

$$T_c \sim \bar{\xi} K_1 \exp\left(-\frac{1}{\lambda - \mu^*}\right), \quad \lambda = \mu\left(1 - \frac{1}{\varepsilon(0, \bar{q})}\right). \tag{2.88}$$

2. The critical temperature depends also on the system's excitation spectrum, which is determined by the equation $\varepsilon(\omega, \bar{q}) = 0$. The Coulomb excitation (plasmon) manifests itself only in the appearance of the Debye screening. The effect on the $T_c$ of the remaining excitations not directly connected with the conduction electrons (the phonons, excitons, etc.) is determined largely by their mean-logarithmic frequency [see (2.66)]

$$\bar{\xi} = \exp\left[\int_0^\infty \frac{dE}{E} \ln E\sigma(E, \bar{q}) \Bigg/ \int_0^\infty \frac{dE}{E}\sigma(E, \bar{q})\right],$$

$$\sigma(\omega, \bar{q}) = -\frac{1}{\pi} \operatorname{Im}\left(\frac{1}{\varepsilon(\omega, \bar{q})}\right)$$

(the frequency of each mode has been represented to the extent of the relative electron-mode coupling constant). The more strongly the vibration mode in question interacts with the conduction electrons, the greater its effectiveness will be. The mean-logarithmic frequency, which can be related with the Grüneisen constant of the material, should neither be too high nor too low: It possesses an optimal value—from the point of view of the raising of $T_c$—[see (2.85)]

$$\bar{\xi}_{\text{opt}} \sim E_F \exp\left(-\frac{1}{\mu}\frac{\varepsilon(0,\bar{q})+1}{\varepsilon(0,\bar{q})-1}\right).$$

The foregoing follows from (2.88) and the dependence

$$\mu^* = \mu/(1+\mu\ln(\xi_2/\bar{\xi})),$$

where $\xi_2 \sim E_F$ [see (2.68)].

These assertions pertain to the case $\bar{\xi}_{\text{opt}} < E_F$. If, on the other hand, we admit the possibility of the inequality $\lambda > 2\mu$, then the optimum value will be $\bar{\xi} \sim E_F$, to which corresponds

$$T_c \sim E_F e^{\frac{\varepsilon(0,\bar{q})}{\mu}}.$$

The foregoing follows from the formula (2.84a).

3. The damping of the excitations adversely affects the critical temperature; the presence of many vibration modes has a similar effect. The most advantageous–from the point of view of the raising of $T_c$–is the case when the entire spectrum of the excitations (exclusing the plasmons) reduces to a single, narrowest possible line located at the optimum frequency (see conclusion 2), and corresponds to the strongest possible interaction with the conduction electrons. This corresponds to $\sigma(\omega,\bar{q}) = A\delta(\omega - \bar{\xi}_{\text{opt}})$, with the highest possible value of the coefficient $A$.

The foregoing follows from the expression for the coefficient $K_1$ in (2.88):

$$K_1 = \exp\left(-\frac{2\mu^2}{g^2}\int_0^\infty \frac{dE\,dE'}{EE'}\left(\frac{E^2+E'^2}{E^2-E'^2}\ln\frac{E}{E'}-1\right)\sigma(E,\bar{q})\,\sigma(E',\bar{q})\right.$$

[see (2.71)], which assumes the value of unity only for a spectrum of the type indicated above.

4. The investigation carried out in this chapter does not give grounds for doubting the possibility in principle of a radical elevation of the critical temperature. The opposite conclusion drawn in [9] is based, in particular, on the inequality $\varepsilon(0,\bar{q}) > 0$, the nonnecessity of which has been established in the present chapter (see Section 2).

These conclusions have been drawn on the basis of an analysis based on a number of limiting assumptions (see Section 1). We are, however, confident that these results will retain their qualitative validity when the weak-interaction assumption is abandoned. As to the assumption that the system is homogeneous and isotropic, the situation may change quite significantly when we go outside its limits. This is true especially in the case of a highly inhomogeneous and anisotropic systems—sandwiches, quasi-one-dimensional and quasi-two-dimensional structures, etc. The corresponding problems are considered in Chapters 6–8. The situation may also change radically when we go over to superconducting systems that are not in a state of thermodynamic equilibrium. Such systems are investigated in Chapter 9.

# 3

# The Electron-Phonon Interaction in Metals and the Problem of Lattice Stability

## 1. Introduction

For the majority of presently known superconductors, the mechanism responsible for the formation of the superconducting state is the electron-phonon interaction. The traditional way of raising the critical temperature, $T_c$, of the superconducting transition involves the use of the potentialities of the phonon mechanism of superconductivity and the choice, with this aim in view, of all new superconducting alloys and compounds.

The highest $T_c$ value attained in this way is itself evidence of great progress: In the compound $Nb_3Ge$ superconductivity exists up to the temperature $T_c$ = 23.2 K, which allows liquid-hydrogen cooling to be employed [1, 2]. This achievement does not, however, solve the problem of high-temperature superconductivity. What then are the causes preventing the production of superconductors with high $T_c$ values? And is it possible to produce superconducting compounds with the ordinary electron-phonon mechanism of interelectron attraction and a critical temperature significantly exceeding 23 K?

The simplest answers to these questions have already been given in Chapter 1. They follow from an analysis of the conventional BCS formulas for the critical temperature [3] or their generalizations to the case of strong electron-phonon coupling [4-6]:

$$T_c = \bar{\omega} e^{-\frac{1+\lambda}{\lambda-\mu^*}}, \tag{3.1}$$

$$\mu^* = \frac{\mu}{1+\mu \ln\left(E_F/\bar{\omega}\right)}. \tag{3.2}$$

It can be seen from these formulas that for values of the electron-phonon coupling constant $\lambda << 1$, the critical temperature is exponentially small in comparison with the characteristic phonon frequencies.

This answer cannot entirely satisfy us for the following reasons: The electron-phonon coupling constant $\lambda$ may not be small (in lead, for example, $\lambda = 1.5$), and in this case it follows even from the formulas (3.1) that the value of $T_c$ can be comparable to $\bar{\omega}$, with the result that since alloys with $\bar{\omega} \approx 10^3$ K are known, values of $T_c \approx 100$ K could have been expected in those of them with $\lambda \approx \frac{1}{2}$ to 1.

And what is more, it follows [6, 7] from the solution of the Éliashberg equations describing superconductors with strong electron-phonon coupling [8] that the formula (3.1) for the critical temperature becomes invalid as $\lambda$ is increased, and that, for $\lambda >> 1$, the quantity $T_c$ can increase with increasing $\lambda$ according to the law

$$T_c \approx \bar{\omega} \sqrt{\bar{\lambda}}. \tag{3.3}$$

This means that, formally, within the framework of the superconductivity theory based on the Éliashberg equations, if we treat $\lambda$, $\mu$, and $\bar{\omega}$ as independent quantities, then there are practically no limitations on the value of the critical temperature of the superconducting transition. But in reality there are, of course, limitations on the possible values of $T_c$ within the framework of the electron-phonon mechanism of superconductivity.

Objections to the phonon mechanism of superconductivity were raised right after its proposal in Fröhlich's first paper [9] by Wentzel [10], according to whom the strong electron-phonon interaction required in [9] would lead not to superconductivity, but to lattice instability. And although, as it subsequently turned out, a weak interelectron interaction is sufficient for superconductivity, for the purpose of raising $T_c$, it is desirable to search for materials with strong electron-phonon interaction, i.e., with a large value of the electron-phonon interaction constant $\lambda$. As a result, there arises again the question of lattice stability or, somewhat more generally, of the stability of the electron-phonon system against a transition into another nonsuperconducting phase (a doubling of the lattice constant, the formation of covalent bonds, etc.).

On the other hand, recently the opinion has come to be fairly widely held that high critical temperatures can be obtained under conditions of the ordinary electron-phonon mechanism of superconductivity precisely in structurally unstable compounds or in materials close to "lattice instability" [11]. This opinion is based to a large extent on the empirical fact that the highest critical temperatures (21 K in the ternary alloy $Nb_3Al_xGe_{1-x}$ and 23 K in $Nb_3Ge$) have been obtained in alloys close to this state of instability. Moreover, as has been shown by Testardi [12], the critical temperature of the superconducting transition depends very strongly on the conditions under which the metal is produced. In particular, the fabrication of a film by sputtering at temperatures close to the structural-transition temperature for the material in question followed by rapid quenching, which enables us to preserve the corresponding structure in the film, significantly raises the superconducting-transition temperature $T_c$.

Undoubtedly of great interest, in the light of all that has been said above, is a detailed elucidation of the following questions: which characteristics of the material determine the value of the critical temperature in real metals? How are these characteristics interrelated? What factors limit $T_c$ in a particular system?

The solution to all these problems requires first and foremost a detailed analysis of the electron-phonon system of a normal metal, which will allow us to describe the interaction responsible for superconductivity. In the course of this analysis, we shall, in particular, attempt to ascertain the limitations imposed on this interaction by the condition that the system should be stable.

The present chapter is devoted to the detailed consideration of all of these problems. Furthermore, the results of the investigations expounded in it will serve as a basis for the subsequent investigation of the superconducting properties of electron-phonon systems and the study of the possibility of raising the critical temperature of the superconducting transition.

The systematic and self-consistent description of the properties of the electron and phonon systems in a metal is one of the central problems of solid state physics. It should be noted right away that there are thus far no clear-cut and precise answers to many questions connected with this problem. In particular, there does not exist a unique and fully substantiated opinion on such an important question as the role of the electron-phonon interaction in the development of lattice instability.

The following three methods are the most widely used in the theory of metals: the Fröhlich model, the adiabatic approximation, and the plasma model. In the Fröhlich model it is assumed that the interaction of an electron with the vibrations of the lattice is described by the following Hamiltonian:

$$H_{\text{int}} = \sum_n [V(\mathbf{r} - \delta\mathbf{R}_n^0 - \delta\mathbf{R}_n) - V(\mathbf{r} - \mathbf{R}_n^0)] = \sum_n \nabla V(\mathbf{r} - \mathbf{R}_n^0)\,\delta\mathbf{R}_n, \tag{3.4}$$

where $V(\mathbf{r} - \mathbf{R}_n)$ is the screened potential of the $n$th ion, $\mathbf{R}_n^0$ is the equilibrium position of the ion, and $\delta\mathbf{R}_n$ is the displacement of the ion from the equilibrium position.

Such an approach to the electron-phonon interaction was first proposed by Bloch for the description of the electrical conductivity of metals [13]. Subsequently, Fröhlich used this method to construct his theory of superconductivity. Since then this model has been widely used to describe diverse properties of electron-phonon systems [8, 14]. As will be seen below, one of the characteristic features of the Fröhlich model is the existence in it of a phonon instability in the presence of a sufficiently strong electron-phonon interaction [10]. This model has, for some time, often been used precisely to describe the structural transitions caused by the electron-phonon interaction [15, 16], especially in one-dimensional and quasi-one-dimensional metallic compounds [17] (for greater details, see Chapters 5 and 7).

The total electron-phonon Hamiltonian used in the Fröhlich model was introduced intuitively. We do not have any rigorous procedure, based on an exact analysis of the electron-ion system of the metal, for deriving this Hamiltonian.

A consistent procedure for analyzing the electron-phonon system was proposed by Born and Oppenheimer [18]. The adiabatic approximation introduced by them was subsequently used to study a number of properties of a metal [19]. This approximation was most systematically developed on the many-particle basis by Brovman and Kagan and was used by them to compute the phonon spectra of metals [20]. Within the framework of this approximation, it was shown, in particular, that contrary to what happens in the Fröhlich model, the renormalization of the adiabatic phonon frequencies as a result of the electron-phonon interaction is negligible—it is of the order of $(m/M)^{1/2}$. This means that the electron-phonon interaction cannot be the direct cause of the development of lattice instability. The electron-phonon interaction matrix element that follows from the adiabatic approximation also differs significantly from the one used in the Fröhlich model [21].

In the third approach (which we shall call the "plasma" approach) to the description of the electron-phonon system of a metal, the investigation of the metal begins with an analysis of the system of ions and electrons interacting according to the Coulomb law. This method allows us to obtain, at least for an isotropic system, clear-cut stability conditions for the system. In contrast to the preceding chapter, where these conditions were formulated in terms of the total permittivity, $\varepsilon(q, \omega)$, of the system, in the present chapter they will be derived from the more generally accepted—in solid-state theory—condition that the squares of the acoustic-vibration frequencies should not possess negative values [i.e., from the condition that $\omega^2(q) > 0$]. Such an approach will in turn allow the elucidation of the nature of the contradiction between the Fröhlich model and the adiabatic approximation with regard to the question of the role of the electron-phonon interaction in the problem of lattice stability.

Using the system of complete self-consistent equations obtained in [22] for the electron-phonon system, we shall briefly consider the changes that allowance for the crystal structure of the metal introduces into the condition for lattice stability obtained in the isotropic case. We shall also briefly discuss the methods of computing the phonon contribution to the self-energy of the electrons.

## 2. The Fröhlich Model

In second-quantization representation the Fröhlich Hamiltonian can be written in the following form:

$$H = \sum_{\mathbf{k}} E_{\mathbf{k}} a_{\mathbf{k}}^{+} a_{\mathbf{k}} + \sum_{\mathbf{q},\lambda} \omega_0 (\mathbf{q}\lambda) \left( b_{\mathbf{q}\lambda}^{+} b_{\mathbf{q}\lambda} + \frac{1}{2} \right) + \\ + \sum_{\mathbf{k},\mathbf{q},\lambda} g_0 (\mathbf{k}, \mathbf{k}+\mathbf{q})\, a_{\mathbf{k}+\mathbf{q}}^{+} a_{\mathbf{k}} \left( b_{\mathbf{q}\lambda} + b_{-\mathbf{q}\lambda}^{\pm} \right). \tag{3.5}$$

Here $E_{\mathbf{k}}$ is the electron energy, $a_{\mathbf{k}}, a_{\mathbf{k}}^{+}$ are electron annihilation and creation operators, $b_{\mathbf{q}\lambda}, b_{\mathbf{q}\lambda}^{+}$ are the corresponding phonon operators, and the $\omega_0(q\lambda)$ are the bare-phonon frequencies, which are assumed in the present model to be acoustic, i.e.,

$$\omega_0 (q\lambda) = cq. \tag{3.6}$$

The quantity $g_0(\mathbf{k}, \mathbf{k} + \mathbf{q})$ is the electron-phonon coupling constant, and, in accordance with the formula (3.4), it is the matrix element of the gradient of the screened electron-ion potential:

$$g_0 (\mathbf{k}, \mathbf{k} + \mathbf{q}) = \frac{1}{\Omega_0} \int_{\Omega} \psi_{\mathbf{k}+\mathbf{q}}^{*} (\mathbf{r})\, \nabla V (\mathbf{r})\, \psi_{\mathbf{k}} (\mathbf{r})\, d^3 r. \tag{3.7}$$

The integration in the formula (3.7) is over a unit cell with volume $\Omega_0$. Often, in the Fröhlich model, the quantity $g_0(\mathbf{k}, \mathbf{k} + \mathbf{q})$ is taken in the form of a constant, i.e., its dependence on the momenta is neglected.

Introducing in the usual manner the electron and phonon Green functions

$$G (\mathbf{k}, t) = -i \langle T a_{\mathbf{k}} (t)\, a_{\mathbf{k}}^{+} (0) \rangle, \tag{3.8}$$

$$D (\mathbf{q}\lambda, t) = -i \langle T b_{\mathbf{q}\lambda} (t)\, b_{\mathbf{q}\lambda}^{+} (0) \rangle, \tag{3.8a}$$

we can write for the Fourier components of these functions the following Dyson equations (Fig. 3.1):

$$G^{-1} (\mathbf{k}, \omega) = G_0^{-1} (\mathbf{k}, \omega) - \Sigma (\mathbf{k}, \omega), \tag{3.9}$$

$$D^{-1} (\mathbf{q}\lambda, \omega) = D_0^{-1} (\mathbf{q}\lambda, \omega) - \Pi (\mathbf{q}\lambda, \omega). \tag{3.10}$$

In these expressions

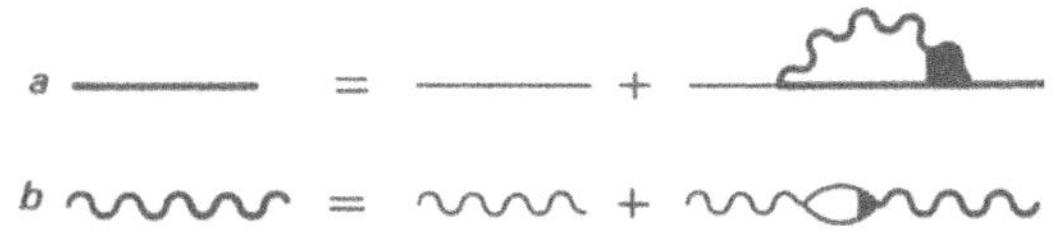


**Figure 3.1. Diagrammatic representation of the Dyson equations for (a) the electron and (b) the phonon, Green function.**

$$G_0(\mathbf{k}, \omega) = \frac{1}{\omega - E_{\mathbf{k}} + i\delta \operatorname{sign}(|k| - k_F)}, \tag{3.11}$$

$$D_0(\mathbf{q}\lambda, \omega) = \frac{\omega_0^2(q\lambda)}{\omega^2 - \omega_0^2(q\lambda)}. \tag{3.12}$$

The self-energy parts, $\Sigma(\mathbf{k}, \omega)$ and $\Pi(\mathbf{k}, \omega)$, of the electron and phonon Green functions can be expressed with the aid of the integral equations

$$\Sigma(p) = ig_0 \int G(p-k) D(k) \Gamma(p-k, p, k) \frac{d^4k}{(2\pi)^4}, \tag{3.13}$$

$$\Pi(k) = -2ig_0 \int G(p) G(p-k) \Gamma(p, p-k, k) \frac{d^4p}{(2\pi)^4}. \tag{3.14}$$

The system of equations (3.9)-(3.14) was first studied in detail by Migdal [14]. It has been shown by him that because of the considerable difference between the electron energies (they are of the order of $E_F$) and the phonon energies, the vertex corrections in Eqs. (3.13) and (3.14) can be neglected, i.e.,

$$\Gamma(p-k, p, k) \approx g_0\left(1 + \sqrt{\frac{m}{M}}\right).$$

A second important fact ascertained in [14] is that $G(\mathbf{k}, \omega)$ differs significantly from $G_0(\mathbf{k}, \omega)$ only in the narrow momentum and frequency ranges $||\mathbf{k}| - k_F| \lesssim \omega_D/v_F$ and $|\omega| \lesssim \omega_D$. Since a much broader range is important in the integral for $\Pi(\mathbf{q}, \omega)$, we can replace $G(p, \omega)$ by $G_0(p, \omega)$ in the integral and obtain for the exact phonon Green function the final expression

$$D(\mathbf{q}\lambda, \omega) = \frac{\omega_0^2(q\lambda)}{\omega^2 - \omega^2(q\lambda)}. \tag{3.15}$$

Here $\omega(q\lambda)$ is the spectrum of the renormalized-phonon excitations. It is determined from the equation

$$\omega^2(q\lambda) = \omega_0^2(q\lambda)\left(1 + g_0^2 \Pi_0(q, 0)\right), \tag{3.16}$$

where $\Pi_0(q, \omega)$ is the polarization operator of the free electrons. In an isotropic three-dimensional system with $\omega < qv_F$,

$$\Pi_0(q, \omega) = -N(0)\left[h\left(\frac{q}{2k_F}\right) + i\pi \frac{m|\omega|}{k_F q} \theta(2k_F - q)\right]; \tag{3.17}$$

here $N(0)$ is the density of states of the electrons at the Fermi surface and

$$h(x) = 1 + \frac{1-x^2}{2x} \ln \left| \frac{1+x}{1-x} \right|, \quad 0(x) = \begin{cases} 1 & \text{for} \quad x > 0, \\ 0 & \text{for} \quad x < 0. \end{cases}$$

For small values of the wave vector $q$ the spectrum of the phonon excitations has the form

$$\omega(q\lambda) = \omega_0(q\lambda)\sqrt{1-2\zeta}. \tag{3.18}$$

Here we have introduced the dimensionless electron-phonon interaction constant

$$\zeta = g_0^2 N(0). \tag{3.19}$$

Before proceeding to discuss the properties of the phonon spectrum given by the Fröhlich model, let us give the results of the computations for the electron Green function. Migdal [14] has shown that the quantity $\Sigma(k, \omega)$ in the narrow momentum and frequency ranges $||\mathbf{k}| - k_F| \lesssim (\omega_D/v_F)$ and $|\omega| \lesssim \omega_D$ (where it is, in fact, important) is a function of only the frequency $\omega$ and therefore can be computed with the aid of Eq. (3.13):

$$\Sigma(\omega) = -\zeta\omega \int_0^{2k_F} \frac{k\,dk}{2k_F^2} \int_0^\infty d\omega' \int_0^\infty d\varepsilon \operatorname{Im} D(k, \varepsilon) \times \frac{1}{(\omega'+\varepsilon)^2 - \omega^2 - i\delta}. \tag{3.20}$$

From (3.20), it is easy to find that for low $\omega$

$$\operatorname{Re}\Sigma(\omega) = -\lambda\omega, \tag{3.21}$$

$$\operatorname{Im}\Sigma(\omega) = -\frac{\pi\zeta}{2} \int_0^{k_2} \frac{k\,dk}{2k_F^2} \frac{\omega_0^2(k)}{\omega^2(k)} \operatorname{sign}\omega, \tag{3.22}$$

where $\lambda$ is the renormalized electron-phonon coupling constant:

$$\lambda = \zeta \int_0^{2k_F} \frac{k\,dk}{2k_F^2} \frac{\omega_0^2(k)}{\omega^2(k)}. \tag{3.23}$$

The upper limit in the formula (3.22) is determined from the condition $\omega(k_2) = \omega$.

It can be seen from these formulas that the electron decrement due to the electron-phonon interaction vanishes near the Fermi surface, and the Green function $G(\mathbf{k}, \omega)$ has the form

$$G(\mathbf{k}, \omega) = \frac{1}{1+\lambda} \frac{1}{\omega - \frac{E_k}{1+\lambda} - i\delta \operatorname{sign}(|\mathbf{k}| - k_F)}. \quad (3.24)$$

Thus, allowing for the fact that $E_k = v_F(|\mathbf{k}| - k_F)$, we see that, as a result of the electron-phonon interaction, the electron velocity at the Fermi surface becomes renormalized:

$$v_F' = \frac{v_F}{1+\lambda}.$$

The residue at the pole of the Green function $G(k, \omega)$ also changes: it decreases from unity to $1/(1 + \lambda)$.

Proceeding to discuss the results obtained in the Fröhlich model, let us note first of all that in accordance with the formula (3.16), the phonon frequencies become unstable when

$$\zeta > \frac{N(0)}{|\Pi_0(q, 0)|} \geqslant \frac{1}{2}. \quad (3.25)$$

In this case the instability first sets in at $q = 0$ when $\zeta = \frac{1}{2}$. For the instability at higher $q$ we need, in accordance with (3.25) and (3.17), larger $\zeta$ values.

Approximating $\Pi_0(q, 0)$ in the form [14]

$$\Pi_0(q, 0) = -N(0)\left[2 - \left(\frac{q}{2k_F}\right)^2\right], \quad (3.26)$$

and using the formulas (3.16), (3.23), and (3.26), we find the value of $\lambda$ near the instability threshold

$$\lambda = \zeta \int_0^{2k_F} \frac{q\,dq}{2k_F^2} \frac{1}{1 - 2\zeta[1 - {}^1/_2 (q/2k_F)^2]} = \zeta \ln \frac{2(1-\zeta)}{1-2\zeta}. \quad (3.27)$$

It is very easy to understand the cause of the appearance in the electronic characteristics of the renormalized coupling constant $\lambda$ in place of the initial $\zeta$: when we allow for the phonon renormalization, the original interelectron interaction $g_0^2 D_0(q, \omega)$ (Fig. 3.2a) goes over into $g_0^2 D(g, \omega)$ (Fig. 3.2b), where $D(q, \omega)$ is the given by the expression (3.15).

**Figure 3.2. The electron-phonon interaction resulting from (a) the bare-phonon Green function and (b) the exact function.**

The true coupling constant is given by the integral of $g_0^2 D(q, \omega)$ over all the momenta at $\omega = 0$, from which we obtain the expressions (3.23) and (3.27). The principal deviation of $\lambda$ from $\zeta$ arises as the result of the change in the phonon frequencies. If we neglect the weak dependence on the momentum of the function $h(q/2k_F)$, then we obtain a relation between $\lambda$ and $\zeta$ in the form

$$\lambda = \zeta \frac{\omega_0^2}{\omega^2} = \frac{\zeta}{1 - 2\zeta}, \tag{3.27a}$$

again first obtained by Fröhlich [9].

It can be seen from the expression (3.27) that within the framework of the Fröhlich model the renormalized electron-phonon interaction constant $\lambda$, which determines, in particular, the change in the electron velocity at the Fermi surface, can be arbitrarily large. As $\zeta$ increases, the renormalized coupling constant $\lambda$ increases continuously, becoming infinite at the very point of instability. Since it is precisely the constant $\lambda$ that enters into the expression for the critical temperature, the stability condition, $\zeta < \frac{1}{2}$, obtained in the Fröhlich model imposes no additional limitations on the value of $T_c$. Moreover, by itself, the condition $\zeta < \frac{1}{2}$ is in a sense essentially empty since there does not exist within the framework of the Fröhlich model itself a method of determining this quantity. It is just the renormalized coupling constant $\lambda$, and not the bare-constant $\zeta$, that is actually observable. If, using (3.27a), we express the bare-constant $\zeta$ in terms of the experimentally observable renormalized constant $\lambda$, then we obtain

$$\zeta = \frac{\lambda}{1 + 2\lambda}.$$

It can be seen from this that for any positive $\lambda$ the bare-constant $\zeta$ satisfies the condition

$$\zeta \leqslant {}^1/_2.$$

There exists in the Fröhlich model another more important reason for the appearance of lattice instability than simply the inequality $\zeta > \frac{1}{2}$. It can be seen from Eq. (3.16), which determines the frequencies of the renormalized phonons, that the square of the phonon frequency becomes negative, with the result that the system becomes unstable when the condition

$$g_0^2 \mid \Pi(q, 0) \mid \geqslant 1 \tag{3.25a}$$

is fulfilled. In one-dimensional systems, or in three-dimensional systems whose Fermi surface has overlapping sections, because of the logarithmic singularities of the polarization operator $\Pi(q, 0)$ at $q = 2k_F$, the condition (3.25a) is fulfilled for all values of the bare-coupling-constant $g_0$ [15–17]. In the section devoted to the plasma model, we shall discuss a similar type of instability in greater detail. We shall obtain an instability condition that is more exact than (3.25a).

## 3. The Adiabatic Approximation

In expounding the adiabatic approximation we shall, in the main, follow Brovman and Kagan's review [20]; therefore, we shall reproduce only the main ideas and not dwell on the computational details.

The adiabatic-approximation idea, which consists in the separation of the fast electronic and the slow phonon variables, was first formulated by Born and Oppenheimer [18]. The main adiabaticity criterion is the nonexistence in the electronic spectrum of excitations with energies of the order of the frequencies of the ionic vibrations. On the face of it, this criterion is clearly violated in metals since electronic excitations with arbitrarily low energies exist near the Fermi surface. The number of such excitations in metals, however, is extremely small in comparison with the total volume of the Fermi sphere. It is this fact that enables us to use the adiabatic approximation to determine quantities integrated over the electron spectrum.

Let us write the total electron–ion Hamiltonian in the form

$$H = H_e(\mathbf{r}) + H_i(\mathbf{R}) + H_{ei}(\mathbf{r}, \mathbf{R}). \tag{3.28}$$

The idea of the adiabatic approach is to find the eigenfunctions of the operator $H_e(\mathbf{r}) + H_{ei}(\mathbf{r}, \mathbf{R})$ for arbitrary positions, $\mathbf{R}_l$, of the ions:

$$[H_e(\mathbf{r}) + H_{ei}(\mathbf{r}, \mathbf{R})]\,\psi_m(\mathbf{r}, \mathbf{R}) = E_m(\mathbf{R})\,\psi_m(\mathbf{r}, \mathbf{R}). \tag{3.29}$$

The eigenfunction of the total Hamiltonian is sought in the form of an expansion in terms of the functions $\psi_m(\mathbf{r}, \mathbf{R})$:

$$\Psi(\mathbf{r}, \mathbf{R}) = \sum_m \chi_m(\mathbf{R})\,\psi_m(\mathbf{r}, \mathbf{R}). \tag{3.30}$$

Then for the function $\chi_m(\mathbf{R})$, which describes the properties of the ion subsystem, we have

$$[H_i(\mathbf{R}) + E_m(\mathbf{R})]\,\chi_m(\mathbf{R}) + \sum_n (A_{mn} + B_{mn})\,\chi_n(\mathbf{R}) = E\chi_m(\mathbf{R}), \quad (3.31)$$

where

$$A_{mn} = \frac{1}{M}\sum_j (\nabla_{\mathbf{R}_j})_{mn}\,\nabla_{\mathbf{R}_j}, \quad B_{mn} = \frac{1}{M}\sum_j \left(\nabla^2_{\mathbf{R}_j}\right)_{mn}, \quad (3.32)$$

$\mathbf{R}_j$ is the coordinate of the $j$th ion, $(\nabla_{\mathbf{R}_j})_{mn}$ is the matrix element of the gradient $\nabla_{\mathbf{R}_j}$ as computed with the functions $\psi_m(\mathbf{r}, \mathbf{R})$.

If in Eq. (3.31) we neglect the off-diagonal terms $A_{mn}$ and $B_{mn}$, then we obtain two independent subsystems: an electron subsystem described by Eq. (3.29) and an ion subsystem described by Eq. (3.31). Then, in Eq. (3.31), which reflects the vibrational properties of the metal, the potential energy $E_m(\mathbf{R})$ is the energy of the electrons in the given static field of the ions. By expanding in the usual manner $H_i(\mathbf{R})$ and $E_m(\mathbf{R})$ around the equilibrium positions, $\mathbf{R}^0_n$, of the ions, i.e., in powers of the deviations $\mathbf{u}_n = \mathbf{R}_n - \mathbf{R}^0_n$, we can derive [20-23] the dynamical matrix of the vibrations:

$$\begin{aligned}\Phi_{\alpha\beta}(\mathbf{q}) = {} & \frac{4\pi e^2 Z^2}{\Omega_0}\sum_{\mathbf{G}} \frac{(\mathbf{q}+\mathbf{G})_\alpha (\mathbf{q}+\mathbf{G})_\beta}{|\mathbf{q}+\mathbf{G}|^2} + \\ & + \frac{1}{\Omega_0}\sum_{\mathbf{G},\mathbf{G}'} (\mathbf{q}+\mathbf{G})_\alpha (\mathbf{q}+\mathbf{G}')_\beta\, V_{ei}(\mathbf{q}+\mathbf{G}) \times \\ & \times \chi(\mathbf{q}+\mathbf{G}, \mathbf{q}+\mathbf{G}')\, V_{ei}(\mathbf{q}+\mathbf{G}').\end{aligned} \quad (3.33)$$

Here $Z$ is the ion charge, $\mathbf{G}$ and $\mathbf{G}'$ are reciprocal-lattice vectors, $V_{ei}(\mathbf{q} + \mathbf{G})$ is the matrix element of the electron-ion interaction, and $\chi(\mathbf{q} + \mathbf{G}, \mathbf{q} + \mathbf{G}')$ is the total susceptibility of the electrons.

The first term in (3.33) is the contribution of the ions of the lattice to the dynamical matrix of the vibrations; the second term, the contribution of the electrons, which is determined by the properties of the electrons in the periodic field of the lattice. The vibration frequencies are determined, as usual, from the equation

$$\mathrm{Det}\,|\bar{\Phi}_{\alpha\beta}(\mathbf{q}) - M\omega^2(\mathbf{q})\,\delta_{\alpha\beta}| = 0,$$

where $\bar{\Phi}_{\alpha\beta}(\mathbf{q}) = \Phi_{\alpha\beta}(\mathbf{q}) - \Phi_{\alpha\beta}(0)$.

In particular, for the longitudinal vibrations in an isotropic system we obtain [20]

$$\omega_l^2(q) = \Omega_i^2(q) + \frac{q^2 V_{ei}(q) N}{M} \chi(q), \tag{3.34}$$

where the $\Omega_i(q)$ are the longitudinal-vibration frequencies of ions.

Within the framework of the adiabatic approximation, the electron-phonon interaction is contained in the off-diagonal $A_{mn}$ and $B_{mn}$ terms of Eq. (3.31). This interaction can be taken into account in the following way. First, let us compute the nonadiabatic correction, $\Delta E$, to the total energy of the system. Then by varying the energy with respect to the corresponding occupation numbers, we obtain the corrections to the phonon and electron energies:

$$\delta\omega_{q\lambda} = \frac{\delta(\Delta E)}{\delta N_{q\lambda}}, \quad \delta\varepsilon_k = \frac{\delta(\Delta E)}{\delta n_k}. \tag{3.35}$$

By taking into account the fact that

$$(\nabla_{R_i})_{mn} = -\frac{[\nabla_{R_i} H_{ei}(\mathbf{r}, \mathbf{R})]_{mn}}{E_m - E_n},$$

and going over to the momentum representation for the electrons and phonons, we can show [20, 21] that

$$\delta\omega_{q\lambda} = \sum_k |M_{\mathbf{kq}\lambda}|^2 \frac{n_{\mathbf{k}} - n_{\mathbf{k+q}}}{E_{\mathbf{k}} - E_{\mathbf{k+q}} - \omega_{\mathbf{q}\lambda} + i\delta} - \sum_k |M_{\mathbf{kq}\lambda}|^2 \frac{n_{\mathbf{k}} - n_{\mathbf{k+q}}}{E_{\mathbf{k}} - E_{\mathbf{k+q}}}, \tag{3.36}$$

where the $M_{\mathbf{kq}\lambda}$ are the Fourier components of the expression

$$\sum_j [\Delta_{R_j} H_{ei}(\mathbf{r}, \mathbf{R})]_{mn}.$$

It can easily be verified from (3.36) that

$$\Delta\omega_{\mathbf{q}\lambda} = \mathrm{Re}\,\delta\omega_{\mathbf{q}\lambda} \approx \omega_{\mathbf{q}\lambda}\left(\frac{\bar{\omega}}{\varepsilon_F}\right)^2, \quad \Gamma_{\mathbf{q}\lambda} = \mathrm{Im}\,\delta\omega_{\mathbf{q}\lambda} \approx \omega_{\mathbf{q}\lambda}\frac{\bar{\omega}}{\varepsilon_F}. \tag{3.37}$$

Here $\bar{\omega}$ is a quantity of the order of the mean phonon frequencies.

For the renormalization of the electron spectrum, we obtain the following expression:

$$\Delta E_{\mathbf{k}} = \sum_{\mathbf{q}\lambda} |M_{\mathbf{kq}\lambda}|^2 \frac{\omega_{\mathbf{q}\lambda}(1 - 2n_{\mathbf{k+q}})}{(E_{\mathbf{k}} - E_{\mathbf{k+q}})^2 - \omega_{\mathbf{q}\lambda}^2}. \tag{3.38}$$

It can easily be verified that the expression (3.38) for $\Delta E_{\mathbf{k}}$ exactly coincides with the expression that can be derived on the basis of the Fröhlich model if in place of the phenomenological electron-phonon interaction constant $g_0^2$ we use the matrix element $|M_{\mathbf{kq}\lambda}|^2$.

On the basis of the adiabatic-approximation results presented here we can draw the following conclusions about the role of the electron-phonon interaction in metals.

1. The phonon frequencies in metals can be computed within the framework of the adiabatic approximation with a high accuracy of the order of $(m/M)^{1/2}$. In this case we obtain the real phonon spectrum right away. The dynamical interaction of the electrons with the adiabatic phonons, which arises from the nonadiabatic corrections $A_{mn}$ and $B_{mn}$, weakly renormalizes the phonon spectrum and clearly cannot lead to lattice instability.

2. The renormalization, due to the nonadiabatic electron-phonon interaction of the electrons in the vicinity of the Fermi surface, is drastic. Qualitatively, these changes in the electron system have the same form as those obtained in the Fröhlich model if we use for the matrix element of the electron-phonon interaction the expression for $M_{\mathbf{kq}\lambda}$ and take the phonon frequencies in the adiabatic approximation.

3. Within the framework of this approximation, the instability of the system manifests itself in the appearance of negative values of $\omega_{q\lambda}^2$ in the adiabatic equation (3.34).

We postpone the concrete discussion of the causes of these instabilities and their possible connection with the electron-phonon interaction to the following section. Here we only note that because of its non-Hamiltonian character and the absence of a convenient diagram technique for the computation of the corresponding corrections, the adiabatic approximation is ill-suited for self-consistent computations of the characteristics of the electron and phonon systems.

## 4. The Plasma Model

We now consider another method, which can be called the plasma model, of describing the electron-phonon system. The description of a metal in this procedure begins with the study of a system of electrons and ions with unscreened Coulomb interaction. The normal modes of such a system are essentially the corresponding plasma oscillations: the subsequent allowance for the electronic screening allows the introduction of ordinary phonons and the description of the resulting electron-phonon interaction. This approach has been well described in a number of monographs [24, 25]; therefore, here we shall indicate the main steps in the calculations, omitting the computational details. In presenting the material, we shall, in the main, follow Schrieffer's book [25].

We write the standard electron-ion Hamiltonian (3.28) in second-quantization representation:

$$H = \sum_{\mathbf{k}} E_{\mathbf{k}} a_{\mathbf{k}}^{+} a_{\mathbf{k}} + \sum_{\mathbf{q}\lambda} \Omega_{\mathbf{q}\lambda} \left( b_{\mathbf{q}\lambda}^{+} b_{\mathbf{q}\lambda} + \frac{1}{2} \right) + \\ + \sum_{\mathbf{k}\mathbf{k}'\lambda} g_{\mathbf{k}\mathbf{k}'\lambda} a_{\mathbf{k}}^{+} a_{\mathbf{k}'} (b_{\mathbf{k}-\mathbf{k}',\lambda} + b_{\mathbf{k}'-\mathbf{k},\lambda}^{+}) + \\ + \frac{1}{2} \sum_{\mathbf{p}\mathbf{k}\mathbf{q}} V_{\mathbf{q}} a_{\mathbf{p}+\mathbf{q}}^{+} a_{\mathbf{k}-\mathbf{q}}^{+} a_{\mathbf{k}} a_{\mathbf{p}}. \tag{3.39}$$

In this expression $E_{\mathbf{k}}$ is the Bloch-electron energy, determined by the expression

$$\left[ \frac{\mathbf{k}^2}{2m} + \sum_{n} V_{ei} (\mathbf{r} - \mathbf{R}_n) + U_{\mathrm{H}} (\mathbf{r}) \right] \psi_{\mathbf{k}} (\mathbf{r}) = E_{\mathbf{k}} \psi_{\mathbf{k}} (\mathbf{r}), \tag{3.40}$$

here $V_{ei}(\mathbf{r} - \mathbf{R}_n)$ is the electron–ion potential and $U_H(\mathbf{r})$ is the Hartree contribution from the interelectron Coulomb interaction; the $\Omega_{\mathbf{q}\lambda}$ are the bare-vibration frequencies of the ionic plasma.

In the simplest jellium model, in which the ions are assumed to form a homogeneous structureless medium,

$$\Omega_{q\lambda}^2 = \frac{4\pi n (Ze)^2}{M}, \tag{3.41}$$

where $n$ is the ion density, $Z$ is the ion charge, and $M$ is the ion mass.

Three excitation branches exist in a crystal. Of them the two transverse ones have an acoustic character, while the bare-longitudinal branch of the excitations has the character of optical plasma oscillations. In this case the following sum rule is satisfied:

$$\sum_{\lambda} \Omega_{q\lambda}^2 = \frac{4\pi n (Ze)^2}{M}. \tag{3.42}$$

The bare-electron–phonon interaction constant $g_{\mathbf{k}\mathbf{k}'\lambda}$ given by the expression

$$g_{\mathbf{k}\mathbf{k}'\lambda} = - \left( \frac{n}{M\Omega_{q\lambda}^2} \right)^{1/2} \langle \mathbf{k}' \,|\, \nabla_i V_{ei} \,|\, \mathbf{k} \rangle \, \mathbf{e}_{\mathbf{q}\lambda} \quad (\mathbf{q} = \mathbf{k}' - \mathbf{k}), \tag{3.43}$$

where the $\mathbf{e}_{\mathbf{q}\lambda}$ are the polarization vectors of the bare phonons.

The fourth term in (3.39) represents the direct interelectron Coulomb interaction:

$$V_q = \frac{4\pi e^2}{q^2}. \tag{3.44}$$

It can easily be shown that in the simplest isotropic system the bare-electron-phonon interaction constant $g^2(\mathbf{k} - \mathbf{k}')$ will have the same type of singularity as the direct interelectron Coulomb interaction:

$$g^2(\mathbf{k} - \mathbf{k}') \approx \frac{1}{(\mathbf{k} - \mathbf{k}')^2}. \tag{3.45}$$

Going over to the renormalized quantities, which is equivalent to taking the screening into account, we retrieve the normal acoustic character of the longitudinal-vibration mode and eliminate the singularity in the interelectron interaction. This procedure for the isotropic system was first carried through by Bohm and Pines [26], and its feasibility in the case of the Bloch electrons of a metal was demonstrated by Hubbard [27]. In our discussion, we shall (following Schrieffer [25]), limit ourselves to the consideration of the quasi-isotropic case within the framework of the random-phase approximation.

First of all, it is necessary to take into account the screening in the Coulomb interaction between the electrons. The diagrams describing this procedure in the random-phase approximation are shown in Fig. 3.3. The integral equation for the screened Coulomb interaction $\tilde{V}(q, \omega)$, depicted in Fig. 3.3, is easily solved, and we obtain

$$\tilde{V}(q, \omega) = \frac{4\pi e^2}{q^2 \varepsilon_e(q, \omega)}, \tag{3.46}$$

where

$$\varepsilon_e(q, \omega) = 1 - \frac{4\pi e^2}{q^2} \Pi(q, \omega) \tag{3.47}$$

is the permittivity of the electron gas, $\Pi(q, \omega)$ is the irreducible polarization operator of the electrons, which is given in the random-phase approximation by the formula (3.17). The screening of the electron-phonon vertex (Fig. 3.4)

$$\tilde{g}(q\lambda) = \frac{g(q\lambda)}{\varepsilon_e(q, \omega)} \tag{3.48}$$

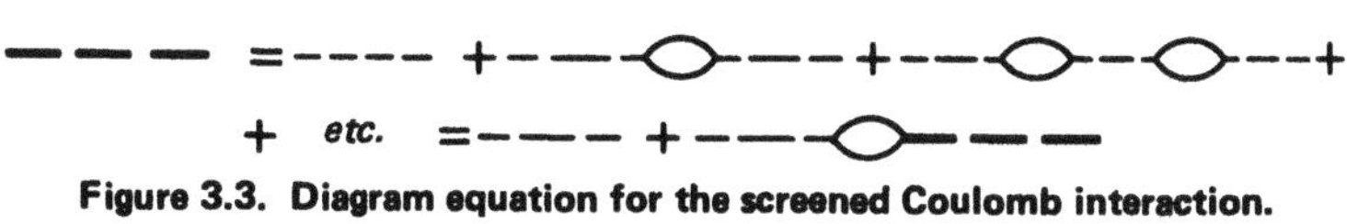


**Figure 3.3. Diagram equation for the screened Coulomb interaction.**

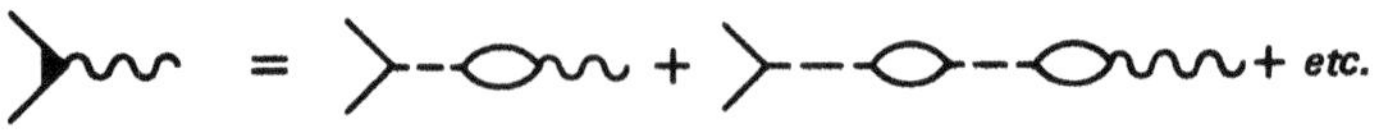


**Figure 3.4. Diagram equation for the screened vertex of the electron–phonon interaction.**

and the phonon spectrum (Fig. 3.5)

$$D^{-1}(q\lambda,\omega) = D_0^{-1}(q\lambda,\omega) - \frac{g^2(q\lambda)}{V(q)}\left(\frac{1}{\varepsilon_e(q,\omega)} - 1\right), \tag{3.49}$$

where

$$D_0(q\lambda,\omega) = \frac{\Omega_{q\lambda}^2}{\omega^2 - \Omega_{q\lambda}^2 + i\delta}, \tag{3.50}$$

occurs in similar fashion. From Eqs. (3.49) and (3.50) we find

$$D(q\lambda,\omega) = \frac{\Omega_{q\lambda}^2}{\omega^2 - \dfrac{g^2(q\lambda)\,\Omega_{q\lambda}^2}{V(q)\,\varepsilon_e(q,0)} - \Omega_{q\lambda}^2\left[1 - \dfrac{g^2(q\lambda)}{V(q)}\right] + i\delta}. \tag{3.51}$$

In writing down (3.51), we neglected the frequency dependence of $\varepsilon(q,\omega)$ since at low $\omega$ of the order of the phonon frequencies this dependence is unimportant. The equality

$$\frac{g^2(q\lambda)}{V(q)} = 1.$$

is valid in the jellium model. In this case $D(q\lambda,\omega)$ can be simplified:

$$D(q\lambda,\omega) = \frac{\Omega_{q\lambda}^2}{\omega^2 - \left(\Omega_{q\lambda}^2/\varepsilon_e(q,0)\right) + i\delta}. \tag{3.52}$$

The poles of this expression determine the frequencies of the renormalized phonons

$$\omega^2(q\lambda) = \frac{\Omega_{q\lambda}^2}{\varepsilon_e(q,0)} \approx \frac{\Omega_{q\lambda}^2}{1 + \left(k_s^2/q^2\right)} = \frac{mZ}{3M}v_F^2 q^2, \tag{3.53}$$

where

$$k_s = \sqrt{\frac{12\pi n e^2}{m v_F^2}}$$

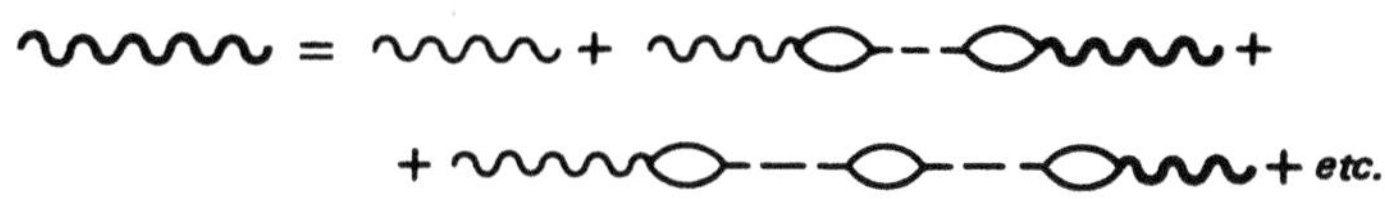


**Figure 3.5.** Diagram equation for the Green function of the screened phonons.

is the Debye screening radius. It can be seen from (3.53) that the renormalized longitudinal phonons obey an acoustic law of dispersion with the sound speed equal to $(mZ/3M)^{1/2}v_F$, where $v_F$ is the Fermi velocity of the electrons. It can easily be verified in a more general (than the "jellium") model that takes account of the deviation of the potential $V_{ei}(q)$ from the purely Coulomb potential that the poles of (3.51) also lead to an acoustic spectrum since it follows from the condition of electrical neutrality that $V_{ei}(q)$ is equal, as before, to $Ze^2/q^2$ at low $q$.

On the basis of the Hamiltonian (3.39), we can also determine the total effective interelectron interaction, which is essential for, for example, the computation of the superconducting properties of a metal (see Chapter 2). It is depicted in Fig. 3.6. Figure 3.6a represents the screened Coulomb interaction; Fig. 3.6b, the electron-phonon interaction with a screened vertex, $\tilde{g}(q\lambda)$.

The total interaction has the form

$$V_{\text{eff}}(q, \omega) = \frac{4\pi e^2}{q^2\varepsilon_e(q, \omega)} + \frac{g^2(q\lambda)}{\varepsilon_e^2(q, 0)} \frac{\Omega_{q\lambda}^2}{\omega^2 - \omega^2(q\lambda)}, \tag{3.54}$$

where $\omega^2(q\lambda)$ is the square of the renormalized-phonon frequency

$$\omega^2(q\lambda) = \Omega_{q\lambda}^2\left[1 - \frac{g^2(q\lambda)}{V(q)}\left(1 - \frac{1}{\varepsilon_e(q, 0)}\right)\right]. \tag{3.55}$$

In the "jellium" model $V_{\text{eff}}(q, \omega)$ reduces to the form

$$V_{\text{eff}}(q, \omega) = \frac{4\pi e^2}{q^2\varepsilon_{\text{eff}}(q, \omega)}, \tag{3.56}$$

where $\varepsilon_{\text{eff}}(q, \omega)$ is the total permittivity of the system:

$$\varepsilon_{\text{eff}}(q, \omega) = \varepsilon_e(q, \omega) - \frac{\Omega_{q\lambda}^2}{\omega^2}. \tag{3.57}$$

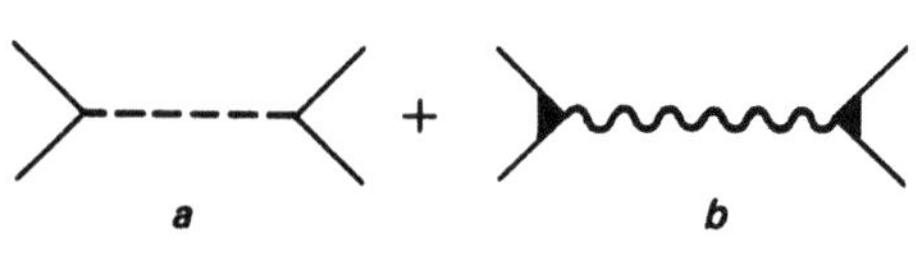


**Figure 3.6.** The total interelectron interaction: (a) the screened Coulomb interaction; (b) the screened electron–phonon interaction.

In the more general case the interelectron interaction also reduces to (3.56), but $\varepsilon_{eff}$ has a form different from (3.57):

$$\varepsilon_{eff}(q, \omega) = \varepsilon_e(q, \omega) - \frac{g^2(q\lambda)}{V(q)} \frac{\Omega_{q\lambda}^2}{\omega^2 - \Omega_{q\lambda}^2 \left[1 - \frac{g^2(q\lambda)}{V(q)}\right]}. \qquad (3.58)$$

Interelectron interactions of the types (3.56) and (3.54) have already been used in Chapter 2 to investigate the superconducting properties of metals. Knowing these interactions, we can also describe the properties of normal metals, in particular, the renormalized excitation spectrum. These computations are carried out in the above-indicated book by Pines and Nozières [24] within the framework of the random-phase approximation, which we have thus far been using, and we shall not dwell on them here. Let us only note that the renormalized electron–phonon interaction constant $\lambda$ is, as can be seen from (3.54), connected with the bare-constant $g^2(q\lambda)$ by the following expression:

$$\lambda = N(0) \int_0^{2k_F} \frac{q\,dq}{2k_F} \tilde{g}^2(q\lambda) \frac{\Omega_{q\lambda}^2}{\omega^2(q\lambda)}. \qquad (3.59)$$

Below we investigate on the basis of the obtained results the problem of lattice stability and the limitations that this condition imposes on the effective interelectron interaction (3.56). (This question has already been considered in Chapter 2 from a slightly different point of view.) The condition for the dynamical stability of a lattice reduces to the requirement that $\omega^2(q\lambda) > 0$. Accordingly, from (3.55) we have

$$1 + \frac{g^2(q\lambda)}{V(q)} \frac{1 - \varepsilon_e(q, 0)}{\varepsilon_e(q, 0)} > 0. \qquad (3.60)$$

If we use the well-known Hubbard approximation for $\varepsilon_e(q, 0)$, then we can rewrite the stability condition in the form

$$g^2(q\lambda) + \frac{U}{2} - V(q) < \frac{1}{|\Pi(q, 0)|}, \qquad (3.61)$$

where $U$ is the interelectron exchange interaction. This condition is formally similar to the one obtained by Chan and Heine [28]. It should, however, be noted that in [28] the condition (3.61) contains the screened electron–phonon interaction constant $\tilde{g}^2(q\lambda)$ instead of the constant $g^2(q\lambda)$ for the interaction with the bare ionic vibrations. This result is clearly erroneous and arose as a result of the incorrect use in [28] of the adiabatic approximation.

Using the formulas (3.58) and (3.61), we can easily verify that the condition for lattice stability in a quasi-isotropic system leads to the following limitations on the effective interelectron interaction

$$V_{\text{eff}}(q, \omega) = \frac{4\pi e^2}{q^2 \varepsilon_{\text{eff}}(q, \omega)}:$$

if

$$\varepsilon_e(q, 0) > 0, \text{ ТО } \varepsilon_{\text{eff}}(q, 0)\left[\frac{g^2(q\lambda)}{V(q)} - 1\right] < 0,$$

if

$$\varepsilon_e(q, 0) < 0, \text{ ТО } \varepsilon_{\text{eff}}(q, 0)\left[\frac{g^2(q\lambda)}{V(q)} - 1\right] > 0.$$

It can be seen from these expressions that, for $\varepsilon_e(q, 0) > 1$, the effective interelectron interaction at zero frequency can be an attractive interaction [i.e., we can have $V_{\text{eff}}(q, 0) < 0$] if the bare electron–ion interaction is stronger than the direct mutual Coulomb repulsion of the electrons. The condition that $V_{\text{eff}}(q, 0)$ be negative can be written together with the lattice stability condition (3.60) in the form

$$\frac{\varepsilon_e(q, 0)}{\varepsilon_e(q, 0) - 1} > \frac{g^2(q\lambda)}{V(q)} > 1. \tag{3.62}$$

Since the permittivity, $\varepsilon_e(q, 0)$, of the electron gas tends to infinity at low $q$, it follows from (3.62) that the effective interelectron interaction cannot become an attractive interaction at low $q$ (more exactly, as $q \longrightarrow 0$). Moreover, it can be rigorously demonstrated on the basis of the electrical-neutrality condition that the ratio $g^2(q\lambda)/V(q)$ should tend at low $q$ to unity:

$$\lim_{q \to 0} \frac{g^2(q\lambda)}{V(q)} = 1.$$

On the other hand, at large $q$ values, which are precisely the most important in superconductivity, the permittivity $\varepsilon(q, 0)$ tends to unity, and the situation is entirely possible in which the ratio $g^2(q\lambda)/V(q)$ is, on the one hand, greater than unity, which makes the effective interelectron interaction attractive but at the same time less than the quantity $\varepsilon_e(q, 0)/[\varepsilon_e(q, 0) - 1]$, which is essential to the stability of the lattice.

The results obtained here are in complete agreement with the results presented in Chapter 2. In particular, as can be seen from the formulas (3.58) and (3.60), the quantity $\varepsilon_{eff}(q, 0)$ in a stable system is subject to the following limitations: either $\varepsilon_{eff}(q, 0) > 1$, or $\varepsilon_{eff}(q, 0) < 0$.

Thus far, we have established only the formal possibility of the existence of attractive interelectron interaction in a stable system. Let us now briefly consider the conditions under which such a situation can actually manifest itself in a quasi-isotropic system. For this purpose, we rewrite (3.62), using for the bare-electron–phonon interaction the expression (3.43):

$$\frac{nq^2 V_{ei}^2(q)}{M\Omega_{q\lambda}^2} > \frac{4\pi e^2}{q^2}. \tag{3.63}$$

Entering into the inequality (3.63) is the plasma frequency, $\Omega_{q\|}$, of the longitudinal vibrations of the ions since in a quasi-isotropic system the electrons interact only with the longitudinal vibrations. In models of the "jellium" type, in which the ions are regarded as forming a structureless continuous medium, there is only one longitudinal mode of ionic vibrations. In this case there obtains the equality

$$\Omega_{q\|}^2 = \frac{4\pi n (Ze)^2}{M}.$$

Now if the electron–ion interaction potential, $V_{ei}(q)$, is a purely Coulomb potential, i.e., if

$$V_{ei}(q) = \frac{4\pi Ze^2}{q^2},$$

then the inequality (3.61) is not fulfilled at any value of the momentum $q$. In such a model, the bare-electron–phonon interaction potential $nq^2 V_{ei}^2(q)/M\Omega_{q\|}^2$ is always equal to the Coulomb potential $V(q)$. In the majority of metals, because of the presence of filled inner electron shells, the interaction of the valence electrons with the ions is not a purely Coulomb interaction but is described by a pseudopotential [29]. At low momenta $q$, because of the electrical-neutrality requirement, the condition

$$\lim_{q\to 0} V_{ei}(q) = \frac{4\pi Z^2 e^2}{q^2}.$$

should also be fulfilled. At large momenta $q$, however, the pseudopotential $V_{ei}(q)$ can differ significantly from the Coulomb potential of the ion. Normally, the pseudopotential is smaller than the Coulomb potential. The case when, as a result of the influence of the polarization of the inner shells, this pseudopotential is greater than the Coulomb potential is, however, entirely possible. It is precisely this situation that is necessary for the existence of attractive interelectron interaction at zero frequency. As the study of the pseudopotentials of real metals shows [29], this possibility is realized, but rarely.

More favorable conditions for the appearance of attractive interelectron interaction arise when allowance is made for the crystalline structure of the ion subsystem. In this case, even if the electron potential is assumed to be a purely Coulomb potential, on account of the sum rule (3.42), the frequency $\Omega_{q\parallel}$ satisfies the inequality

$$\Omega_{q\parallel}^2 < \frac{4\pi Z^2 e^2 n}{M}.$$

Because of this condition, the inequality (3.63) is in fact fulfilled at any $q \neq 0$. In that case the effective interelectron interaction at zero frequency, $V_{\mathrm{eff}}(q, 0)$, is a purely attractive interaction for all momenta. As has already been emphasized in Chapter 2, this circumstance significantly improves the conditions for the onset of the superconducting state in such a system. Another factor that is favorable to the appearance of interelectron attraction in a crystal lattice is the effect of the Umklapp processes.

Coherent computations of $\varepsilon_{\mathrm{eff}}(q, 0)$ have been carried out for a number of simple metals by Dolgov and Maksimov [30]. As is well known, because of the smallness of the ion pseudopotential [29], these metals can, to a high degree of accuracy, be regarded as quasi-isotropic. It is shown in [30] that, in the alkali metals, because of the smallness of the pseudopotential $V_{ei}(q)$, the increase in the quantity $nq^2 V_{ei}^2(q)/M\Omega_{q\parallel}^2$ resulting from the deviation of the quantity $\Omega_{q\parallel}^2$ from the square of the plasma frequency is not sufficient for the condition (3.63) to be fulfilled. An in metals such as Al and Pb, in which the pseudopotential $V_{ei}(q)$ is not so small, the effective permittivity $\varepsilon_{\mathrm{eff}}(q, 0)$ is a negative quantity at low $q$ but becomes positive at large $q$ because of the significant decrease that then occurs in the pseudopotential $V_{ei}(q)$ as compared to the Coulomb potential of a point ion. It can be shown [30] on the basis of the phonon-spectrum calculations carried out in [31] that for metallic hydrogen, in which the electron-ion interaction is described by the simple Coulomb law, the effective permittivity $\varepsilon_{\mathrm{eff}}(q, 0)$ is negative at all momenta $q$.

Thus far, the entire investigation has been based on the random-phase approximation. Allowance for the Coulomb interaction in the higher orders of perturbation theory for an isotropic electron-phonon system has been made by Batyev and Pokrovskii [32]. We shall consider certain characteristic features that arise in this case in the following section. Here, jumping ahead, we note that when precise allowance is made for the Coulomb interaction, the effective interelectron interaction no longer reduces to the function $V_{\mathrm{eff}}(q, \omega)$, which depends only on the momentum transfer $q$ and the energy $\omega$. A comparison of the strengths of the electron-phonon and Coulomb interactions then becomes difficult and has, in fact, thus far not been carried out to any degree of accuracy.

This question is considered within the framework of the local approximation for the exchange-correlation interaction by Dolgov, *et al.* in [33]. They show that in this approximation the effective interaction $V_{eff}(q, \omega)$ no longer reduces to the formula (3.56) but that the sign of this interaction at $\omega = 0$ is determined as before by the sign of the effective permittivity $\varepsilon_{eff}(q, 0)$.

As to the stability condition for the isotropic system, (3.61) obtained by us in the random-phase approximation, since the lattice dynamics is determined only by the value of the permittivity, $\varepsilon_e(q, 0)$, of the electrons, it remains formally valid when the Coulomb interaction is taken into account in all orders of perturbation theory [22]. In this case the permittivity, $\varepsilon_e(q, \omega)$, should be computed not in the random-phase approximation, but with complete allowance for the Coulomb interaction.

The investigation of the plasma model also allows us to gain an understanding of the essence of the contradictions that arise between the results obtained in the adiabatic approximation and those obtained in the Fröhlich model. Let us, first, note that the expression for the phonon frequencies of the isotropic system that follows from the adiabatic approximation in accordance with the formula (3.34) actually coincides entirely with the value of $\omega^2(q\lambda)$ obtained in the plasma model [see (3.55)]. Within the framework of the plasma model, the formula (3.55) for the squares of the phonon frequencies can be written somewhat more generally in the form

$$\omega^2 (q\lambda) = \Omega_{q\lambda}^2 [1 + g^2 (q\lambda) \chi_e (q, \omega (q\lambda))]. \tag{3.64}$$

In writing down (3.64), we introduced the electron susceptibility $\chi_e(q, \omega)$, which is connected with the permittivity $\varepsilon_e(q, \omega)$ by the relation

$$\chi_e (q, \omega) = \frac{1}{V(q)} \left( \frac{1}{\varepsilon_e (q, \omega)} - 1 \right),$$

and we also took into account the dependence of this function on the frequency $\omega$.

Earlier, in computing the phonon frequencies within the framework of the plasma model, we neglected the dependence of the function $\chi_e(q, \omega)$ on the frequency $\omega$. This approximation constitutes the adiabatic approximation for the plasma model. The physical meaning of this approximation is also fairly simple. It can be seen from the expression $\omega_{q\lambda}^2 = \Omega_{q\lambda}^2 / \varepsilon_e(q, 0)$ for the phonon frequencies that the lattice dynamics is determined by the inter-ion interaction screened by the static permittivity. A precisely similar situation is obtained in the conventional Born–Oppenheimer approximation [see Eq. (3.33) for the dynamical vibration matrix]. The electron-phonon-interaction-induced nonadiabatic correction to the phonon frequencies can be computed by solving the transcendental equation (3.64). It can easily be shown that because of the smallness of the ion velocity as compared to the Fermi velocities of the electrons, allowance for the frequency dependence of the function $\chi_e(q, \omega)$ will result in the phonon frequencies' changing by a small amount of the order of $(m/M)^{1/2}$.

It is clear from the foregoing that the "softening" of the frequencies of the actually observable phonons and the lattice instability (the vanishing of the phonon frequencies in the adiabatic approximation or in the plasma model) can be related to the change that occurs in the effective inter-ion interaction as a result of the change in the static permittivity of the electrons. Such is the situation in, for example, one-dimensional or quasi-one-dimensional metallic systems, in which, as is well known [17], the quantity $\varepsilon_e(q, 0)$ at $T = 0$ has a logarithmic singularity and becomes infinite for momenta $q = 2k_F$. In this case the phonon frequency and the Fourier transform of the effective inter-ion interaction vanish at $q = 2k_F$.

On the other hand, as can also be seen from the plasma model, another interpretation of the lattice instability is also possible. We can introduce "bare" phonons, e.g., plasma oscillations of the ions. Then the subsequent allowance for the interaction of the "bare" phonons with the electrons leads to their renormalization, which yields precisely the real spectrum of the phonon excitations of the metal. The onset of lattice instability and the appearance of negative values of $\omega^2(q)$ can thus be regarded as the result of the effect of the electron-phonon interaction on the spectrum of the bare plasma oscillations.

It is clear from this that the main discrepancy between the adiabatic approximation and the Fröhlich model does not lie in the question of whether the electron-phonon interaction renormalizes the vibration frequencies drastically or slightly but whether this interaction can be the cause of the lattice instability. These, as we have already seen, are purely terminological questions, and are connected with the choice of the basic approximation for the phonons and electrons.

The main shortcoming of the Fröhlich model as compared with the adiabatic approximation is connected with the absence in the Fröhlich model of a consistent method of allowing for the interelectron Coulomb interaction. It is usually assumed when the Fröhlich Hamiltonian is being written down that the electron-phonon interaction constant and the vibration frequencies of the bare phonons should be taken with allowance for the screening of the electron-ion and ion-ion interactions by the conduction electrons. However, there does not exist any consistent method of division of the Coulomb interaction into the part that has already been taken into account in the bare values of the constants of the Fröhlich Hamiltonian and the part that subsequently manifests itself in the renormalization of the bare phonons as a result of the electron-phonon interaction. Thus, the conclusion drawn in the Fröhlich model that the bare *longitu-*

*dinal acoustic vibrations* are unstable because of the electron–phonon interaction is, as is rightly pointed out in [20], actually due to the fact that the electron–ion interaction is taken into account twice: first, in the choice of the acoustic spectrum for the bare phonons; second, in the computation of the renormalization of the sound.

Even if we drop the acoustic law of dispersion of the bare phonons in the Fröhlich model, the absence in this model of the interelectron Coulomb interaction will all the same lead to an incorrect criterion for lattice stability. Indeed, for an isotropic system, the stability criterion in the Fröhlich model can be written in the form

$$g_0^2 \,|\, \Pi\,(q,\,0)\,| < 1,$$

whereas the correct stability criterion, obtained in the plasma model or in the adiabatic approximation, can be written in accordance with the formulas (3.61) as follows:

$$\frac{g^2\,(q\lambda)\,|\,\Pi\,(q,\,0)\,|}{1+V\,(q)\,|\,\Pi\,(q,\,0)\,|} < 1.$$

It can be seen from a comparison of these formulas that, within the framework of the Fröhlich model, lattice instability can, in the presence of singularities in $\Pi(q, 0)$, arise at any value of the electron–bare-phonon interaction constant, whereas according to the exact stability criterion, the instability arises only when the interaction of the electrons with the bare phonons is stronger than the Coulomb interaction between the electrons. The result that the interelectron Coulomb interaction has a stabilizing influence on the lattice has also been obtained by a number of authors [34, 35] with models significantly different from the plasma model.

The availability of a convenient diagram technique makes the plasma model, as opposed to the adiabatic approximation, extremely attractive for the consistent description of the electrons and phonons, not only in an isotropic system but also in real metals with a crystal lattice. The complete system of equations in the plasma model that take account of the crystal structure of the metal has been formally obtained by Garland [36]. Unfortunately, the introduction from the very beginning of the plasma vibrations of the crystal lattice leads to the hybridization of the various vibration modes when allowance is made for the interaction of these vibrations with the conduction electrons. The determination of the true normal modes of the lattice then becomes an extremely difficult problem. Much more convenient is the use from the very beginning of the coordinate representation for the electron and ion Green functions, with their subsequent diagonalization in the appropriate basis.

## 5. Self-Consistent Description of the Electron-Phonon System of a Metal

The first attempt to construct a self-consistent theory of the electron-phonon system was made by Baym [37]. Subsequently, this approach was developed in [38–40]. There is no doubt that the self-consistent equations for the electron and phonon subsystems will be extremely complex. Unfortunately, in some of the indicated papers these equations are, moreover, unnecessarily overcomplicated. In [37, 38], the rigorous sum rules, which allow the simplification of the obtained expressions, are not used. In [40], instead of a single permittivity function, Rajagopal and Cohen introduce several types of such functions. In this section we present a system of exact self-consistent equations, first obtained in [22], for the electron and phonon Green functions. In our opinion, this system of equations is the most convenient system for a concrete investigation of the properties of metals.

Let us begin the investigation of the electron-phonon system with the introduction of the total electron-ion interaction Hamiltonian:

$$H = H_e + H_{ei} + H_i, \tag{3.65}$$

$$H_e = \int d^3r\, \psi^+(\mathbf{r}, t)\left(-\frac{\nabla^2}{2m} - \mu\right)\psi(\mathbf{r}, t) + \\ + \frac{1}{2}\int d^3r_1 d^3r_2 \psi^+(\mathbf{r}_1, t)\,\psi^+(\mathbf{r}_2, t)\frac{e^2}{|r_1 - r_2|}\psi(\mathbf{r}_2, t)\psi(\mathbf{r}_1, t), \tag{3.66}$$

$$H_{ei} = \sum_n \int d^3r \psi^+(\mathbf{r}, t)\,\psi(\mathbf{r}, t)\,V_{ei}(\mathbf{r} - \mathbf{R}_n^0) - \\ - \sum_n \int d^3r \psi^+(\mathbf{r}, t)\,\psi(\mathbf{r}, t)\,u_n^\alpha\,\nabla_\alpha V_{ei}(\mathbf{r} - \mathbf{R}_n^0) + \\ + \frac{1}{2}\sum_n \int d^3r \psi^+(\mathbf{r}, t)\,\psi(\mathbf{r}, t)\,u_n^\alpha u_n^\beta \nabla_\alpha \nabla_\beta V_{ei}(\mathbf{r} - \mathbf{R}_n^0), \tag{3.67}$$

$$H_i = \frac{1}{2}\sum_{n,\alpha} M\,[\dot{u}_n^\alpha]^2 + \frac{1}{2}\sum_{n,n'} V_{ii}(\mathbf{R}_n^0 - \mathbf{R}_{n'}^0) + \\ + \frac{1}{2}\sum_{n,n'}(u_n^\alpha - u_{n'}^\alpha)\,\nabla_\alpha V_{ii}(\mathbf{R}_n^0 - \mathbf{R}_{n'}^0) + \\ + \frac{1}{4}\sum_{n,n'}(u_n^\alpha - u_{n'}^\alpha)(u_n^\beta - u_{n'}^\beta)\,\nabla_\alpha\nabla_\beta V_{ii}(\mathbf{R}_n^0 - \mathbf{R}_{n'}^0). \tag{3.68}$$

In writing down the Hamiltonian, we limited ourselves to the harmonic approximation by expanding all the quantities to second order in the small displacements of the ions from the equilibrium positions:

$$\mathbf{u}_n = \mathbf{R}_n - \mathbf{R}_n^0.$$

For the electrons we use the second-quantization representation, while the quantities describing the ion subsystem have been written in first-quantized form; $V_{ii}$ is the ion–ion interaction potential, while $V_{ei}$ is the electron–ion interaction potential.

Let us introduce in the usual manner the temperature Green function for the electrons

$$G(\mathbf{r}\tau, \mathbf{r}'\tau') = -i\langle T_\tau \psi(\mathbf{r}, \tau)\psi^+(\mathbf{r}', \tau')\rangle \tag{3.69}$$

and the Green function for the ion displacements

$$D_{nn'}^{\alpha\beta}(\tau) = -i\left\langle T_\tau\left(u_n^\alpha(\tau) - \left\langle u_n^\alpha\right\rangle\right), \quad \left(u_{n'}^\beta(0) - \left\langle u_{n'}^\beta\right\rangle\right)\right\rangle. \tag{3.70}$$

In an ideal crystal, the function $D_{nn'}^{\alpha\beta}(\tau)$ goes over into the normal phonon Green function if we use the phonon representation for the ion displacements:

$$\mathbf{u}_n^\alpha = \sum_{\mathbf{q},\lambda} e^{i\mathbf{q}\mathbf{R}_n^{(0)}} \mathbf{e}_{q\lambda}\left(b_{q\lambda} + b_{-q\lambda}^+\right).$$

As we have already pointed out, it will be more consistent to work with these equations in coordinate space; therefore, we shall use the Green function $D_{nn'}^{\alpha\beta}(\tau)$ directly and not go over to the phonon representation right up to the derivation of the final formulas. For the derivation of the equations satisfied by the functions $G(\mathbf{r}\tau, \mathbf{r}'\tau')$ and $D_{nn'}^{\alpha\beta}(\tau)$, we shall use the method of functional differentiation [41]. For this purpose, let us add to the Hamiltonian (3.65) the external sources

$$H' = \int d^3r\psi^+(\mathbf{r}, t)\,\psi(\mathbf{r}, t)\,U(\mathbf{r}, t) + \sum_{n,\alpha} \mathcal{I}_n^\alpha(t)\,u_n^\alpha(t).$$

Let us write down the equations for the operators $\psi(\mathbf{r}, \tau), u_n^\alpha(\tau)$:

$$i\,\frac{\partial\psi(\mathbf{r}, \tau)}{\partial\tau} = \frac{\delta H}{\delta\psi^+(\mathbf{r}, \tau)}, \qquad -M\,\frac{\partial^2 u_n^\alpha(\tau)}{\partial\tau^2} = \frac{\delta H}{\delta u_n^\alpha}. \tag{3.71}$$

Averaging the equation for $u_n^\alpha(\tau)$ and taking into account the fact that

$$\frac{\delta\left\langle u_n^\alpha\right\rangle}{\delta\mathcal{J}_{n'}^\beta(\tau)} = D_{nn'}^{\alpha\beta}(\tau), \tag{3.72}$$

we obtain the equation for the function $D_{nn'}^{\alpha\beta}(\tau)$:

$$\begin{aligned} M\frac{\partial^2}{\partial\tau^2}D_{nn'}^{\alpha\beta}(\tau) &= \frac{\delta}{\delta\mathcal{J}_{n'}^\beta(\tau)}\int d^3r\,\langle\psi^+(\mathbf{r},\tau)\,\psi(\mathbf{r},\tau)\rangle\,\nabla_\alpha V_{ei}\left(\mathbf{r}-\mathbf{R}_n^0\right) - \\ &- \frac{\delta}{\delta\mathcal{J}_{n'}^\beta(\tau)}\int d^3r\left\langle\psi^+(\mathbf{r},\tau)\,\psi(\mathbf{r},\tau)\,u_n^\gamma(\tau)\right\rangle\times \\ &\times\nabla_\alpha\nabla_\gamma V_{ei}\left(\mathbf{r}-\mathbf{R}_0^n\right) + \sum_{n''\neq n}\nabla_\gamma\nabla_\alpha V_{ii}\left(\mathbf{R}_n^0-\mathbf{R}_{n''}^0\right)D_{nn'}^{\gamma\beta}(\tau) - \\ &- \sum_{n''\neq n}\nabla_\alpha\nabla_\gamma V_{ii}\left(\mathbf{R}_n^0-\mathbf{R}_{n'}^0\right)D_{n''n'}^{\gamma\beta}(\tau) - \delta_{nn'}^{\alpha\beta}\,\delta(\tau), \end{aligned} \tag{3.73}$$

and, correspondingly, for $G(\mathbf{r}\tau, \mathbf{r}'\tau')$:

$$\begin{aligned} &\left[i\frac{\partial}{\partial\tau}+\left(\frac{\nabla^2}{2m}+\mu\right)-U(x)+\sum_n V_{ei}\left(\mathbf{r}-\mathbf{R}_n^0\right)\right]G(x,x') = \\ &= \delta(x-x') + i\sum_n\nabla_\alpha V_{ei}\left(\mathbf{r}-\mathbf{R}_n^0\right)\left\langle T_\tau u_n^\alpha(\tau)\,\psi(x)\,\psi^+(x')\right\rangle - \\ &- \frac{i}{2}\sum_n\nabla_\alpha\nabla_\beta V_{ei}\left(\mathbf{r}-\mathbf{R}_n^0\right)\left\langle T_\tau u_n^\alpha u_n^\beta\psi(x)\,\psi^+(x')\right\rangle - \\ &- ie^2\int d^3r''\,\frac{1}{|\mathbf{r}''-\mathbf{r}|}\,\langle T_\tau\psi^+(\mathbf{r}''\tau)\,\psi(\mathbf{r}''\tau)\,\psi(x)\,\psi^+(x')\rangle \end{aligned} \tag{3.74}$$

[in Eq. (3.74) we have introduced the notation $x = (\mathbf{r}\tau)$ and $x' = (\mathbf{r}'\tau')$].

The use of the rules of functional differentiation allows us to significantly simplify Eqs. (3.73) and (3.74). We do not give details here of the subsequent calculations (they can be found in [22]) but write down immediately the final form of these equations:

$$\left[D_{nn'}^{\alpha\beta}(\tau)\right]^{-1} = \left[D_{0nn'}^{\alpha\beta}(\tau)\right]^{-1} - M_{nn'}^{\alpha\beta}(\tau). \tag{3.75}$$

In this formula the matrix $D_{0nn'}^{\alpha\beta}(\tau)$ is determined by the dynamics of the vibration of the ion lattice:

$$\Big[M\frac{\partial^2}{\partial\tau^2}+\delta_{n''n}\sum_{n'''\neq n}\nabla_\alpha\nabla_\gamma V_{ii}(\mathbf{R}_n^0-\mathbf{R}_{n'''}^0)-$$

$$-\sum_{n''\neq n}\nabla_\alpha\nabla_\gamma V_{ii}(\mathbf{R}_n^0-\mathbf{R}_{n''}^0)\Big]D_{n''n'}^{\gamma\beta}(\tau)=-\delta(\tau)\,\delta_{nn'}^{\alpha\beta}. \tag{3.76}$$

The longitudinal vibration mode in $D_{0nn'}^{\alpha\beta}$ is the ion-plasma vibration; the charge-density preserving transverse modes, on the other hand, are of acoustic nature even in the ion system. The quantity $M_{nn'}^{\alpha\beta}$ is the contribution of the electrons to the lattice dynamics:

$$M_{nn'}^{\alpha\beta}(\tau)=\delta_{nn'}\sum_{n''}\int dx''\,d^3r\chi_e(x,\mathbf{r}''\tau^+)\times$$
$$\times\nabla_\alpha V_{ei}(\mathbf{r}-\mathbf{R}_n^0)\nabla_\beta V_{ei}(\mathbf{r}-\mathbf{R}_{n''}^0)-$$
$$-\int dx''\,d^3r\chi_e(x,x'')\nabla_\alpha V_{ei}(\mathbf{r}-\mathbf{R}_n^0)\nabla_\beta V_{ei}(\mathbf{r}-\mathbf{R}_{n'}^0). \tag{3.77}$$

The total susceptibility of the electrons, $\chi_e(x,x')$, figures in equation (3.77). The nonadiabaticity of the electron contribution to the lattice dynamics is contained only in the second term in $M_{nn'}^{\alpha\beta}$; the electron susceptibility, $\chi_e$, entering into play in the first term is the equal-time susceptibility. If we go over to the phonon representation in Eqs. (3.75)-(3.77) and neglect the nonadiabaticity in the second term, i.e., if we set

$$\chi_e(x,x')=\delta(\tau-\tau')\chi_e(r,r'), \tag{3.78}$$

we naturally arrive at the standard formulas of the adiabatic approximation for the phonon system [see (3.33)].

Simple, but long computations [22] for the electron Green function $G(x,x')$ lead to the following results:

$$G^{-1}(x,x)=i\frac{\partial}{\partial\tau}+\left[\frac{\nabla^2}{2m}+\mu\right]-U_{\text{eff}}(x)-\Sigma(x,x'). \tag{3.79}$$

In formula (3.79)

$$U_{\text{eff}}(x)=U(x)+e^2\int d^3r''\frac{\langle\rho_e(r'',\tau)\rangle}{|\mathbf{r}''-\mathbf{r}|}+$$

$$+\sum_n V_{ei}(\mathbf{r}-\mathbf{R}_n^0)-\sum_n\nabla_\alpha V_{ei}(\mathbf{r}-\mathbf{R}_n^0)\langle u_n^\alpha\rangle. \tag{3.80}$$

After switching off the external source, i.e., after setting $U(x) = 0$, the quantity $U_{eff}(x)$ goes over into the standard Hartree potential for the electrons in the periodic field of the ions. The self-energy part of the electron Green function has the form

$$\Sigma(x, x') = -i\int dx_1\Big\{\int d^3r'' \frac{e^2}{|\mathbf{r}'' - \mathbf{r}|}\varepsilon_e^{-1}(\mathbf{r}'', \tau, x_1) + \\ + \sum_{n,n'}\int dx_2 dx_3 dx_4 \varepsilon_e^{-1}(\mathbf{r}\tau, x_3)\nabla_\alpha V_{ei}(\mathbf{r}_3 - \mathbf{R}_n^0)\varepsilon_e^{-1}(x_1, x_4)\times \\ \times \nabla_\beta V_{ei}(\mathbf{r}_4 - \mathbf{R}_{n'}^0) D_{nn'}^{\alpha\beta}(\tau_4 - \tau_3)\Big\}\Gamma(x_2, x', x_1)G(x, x_2). \quad (3.81)$$

The first term in the parentheses is connected with the contribution of the screened Coulomb interaction to the self-energy of the electrons. The second term is the phonon contribution to the self-energy. In fact, the expression in the parentheses is the total interelectron interaction $V_{eff}(x, x_1)$ and is the sum of the screened Coulomb interaction (the first term) and the electron-phonon interaction (the second term) [cf. (3.57)].

The vertex function $\Gamma(x_2, x', x_1)$ is defined as follows:

$$\Gamma(x_2, x', x_1) = \frac{\delta G^{-1}(x_2, x')}{\delta U_{eff}(x_1)}. \quad (3.82)$$

We can, using the method of functional differentiation, write the following equation for it:

$$\Gamma(x_2, x', x_1) = \delta(x_2 - x')\delta(x_2 - x_1) - \int dx_3 dx_4 dx_5 dx_6 \times \\ \times \frac{\delta\Sigma(x_2, x')}{\delta G(x_3, x_4)} G(x_3, x_5)\Gamma(x_5, x_6, x_1)G(x_6, x_4). \quad (3.83)$$

The simplest diagrams for the vertex function $\Gamma(x_2, x', x_1)$ are shown in Fig. 3.7a-f.

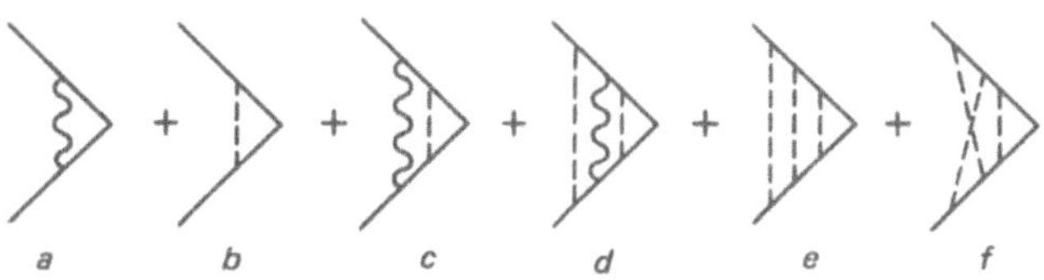


**Figure 3.7. The simplest diagrams for the vertex function $\Gamma(x_2, x', x_1)$. A wavy line represents the exact electron-phonon interaction; a dashed line, the screened Coulomb interaction between the electrons.**

We shall discuss the methods of computing $\Gamma(x_2, x', x_1)$ later; right now we will determine the remaining functions entering into the equations obtained by us.

The electron susceptibility $\chi_e(x, x')$ entering into the equation for the phonon Green function is determined by the following equality:

$$\chi_e = \int dx_1 \varepsilon_e^{-1}(x, x_1)\, \Pi(x_1, x'), \tag{3.84}$$

where $\Pi(x_1, x')$ is the irreducible polarization operator of the electrons:

$$\Pi(x, x') = \frac{\delta\langle\rho_e(x)\rangle}{\delta U_{\mathrm{eff}}(x')} = -\int dx_1 dx_2 G(x, x_1)\, G(x_2, x)\, \Gamma(x_1, x_2, x'), \tag{3.85}$$

The quantity $\varepsilon_e^{-1}(x, x_1)$ is the inverse permittivity of the system, determined by the equation

$$\varepsilon_e^{-1}(x, x_1) = e^2 \int dx'' d^3 r_1 \frac{1}{|\mathbf{r}_1 - \mathbf{r}|}\, \Pi(x, x'')\, \varepsilon_e^{-1}(x'', x_1) + \delta(x - x_1). \tag{3.86}$$

It is practically impossible at present to obtain the exact solution to the complete system of self-consistent equations for the electrons and phonons. The main, fundamental difficulty lies in the computation of the vertex function $\Gamma(x_2, x', x_1)$, which enters into both the expression for the self-energy part, $\Sigma(x, x')$, of the electron Green function and the expression for the exact electron polarization operator $\Pi(x, x')$, which determines, in particular, the contribution of the electrons to the phonon frequencies.

As has been further shown by Migdal [14], the computation of the phonon corrections to the vertex function leads to quantities that are smaller than the bare vertex function in the ratio $(m/M)^{1/2}$. Indeed, it can easily be verified that the contribution of the diagram in Fig. 3.7a is smaller than that of the diagram in Fig. 3.7b precisely by a factor of $(M/m)^{1/2}$. The diagrams 3.7d and 3.7e are related in the same way. From this it apparently follows that in computing the vertex function $\Gamma(x_2, x', x_1)$, we can neglect all the diagrams with phonon lines. It can be shown that we can, in fact, do this when computing the polarization operator. At least at temperatures lower than the characteristic phonon frequencies we can use in the formula (3.85) in place of the complete vertex $\Gamma(x_2, x', x_1)$ the function $\Gamma_0(x_2, x', x_1)$, which is determined by only the screened Coulomb interaction (the sum of diagrams of the type b, e, and a in Fig. 3.7.)

In metals the Coulomb interaction does not have any smallness parameter; therefore, for all intents and purposes there is no consistent procedure that enables us to select the most important diagrams for the computation of $\Gamma_0(x_2, x', x_1)$. The problem of the Coulomb correlations at electron densities corresponding to real metals has in recent years become one of the most pressing problems of solid state physics [20, 42–45].

Returning to the determination of the self-energy part of the electron Green function (3.81), we note that in this case the total neglect of the phonon corrections to the vertex function (as was done by Heine *et al.* [46]) is inadmissible. In order to gain an understanding of the cause of this, we rewrite schematically the expression for $\Sigma(x, x')$ in the form of a sum of two terms:

$$\Sigma(x, x') = -i\int \widetilde{V}_{ee} G\Gamma - i\int \nabla_\alpha \widetilde{V}_{ei} D^{\alpha\beta} \nabla_\beta \widetilde{V}_{ei} \Gamma G, \tag{3.87}$$

where we have denoted by $\widetilde{V}_{ee}$ the screened interelectron Coulomb interaction and by $\widetilde{V}_{ei}$ the screened electron–ion interaction.

It is well known that the self-energy part of the electron Green function that arises as a result of the interaction with the phonons is proportional to the mean phonon frequency: $\Sigma_{\text{ph}} \approx \omega_D$. The Coulomb contribution to the self-energy of the electrons of a metal is proportional to the Fermi energy: $\Sigma_{\text{Coul}} \approx E_F$. Therefore, without allowance for the vertex corrections, the first term in (3.87) is larger than the second by a factor of $E_F/\omega_D$.

The necessity to take the phonon contribution to $\Sigma(x, x')$ into account is because it is localized largely in the vicinity of the Fermi surface and is, therefore, not small compared to the electron energy. Since each auxiliary phonon line in the vertex function gives rise to a small quantity of the order of $(m/M)^{1/2}$, we can replace the vertex function in the second term in the formula (3.87) by $\Gamma_0$, neglecting the phonon corrections. This cannot, however, be done in the first term.

Let us consider the series, shown in Fig. 3.8, of simplest diagrams for the first term in the formula (3.87). It can easily be verified that diagrams of types

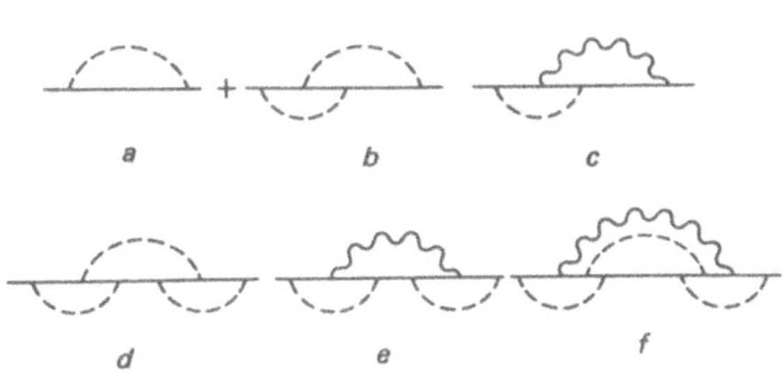


**Figure 3.8. The simplest diagrams for the first term in the formula (3.87) for the self-energy part of the electron Green function.**

**Figure 3.9. Skeleton diagrams for the self-energy part, Σ, of the electron Green function. The circles and open and crossed rectangles denote, respectively, vertices with different numbers of Coulomb-interaction lines going out of them.**

c and e in Fig. 3.8 are not small compared to the phonon contribution arising from the second term and, above all, they are also localized in the vicinity of the Fermi surface.

The whole sum of the diagrams arising from the first term with allowance for a single phonon line in the vertex function and from the second term can be reduced to the skeleton diagrams shown in Fig. 3.9.

It can be shown (this has been done for $b$ by Batyev and Pokrovskii [32)] that $b$ and $c$ are smaller than the direct Coulomb diagram shown in Fig. 3.8a in the ratio $(m/M)^{1/2}$; they have the same energy dependence possessed by the simple Coulomb diagram, i.e., they are not important in so far as the determination of the electron spectrum is concerned.

Taking all of the foregoing into account, we can rewrite the schematic equation (3.87) for the self-energy part of the electron Green function in the following form:

$$\Sigma(x, x') = -i\int \widetilde{V}_{ee} G \Gamma_0 - i\int \nabla_\alpha \widetilde{V}_{ei} \Gamma_0 D^{\alpha\beta} \nabla_\beta \widetilde{V}_{ei} \Gamma_0 G. \tag{3.88}$$

Let us now consider the methods of computing the electronic properties of metals on the basis of the above-presented equations for the electron Green function (3.79) and the self-energy part $\Sigma(x, x')$ (3.88). We shall also discuss briefly the changes that allowance for the crystalline structure of metals effects in the condition for lattice stability obtained in the preceding section.

Proceeding to the investigation of the electronic properties of metals, first, we note that the conventional band-structure picture obtained for the electronic spectrum on the basis of an analysis of the single-particle Hamiltonian with a periodic external field clearly loses its meaning when a consistent allowance is made for the interelectron Coulomb interaction. Because the single-particle excitations in an interacting system are damped, we cannot even introduce such quantities as the function $E_{\mathbf{k}}$ for the entire bulk of electronic states lying far from the Fermi surface. The consistent determination of the electronic spectrum is possible only for the quasiparticle excitations close to the Fermi surface. We can, in accordance with Eqs. (3.79) and (3.88), write down the equation for the determination of this spectrum:

$$\left[-\frac{\nabla^2}{2} - \mu + \sum_n V_{ei}(\mathbf{r} - \mathbf{R}_n^0) + U_{\mathrm{H}}(\mathbf{r})\right] f_l(\mathbf{r}) + \\ + \int d^3r'' \Sigma(\mathbf{r}, \mathbf{r}'', E_l) f_l(\mathbf{r}'') = E_l f_l(\mathbf{r}), \tag{3.89}$$

Here the $f_i$ are electron wave functions and $U_H(\mathbf{r})$ is the normal Hartree potential for the electrons, which is determined by the Eqs. (3.80).

Because of the presence of an imaginary part in the quantity $\Sigma(\mathbf{r}, \mathbf{r}'', E_i)$, the eigenfunctions, $f_i(\mathbf{r})$, of Eq. (3.89) do not, in the general case, possess the properties of the normal Bloch wave functions. First, the $f_i(\mathbf{r})$ are not orthonormalized functions, which greatly complicates all of computations based on their use.

Because of the strong damping of the electrons near the Fermi surface as a result of the electron-phonon interaction, it is more convenient to split the problem of determining the electronic structure of a metal into two parts. To begin with, we should compute the Green function and the excitation spectrum of the electrons, allowing only for the interaction of the electrons with the rigid lattice and their Coulomb interaction with each other. In this case it is necessary to solve Eq. (3.89), using as the function $\Sigma(\mathbf{r}, \mathbf{r}'', E_i)$ only the first term of the expression (3.88). Naturally, the function $\Sigma(\mathbf{r}, \mathbf{r}'', E_i)$ should be determined in a self-consistent fashion. After this, it is necessary to take into account the changes that arise in the electronic spectrum as a result of the electron-phonon interaction.

Detailed results of such calculations for a number of model systems with the use of the simplest approximations for the self-energy part $\Sigma(\mathbf{r}, \mathbf{r}'', E_i)$ are given in Hedin and Lundqvist's review article [47], and we shall not touch upon them here. Let us only note that it has recently been shown [48, 49] that such an approach, even if for $\Sigma(\mathbf{r}, \mathbf{r}'', E_i)$ we use an approximation that completely neglects the vertex functions in Eq. (3.88), leads to a good description of the Fermi surface in such metals as lithium and aluminum.

To determine the exact changes caused by the electron-phonon interaction, it is necessary to compute not only the excitation spectrum $E_i$, but also the functions $f_i(\mathbf{r})$. Unfortunately, such calculations have thus far not been carried out. All this does not allow us to answer within the framework of the present approach the following very important question: How is the potential determining the electronic band structure related to the potential whose gradient enters into the matrix element of the electron-phonon interaction? As has already been noted, these potentials are assumed to be the same in the Fröhlich model.

Calculations of the electron-phonon interaction constant in metals have recently been published [50-53] that are based to a very large extent on the above-indicated property of the Fröhlich model. In these calculations the authors use for the description of the electron-phonon interaction a model potential whose parameters are determined by fitting data on the shape of the electron Fermi surface. If, on the other hand, we proceed from Eq. (3.89), which determines the band spectrum of the quasiparticle excitations, and Eq. (3.88) for the self-energy part $\Sigma(x, x')$, according to which the matrix element of the electron-phonon interaction is related with the gradient of the potential

$$\widetilde{\widetilde{V}}_{ei} \approx \Gamma_0 \widetilde{V}_{ei},$$

then, generally speaking, it seems impossible to establish any connection between these potentials.

In order to gain an understanding of this problem, we use the method proposed for the isotropic system in [24, 54]. Let us limit ourselves to the study of a system of quasifree electrons with a weak electron-ion interaction, which allows us to take it into account within the framework of perturbation theory. To begin with, let us compute the Green function for an interacting homogeneous electron gas:

$$G_c(k, \omega) = \frac{1}{\omega - E_k - \Sigma_c(k, \omega)}. \tag{3.90}$$

Expanding the function $\Sigma_c(k, \omega)$ for small $\omega$ and $k$ of the order of $k_F$:

$$\Sigma_c(k, \omega) = \Sigma_c + E_k \frac{1}{v_F} \frac{\partial \Sigma_c}{\partial k} + \omega \frac{\partial \Sigma_c}{\partial \omega}, \tag{3.91}$$

we can write the Green function $G_c(k, \omega)$ in the vicinity of the Fermi surface in the form

$$G_e(k, \omega) = \frac{1}{Z_c} \frac{1}{\omega - \widetilde{E}_k + i\delta}. \tag{3.92}$$

In this expression $\widetilde{E}_{\mathbf{k}}$ is the renormalized quasiparticle energy:

$$\widetilde{E}_k = E_k Z_c^{-1} \left(1 + \frac{1}{v_F} \frac{\partial \Sigma_c}{\partial k}\right)_{\substack{k=k_F,\\ \omega=0}}. \tag{3.93}$$

Introducing the renormalized mass $m^*$, let us write:

$$\widetilde{E}_k = \frac{m}{m^*} E_k.$$

The quantity $Z_c$ is the renormalization factor, which indicates the change occurring in the magnitude of the residue at the pole of the Green function as a result of the interelectron interaction:

$$Z_c = \left(1 - \frac{\partial \Sigma_c}{\partial \omega}\right)_{\substack{k=k_F,\\ \omega=0}}. \tag{3.94}$$

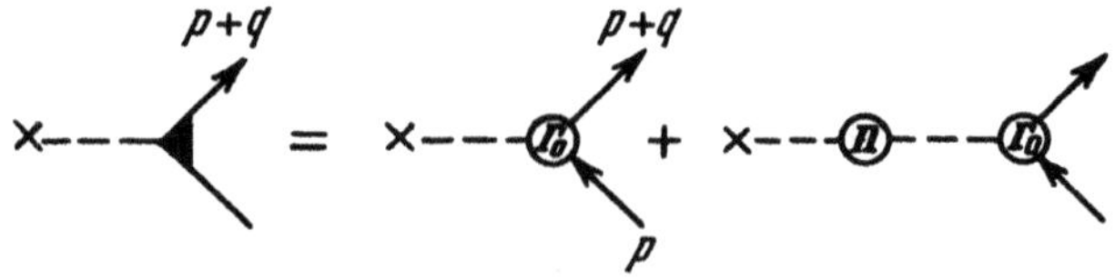


**Figure 3.10. The diagram equation for the effective electron interaction.**

The effective electron-ion interaction can be represented by the diagrams shown in Fig. 3.10 and can be written in the form

$$\tilde{\tilde{V}}_{ei} = \frac{V_{ei}(q)}{\varepsilon_e(q, 0)} \Gamma_0(k, q). \tag{3.95}$$

Here $V_{ei}(q)$ is the matrix element of the bare, unscreened electron-ion interaction and $\Gamma_0(k, q)$ is the vertex function of the electron-electron Coulomb inteaction. In the second-order perturbation theory in terms of the $\tilde{\tilde{V}}_{ei}$ for the electrons near the Fermi surface, we can write for the Green function $\tilde{G}(\mathbf{k}, \omega)$ the equation

$$\tilde{G}(\mathbf{k}, \omega) = G_c(\mathbf{k}, \omega) + G_c(\mathbf{k}, \omega) \sum_{\mathbf{G}} \left|\tilde{\tilde{V}}_{ei}(k_F, \mathbf{G})\right|^2 \times G_c(\mathbf{k}+\mathbf{G}) \tilde{G}(\mathbf{k}, \omega), \tag{3.96}$$

or

$$\tilde{G}(\mathbf{k}, \omega) = \frac{1}{G_c^{-1}(\mathbf{k}, \omega) - \sum_{\mathbf{G}} \left|\tilde{\tilde{V}}_{ei}(k_F, \mathbf{G})\right|^2 G_0(\mathbf{k}+\mathbf{G}, \omega)}, \tag{3.97}$$

where $\mathbf{G}$ is a reciprocal-lattice vector.

From Eq. (3.97) we obtain a condition determining the band spectrum of the electrons:

$$(\omega - \tilde{E}_{\mathbf{k}})(\omega - \tilde{E}_{\mathbf{k}+\mathbf{G}}) - \left|\tilde{\tilde{V}}_{ei}(k_F, G)\right|^2 / Z_c^2 = 0. \tag{3.98}$$

In writing down (3.98), we limited ourselves to the consideration of one of the reciprocal-lattice vectors.

As can be seen from Eq. (3.98), the periodic potential, $\tilde{\tilde{V}}_{ei}$, determining the band structure of the electrons has the form

$$\tilde{\tilde{V}}_{ei} = \frac{V_{ei}(q)}{\varepsilon_e(q, 0)} \frac{\Gamma_0(k_F, q)}{Z_c}. \tag{3.99}$$

Luttinger and Nozières, [55] have obtained for the system in question the Ward identity, which allows to show that

$$\lim_{q \to 0} \frac{V_{ei}(q)}{\varepsilon_e(q, 0)} \frac{\Gamma_0(k_F, q)}{Z_c} = \frac{V_{ei}(q)}{\varepsilon_e(q, 0)} = -\frac{Z}{2N(0)}, \tag{3.100}$$

where $N(0)$ is the density of electron states.

As can now be verified from a comparison of the formulas (3.88) and (3.89), the periodic potential, $\tilde{\tilde{V}}_{ei}$, figures in the definitions of the band structure and the matrix potential of the electron-phonon interaction. In this sense, Bloch's idea, expressed in the Fröhlich model, that these potentials coincide is justified to some extent. Let us, nevertheless, emphasize that the deviation of the electron band spectrum from the free-electron spectrum $E_{\mathbf{k}} = (k^2/2m) - \mu$ is connected not only with the potential $\tilde{\tilde{V}}_{ei}$; As can be seen from the formula (3.98), the interelectron Coulomb interaction also makes a direct contribution to the band energy, which amounts to the replacement of the quantity $E_{\mathbf{k}}$ by the renormalized energy $\tilde{E}_{\mathbf{k}}$.

In simple metals, as calculations by Rice [56] have shown, this renormalization is small. Although the Coulomb interaction in metals is itself not weak, there occurs in the determination of $\tilde{E}_{\mathbf{k}}$ in accordance with the formula (3.93) a substantial compensation of the quantities $Z_c^{-1}$ and

$$\left(1 + \frac{1}{v_F}\frac{\partial \Sigma_c}{\partial k}\right)_{k=k_F}.$$

Therefore, in simple metals, the potential $\tilde{\tilde{V}}_{ei}$, as determined from the formula (3.98) with the energy $\tilde{E}_{\mathbf{k}}$ replaced by the free-electron energy $E_{\mathbf{k}}$, is not very different from its true value. This is the explanation for the success of Allen's [50] and Das's [51] calculations of the electron-phonon interaction constant, which are based precisely on this definition of the potential $\tilde{\tilde{V}}_{ei}$.

The effect of the Coulomb interaction on the electronic spectrum of the transition metals has not been studied theoretically. There is no reason whatsoever to expect *a priori* that the effect of this interaction will be weak. All the more interesting are the results obtained in [52, 53], where calculations of the electron-phonon coupling constant for the transition metals are carried out in the same fashion, and a good agreement with the experimental data is obtained.

We shall not at the moment consider in detail the computation of the changes that occur in the electron spectrum as a result of the electron-phonon interaction. This will be done in the following chapter. Right now, let us briefly discuss certain problems concerning the lattice dynamics of metals.

With that end in view, let us rewrite the expression for the dynamical vibration matrix that follows from (3.75)-(3.77) [see also (3.33)] in the form

$$\Phi_{\alpha\beta}(\mathbf{q}) = \frac{4\pi Z^2 e^2}{\Omega_0} \sum_{\mathbf{G}} \frac{(\mathbf{q}+\mathbf{G})_\alpha (\mathbf{q}+\mathbf{G})_\beta}{|\mathbf{q}+\mathbf{G}|^2} +$$
$$+ \frac{1}{\Omega_0} \sum_{\mathbf{G},\mathbf{G}'} (\mathbf{q}+\mathbf{G})_\alpha V_{ei}(\mathbf{q}+\mathbf{G}) (\mathbf{q}+\mathbf{G}')_\beta V_{ei}(\mathbf{q}+\mathbf{G}') \times$$
$$\times \frac{|\mathbf{q}+\mathbf{G}|^2}{4\pi e^2} \left[\varepsilon_e^{-1}(\mathbf{q}+\mathbf{G}, \mathbf{q}+\mathbf{G}') - \delta_{\mathbf{G}\mathbf{G}'}\right]. \qquad (3.101)$$

In writing down this formula, we have used as the ion-ion interaction potential the purely Coulomb potential and have expressed the electronic susceptibility in terms of the inverse permittivity matrix $\varepsilon_e^{-1}(\mathbf{q}+\mathbf{G}, \mathbf{q}+\mathbf{G})$. If the electron–ion interaction is also assumed to be a purely Coulomb interaction, then the dynamical vibration matrix can be written in the following very graphic form:

$$\Phi_{\alpha\beta}(\mathbf{q}) = \frac{4\pi Z^2 e^2}{\Omega_0} \sum_{\mathbf{G},\mathbf{G}'} \frac{(\mathbf{q}+\mathbf{G})_\alpha (\mathbf{q}+\mathbf{G}')_\beta}{|\mathbf{q}+\mathbf{G}|^2} \varepsilon_e^{-1}(\mathbf{q}+\mathbf{G}, \mathbf{q}+\mathbf{G}'). \quad (3.102)$$

It can be seen from (3.102) that the dynamical matrix is the second derivative of the screened inter-ion Coulomb interaction. In an isotropic system the permittivity at zero frequency is a function only of the momentum transfer, and therefore the effective screened inter-ion interaction is a central pair interaction, i.e., it depends only on the distances between the ions.

The presence of singularities in the permittivity leads to the appearance of long-range inter-ion interactions. In the isotropic three-dimensional case these are the well-known Friedel oscillations in the inter-ion interaction potential. The presence of such singularities in the inter-ion interaction potential in turn leads to singularities in the phonon spectra. In the three-dimensional case for almost free electrons, these singularities are weak and manifest themselves only in the fine structure of the phonon spectra (as Kohn anomalies). In the quasi-one-dimensional case the singularity in the permittivity can, as has already been discussed in the preceding section (see also Chapter 7), lead to drastic changes in the phonon spectrum.

In a crystal, the permittivity is a matrix in the reciprocal-lattice vectors. As has been shown in a number of papers [57, 58], allowance for the off-diagonal elements of $\varepsilon_e^{-1}(\mathbf{q}+\mathbf{G}, \mathbf{q}+\mathbf{G}')$ with $\mathbf{G} \neq \mathbf{G}'$ corresponds to the consideration of the corrections for the deviation of the local field in the crystal from the mean field. The corrections are fairly small in the simple metals. But even in them, allowance for the off-diagonal elements of $\varepsilon_e^{-1}(\mathbf{q}+\mathbf{G}, \mathbf{q}+\mathbf{G}')$ leads to the appearance of effective many-ion noncentral interactions. These interactions have a very considerable effect on the lattice dynamics of many multivalent metals, in particular, those like lead and tin [20]. Without allowance for such interactions, the lattice of tin would simply be unstable.

Even more important is the consideration of these off-diagonal matrix elements in the transition metals, in which the conduction electrons include both almost free *s* electrons and strongly coupled *d* electrons. The contribution of the latter electrons to the off-diagonal elements of $\varepsilon_e^{-1}(\mathbf{q}+\mathbf{G}, \mathbf{q}+\mathbf{G}')$ can be very considerable, especially in the presence of a partially covalent or an ionic bond, as obtains in a number of compounds of the transition metals.

Papers have recently been published [23, 59] in which methods for computing the function $\varepsilon_e(\mathbf{q}+\mathbf{G}, \mathbf{q}+\mathbf{G}')$ are proposed that allow us to carry out easily the inversion of this matrix and determine the matrix, $\varepsilon_e^{-1}(\mathbf{q}+\mathbf{G}, \mathbf{q}+\mathbf{G}')$, necessary for the computation of the phonon frequencies. Let us, as usual [23], write the matrix $\varepsilon_e(\mathbf{q}+\mathbf{G}, \mathbf{q}+\mathbf{G}')$ in the following form:

$$\varepsilon_e(\mathbf{q}+\mathbf{G},\mathbf{q}+\mathbf{G}') = \delta_{\mathbf{G}\mathbf{G}'} - V(\mathbf{q}+\mathbf{G})\,\Pi(\mathbf{q}+\mathbf{G},\mathbf{q}+\mathbf{G}'), \tag{3.103}$$

where

$$V(\mathbf{q}+\mathbf{G}) = \frac{4\pi e^2}{|\mathbf{q}+\mathbf{G}|^2},$$

$\Pi(\mathbf{q}+\mathbf{G},\mathbf{q}+\mathbf{G}')$ is the generalized polarization operator of the electrons, which, in the simplest approximation, can be written as

$$\Pi(\mathbf{q}+\mathbf{G},\mathbf{q}+\mathbf{G}') = \sum_{\substack{n,n'\\ \mathbf{k},\mathbf{k}}} \frac{f(n,\mathbf{k})-f(n',\mathbf{k}')}{E_n(\mathbf{k})-E_{n'}(\mathbf{k}')}\times$$
$$\times\langle n,\mathbf{k}|e^{-i(\mathbf{q}+\mathbf{G})\mathbf{r}}|n',\mathbf{k}'\rangle\langle n',\mathbf{k}'|e^{i(\mathbf{q}+\mathbf{G}')\mathbf{r}}|n,\mathbf{k}\rangle. \tag{3.104}$$

Here $f(n, k)$ is the occupation number for the electrons in a state with a given momentum $\mathbf{k}$ in the $n$th band, $E_n(\mathbf{k})$ is the energy of the Bloch electrons, and $\langle n,\mathbf{k}|e^{-i(\mathbf{q}+\mathbf{G})\mathbf{r}}|n',\mathbf{k}'\rangle$ is a matrix element, computed with the wave functions of the Bloch electrons.

By using the expansion of the Bloch wave functions $\psi_{n\mathbf{k}}(\mathbf{r})$ in terms of the localized Wannier functions $\Phi_n(\mathbf{r}-\mathbf{R}^l)$,

$$\psi_{n\mathbf{k}}(\mathbf{r}) = \frac{1}{\sqrt{N}}\sum_{\mathbf{R}^l} e^{i\mathbf{k}\mathbf{R}^l}\Phi_n(\mathbf{r}-\mathbf{R}^l),$$

we can represent the polarization operator of the electrons in a form that is factorized with respect to the vectors $\mathbf{q}+\mathbf{G}$ and $\mathbf{q}+\mathbf{G}'$:

$$\Pi(\mathbf{q}+\mathbf{G},\mathbf{q}+\mathbf{G}') = \sum_{\substack{n,n'\\ l,l'}} A^l_{nn'}(\mathbf{q}+\mathbf{G})\,N^{ll'}_{\substack{nn'\\ nn'}}(\mathbf{q})\,A^{l'*}_{nn'}(\mathbf{q}+\mathbf{G}'). \tag{3.104a}$$

Here

$$A^l_{nn'} = \int_{\Omega_0}\Phi^*_n(\mathbf{r})\,e^{-i(\mathbf{q}+\mathbf{G})\mathbf{r}}\Phi_{n'}(\mathbf{r}+\mathbf{R}^l)\,d^3r \tag{3.105}$$

and

$$N^{ll'}_{\substack{nn'\\ nn'}}(\mathbf{q}) = \sum_{k}\frac{f(n,\mathbf{k})-f(n',\mathbf{k}')}{E_n(\mathbf{k})-E_{n'}(\mathbf{k}')}\,e^{-i(\mathbf{k}+\mathbf{q})(\mathbf{R}^l-\mathbf{R}^{l'})}. \tag{3.106}$$

In matrix form, the expression for $\hat{\varepsilon}$ can be written in the form

$$\hat{\varepsilon}_e = 1 - A\hat{N}A^+. \tag{3.107}$$

Because the factorization of the matrix $\Pi(\mathbf{q}+\mathbf{G}, \mathbf{q}+\mathbf{G}')$, formally, the inversion of $\hat{\varepsilon}_e$ is extremely simple, and the matrix $\hat{\varepsilon}_e^{-1}$ can be written in the form

$$\hat{\varepsilon}_e^{-1}(\mathbf{q}+\mathbf{G}, \mathbf{q}+\mathbf{G}') = \\ = \delta_{\mathbf{G}\mathbf{G}'} + \sum_{\substack{m,m'\\ n,n'\\ l,l'}} A^{l}_{mm'}(\mathbf{q}+\mathbf{G})\, S^{ll'}_{\substack{mm'\\ nn'}}(\mathbf{q})\, A^{l'*}_{nn'}(\mathbf{q}+\mathbf{G}'), \tag{3.108}$$

where the matrix $\hat{S}$ is defined as

$$\hat{S} = (\hat{N}^{-1} - \hat{V})^{-1} \tag{3.109}$$

and

$$V^{ll'}_{\substack{mm'\\ nn'}} = \sum_{G''} A^{l*}_{mm'}(\mathbf{q}+\mathbf{G}'')\, A^{l'}_{nn'}(\mathbf{q}+\mathbf{G}'')\, V(\mathbf{q}+\mathbf{G}''). \tag{3.110}$$

As can be seen from the obtained formulas, the possibility of actually accomplishing this inversion operation depends significantly on the dimensionality of the matrix $\hat{N}$. It is clear from the formulas (3.105) and (3.106) that the dimensionality of $\hat{N}$ is determined by the number of Bloch bands that must be taken into account in the computation of the electronic properties of the metal, as well as by the number of lattice vectors, $\mathbf{R}^l$, $\mathbf{R}^{l'}$, determining the quantity $A^l_{nn'}(\mathbf{q}+\mathbf{G})$. It is clear that, because of the energy denominators in the formula (3.106), we can, in computing the matrix $\hat{N}$, limit ourselves to a few low-lying Bloch bands.

On the other hand, because of the good localization of the Wannier function, it is sufficient to allow for only the overlap of the nearest neighbors in the formula (3.105).

With metals, due particularly to the presence in them of almost free $s$ electrons, it is more convenient from the very outset to split the permittivity matrix into a diagonal part, stemming from the intraband transitions in the unfilled $s$ bands, and into a nondiagonal part, connected with the contribution of the well-localized $d$ electrons:

$$\hat{\varepsilon}_e(\mathbf{q}+\mathbf{G}, \mathbf{q}+\mathbf{G}') = \varepsilon_0(\mathbf{q}+\mathbf{G})\delta_{GG'} - \\ -V(\mathbf{q}+\mathbf{G}') \sum_{\substack{m,m'\\ l,l'}} A^{l}_{mm'}(\mathbf{q}+\mathbf{G})\, N^{ll'}_{\substack{mm'\\ mm'}}(\mathbf{q})\, A^{l'*}_{mm'}(\mathbf{q}+\mathbf{G}'). \tag{3.111}$$

Then the inverse permittivity matrix can be written in the form

$$\hat{\varepsilon}_e^{-1}(\mathbf{q}+\mathbf{G},\mathbf{q}+\mathbf{G}') = \frac{1}{\varepsilon_0(\mathbf{q}+\mathbf{G})}\Bigg[\delta_{GG'} + \frac{V(\mathbf{q}+\mathbf{G})}{\varepsilon_0(\mathbf{q}+\mathbf{G})} \times \times \sum_{\substack{l,l'\\ n,n'\\ m,m'}} A^l_{mm'}(\mathbf{q}+\mathbf{G})\, S^{ll'}_{\substack{mm'\\ nn'}}(\mathbf{q})\, A^{l'*}_{nn'}(\mathbf{q}+\mathbf{G}')\Bigg], \tag{3.112}$$

where the matrix $S$ is given, as before, by the formula (3.109), while the matrix $\hat{V}$ has the form

$$V^{ll'}_{\substack{mm'\\ nn'}}(\mathbf{q}) = \sum_{G''} A^l_{mm'}(\mathbf{q}+\mathbf{G}'')\, A^{l'*}_{nn'}(\mathbf{q}+\mathbf{G}'')\,\frac{V(\mathbf{q}+\mathbf{G}'')}{\varepsilon_0(\mathbf{q}+\mathbf{G}'')}. \tag{3.113}$$

Knowing now the expression for the matrix $\hat{\varepsilon}^{-1}$, we can, in accordance with the formula (3.102), write down the complete dynamical vibration matrix:

$$\begin{aligned}\Phi_{\alpha\beta}(\mathbf{q}) &= \frac{4\pi Z^2 e^2}{\Omega_0}\sum_G \frac{(\mathbf{q}+\mathbf{G})_\alpha(\mathbf{q}+\mathbf{G})_\beta}{|\mathbf{q}+\mathbf{G}|^2} + \\ &+ \frac{1}{\Omega_0}\sum_G (\mathbf{q}+\mathbf{G})_\alpha V_{ei}(\mathbf{q}+\mathbf{G})(\mathbf{q}+\mathbf{G})_\beta V_{ei}(\mathbf{q}+\mathbf{G}) \times \\ &\qquad \times \frac{|\mathbf{q}+\mathbf{G}|^2}{4\pi e^2}\left[\varepsilon_0^{-1}(\mathbf{q}+\mathbf{G}) - \delta_{GG'}\right] + \\ &+ \frac{1}{\Omega_0}\sum_{\substack{G,G'\\ nn'\\ mm'\\ ll'}} (\mathbf{q}+\mathbf{G})_\alpha \frac{V_{ei}(\mathbf{q}+\mathbf{G})}{\varepsilon_0(\mathbf{q}+\mathbf{G})}(\mathbf{q}+\mathbf{G}')_\beta \frac{V_{ei}(\mathbf{q}+\mathbf{G}')}{\varepsilon_0(\mathbf{q}+\mathbf{G}')} \times \\ &\qquad \times A^l_{nn'}(\mathbf{q}+\mathbf{G})\, S^{ll'}_{\substack{mm'\\ nn'}}(\mathbf{q})\, A^{l*}_{mm'}(\mathbf{q}+\mathbf{G}'),\end{aligned} \tag{3.114}$$

or in matrix form

$$\hat{\Phi} = \hat{C} + \hat{E} + \sum_{s,s'} U(s)\,\hat{S}(s,s')\,U^+(s'). \tag{3.115}$$

In this formula we have denoted the indices $(l, n, n')$ by the letter $s$ and introduced the quantity

$$U_\alpha(s) = \sum_G (\mathbf{q}+\mathbf{G})_\alpha A^l_{nn'}(\mathbf{q}+\mathbf{G})\,\frac{V_{ei}(\mathbf{q}+\mathbf{G})}{\varepsilon_0(\mathbf{q}+\mathbf{G})}. \tag{3.116}$$

In accordance with the formula (3.115), the dynamical vibration matrix is determined by the inter-ion Coulomb interaction matrix $\hat{C}$ and the matrix $\hat{E}$ that describes the screening of the inter-ion Coulomb interaction as a result of the presence of the almost free electrons. The third term in the dynamical vibration matrix arises on account of the contribution of the distributed dipoles connected with the highly localized electrons of the filled bands.

In simple metals only the contributions from the matrices $\hat{C}$ and $\hat{E}$ are normally taken into account, although, as has already been noted, this is clearly inadequate for the multivalent metals. In dielectrics, in which $\varepsilon_0(\mathbf{q} + \mathbf{G}) = 1$, only the first and third terms in the formula (3.115) contribute to the vibration matrix. In the covalent semiconductors and transition metals the contributions from the three terms to the dynamical vibration matrix are all very important.

In many transition metals considerable concellation of the contributions to the lattice dynamics from the diagonal and off-diagonal elements of the matrix $\hat{\varepsilon}_e^{-1}(\mathbf{q} + \mathbf{G}, \mathbf{q} + \mathbf{G}')$ occurs, so that the characteristic phonon frequencies of these metals are much closer to the frequencies of the purely ionic lattice defined by the matrix $\hat{C}$ than is the case in the simple multivalent metals.

The problem of the anomalies of the phonon spectra in the transition metals and their compounds has in the last few years become one of the most pressing and often discussed problems in solid-state physics (see [33, 60-68]). The most widely accepted idea about the nature of these anomalies is connected with the possibility of the appearance of charge-density waves (CDW) in the electron subsystem of the metal. To such a possibility corresponds the vanishing of the denominator in the matrix $\hat{S} = (\hat{N}^{-1} - \hat{V})^{-1}$, which has the form

$$N^{-1}(q) - A(q)\,\frac{V(q)}{\varepsilon_0(q)}\,A^*(q). \tag{3.117}$$

It is natural that matters do not, in fact, get to the point where CDW actually appear in the electron subsystem. As can be seen from the formulas (3.114) and (3.115), as the electron subsystem approaches the point of instability, the electron contribution to the dynamical vibration matrix increases, and the corresponding phonon frequencies soften drastically or even vanish. This approach to the explanation of the anomalies in the phonon spectra and of structural transitions has been most systematically developed by Sinha [60, 63], Hanke [65], and a number of other investigators (see [63]).

There is another point of view regarding the nature of the anomalies in the phonon spectra in metals. It is also connected with the influence of the electron subsystem on the phonon frequencies. In this case, however, the main attention is focused not on the properties of the electron subsystem alone, a subsystem that may not have any tendency whatsoever toward instability, but on the properties of the electron-phonon interaction matrix elements. According to the results of calculations carried out in [66-68], the anomalies in the phonon spectra in pure niobium, as well as in a number of niobium compounds, are due to the large magnitudes of the electron-phonon interaction matrix elements. In these cases the expression (3.117) is not at all small, and there is no tendency toward the appearance of CDW in the electron subsystem alone.

On the whole, this problem is still very far from the final solution, and, therefore, we limit ourselves here to only some remarks apropos of the two above-indicated approaches to the elucidation of the nature of the anomalies in the phonon spectra. The microscopic computations of the function $N(q)$ for Nb [66, 69], Sc [70], and a number of other metals [69] do not reveal any serious anomalies that could be related to the experimentally observed anomalies in the phonon spectra in these metals. This circumstance would seem to support the idea that the electron–phonon interaction matrix elements are important for the description of the nature of these anomalies. But then there arises the important question of the correct computation of these matrix elements since the calculations in [66–68] were, in fact, carried out within the framework of the Fröhlich model. As has repeatedly been pointed out in this book, such an approach to the problem of the phonon spectra is certainly not a consistent one.

In view of this, the consistent allowance for the off-diagonal elements of the matrix $\varepsilon^{-1}(\mathbf{q}+\mathbf{G}, \mathbf{q}+\mathbf{G}', 0)$, which are normally described as local-field effects, acquires great significance in the analysis of the nature of the anomalies in the phonon spectra. It should be noted that, unfortunately, virtually no investigation has so far been carried out on the role of the electron correlations, which are extremely important precisely in the problem of local-field effects [71].

# 4

# Superconductivity in Three-Dimensional Quasiisotropic Systems

## 1. Introduction

Various proposals concerning the synthesis and investigation of special structures (quasi-one-dimensional metallic systems, systems of the "sandwich" type, etc.), as well as new nonphonon mechanisms of superconductivity in such systems have been put forward in discussions of the problem of high-temperature superconductivity. We cannot, however, regard as having been exhausted the more traditional way of raising the critical temperature, $T_c$, of superconductors, namely, the synthesis of ordinary quasi-isotropic three-dimensional systems (metals, alloys, compounds), in which the phonon mechanism of interelectron attraction is, as a rule, responsible for superconductivity. The record $T_c$ values that we have now ($T_c = 23.2$ K in $Nb_3Ge$) were attained in precisely this way. The extrapolation of the presently available data allows us to hope for a value of $T_c \approx 30$ K in the A-15 structure compound $Nb_3Si$. In principle, higher critical-temperature values are possible in materials with the phonon mechanism of superconductivity.

From this point of view, the analysis of such systems with the object of finding out how their superconducting characteristics depend on the properties of the normal metal and what causes limit the value of the critical temperature is of great interest. The investigation of these problems may also turn out to be more generally useful. For one thing, we can, on the basis of the considerable factual data that have been accumulated for three-dimensional systems with the normal electron–phonon interaction, draw some general conclusions that will also be applicable to nonphonon mechanisms of interelectron attraction. For another, since in any system, and for any mechanism of superconductivity, the electrons inevitably interact with the lattice, the possible limitations on the $T_c$ value that are connected with the electron–phonon interaction may turn out to be important in these cases also. Moreover, many general relations connecting the $T_c$ value, the coupling constant $\lambda$, and the electronic and phonon characteristics of the metal (such characteristics as, for example, the density of states at the Fermi surface and the characteristic phonon frequencies) are easily carried over to the case of the electron–exciton interaction.

The investigation of superconductivity in three-dimensional systems in which the phonon mechanism of interelectron attraction predominates is the purpose of the present chapter. We shall, to begin with, derive the basic equations of the theory of superconductivity in a quasi-isotropic three-dimensional system of interacting electrons. We shall, in deriving these equations, endeavor to consider exactly not only the electron–phonon interaction but also the interelectron Coulomb repulsion. Unfortunately, a rigorous quantitative analysis of the obtained equations is impossible because of the absence of methods of computing the Coulomb interaction at the electron densities existing in real metals.

It was shown in Chapter 2 that in the weak-coupling approximation, it is not the Coulomb-interaction constant, $\mu$, itself that enters into the superconducting characteristics of a metal but a significantly smaller quantity: the pseudopotential $\mu^*$:

$$\mu^* = \frac{\mu}{1 + \mu \ln\left(E_F/\overline{\omega}\right)}. \tag{4.1}$$

Here $\overline{\omega}$ is the characteristic phonon frequency. It turns out that, in the case of strong coupling, it is impossible to obtain a concrete expression for the quantities $\mu$ and $\mu^*$, but it can be shown that, as before, the role of the direct interelectron Coulomb repulsion is suppressed because of the appreciable difference between the magnitudes of the electron energy and the characteristic phonon energies.

In the majority of superconductors, especially those with high $T_c$ values, the electron–phonon coupling constant $\lambda$ is, on the average, of the order of unity. As McMillan's analysis [1] of the experimental data shows, the Coulomb constant $\mu$ and, consequently, the pseudopotential $\mu^*$ in these materials are certainly less than $\lambda$. Thus, in the transition metals, $\mu^*$ is, on the average, of the order of 0.1–0.2. Because of the relatively small contribution of the direct interelectron Coulomb interaction as compared to the contribution of the electron–phonon interaction, it is the normal practice in the theory of superconductivity to consider a simplified model for the interelectron Coulomb interaction, specifically, the model in which this interaction is described by some constant.

Within the framework of such a model, it is, as has been shown by McMillan [1], possible to express the critical temperature of the superconducting transition directly in terms of the characteristic phonon frequencies $\omega_D$, the electron-phonon coupling constant $\lambda$, and the pseudopotential $\mu^*$ [see (1.32)]:

$$T_c = \frac{\omega_D}{1.45} \exp\left[-\frac{1{,}04\,(1+\lambda)}{\lambda - \mu^*\,(1+0.62\lambda)}\right]. \tag{4.2}$$

Concrete values were obtained in [1] for the numerical coefficients in this formula from a numerical solution of the superconductivity equations for a system with quite a specific phonon spectrum. A number of attempts were later made [2-5] to improve the formula (4.2) and make it more universal. In this chapter we analyze the superconductivity equations on the basis of the results obtained in [6] with the object of making the derivation of the formula (4.2) itself more rigorous and the meaning of the numerical coefficients appearing in this formula more precise. We shall satisfy ourselves that the values of these coefficients are connected with the form of the phonon spectrum and can vary from metal to metal as this spectrum is varied.

Next, we shall derive, largely on the basis of the results presented in the preceding chapter, formulas for the electron-phonon coupling constant $\lambda$. On their basis, we shall study the dependence of this quantity on the electronic characteristics of a metal and on the properties of the phonon subsystem. In the process, we shall also discuss the empirical laws found as a result of an analysis of the experimental data to govern the variation of the constant $\lambda$ as the mean valency or the structure of metals and alloys is varied. We shall atempt to establish the possibility of a theoretical explanation of some of these laws. We shall also consider in detail the correlation between superconductivity with high $T_c$ values and lattice instability. In conclusion, we shall briefly discuss the problems connected with the possibility of realizing the nonphonon mechanism of superconductivity in three-dimensional systems.

On the whole, the investigation carried out in this chapter shows that the existing theoretical methods provide, at least for an electron-phonon system, an entirely satisfactory description of superconductivity. They allow a qualitative, and often a quantitative, determination of which of the characteristics of the normal metal the critical temperature of the superconducting transition depends on and how it depends on them. Thus, the complexity of the prediction of the highest $T_c$ values for metals and alloys and of the actual production of such materials is connected not with the specific nature of the phenomenon of superconductivity itself, but with the imperfect understanding at present of the properties of normal metals. The stability of one or another structure and the feasibility of the production of alloys and compounds with prescribed parameters are all problems whose proper study are only now beginning.

## 2. Description of the Superconducting State in a Quasiisotropic Three-Dimensional System

As is well known, the superconducting state is characterized by the appearance in the system of a coherent condensate of Cooper pairs and related anomalous mean quantities of the type $\langle\psi_\uparrow^\dagger\psi_\downarrow^\dagger\rangle$, where $\psi_\uparrow^\dagger$ and $\psi_\downarrow^\dagger$ are the creation operators for electrons with oppositely directed spins. The equations for the electron and phonon Green functions that take account of the existence of the anomalous means were first derived by Éliashberg [7] within the framework of the Fröhlich model.

As we have noted, the Fröhlich model is, for a number of reasons, not very suitable for the investigation of such questions as the dependence of the electron-phonon coupling constant on the properties of the normal metal, the correlation between lattice stability and the strength of the electron-phonon interaction, etc. The rigorous investigation of all these questions would require the generalization, with allowance for the existence of the anomalous means, of the exact self-consistent equations obtained for the Green functions of the electron-phonon system in the preceding chapter. We shall, however, not perform such an investigation but shall limit ourselves to the consideration of the superconductivity equations for a quasi-isotrpic three-dimensional system within the framework of the plasma model. This approximation allows us to answer many questions that arise in the investigation of the superconducting properties of metals.

The superconductivity of such systems has been studied in detail in a number of papers [8–10]. In expounding the material of this section, we shall, in the main, adhere to the scheme suggested in Scalapino's review article [9].

To begin with, let us note that in deriving the equations of superconductivity, it is convenient to use the Nambu technique and introduce for the electrons the two-component spinor operators:

$$\psi_{\mathbf{k}} = \begin{vmatrix} a_{\mathbf{k}\uparrow} \\ a^{+}_{-\mathbf{k}\downarrow} \end{vmatrix}, \qquad \psi^{+}_{\mathbf{k}} = \begin{vmatrix} a^{+}_{\mathbf{k}\uparrow}, & a_{-\mathbf{k}\downarrow} \end{vmatrix}. \tag{4.3}$$

With the aid of these operators, the Hamiltonian of the plasma model [see formula (3.39)] can be rewritten in the following form:

$$\begin{aligned} H = {} & \sum_{\mathbf{k}} E_{\mathbf{k}} \psi^{+}_{\mathbf{k}} \tau_3 \psi_{\mathbf{k}} + \sum_{\mathbf{q}\lambda} \Omega_{\mathbf{q}\lambda} \left( b^{+}_{\mathbf{q}\lambda} b_{\mathbf{q}\lambda} + \frac{1}{2} \right) + \\ & + \sum_{\mathbf{k},\mathbf{k}',\lambda} g_{\mathbf{k}\mathbf{k}'\lambda} \psi^{+}_{\mathbf{k}'} \tau_3 \psi_{\mathbf{k}} \left( b_{\mathbf{k}'-\mathbf{k},\lambda} + b^{+}_{\mathbf{k}-\mathbf{k}',\lambda} \right) + \\ & + \frac{1}{2} \sum_{\mathbf{k},\mathbf{k}',\mathbf{q}} \frac{4\pi e^2}{q^2} \left( \psi^{+}_{\mathbf{k}-\mathbf{q}} \tau_3 \psi_{\mathbf{k}} \right) \left( \psi^{+}_{\mathbf{k}'+\mathbf{q}} \tau_3 \psi_{\mathbf{k}'} \right). \end{aligned} \tag{4.4}$$

Here and below the $\tau$'s are the ordinary Pauli matrices:

$$\tau_1 = \begin{vmatrix} 0 & 1 \\ 1 & 0 \end{vmatrix}, \quad \tau_2 = \begin{vmatrix} 0 & -i \\ i & 0 \end{vmatrix}, \quad \tau_3 = \begin{vmatrix} 1 & 0 \\ 0 & -1 \end{vmatrix}, \quad \tau_4 = \begin{vmatrix} 1 & 0 \\ 0 & 1 \end{vmatrix}. \tag{4.5}$$

Let us introduce for the electrons the thermodynamic Green function, $G(\mathbf{k}, \tau)$:

$$G(\mathbf{k}, \tau) = -\langle T_\tau \psi_\mathbf{k}(\tau) \psi_\mathbf{k}^+(0) \rangle. \tag{4.6}$$

This Green function can be written in the matrix form:

$$G(\mathbf{k}, \tau) = \begin{vmatrix} -\langle T a_{\mathbf{k}\uparrow}(\tau) a_{\mathbf{k}\uparrow}^+(0) \rangle & -\langle T a_{\mathbf{k}\uparrow}(\tau) a_{-\mathbf{k}\downarrow}(0) \rangle \\ -\langle T a_{-\mathbf{k}\downarrow}^+(\tau) a_{\mathbf{k}\uparrow}^+(0) \rangle & -\langle T a_{-\mathbf{k}\downarrow}^+(\tau) a_{-\mathbf{k}\downarrow}(0) \rangle \end{vmatrix}. \tag{4.7}$$

It can be seen from expression (4.7) that the diagonal matrix elements of the $G(\mathbf{k}, \tau)$ function are the normal electron Green functions, while the off-diagonal matrix elements are the anomalous Green functions, because of the presence in the system of a condensate of Cooper pairs.

Introducing in the usual manner the phonon Green function $D_\lambda(\mathbf{q}, \tau)$, we can write down the Fourier expansions of the $G(\mathbf{k}, \tau)$ and $D_\lambda(\mathbf{q}, \tau)$ functions:

$$\begin{aligned} G(\mathbf{k}, \tau) &= T \sum_{\omega_n} \exp[-i\omega_n \tau]\, G(\mathbf{k}, i\omega_n), \\ D_\lambda(\mathbf{q}, \tau) &= T \sum_{\omega_\nu} \exp[-i\omega_\nu \tau]\, D_\lambda(\mathbf{q}, i\omega_\nu). \end{aligned} \tag{4.8}$$

In the formula (4.8) the summation for the function $G(\mathbf{k}, \tau)$ is over $\omega_n = (2n + 1)\pi T$, while the summation for $D_\lambda(\mathbf{q}, \tau)$ is over $\omega_\nu = 2\pi\nu T$.

Below we shall not consider the equations for the phonon Green function since these equations do not differ from the equations obtained earlier for the normal metal. As for the electron Green function, we can write the matrix Dyson equation:

$$G^{-1}(\mathbf{k}, i\omega_n) = i\omega_n \tau_4 - E_k \tau_3 - \Sigma(\mathbf{k}, i\omega_n), \tag{4.9}$$

where the self-energy part $\Sigma(\mathbf{k}, i\omega_n)$ can be expanded in terms of the Pauli matrices:

$$\begin{aligned} \Sigma(\mathbf{k}, i\omega_n) = [1 - Z(\mathbf{k}, i\omega_n)]\, i\omega_n \tau_4 + \Phi(\mathbf{k}, i\omega_n)\, \tau_1 + \\ + \Phi_1(\mathbf{k}, i\omega_n)\, \tau_2 + \chi(\mathbf{k}, i\omega_n)\, \tau_3. \end{aligned} \tag{4.10}$$

In this expression the function $\Phi(\mathbf{k}, i\omega_n)$ is the order parameter of the superconducting state, while the functions $Z(\mathbf{k}, i\omega_n)$ and $\chi(\mathbf{k}, i\omega_n)$ represent the

**Figure 4.1. Diagrammatic representation of the self-energy part for the matrix Green function G($k, i\omega_n$).**

renormalization of the electron spectrum and change little when the material goes over from the normal into the superconducting state.

Limiting ourselves to the consideration of homogeneous superconductors in the absence of a magnetic field, we can choose $\Phi_1(\mathbf{k}, i\omega_n) = 0$.

The diagrams for the self-energy part $\Sigma(\mathbf{k}, i\omega_n)$ are shown in Fig. 4.1 and consist of a sum of two contributions: the Coulomb and electron-phonon contributions. Analytically, they can be written in the following form:

$$\Sigma_c(\mathbf{k}, i\omega_n) = -T \sum_{\mathbf{k}', \omega_m} \widetilde{V}_{ee}(\mathbf{k}-\mathbf{k}', i\omega_n - i\omega_m) \times$$
$$\times \Gamma(\mathbf{k}, \mathbf{k}', i\omega_n, i\omega_m)\, \tau_3 G(\mathbf{k}', i\omega_m)\, \tau_3, \tag{4.11}$$

$$\Sigma_{ph}(\mathbf{k}, i\omega_n) = -T \sum_{\mathbf{k}', \omega_m} \left| \frac{\Gamma(\mathbf{k}, \mathbf{k}')\, g_\lambda(\mathbf{k}-\mathbf{k}')}{\varepsilon(\mathbf{k}-\mathbf{k}')} \right|^2 \times$$
$$\times D_\lambda(\mathbf{k}-\mathbf{k}', i\omega_n - i\omega_m)\, \tau_3 G(\mathbf{k}', i\omega_m)\, \tau_3. \tag{4.12}$$

In these formulas $\widetilde{V}_{ee}(\mathbf{k} - \mathbf{k}', i\omega_n - i\omega_m)$ is the screened interelectron Coulomb interaction:

$$\widetilde{V}_{ee}(\mathbf{k}-\mathbf{k}', i\omega_n - i\omega_m) = \frac{4\pi e^2}{(\mathbf{k}-\mathbf{k}')^2\, \varepsilon(\mathbf{k}-\mathbf{k}', i\omega_n - i\omega_m)}. \tag{4.13}$$

The quantity $\Gamma(\mathbf{k}, \mathbf{k}', i\omega_n, i\omega_m)$ is the vertex function of the Coulomb interaction. The equations for the functions $\varepsilon(\mathbf{q}, i\omega_n)$ and $\Gamma(\mathbf{k}, \mathbf{k}', i\omega_n, i\omega_m)$ coincide with the corresponding equations for the normal metal, and we shall not at the moment concern ourselves with their computation. In writing down (4.12), we neglected the frequency dependence of the functions $\varepsilon(\mathbf{q}, i\omega_n)$ and $\Gamma(\mathbf{k}, \mathbf{k}', i\omega_n, i\omega_m)$ since in this case we are interested in small values of the frequencies $i\omega_n$ and $i\omega_m$, i.e., values of the order of the mean phonon frequency $\overline{\omega}$.

Using the representation (4.10) for the function $\Sigma(\mathbf{k}, i\omega_n)$, we can write for the Coulomb part $\Sigma_c(\mathbf{k}, i\omega_n)$ three equations for the functions $\Phi_c(\mathbf{k}; i\omega_n)$, $Z_c(\mathbf{k}, i\omega_n)$, and $\chi_c(\mathbf{k}, i\omega_n)$:

$$\Phi_c(\mathbf{k}, i\omega_n) = T \sum_{\mathbf{k}',\omega_m} \frac{\widetilde{V}_{ee}(\mathbf{k}-\mathbf{k}', i\omega_n - i\omega_m)\,\Gamma\Phi(\mathbf{k}', i\omega_m)}{[Z(\mathbf{k}', i\omega_m)\,i\omega_m]^2 - \widetilde{E}^2_{\mathbf{k}'} - \Phi^2(\mathbf{k}', i\omega_m)}, \tag{4.14}$$

$$[1 - Z_c(\mathbf{k}, i\omega_n)]\,i\omega_n = -T \sum_{\mathbf{k}',\omega_m} \frac{\widetilde{V}_{ee}(\mathbf{k}-\mathbf{k}', i\omega_n - i\omega_m)\,\Gamma Z(\mathbf{k}', i\omega_m)\,i\omega_m}{[Z(\mathbf{k}', i\omega_m)\,i\omega_m]^2 - \widetilde{E}^2_{\mathbf{k}'} - \Phi^2(\mathbf{k}', i\omega_m)}, \tag{4.15a}$$

$$\chi_c(\mathbf{k}, i\omega_n) = -T \sum_{\mathbf{k}',\omega_m} \frac{\widetilde{V}_{ee}(\mathbf{k}-\mathbf{k}', i\omega_n - i\omega_m)\,\Gamma\widetilde{E}_{\mathbf{k}'}}{[Z(\mathbf{k}', i\omega_m)\,i\omega_m]^2 - \widetilde{E}^2_{\mathbf{k}'} - \Phi^2(\mathbf{k}', i\omega_m)}. \tag{4.15b}$$

Here $\widetilde{E}_{\mathbf{k}} = E_{\mathbf{k}} + \chi(\mathbf{k}, i\omega)$. In solving the equations for the functions $Z_c(\mathbf{k}, i\omega_n)$ and $\chi_c(\mathbf{k}, i\omega_n)$, we can neglect the superconducting correlations, i.e., we can set $\Phi(\mathbf{k}', i\omega_n)$ equal to zero. We shall return a little later to the discussion of the equation for $\Phi_c(\mathbf{k}, i\omega_n)$. Right now we proceed to consider the phonon contribution to the self-energy part of the electron Green function.

Using the spectral representations for the electron and phonon Green functions:

$$G(\mathbf{k}, i\omega_n) = -\int_{-\infty}^{\infty} \frac{d\omega'}{\pi} \frac{\operatorname{Im} G(\mathbf{k}, \omega' + i\delta)}{i\omega_n - \omega'}, \tag{4.16}$$

$$D_\lambda(\mathbf{q}, i\omega_\nu) = \int_0^{\infty} d\Omega\, B_\lambda(\mathbf{q}, \Omega) \left[\frac{1}{i\omega_n - \Omega} - \frac{1}{i\omega_n + \Omega}\right], \tag{4.17}$$

where

$$B_\lambda(\mathbf{q}, \Omega) = -\frac{1}{\pi} \operatorname{Im} D_\lambda(\mathbf{q}, \Omega + i\delta),$$

we can derive for the analytic continuation of $\Sigma_{\text{ph}}(\mathbf{k}, \omega)$ the following equation:

$$\Sigma_{\text{ph}}(\mathbf{k},\omega)=\int\limits_{-\infty}^{\infty}\frac{d\omega'}{\pi}\int\limits_{0}^{\infty}d\Omega\int\frac{d^3k'}{(2\pi)^3}\tau_3\,\text{Im}\,[\mathrm{G}(\mathbf{k}',\omega'+i\delta)]\,\tau_3\times$$
$$\times\sum_{\lambda}B_{\lambda}(\mathbf{k}-\mathbf{k}',\Omega)\left|\frac{\Gamma(\mathbf{k},\mathbf{k}')\,g_{\lambda}(\mathbf{k},\mathbf{k}')}{\varepsilon(\mathbf{k},\mathbf{k}')}\right|^2\times$$
$$\times\left[\frac{f(-\omega')+N(\Omega)}{\omega'+\Omega-\omega}+\frac{f(\omega')+N(\Omega)}{\omega'-\Omega-\omega}\right]. \tag{4.18}$$

Here $f(\omega)$ and $N(\Omega)$ are, respectively, the Fermi and Boson distribution functions.

Since in the phonon part of the function $\Sigma(\mathbf{k},\omega)$ and in the integral over $\omega'$, only the values of $\omega$ and $\omega'$ of the order of the mean phonon frequencies are important, the Green function $\mathbf{G}(\mathbf{k},\omega)$ can in this case be written in the form

$$\mathrm{G}^{-1}(\mathbf{k},\omega)=Z_c(k_F)\big[\widetilde{Z}_{\text{ph}}(\mathbf{k},\omega)\,\omega\tau_4-$$
$$-\big(\widetilde{E}_k-\widetilde{\chi}_{\text{ph}}(\mathbf{k},\omega)\big)\,\tau_3-\widetilde{\Phi}(\mathbf{k},\omega)\,\tau_1\big], \tag{4.19}$$

where

$$1-\widetilde{Z}_{\text{ph}}=\frac{1-Z_{\text{ph}}}{Z_c},\quad \widetilde{E}_k=\frac{E_k}{Z_c}\left(1+\frac{1}{v_F}\frac{\partial\chi_c}{\partial k}\right),$$
$$\chi_{\text{ph}}=\frac{\chi_{\text{ph}}}{Z_c},\quad \widetilde{\Phi}=\frac{\Phi}{Z_c}. \tag{4.20}$$

Using the expressions (4.19) and (4.20), we can rewrite Eq. (4.18) for the reduced function

$$\widetilde{\Sigma}_{\text{ph}}(\mathbf{k},\omega)=\int\limits_{-\infty}^{\infty}\frac{d\omega'}{\pi}\int\limits_{0}^{\infty}d\Omega\int\frac{d^3k'}{(2\pi)^3}\tau_3\,\text{Im}\big[\widetilde{\mathrm{G}}(\mathbf{k}',\omega'+i\delta)\big]\,\tau_3\times$$
$$\times\sum_{\lambda}B_{\lambda}(\mathbf{k}-\mathbf{k}',\Omega)\,|\,\widetilde{g}_{\lambda}(\mathbf{k},\mathbf{k}')\,|^2\times$$
$$\times\left[\frac{f(-\omega')+N(\Omega)}{\omega'+\Omega-\omega}+\frac{f(\omega')+N(\Omega)}{\omega'-\Omega-\omega}\right]. \tag{4.21}$$

In this equation

$$\widetilde{G}^{-1}(\mathbf{k}, \omega) = \widetilde{Z}_{ph}(\mathbf{k}, \omega)\,\omega\tau_4 - (\widetilde{E}_k + \chi_{ph})\,\tau_3 - \widetilde{\Phi}\tau_1, \tag{4.22}$$

$$\widetilde{g}_\lambda(\mathbf{k}, \mathbf{k}') = \frac{\Gamma(\mathbf{k}, \mathbf{k}')\, g_\lambda(\mathbf{k}, \mathbf{k}')}{Z_c \varepsilon(\mathbf{k}, \mathbf{k}')}. \tag{4.23}$$

By comparing the formula, (4.23), for $\widetilde{g}_\lambda(\mathbf{k}, \mathbf{k}')$ with the formula, (3.99), for the matrix element of the periodic potential, we can verify the validity of the conclusion drawn in the preceding chapter about the connection between these two quantities.

By writing in the usual manner the integration over the momenta $\mathbf{k}'$ as integration over a surface with a given energy and over the energy $E_{k'}$:

$$\int \frac{d^3k'}{(2\pi)^3} = \int \frac{dS_{k'}}{v_F} \int dE_{k'}$$

(here $dS_{\mathbf{k}'}$ is an element of the energy surface and $v_F$ is the electron velocity at the Fermi surface), we can perform the integration over $F_{\mathbf{k}'}$ in Eq. (4.21):

$$\int_{-\infty}^{\infty} dE_{k'}\widetilde{G}(\mathbf{k}', \omega') = -\frac{i\pi\,[\omega'\widetilde{Z}(\omega')\,\tau_4 + \widetilde{\Phi}(\omega')\,\tau_1]}{[\omega'^2\widetilde{Z}^2(\omega') - \widetilde{\Phi}^2(\omega')]}. \tag{4.24}$$

In deriving (4.24), we used the fact that, as in the normal metal, the phonon contribution to the self-energy part of the electron Green function depends only on the energy $\omega$ and does not depend on the momenta.

Finally, the equation for the phonon part of the function $\Sigma(\mathbf{k}, \omega)$ assumes the form

$$\widetilde{\Sigma}_{ph}(\omega) = \int_{-\infty}^{\infty} d\omega' \int_0^{\infty} d\Omega\,\alpha^2(\Omega)\,g(\Omega)\times$$

$$\times \operatorname{Re}\left\{\frac{\tau_3\,[\widetilde{Z}_{ph}(\omega')\,\omega'\tau_4 - \widetilde{\Phi}\tau_1]\,\tau_3}{\sqrt{\omega'^2\widetilde{Z}^2(\omega') - \widetilde{\Phi}^2(\omega')}}\right\}\left[\frac{f(-\omega') + N(\Omega)}{\omega' + \Omega - \omega} + \frac{f(\omega') + N(\Omega)}{\omega' - \Omega - \omega}\right]. \tag{4.25}$$

In Eq. (4.25) we have introduced the notation

$$\alpha^2(\Omega)\,g(\Omega) = \int \frac{dS_k}{v_F} \int \frac{dS_{k'}}{v_F} \sum_\lambda |\widetilde{g}_\lambda(\mathbf{k}, \mathbf{k}')|^2 B_\lambda(\mathbf{k}, \mathbf{k}', \Omega) \Big/ \int \frac{dS_k}{v_F}. \tag{4.26}$$

The function $\alpha^2(\Omega)g(\Omega)$ is the central function in the entire problem of the effect of the electron-phonon interaction on the electronic properties of metals.

In the following sections we shall discuss in greater detail the form of this function and the methods of computing it; here let us only note that this function can be determined from tunneling experiments [11]. At the present moment the form of this function is well known for a very large number of superconducting metals and alloys.

Returning to the determination of $\Sigma_c(\mathbf{k}, \omega)$, the Coulomb contirbution to the self-energy part, let us note that for this function it is not possible to obtain a one-dimensional integral equation similar to (4.25). The point is that a broad region of the $\omega$- and $\omega'$-energy values of the order of the Fermi energies, $E_F$, of the metal is important in the Eqs. (4.15), and it is absolutely impossible to neglect the dependence of $\Sigma_c(\mathbf{k}, i\omega_n)$ on the momentum **k**. For this same reason, in Eqs. (4.15) we cannot neglect the dependence of the functions $\varepsilon(\mathbf{k}, i\omega_n)$ and $\Gamma(\mathbf{k}, \mathbf{k}', i\omega_n, i\omega_{n'})$ on the frequency. On account of this, a rigorous, quantitative analysis of the Eqs. (4.15) is extremely complicated and has, in fact, not been performed by anybody. Qualitative estimates made in [8] show that, in the strong-coupling case, the role of the direct Coulomb repulsion in the determination of the superconducting characteristics of metals is greatly suppressed. In particular, when the frequency dependence of the functions $\Gamma(\mathbf{k}, \mathbf{k}', i\omega_n, i\omega_{n'})$ and $\varepsilon(\mathbf{k} - \mathbf{k}', i\omega_m)$ is neglected, the equation, (4.14), for the Coulomb contribution to the order parameter $\Phi_c(\mathbf{k}, i\omega_n)$ can easily be reduced to a pseudopotential form close to the form obtained in the weak-coupling case [9].

For the majority of superconductors, the neglect of the frequency dependence of the functions $\Gamma(\mathbf{k}, \mathbf{k}', i\omega_n, i\omega_{n'})$ and $\varepsilon(\mathbf{k}, \mathbf{k}', i\omega_m)$, and the replacement of the kernel of the integral equation for $\Phi_c(\mathbf{k}, i\omega_n)$ by some constant will by themselves hardly lead to serious errors in the computation of their properties. Unfortunately, the consistent determination of the very magnitude of this constant is absolutely impossible in such an approach. We can, to be sure, attempt to establish its value by comparing the experimental data with the computations performed in such a model. It is, perhaps, worth noting that the replacement of the kernel in Eq. (4.14) by a constant quantity does not at all imply, in general, the replacement of the entire Coulomb interaction by a constant. Such a replacement of that would have led in the other equations, (4.15a) and (4.15b), for the renormalization functions $Z(\mathbf{k}, i\omega_n)$ and $\chi(\mathbf{k}, i\omega_n)$ to trivial answers: $Z(\mathbf{k}, i\omega_n) = 1, \chi(\mathbf{k}, i\omega_n) = 0$.

For certain transition metals, especially at the end of the transition-metal series, where an important role is playing by the Coulomb correlations, which are responsible for the appearance of ferromagnetism, the neglect of the frequency dependence of the vertex function and in the equation for the order parameter $\Phi_c(\mathbf{k}, i\omega_n)$ turns out to be inadmissible. In this case spin fluctuations, which lead to a strong dependence of the vertex function on the frequency $\omega$ even at fairly low $\omega$ values, arise in the system. As has been shown by

Berk and Shrieffer [12], these effects completely suppress the superconductivity in such metals as platinum and palladium.

Below we shall practically not touch upon questions connected with the effect of spin fluctuations on superconductivity; they are considered in detail in [12]. On the other hand, in computing the critical temperature of superconductors and in analyzing its dependence on the properties of the normal metal, we shall use the above-indicated simple model in which the kernel in Eq. (4.14) is replaced by a constant quantity $\mu$. Then the Eqs. (4.15) and (4.25) unite into the following single equation for the order parameter $\Phi(\omega) = Z(\omega)\Delta(\omega)$:

$$Z(\omega)\,\Delta(\omega) = \int_{-\infty}^{\infty} d\omega'\,\mathrm{Re}\left\{\frac{\Delta(\omega')}{[\omega'^2-\Delta^2(\omega')]^{\frac{1}{2}}}\right\}\times$$

$$\times\int_0^{\infty} d\Omega\,\alpha^2(\Omega)\,g(\Omega)\left[\frac{f(-\omega')+N(\Omega)}{\omega'+\Omega-\omega}+\frac{f(\omega')+N(\Omega)}{\omega'-\Omega-\omega}\right]-$$

$$-\mu\int_0^{E_F} d\omega'\,\mathrm{Re}\left\{\frac{\Delta(\omega')}{[\omega'^2-\Delta^2(\omega')]^{\frac{1}{2}}}\right\}\tanh\frac{\omega'}{2T}. \qquad (4.27)$$

The renormalization function $Z(\omega)$ is determined in this model by only the electron-phonon interaction, and the equation for $Z(\omega)$ has the form

$$[1-Z(\omega)]\,\omega = \int_0^{\infty} d\omega'\,\mathrm{Re}\left\{\frac{\omega'}{[\omega'^2-\Delta^2(\omega')]^{\frac{1}{2}}}\right\}\times$$

$$\times\int_0^{\infty} d\Omega\,\alpha^2(\Omega)\,g(\Omega)\left[\frac{f(-\omega')+N(\Omega)}{\omega'+\Omega+\omega}-\frac{f(-\omega')+N(\Omega)}{\omega'+\Omega-\omega}+\right.$$

$$\left.+\frac{f(\omega')+N(\Omega)}{-\omega'+\Omega+\omega}-\frac{f(\omega')+N(\Omega)}{-\omega'+\Omega-\omega}\right]. \qquad (4.28)$$

## 3. The Critical Temperature of Superconductors with Strong Electron-Phonon Interaction

The critical temperature, $T_c$, of the superconducting transition is determined from the solution to the Eqs. (4.27) and (4.28) as the temperature at which $\Delta(\omega) \to 0$. In this case Eq. (4.28) becomes virtually an equality directly determining the function $Z(\omega)$:

$$[1 - Z(\omega)]\,\omega = \int_0^\infty d\omega' \int_0^\infty d\Omega\, \alpha^2(\Omega)\, g(\Omega)\, \{[f(-\omega') + N(\Omega)] \times$$
$$\times K_-(\omega', \Omega, \omega) + [f(\omega') + N(\Omega)]\, K_-(-\omega', \Omega, \omega)\}. \quad (4.29)$$

On the other hand, the equation for the order parameter $\Delta(\omega)$ can be rewritten in the form

$$Z(\omega)\,\Delta(\omega) =$$
$$= \int_0^\infty d\omega' \operatorname{Re} \Delta(\omega') \int_0^\infty d\Omega\, \alpha^2(\Omega)\, g(\Omega)\, \{[f(-\omega') + N(\Omega)] \times$$
$$\times K_+(\omega', \Omega, \omega) - [f(\omega') + N(\Omega)]\, K_+(-\omega', \Omega, \omega)\} -$$
$$- \int d\omega' \operatorname{Re} \Delta(\omega')\, V_c(\omega, \omega') \tanh \frac{\omega'}{2T}. \quad (4.30)$$

In writing down the expressions (4.29) and (4.30), we introduced the notation

$$K_\pm(\omega', \Omega, \omega) = \frac{1}{\omega' + \Omega + \omega + i\delta} \pm \frac{1}{\omega' + \Omega - \omega - i\delta},$$
$$V_c(\omega, \omega') = \mu\theta(E_F - \omega)\,\theta(E_F - \omega').$$

To begin with, we shall be occupied with the computation of the function $Z(\omega)$ and the elucidation of its physical meaning. As can be seen from Eq. (4.22), in the normal metal the Green function for the electrons interacting with the phonons can be written in the form

$$G(k, \omega) = \frac{1}{Z(\omega)\,\omega - E_k}. \quad (4.31)$$

As follows from this expression, in the vicinity of the Fermi surface, the renormalized excitation spectrum of the electrons has the form

$$\tilde{E}_k = \frac{E_k}{\operatorname{Re} Z(0)}. \quad (4.32)$$

If we take account of the fact that in the vicinity of the Fermi surface

$$E_k = v_F(k - k'),$$

then we see that the electron-phonon interaction renormalizes the electron velocity, or, equivalently, the electron mass:

$$m^* = m \operatorname{Re} Z(0). \tag{4.33}$$

As can be seen from the expression (4.29), the renormalization of the electron mass depends essentially on the temperature of the system. At temperatures significantly lower than the characteristic phonon frequencies, Re $Z(0)$ can be written in the form

$$\operatorname{Re} Z(0) = 1 + 2\int_0^\infty \frac{d\Omega}{\Omega}\alpha^2(\Omega)\, g(\Omega)\left[1 + \frac{\pi^2}{3}\left(\frac{T}{\Omega}\right)^2\right], \tag{4.34}$$

from which it follows that For $T \ll \overline{\Omega}$

$$\operatorname{Re} Z(0) = 1 + 2\int_0^\infty \frac{d\Omega}{\Omega}\alpha^2(\Omega)\, g(\Omega) = 1 + \lambda,$$

$$\lambda = 2\int_0^\infty \frac{d\Omega}{\Omega}\alpha^2(\Omega)\, g(\Omega). \tag{4.35}$$

The above-introduced quantity $\lambda$ is usually called the electron-phonon coupling constant and characterizes, in particular, the renormalization of the electron mass at zero temperature. As will be verified a little later, it is precisely this quantity $\lambda$ that enters into the expression for the critical temperature of the superconducting transition.

On the other hand, at high temperatures $T \ll \overline{\Omega}$, the electron mass is not renormalized, and the quantity Re $Z(0)$ tends to unity like

$$\operatorname{Re} Z(0) = 1 + \frac{7\zeta(3)}{\pi^2 T^2}\int_0^\infty d\Omega\, \alpha^2(\Omega)\, g(\Omega)\, \Omega.$$

The measurement of the effective mass of the electron at low temperatures $T \ll \overline{\Omega}$ and at high $T \gg \Omega$ in principle allows the determination of the electron-phonon coupling constant $\lambda$. Investigations have recently been published in which the variation of the effective electron mass with temperature was determined from electronic-thermal-capacity [13] and cyclotron-resonance [14] measurements. The $\lambda$ values obtained from these experiments are in good agreement with the theoretical calculations.

The imaginary part of the function $Z(\omega)$ determines the electron damping as a result of the electron-phonon interaction. The magnitude, $\Gamma(\omega)$, of this damping is given by the expression

$$\Gamma(\omega) = \omega \operatorname{Im} Z(\omega) = 2\pi \int_0^\infty d\Omega\, \alpha^2(\Omega)\, g(\Omega)\, [N(\Omega) + \\ + \frac{1}{2}(f(\Omega+\omega) + f(\Omega-\omega))]. \tag{4.36}$$

Unfortunately, the exact solution of Eq. (4.30) for the order parameter $\Delta(\omega)$ for any relation between the critical temperature $T_c$ and the mean phonon frequencies $\overline{\omega}$ is extremely difficult. In the majority of the papers [1 - 5] in which the critical temperature is computed, the extremely simple variational function $\Delta(\omega)$ is used in the form of two steps:

$$\Delta = \begin{cases} \Delta_0, & \omega < \omega_c, \\ \Delta_\infty, & E_F > \omega > \omega_c. \end{cases} \tag{4.37}$$

It can be shown [6] that, strictly speaking, such an approximation for $\Delta(\omega)$ is very far from the actual behavior of the function $\Delta(\omega)$ at nonzero temperatures. In fact, the function $\Delta(\omega)$ is complex, and, besides, its real part at $T \neq 0$ is, in contrast to the formula (4.37), not a constant at low $\omega$ but tends to zero. Indeed, using Eq. (4.30), we can easily verify that

$$\operatorname{Re}[Z(\omega)\Delta(\omega)] = \Sigma_1 \neq 0, \quad \operatorname{Im}[Z(\omega)\Delta(\omega)] = \omega\Sigma_2, \tag{4.38}$$

where $\Sigma_1$ and $\Sigma_2$ are some constants, the exact values of which are not important to us at the moment. Further, it follows from (4.38), (4.34), and (4.36) that

$$\operatorname{Re}\Delta(\omega) = \operatorname{Re}\frac{\Sigma_1 + i\omega\Sigma_2}{Z(\omega)} \approx \omega^2 \frac{\lambda\Sigma_1 + \Gamma\Sigma_2}{\Gamma^2},$$

$$\operatorname{Im}\Delta(\omega) = \operatorname{Im}\frac{\Sigma_1 + i\omega\Sigma_2}{Z(\omega)} \approx \omega\frac{\Sigma_1}{\Gamma}. \tag{4.39}$$

Such behavior of the function $\Delta(\omega)$ also does not allow us to use for the solution of Eq. (4.30) the method of logarithmic-singularity separation expounded in Chapter 2.

It is shown in [6] that all these complications arise largely because of the presence at $T \neq 0$ of the decrement $\Gamma(0)$ for the conduction electrons directly at the Fermi surface:

$$\Gamma(0) = 2\pi \int_0^\infty d\Omega\, \alpha^2(\Omega)\, g(\Omega)\, [N(\Omega) + f(\Omega)]. \tag{4.40}$$

If the characteristic phonon frequencies $\overline{\omega}$ are higher that the given temperature $T$, then the quantity $\Gamma(0)$ is small:

$$\Gamma(0) \approx \lambda T \left(\frac{T}{\bar{\omega}}\right)^2.$$

In the majority of real superconductors, the critical temperature is much lower than the value of $\bar{\omega}$. For them, we can rewrite, neglecting the quantities of the order of $(T_c/\omega)^2$, the equations determining $T_c$ in the following form:

$$Z(\omega)\Delta(\omega) = \int_0^\infty \frac{d\omega'}{\omega'} \operatorname{Re}\Delta(\omega') \operatorname{th}\frac{\omega'}{2T_c}\left[\int_0^\infty d\Omega\, \alpha^2(\Omega) g(\Omega) K_+(\omega, \Omega, \omega') - V_c(\omega, \omega')\right], \tag{4.41}$$

$$\operatorname{Re} Z(\omega) = 1 - \frac{1}{\omega}\int_0^\infty d\Omega\, \alpha^2(\Omega) g(\Omega) \int_{-\infty}^\infty d\omega' f(\omega') K_-(\omega, \Omega, \omega'), \tag{4.42}$$

$$\operatorname{Im} Z(\omega) = 0. \tag{4.43}$$

The influence on the critical temperature of the phonons with low frequencies $\omega \lesssim T_c$, as well as the behavior of superconductors with a very large coupling constant $\lambda \gg 1$, in which case the consideration of the damping of the electrons at the Fermi surface is essential, will be discussed at the end of this section. Let us rewrite the equation (4.41) for the function $\phi(\omega) = \operatorname{Re}\Delta(\omega)$:

$$\varphi(\omega) = \int_0^\infty \frac{d\omega'}{\omega'} \varphi(\omega') \tanh\frac{\omega'}{2T_c}[L(\omega', \omega) - \tilde{V}_c(\omega', \omega)], \tag{4.44}$$

where

$$L(\omega', \omega) = \frac{1}{Z(\omega)}\int_0^\infty d\Omega\, \alpha^2(\Omega) g(\Omega)\left[\frac{1}{\omega' + \Omega + \omega} + \frac{1}{\omega' + \Omega - \omega}\right],$$

$$L(0, 0) = \frac{\lambda}{Z(0)}, \tag{4.45}$$

$$\tilde{V}_c(\omega', \omega) = \frac{V_c(\omega', \omega)}{Z(\omega)}. \tag{4.46}$$

Equation (4.44) has a logarithmic singularity with respect to $T_c$ and can now be solved by the method developed in [15] and expounded for the weak-coupling case in Chapter 2. We reproduce only the most important points of the solution, not dwelling on the details. Let us write $T_c$ in the form

$$T_c = 1{,}14\bar{\omega} \exp\left[-\frac{1}{\chi(0)}\right], \tag{4.47}$$

where

$$\chi(\omega) = [L(0,\omega) - \widetilde{V}_c(0,\omega)] - A_{\omega'}\{[L(\omega',\omega) - \widetilde{V}_c(\omega',\omega)]\,\chi(\omega')\}. \tag{4.48}$$

Here we have introduced, as in Chapter 2, the notation

$$A_{\omega'}\{f(\omega')\} = \int\limits_0^\infty d\omega' \ln\frac{\omega'}{\bar{\omega}}\frac{d}{d\omega'} f(\omega').$$

The quantity $\bar{\omega}$ is, for the time being, an arbitrary frequency whose value is comparable to the mean phonon frequency. Let us write the function $\chi(\omega)$ in the form

$$\chi(\omega) = \chi_{ph}(\omega) + \chi_c(\omega), \tag{4.49}$$

and set

$$\chi_{ph} = L(0,\omega) - A_{\omega'}\{L(\omega',\omega)\,\chi(\omega')\}. \tag{4.50}$$

Then for the function $\chi_c(\omega)$ we have

$$\chi_c(\omega) = -\widetilde{V}_c(0,\omega)\,[1 - A_{\omega'}\{\chi_{ph}(\omega')\}] - \\ - A_{\omega'}\{[L(\omega',\omega) - \widetilde{V}_c(\omega',\omega)]\,\chi_c(\omega')\}. \tag{4.51}$$

From Eq. (4.52) we can easily derive an expression for $\chi_c(0)$:

$$\chi_c(0) = -\frac{1}{Z(0)}\frac{\mu\,[1 - A_{\omega'}\{\chi_{ph}(\omega')\}]}{1 + \mu\ln\frac{E_F}{\bar{\omega}} + A_{\omega'}\{L(\omega',0)\}}. \tag{4.52}$$

Returning to the equation, (4.50), defining the function $\chi_{ph}(\omega)$, we note that this equation can be solved by the method of iterations. The expansion parameter here will be the quantity $L(0,0) = \lambda/(1+\lambda)$. As shown in [6], even for coupling constants, $\lambda$, of the order of unity, it is sufficient to limit ourselves to the lowest approximation for $\chi_{ph}(\omega)$. Then

$$\chi_{ph}(\omega) = L(0,\omega) - A_{\omega'}\{L(\omega',\omega)\,L(0,\omega')\}. \tag{4.53}$$

Finally, the expression for the critical temperature assumes the form

$$T_c = 1.14\bar{\omega} \exp\left(-\frac{1+\lambda}{a}\right),$$

where

$$a = \lambda - \frac{\mu [1 - A_{\omega'} \{\chi_{\text{ph}}(0, \omega')\}]}{1 + \mu \ln (E_F/\omega) + A_{\bar{\omega}'} \{L(\bar{\omega}', 0)\} - (1+\lambda) A_{\omega'} \{L(\omega', 0) L(0, \omega')\}}. \quad (4.54)$$

Thus far, the quantity $\bar{\omega}$ has not been precisely defined. Let us define it in such a way that the Coulomb contribution to the critical temperature has the conventional form of the Coulomb pseudopotential. For this purpose, let us define $\bar{\omega}$ by the condition

$$A_{\omega'} \{L(\omega', 0)\} = 0. \quad (4.55)$$

Then it follows from (4.52) that

$$\chi_c(0) = -\frac{\mu^* (1+\alpha)}{1+\lambda}, \quad (4.56)$$

where the Coulomb pseudopotential $\mu^*$ is equal to

$$\mu^* = \frac{\mu}{1 + \mu \ln (E_F/\bar{\omega})}, \quad (4.57)$$

$$\alpha = A_{\omega'} A_{\omega} \{L(\omega', \omega) L(0, \omega')\}. \quad (4.58)$$

Using the expression (4.55), we can easily show that the pre-exponential function is expressible in terms of the mean-logarithmic frequency

$$\ln \bar{\omega} = \ln \omega_{\log} = \frac{2}{\lambda} \int_0^\infty \frac{d\Omega}{\Omega} \alpha^2(\Omega) g(\Omega) \ln \Omega. \quad (4.59)$$

Writing the expression for $T_C$ with the use of the condition (4.55), we obtain

$$T_c = 1.14\omega_{\log} \times$$
$$\times \exp\left[-\frac{1+\lambda}{\lambda - \mu^*(1+\alpha) - (1+\lambda) A_{\omega'} \{L(\omega', 0) L(0, \omega')\}}\right]. \quad (4.60)$$

Generally speaking, the formula, (4.60), for the critical temperature differs significantly from both the McMillan formula (4.2) and the formulas proposed in [1-5]. The difference lies largely in the third term of the index of the exponential function. Although the magnitude of this term is proportional to $\lambda\,[\lambda/(1+\lambda)]$ and is less than the first term for any $\lambda$, it cannot at all be neglected. It can easily be verified that allowance for this quantity for any $\lambda$ leads at least to the appearance of the numerical coefficient of the order of unity in the expression for $T_c$. Expanding the index of the exponential function in a power series in the small quantity $(1+\lambda)A_{\omega'}\{L(\omega',0)L(0,\omega')\}$, we obtain

$$T_c = 1.14\omega_{\log} e^K \exp\left[-\frac{1+\lambda}{\lambda-\mu^*(1+\alpha)}\right], \tag{4.61}$$

where

$$K = \frac{(1+\lambda)^2 A_{\omega'}\{L(\omega',0)L(0,\omega')\}}{[\lambda-\mu^*(1+\alpha)]^2}. \tag{4.62}$$

Taking account of the fact that the condition $\mu^* \ll \lambda$ is normally fulfilled, we obtain

$$K = \frac{A_{\omega'}\{L(\omega',0)L(0,\omega')\}}{L^2(0,0)} = \frac{2}{\lambda^2}\int_0^\infty \frac{d\Omega_1}{\Omega_1}\alpha^2(\Omega)g(\Omega_1)\times$$
$$\times \ln\frac{\Omega_1^2}{\Omega_{\log}^2}\int_0^\infty d\Omega_2\,\alpha^2(\Omega_2)g(\Omega_2)\left[\frac{1}{\Omega_2+\Omega_1}-\frac{1}{\Omega_2-\Omega_1}\right]. \tag{4.63}$$

It can be seen from this expression that $K$ virtually does not depend on the magnitude of the entire coupling constant $\lambda$ but depends on the form of the function $\alpha^2(\Omega)g(\Omega)$.

Let us consider as an example the expression for $T_c$ in the case when the function $\alpha^2(\Omega)g(\Omega)$ can be represented in the form of a sum of $\delta$ functions. Let

$$\alpha^2(\Omega)g(\Omega) = \frac{\lambda_1\omega_1}{2}\delta(\Omega-\omega_1) + \frac{\lambda_2\omega_2}{2}\delta(\Omega-\omega_2), \tag{4.64}$$

then

$$\lambda = \lambda_1+\lambda_2, \quad \omega_{\log} = \omega_1^{\lambda_1/\lambda}\omega_2^{\lambda_2/\lambda},$$

$$K = -\frac{1}{2} + \frac{\lambda_1\lambda_2}{2\lambda^2}\left[1 - \frac{\omega_1^2+\omega_2^2}{\omega_1^2-\omega_2^2}\ln\frac{\omega_2}{\omega_1}\right].$$

If the frequencies of the two peaks do not differ very much from each other, then the second term in the formula for $K$ is small to the degree of the smallness of $(\omega_2 - \omega_1)^2/\omega_2^2$, and it can be neglected. In this case $K = -\frac{1}{2}$, and the expression for the critical temperature can be written in the form

$$T_c = \frac{\omega_{\log}}{1.43} \exp\left[-\frac{1+\lambda}{\lambda - \mu^*(1+\alpha)}\right]. \tag{4.65}$$

It is evident from a comparison of the formulas (4.2) and (4.65) that for this case the two formulas coincide virtually totally if we assume that the quantitv $\omega_D$ figuring in the McMillan formula (4.2) is equal to $\omega_{\log}$. This coincidence of the formulas is not surprising since McMillan obtained the numerical coefficients in his formula from the solution of the Éliashberg equations for the specific $\alpha^2(\Omega)g(\Omega)$ function corresponding to metallic niobium, in which this function comprises two narrow and close peaks. Moreover, calculations carried out in [6] show that in the majority of the superconductors for which the function $\alpha^2(\Omega)g(\Omega)$ is known, the quantity $K$ differs very little from $\frac{1}{2}$ and, therefore, $T_c$ computations performed with the aid of the formula (4.65) yield results that are in good agreement with the experimental data. A similar inference is made in [16], where it is shown that a formula practically coinciding with (4.65) yields the correct value for $T_c$ right up to $\lambda \approx 1.5$.

From the derivation presented here of the formulas, (4.60) and (4.65), for $T_c$, we can also see the limitedness of these formulas as well as of the McMillan formula and all the other approximate expressions for $T_c$. Hence it is also clear that the determination of the Coulomb pseudopotential $\mu^*$ and, consequently, the Coulomb constant $\mu$ on the basis of the McMillan formula is highly conditional. Indeed, if $\mu^*$ is significantly smaller than $\lambda$, as is often the case in superconductors, then the third iteration for $\chi_{\text{ph}}(\omega)$ that was neglected in the derivation of the formula (4.65) may prove to be entirely comparable to $\mu^*$, although it also will, of course, be much smaller than the quantity $\lambda$.

In [17] an expression for $T_c$ is obtained that formally takes all the iterations in the determination of the functions $\chi_{\text{ph}}(\omega)$ and $\chi_c(\omega)$ into account. In our opinion, this is not at all necessary since the use of only two iterations furnishes sufficient accuracy right up to $\lambda = 1.5$ [6, 16], while at large values of $\lambda$ the equations (4.42) and (4.43) themselves are unsuitable and the more exact Éliashberg equations (4.29) and (4.30) should be used. The use of these equations is also necessary in the consideration of the question of the effect of the low-frequency phonons, with $\omega \lesssim T_c$ on $T_c$.

In the presence of low-frequency phonons, or at large values of $\lambda \gg 1$, it is virtually impossible to obtain the exact analytic solution to Eqs. (4.29) and (4.30) that takes into account both the existence of thermal phonons and the damping of the order parameter $\Delta(\omega)$ at low $\omega$ because of these phonons. Because of the oscillatory dependence of $\Delta(\omega)$ on the frequency $\omega$, the numerical solution of these equations leads to large errors. It is much more convenient in this case to use the Éliashberg equations (4.29), (4.30) in the temperature technique:

$$Z(i\omega_n)\Delta(i\omega_n) = \pi T_c \sum_{\omega_m=-\omega_{\max}}^{\omega_{\max}} [L(\omega_n-\omega_m)-\mu^*]\frac{\Delta(i\omega_m)}{|\omega_m|}, \tag{4.66}$$

$$[1-Z(i\omega_n)]\, i\omega_n = -\pi T_c \sum_{\omega_m} \operatorname{sign}\omega_m L(\omega_n-\omega_m),$$

$$L(\omega_n-\omega_m) = 2\int_0^\infty d\Omega\, \alpha^2(\Omega)\, g(\Omega)\frac{\Omega}{\Omega^2+(\omega_n-\omega_m)^2},$$

$$\omega_n = (2n+1)\pi T_c. \tag{4.67}$$

The summation in Eq. (4.66) is performed up to the maximum frequency $\omega_{max}$, which is of the order of $10\omega_D$ ($\omega_D$ is the Debye frequency of the metal). The truncation of the summation is introduced for the purpose of using the Coulomb pseudopotential $\mu^*$ instead of the entire Coulomb interaction $\mu$. Let us first of all consider the effect on $T_c$ of the low-frequency phonons with $\Omega \ll T_c$ and let us use for the function $\alpha^2(\Omega)g(\Omega)$ the expression (4.64), assuming that $\omega_1 \ll T_c$. The kernel, $L(\omega_n - \omega_m)$, of the equation can also be written in the form

$$L(\omega_n-\omega_m) = L_1(\omega_n-\omega_m) + L_2(\omega_n-\omega_m).$$

As can be seen from the formula (4.67), for $\omega_1 \ll T_c$, we can, in the expression for $L_1(\omega_n - \omega_m)$, limit ourselves to the one term with $\omega_n = \omega_m$:

$$L_1(0) = \lambda_1 = 2\int_0^\infty \frac{d\Omega}{\Omega}\alpha_1^2(\Omega)\, g_1(\Omega).$$

With the aid of (4.67), we can easily verify that $Z(i\omega_n) = 1 + f_1(i\omega_n) + f_2(i\omega_m)$, where $f_1(i\omega_n) = \lambda_1 \pi T_c/|\omega_n|$. Substituting these expressions for $L$ and $Z$ into (4.66), we finally have

$$[1+f_2(i\omega_n)]\Delta(i\omega_n) + \frac{\pi T_c\lambda_1}{|\omega_n|}\Delta(i\omega_n) = \frac{\pi T_c\lambda_1}{|\omega_n|}\Delta(i\omega_n) + \pi T_c \sum_{\omega_m}{}' [L_2(\omega_n-\omega_m)-\mu^*]\frac{\Delta(i\omega_m)}{|\omega_m|}. \tag{4.68}$$

Hence it can be seen that all the terms containing $\lambda_1$ in this equation cancel out, and $T_c$ is determined only by the contribution of the phonons with the frequency $\omega_2$. Thus, the phonons with frequencies that are low compared to $T_c$ have almost no effect on the critical temperature itself. This fact is easy to understand on the basis of simple physical arguments. The point is that the phonon excitations with energies lower than the temperature of the electron gas are perceived by the electrons essentially as static impurities. And, as is well known, static impurities exert no direct influence on the quantity $T_c$. Formally, by letting $\omega_1$ tend to zero and using the formula (4.65), we can arrive at the incorrect answer that the critical temperature in this case also tends to zero.

In deriving Eq. (4.68), we discarded the terms of the type $\lambda_1(\omega_1/T_c)^2$. If as $\omega_1$ tends to zero the coupling constant $\lambda_1$ remains constant or decreases, then these terms are indeed small. If, on the other hand, $\lambda_1$ in this case increases with decreasing $\omega_1$ according to the law $\lambda_1 \sim \omega_1^{-2}$, then such low-frequency vibrations make a contribution to $T_c$ even at $\omega = 0$. We can hardly attach any physical meaning to such behavior of the constant $\lambda$.

A quite curious circumstance arising from Eqs. (4.60) - (4.68) is the possibility of the existence in these equations at large $\lambda$ of $T_c$ values exceeding the mean phonon frequencies. For $T_c \gg \overline{\omega}$, there occurs in Eq. (4.66), as before and as in the derivation of (4.68), a canceling out of the dominant terms with $\omega_n = \omega_m$, so that only the terms with $\omega_m = \omega_n \pm 1$, which are of the order of $\lambda(\overline{\omega}/T_c)^2$, remain. The condition for the solvability of the obtained equations reduces essentially to the relation

$$\lambda\left(\frac{\overline{\omega}}{T_c}\right)^2 \approx 1.$$

Hence it follows that†

$$T_c \approx \overline{\omega}\sqrt{\lambda}. \tag{4.69}$$

More accurate calculations carried out in [16] show that the values of $T_c$ approach a limiting value given by an expression of the type (4.69) only when $\lambda > 10$. The possibility of the existence in real metals of such large values of $\lambda$ seems to be highly questionable. The highest $\lambda$ values are attained [16] in amorphous Pb-Bi alloys, in which $\lambda = 1.5$. A specific analysis of the behavior of the critical temperature as a function of the frequencies of, or the electron-phonon coupling constant in, the metal will be carried out in subsequent sections of this chapter. For the moment let us proceed to study the electron-phonon coupling constant and its dependence on the properties of the normal metal.

†The result (4.69) may turn out to be incorrect if the quantities $\lambda$ and $\overline{\omega}$ are not independent (see [100]).

## 4. Computation of the Electron-Phonon Coupling Constant in Metals

As follows from the formula (4.65), the critical temperature of the superconducting transition in a metal is determined largely by two quantities characterizing a metal: the mean-logarithmic frequency and the electron-phonon coupling constant. Both of these quantities can easily be computed from the formulas (4.35) and (4.59) if the function $\alpha^2(\omega)g(\omega)$ is known. The study of the tunneling characteristics of the superconductor-normal metal transitions enables us to establish the form of this function; at present the form of this function is known for a very large number of metals and alloys [11].

Figure 4.2 shows plots of the function $\alpha^2(\omega)g(\omega)$ and the density-ot-states function, $g(\omega)$, for lead. It can be seen that the two curves are very similar in shape. This shows that the function $\alpha^2(\omega)$ depends very weakly on the energy at least in the region of the principal peaks in the density of phonon states.

In the general case the function $\alpha^2(\omega)$ apparently does not change by more than a factor of 2 in the region from low energies to high $\omega$ [18,19]. Only the region of very low $\omega$, where the function $\alpha^2(\omega)g(\omega)$ behaves like $1/\omega$, is excluded. This region may turn out to be quite broad in disordered alloys [20] or in certain compounds of the transition metals [21], in which the role of the Umklapp processes is very important.

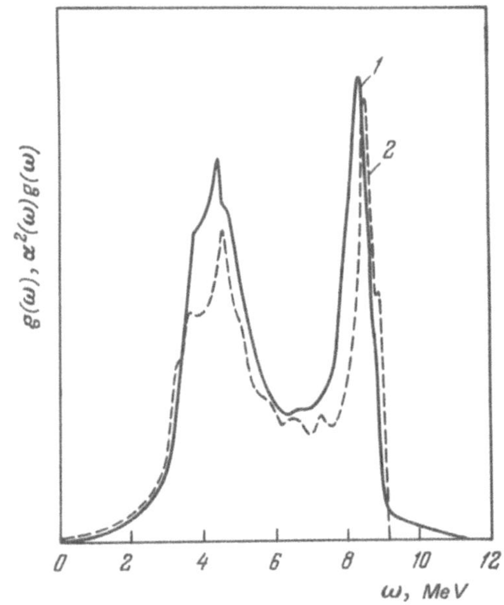


**Figure 4.2. Plots of the functions: (1) $g(\omega)$, obtained from neutron scattering measurements and (2) $\alpha^2(\omega)g(\omega)$, found from tunneling measurements for Pb.**

Unfortunately, the method of tunneling spectroscopy allows the measurement of the function $\alpha^2(\omega)g(\omega)$ only in superconducting metals and alloys. This does not make it possible for us to carry out a detailed analysis of this function in all metals and alloys and determine exactly which form of this function is favorable to the existence of high $T_c$ values. Recently, a method was proposed that allows the determination of the function $\alpha^2(\omega)g(\omega)$ even for normal metals from the investigation of the current-voltage characteristics of point junctions [22]. Undoubtedly, with increase of the accuracy of such measurements, this method will become a very effective means of studying the electron-phonon interaction.

It should be emphasized that a change in the shape of the function $\alpha^2(\omega)g(\omega)$ does not always lead to a change in the electron-phonon coupling constant $\lambda$. In particular, the smearing out of the narrow peaks in the function $\alpha^2(\omega)g(\omega)$ upon the isotopic disordering of a metal does not at all alter the values of $\lambda$. Moreover, it can be rigorously demonstrated [23] that the electron-phonon coupling constant does not depend on the mass of the atoms of the alloy in both the ordered and disordered states.

Thus far, this function has been computed only for some simple metals and their alloys [19, 24, 25].†

In the absence of such calculations, the analysis of the dependence of the coupling constant $\lambda$ on the properties of the normal metal is very often performed, using the empirical formula proposed by McMillan [1] for $\lambda$:

$$\lambda = \frac{N(0)\langle I^2\rangle}{M\langle\omega^2\rangle}. \qquad (4.70)$$

Here $N(0)$ is the density of states of the electrons at the Fermi surface, $I$ is the electron-ion interaction matrix element, $\langle\omega^2\rangle$ is the mean square of the phonon frequency, and $M$ is the mass of the ions of the metal. McMillan has shown on the basis of an analysis of a large number of experimental data for the transition metals and their alloys that the quantity $N(0)\langle I^2\rangle$ is roughly a constant for all the materials investigated by him and, consequently, that the change that occurs in $\lambda$ when we go from one material to another is largely due to the change in the quantity $M\langle\omega^2\rangle$. Later, Hopfield [30] produced a theoretical justification of the

†Computations of this function have also been carried out recently for some transition metals. In particular, computations for Nb are carried out in [26–28]. The quantitative agreement of the obtained results with the experimental data is not quite satisfactory. The computed values of the function $\alpha^2(\omega)g(\omega)$, especially for large energy values, significantly exceed the experimental values. The cause of these discrepancies is not clear. It is possible that they are due to errors in the experimental data since tunneling measurements in the transition metals present considerable difficulties. A more detailed discussion of this problem can be found in [26–28] as well as in the published proceedings of the second Rochester conference on superconductivity in *d*- and *f*-band metals [29].

nondependence of $N(0)\langle I^2\rangle$ on the density of electron states in the transition metals.

Unfortunately, in writing down the formula (4.70), McMillan did not give an explicit expression for the quantity $M\langle\omega^2\rangle$. In the majority of papers, estimates of the quantity $\langle\omega^2\rangle$ are made from measurements of the Debye frequency, $\omega_D$, of the metal. Even in a one-component metal the replacement of $\langle\omega^2\rangle^{1/2}$ by the quantity $\omega_D$ can be erroneous. As for the multicomponent alloys, especially those whose components have essentially different masses, it is generally difficult to make any choice for this quantity.

In order to get rid of the uncertainty in the choice of the quantity $M\langle\omega^2\rangle$, we shall present a simplified derivation of the formula (4.70). To do this, let us, first, rewrite the expression, (4.26), for the function $\alpha^2(\omega)g(\omega)$ in the form proposed in [31, 32]:

$$\alpha^2(\omega)\,g(\omega)=\frac{1}{N\pi^3N(0)}\sum_{n,n'}\int d^3r\int d^3r'\,\mathrm{Im}\,[G(\mathbf{r},\mathbf{r}',E_F)]\times$$
$$\times\nabla_\alpha\tilde{\tilde{V}}_{ei}(\mathbf{r}-\mathbf{R}_n)\,\mathrm{Im}\left[D^{\alpha\beta}_{nn'}(\omega)\right]\nabla_\beta\tilde{\tilde{V}}_{ei}(\mathbf{r}'-\mathbf{R}_{n'})\times$$
$$\times\,\mathrm{Im}\,[G(\mathbf{r},\mathbf{r}',E_F)]. \qquad (4.71)$$

Here $N$ is the number of ions, $\tilde{\tilde{V}}_{ei}(\mathbf{r}-\mathbf{R}_n)$ is the screened electron–ion interaction potential,

$$-\frac{1}{\pi}\,\mathrm{Im}\,G(\mathbf{r},\mathbf{r}',\varepsilon_F)=\frac{\Omega}{(2\pi)^3}\int_\Omega d^3k\,\delta(E_\mathbf{k}-E_F)\psi^*_\mathbf{k}(\mathbf{r})\,\psi_\mathbf{k}(\mathbf{r}'). \qquad (4.72)$$

The Bloch wave functions $\psi_\mathbf{k}(\mathbf{r})$ have been chosen such that they are normalized in the volumn, $\Omega$, of one unit cell; $D^{\alpha\beta}_{nn'}(\omega)$ is the Green function for the ion displacements.

Using for the coupling constant $\lambda$ the expression (4.35), and taking into account the fact that

$$2\int\frac{d\omega}{\omega}\,\mathrm{Im}\,D^{\alpha\beta}_{nn'}(\omega)=\pi D^{\alpha\beta}_{nn'}(0),$$

we find that

$$\lambda=\frac{1}{N\pi^2N(0)}\sum_{n,n'}\int d^3r\int d^3r'\,\mathrm{Im}\,[G(\mathbf{r},\mathbf{r}',E_F)]\times$$
$$\times\nabla_\alpha\tilde{\tilde{V}}_{ei}(\mathbf{r}-\mathbf{R}_n)\,D^{\alpha\beta}_{nn'}(0)\,\nabla_\beta\tilde{\tilde{V}}_{ei}(\mathbf{r}'-\mathbf{R}_{n'})\,\mathrm{Im}\,G(\mathbf{r},\mathbf{r}',E_F). \qquad (4.73)$$

On the basis of (4.73), we can rigorously prove that the electron–phonon coupling constant does not depend on the ion mass. It follows directly from the theory of lattice vibrations, as well as from the formulas obtained in the preceding chapter for the phonon Green function, that

$$D^{\alpha\beta}_{nn'}(0) = (\Phi^{\alpha\beta}_{nn'})^{-1}, \tag{4.74}$$

where $\Phi^{\alpha\beta}_{nn'}$ is the dynamical vibration matrix, determined only by the screened inter-ion interaction.

The dominant contribution to the coupling constant is made by the interactions with the phonons with large momenta, and it is these same phonons that make the greatest contribution to the quantity $D^{\alpha\beta}_{nn'}(0)$. As shown in [33] by Appel and Kohn, because these phonons are nearly localized vibrations, we can assume with a high degree of accuracy that

$$D^{\alpha\beta}_{nn'}(0) = D^{\alpha\beta}_{nn}(0)\,\delta_{nn'}. \tag{4.75}$$

Substituting the formulas (4.74) and (4.75) into (4.73), we obtain

$$\lambda = \frac{1}{\pi^2 N(0)}\frac{1}{N}\sum_n \int d^3r \int d^3r' \,\mathrm{Im}\,[G(\mathbf{r}, \mathbf{r}', E_F)] \times$$

$$\times \nabla_\alpha \tilde{\tilde{V}}_{ei}(\mathbf{r}-\mathbf{R}_n)\,\nabla_\beta \tilde{\tilde{V}}_{ei}(\mathbf{r}'-\mathbf{R}_n)\,\mathrm{Im}\,[G(\mathbf{r}, \mathbf{r}', E_F)]\,(\Phi^{\alpha\beta}_{nn})^{-1}. \tag{4.76}$$

The expression (4.76) can be rewritten in the form†

$$\lambda = \frac{1}{N}\sum_n N(0)\,\langle I^2_{n,\alpha\beta}\rangle\,(\Phi^{\alpha\beta}_{nn})^{-1}, \tag{4.77}$$

where $N(0)\langle I^2\rangle$ is the matrix element of the electron–ion interaction in the $n$th unit cell:

$$N(0)\langle I^2_{n,\alpha\beta}\rangle = \frac{1}{\pi^2 N(0)}\int_{\Omega_n} d^3r \int_{\Omega_n} d^3r' \,\mathrm{Im}\,[G(\mathbf{r}, \mathbf{r}', E_F)] \times$$

$$\times \nabla_\alpha \tilde{\tilde{V}}_{ei}(\mathbf{r})\,\nabla_\beta \tilde{\tilde{V}}_{ei}(\mathbf{r}')\,\mathrm{Im}\,[G(\mathbf{r}, \mathbf{r}', E_F)]. \tag{4.78}$$

†As has been shown by Dolgov and Mazin [34], the relation (4.75) is clearly inexact. Moreover, the properties of the function $D^{\alpha\beta}_{nn'}(0)$ cannot by themselves serve as sufficient justification for the use of the approximation (4.76). A more important circumstance that leads to the indicated separation of the constant into two cofactors is, in fact, not the smallness of the function $D^{\alpha\beta}_{nn'}(0)$ with $n \neq n'$ as compared to $D^{\alpha\beta}_{nn}(0)$, but the existence in the matrix elements of interference cofactors of the type $\exp[i(\mathbf{k}-\mathbf{k}')\cdot(\mathbf{R}^0_n-\mathbf{R}^0_{n'})]$. Since the most important momentum transfers $\mathbf{q} = \mathbf{k} - \mathbf{k}'$ have magnitudes of the order of $2k_F$, the effect of these interference cofactors can lead to the smallness of the contribution to the constant $\lambda$ of the terms with $n \neq n'$. But this conclusion turns out to be incorrect, in particular, for the alkali metals, in which the characteristic momentum transfers $q < 2k_F$. A similar result has been obtained by Rietschel [35].

In a one-component metal we have

$$\frac{1}{3N}\sum_n (\Phi_{nn}^{\alpha\beta})^{-1} = \frac{1}{M}\int \frac{g(\omega)\,d\omega}{\omega^2} = \frac{1}{M}\left\langle \frac{1}{\omega^2} \right\rangle. \tag{4.79}$$

By comparing the formula (4.77) with the expression (4.70), proposed by McMillan for the coupling constant λ, we can now give real meaning to the quantities contained in it for both single-component metals and alloys.† In particular, for an alloy of elements, A and B, with significantly different masses, the formula (4.77) coincides with the formula proposed earlier by Phillips [36]:

$$\lambda = \frac{N_{\rm A}(0)\langle I_{\rm A}^2\rangle}{M_{\rm A}\langle \omega_{\rm A}^2\rangle} + \frac{N_{\rm B}(0)\langle I_{\rm B}^2\rangle}{M_{\rm B}\langle \omega_{\rm B}^2\rangle}, \tag{4.80}$$

where the density of electron states for a given atom is defined as

$$N_A(0) = -\frac{1}{\pi}\int_{\Omega_{\rm A}} d^3r \,\mathrm{Im}\, G(\mathbf{r}, \mathbf{r}', E_F).$$

Proceeding to the consideration of the quantity $N(0)\langle I^2\rangle$, let us first of all note that its behavior, specifically its dependence on a valency of the metal, in the case of the simple metals is very different from what obtains in the transition metals. For the simple metals, we can, when computing this quantity, limit ourselves to the nearly free-electron approximation. Then

$$N(0)\langle I^2\rangle = \frac{N}{v_F}\frac{1}{(2\pi)^2}\int_0^{2k_F} q^3V^2(q)\,dq, \tag{4.81}$$

where $V(\beta)$ is the electron–ion interaction pseudopotential. Introducing the averaged pseudopotential value

$$\overline{V^2(q)} = \int_0^{2k_F} q^3V^2(q)\,dq \Big/ \int_0^{2k_F} q^3V^2(0)\,dq,$$

and taking into account the fact that

$$V(0) = -\frac{Z}{2N(0)},$$

†It can be seen from this, in particular, that it is the quantity $\langle 1/\omega^2\rangle$ and not $1/\langle\omega^2\rangle$ that, strictly speaking, enters into the expression for the constant λ. These quantities normally do not differ very significantly from each other, and below wherever the difference between them does not play a fundamental role we shall neglect it.

where $Z$ is the valency of the metal, we have

$$N(0)\langle I^2\rangle = \frac{4\pi^2 E_F N Z^2}{k_F}\overline{V^2(q)}. \tag{4.82}$$

Finally, for the simple metals, it is convenient to write the expression for $\lambda$ using the formula (4.79) in the following form:

$$\lambda = \pi \frac{E_F}{e^2 k_F}\left\langle \frac{\Omega_{\text{pl}}^2}{\omega^2}\right\rangle \overline{V^2(q)}, \tag{4.83}$$

where we have introduced the notation

$$\Omega_{\text{pl}} = \frac{4\pi N Z^2 e^2}{M}. \tag{4.84}$$

For the transition metals, the nearly free-electron approximation is absolutely inapplicable, and a more rigorous approach to the computation of the quantity $N(0)\langle I^2\rangle$ is necessary. With that end in view, let us, using (4.78) and (4.72), rewrite the expression for $N(0)\langle I^2\rangle$ in the form

$$N(0)\langle I^2\rangle = \frac{1}{N(0)}\int \frac{dS_k}{v_k}\int \frac{dS_{k'}}{v_{k'}} |\langle \mathbf{k}|\nabla \tilde{\tilde{V}}_{ei}|\mathbf{k}'\rangle|^2. \tag{4.85}$$

In writing down this formula, we used the fact that the screened electron-ion interaction potential is a local potential, and that it (or, more precisely, its gradient) can be assumed to be concentrated in one unit cell. Th matrix element of the gradient of the potential has the standard form:

$$\langle \mathbf{k}|\nabla \tilde{\tilde{V}}_{ei}|\mathbf{k}'\rangle = \frac{1}{\Omega_0}\int_{\Omega_0} d^3r\, \psi^*_{\mathbf{k}'}(\mathbf{r})\, \nabla V_{ei}(\mathbf{r})\, \psi_{\mathbf{k}}(\mathbf{r}). \tag{4.86}$$

The Bloch function $\psi_{\mathbf{k}}(r)$ in the unit cell in question can be represented in the form of an expansion in terms of the spherical harmonics $Y_{lm}(\theta, \phi)$:

$$\psi_k(\mathbf{r}) = \sum_{lm} Y_{lm}(\theta, \varphi)\, g_{lmk}(r).$$

The density of electron states with a given $l$ at the Fermi surface is defined as

$$N_l(0) = \lim_{\Delta \to 0} \sum_{\substack{k \\ E_F < E_k < E_F + \Delta}} \sum_{m} \int_0^{r_0} |g_{lmk}(r)|^2 r^2 \, dr,$$

where $r_0$ is the radius of the unit cell. The total density of states, $N(0)$, is equal to

$$N(0) = \sum_l N_l(0).$$

Let us substitute the expression for $\psi_{\mathbf{k}}(r)$ into the formula (4.86). Because of the spherical symmetry of the potential $\tilde{\tilde{V}}_{ei}(r)$, only the transitions between states with $l$ and $l \pm 1$ contribute to the matrix element $\langle \mathbf{k} | \nabla \tilde{\tilde{V}}_{ei} | \mathbf{k}' \rangle$. In the transition metals, the most important transitions are the transitions between the $p$ and $d$ states, i.e., the states with $l = 1$ and $l = 2$.† In this case the expression for $N(0)\langle I^2 \rangle$ can be rewritten [30] in the form

$$N(0)\langle I^2 \rangle = \left( \frac{d\tilde{\tilde{V}}_{ei}}{dz} \right)^2 \frac{N_p(0)\, N_d(0)}{N(0)}, \tag{4.87}$$

where

$$\left( \frac{d\tilde{\tilde{V}}_{ei}}{dz} \right)^2 = \frac{\left| \int_0^{r_0} r^2 g_{20k}^* \left[ \nabla_z V_{ei}(r) \right] g_{10k'} dr \right|^2}{\int_0^{r_0} r^2 |g_{10k'}|^2 dr \int_0^{r_0} r^2 |g_{20k}|^2 dr}.$$

As has been shown by Hopfield [30], the quantity $(d\tilde{\tilde{V}}_{ei}/dz)^2$ is actually an atomic characteristic of the material and depends little on its crystallographic properties. Further, since in the transition metals $N_d(\mathrm{o}) >> N_p(0)$ and $N_p(0)$

†Recent work on the microscopic calculation of the quantity $N(0)\langle I^2 \rangle$ has shown that large contributions are made to the value of $N(0)\langle I^2 \rangle$ not only by the $p$–$d$ transitions, as was assumed by Hopfield [30], but also by the $d$–$f$ transition, i.e., by the transitions from states with $l = 2$ into states with $l = 3$. Since in the transition metals the $f$ component of the wave function of an electron at a given site arises largely on account of the tails of the $d$ wave functions from other unit cells, the density of states of the $f$ component turns out to be proportional to the density of states of the $d$ electrons. This leads to the direct dependence of the quantity $N(0)\langle I^2 \rangle$ on the density of electron states. The coefficient of proportionality in this case depends on the location of the Fermi level in the Brillouin zone, on the series number of the transition metals, etc. A detailed discussion of the present situation regarding the understanding of the microscopic nature of the quantity $N(0)\langle I^2 \rangle$ can be found in [37–40], as well as in Allen's and Grimvall's review papers [41].

has the same order of magnitude as in the simple metals, we obtain

$$N(0)\langle I^2\rangle = N_p(0)\left(\frac{d\tilde{\tilde{V}}_{ei}}{dz}\right)^2. \tag{4.88}$$

The coupling constant λ for the transition metals can be written in the form

$$\lambda = N_p(0)\left(\frac{d\tilde{\tilde{V}}_{ei}}{dz}\right)^2 \frac{1}{M}\left\langle\frac{1}{\omega^2}\right\rangle. \tag{4.89}$$

The formula (4.89) thus obtained can serve as a justification of the nondependence, observed by McMillan [1], of the coupling constant λ on the density of states in the transition metals since $N_p(0)$ is nearly the same for all the transition metals.

Of course, (4.89) is an approximate formula, and, for the exact computation of λ, it is necessary to take all the scattering channels into account. The corresponding formulas are given in [31, 32, 42]. It is shown in these papers that an expression for the quantity $N(0)\langle I^2\rangle$ can be written down, using the phase shifts connected with the scattering of the electrons by the potential $\tilde{\tilde{V}}_{ei}(r)$, in the form

$$N(0)\langle I^2\rangle = \frac{E_F}{\pi^2 N^2(0)} \sum_{l=0}^{\infty} \frac{2(l+1)\sin^2(\delta_{l+1}-\delta_l)\, N_{l+1}N_l}{N_{l+1}^1 N_l^1}. \tag{4.90}$$

Here $N_l$ is the exact density of electron states with a given angular momentum $l$ at the Fermi surface, $\delta_l$ is the scattering phase, and $N_l^1$ is the component of the density of states with allowance for scattering by only one center.

Calculations of the cupling constant λ, performed with the aid of the formula (4.90) for certain transition and simple metals, lead to surprisingly good agreement with the experimental data [31, 32, 42].† We shall discuss these calculations in greater detail in the following section when considering the critical temperature of the superconducting transition.

## 5. Dependence of the Critical Temperature on the Properties of the Normal Metal

Correlations have time and again been established in investigations of superconducting materials between the values of $T_c$ and a number of the properties of

†A large number of papers devoted to various aspects of the microscopic computation of the electron–phonon coupling constant λ have recently been published. A detailed discussion of them can be found in Allen's and Grimvall's review articles [41] as well as in the book by Vonsovskii *et al.* [43]. A great deal of material can also be found on this subject in the reports of the recent conferences [29, 44].

the corresponding metals. Thus, the quantity $T_c$ has been observed to correlate with the melting point and the strength of the metal [45], with the thermal coefficient of resistance $d\rho/dT$ at high temperatures [23], etc. The best known correlation is the one enunciated by Matthias's rule [46], which relates the value of the critical temperature $T_c$ with the number, $Z$, of valence electrons of the metal in question.

Proceeding to study the connection between the quantity $T_c$ and the properties of the normal metal, let us, to begin with, consider in greater detail the properties of the principal quantity determining $T_c$: the coupling constant $\lambda$. For this purpose, let us rewrite the expression, (4.70), for this quantity:

$$\lambda = \frac{N(0)\langle I^2\rangle}{M}\left\langle\frac{1}{\omega^2}\right\rangle. \tag{4.70a}$$

Let us emphasize right away that in this expression not only the factor $N(0)\langle I^2\rangle$, but also the quantity $\langle 1/\omega^2\rangle$, is determined by the electronic properties of the metal (i.e., by the valency of the metal, the strength of the electron interaction, etc.).

In accordance with the results of Chapter 3, the phonon frequencies can be written in the form

$$\omega^2 = \Omega_i^2 - \omega_e^2, \tag{4.91}$$

where the $\Omega_i$ are the bare frequencies of the ionic lattice and $\omega_e$ is the contribution of the electrons to the phonon frequencies, a contribution that is proportional to the electronic susceptibility and the electron-ion interaction potential.

As the electronic properties of the metal, e.g., its valency and density of electron states, are varied, both the electron-ion interaction matrix element $N(0)\langle I^2\rangle$ and the mean square of the phonon frequencies $\langle 1/\omega^2\rangle$, generally speaking, vary. Therefore, the question arises: the variation of which of the quantities, $N(0)\langle I^2\rangle$ or $\langle 1/\omega^2\rangle$, determines the variation of the coupling constant $\lambda$ to a greater degree? Below we shall discuss this question on the basis of both theoretical arguments and the experimental data. It should, however, be said that there is no unique answer to it. Either factor can play the more important role, depending on the circumstance and the specific system. On the whole, there is hardly any sense in carrying out a comparison of the data for very different materials on the basis of the formulas (4.70a) and 4.91). It is more reasonable to consider a number of similar materials; specific relationships may then be identified.

Let us begin our investigation with the simple *sp* metals and write for them the expression (4.83) for the constant $\lambda$ in the following form:

$$\lambda = \frac{1.51}{r_s} \overline{V^2(q)} \left\langle \frac{\Omega^2_{\text{пл}}}{\omega^2} \right\rangle. \tag{4.83a}$$

In writing does this formula, we used the equality

$$\frac{E_F}{e^2 k_F} = \frac{0{,}96}{r_s},$$

where $r_s$ is the ratio of the interelectron distance to the Bohr radius of the electron. In the simple metals $r_s$ varies from $r_s \sim 4$–$5$ for the alkali metals to $r_s \sim 2$ in the multivalent metals.

In the simple metals the electron-ion interaction can be described [9] by a pseudopotential of the following form:

$$V(q) = \frac{-(4\pi e^2/q^2)\cos R_c q}{\varepsilon_e(q,0)}, \tag{4.92}$$

where $R_c$ is the radius of the ion shell and $\varepsilon_e(q, 0)$ is the permittivity of the electron gas. As can be seen from (4.92), the pseudopotential is smaller than the potential of the direct electron Coulomb interaction at large momenta $q$ and vanishes when the momentum is $q_0 = \frac{1}{2}\pi R_c^{-1}$. Calculations carried out in [9] for lead with the use of the value $R_c = 0.57$ Å lead to a very small value for $\overline{V^2(q)}$:

$$\overline{V^2(q)} = 0.04.$$

In the alkali metals, in which $q_0 \approx \mathrm{w} k_F$, the quantity $\overline{V^2(q)}$ is even smaller. Because of the smallness of the mean pseudopotential, the electron-phonon coupling constant in the simple metals will be of the order of unity only when the mean phonon frequencies are extremely small compared to $\Omega_{\text{pi}} = (4\pi N Z^2 e^2/M)^{1/2}$. The phonon frequencies in the simple metals have been repeatedly computed, and the following fact has been discovered. In the univalent metals, the high-frequency vibrations differ little from the vibrations of the purely ionic lattice. In this case the ratio $\langle \Omega^2_{\text{pi}}/\omega^2 \rangle$ is not too high, and the coupling constant $\lambda$ in these metals is of the order of 0.1-0.3.

As the valency of the metal increases, the contribution of the electrons to the phonon frequencies $\omega_e^2$ increases, and, as a result, the ratio $\Omega^2_{\text{pi}}/\omega^2$ increases continuously. This can clearly be seen from Table 4.1, which was taken from [47]. In it are given, in units of the plasma frequency $\Omega_{\text{pi}}$, the vibration frequencies at certain points near the boundaries of the Brillouin zones for Na,

TABLE 4.1

| | Na | Al | | Pb | |
|---|---|---|---|---|---|
| | $L, T(H_{15})$ | $L(X_1)$ | $T(X_5)$ | $L(X_1)$ | $T(X_5)$ |
| $(\Omega_i/\Omega_{\text{pl}})^2$ | 0.333 | 0.678 | 0.161 | 0.678 | 0.161 |
| $(\omega/\Omega_{\text{pl}})^2$ | 0.248 | 0.105 | 0.037 | 0.031 | 0.007 |

Al, and Pb; $\Omega_i$ is the frequency of the ionic lattice and $\omega$ is the true vibration frequency.

It can be seen from the table that the ratio $\Omega_{\text{pl}}^2/\omega^2$ is equal to four for the $L$ and $T(H_{15})$ vibrations in Na. In Al, this ratio already has, for the two vibrations $L(X_1)$ and $T(X_5)$, values, respectively, equal to 9.54 and 27. In Pb, the ratio $\Omega_{\text{pl}}^2/\omega^2$ possesses even greater values: 32.2 and 142. It can be seen from an analysis of these data that the greatest contribution to the coupling constant $\lambda$ in the multivalent metals is made by the transverse phonons [in our case these are the $T(X_5)$ phonons], for which the ratio $\Omega_{\text{pl}}^2/\omega^2$ is significantly larger than for the longitudinal vibrations.

The increase of the critical temperature of the superconducting transition in the simple metals with increasing valency of the metal is a long-established fact [46]. The causes of the increase of the coupling constant $\lambda$ and the critical temperature $T_c$ with increasing valency of the metal have been thoroughly investigated by Dynes and Rowell [48] and Allen and Dynes [16, 49]. They have shown that the main cause of this phenomenon is connected not with the decrease of the absolute values of the phonon frequencies $\omega^2$, but with the growth of $\Omega_{\text{pl}}^2$. This means that the simplest formula

$$\lambda = \frac{C}{M\langle\omega^2\rangle}, \tag{4.93}$$

where $C$ has the same value for all metals, cannot be used for the coupling constant $\lambda$ in the simple metals.

It should, however, be emphasized that the formula (4.93) turns out to be quite adequate for the description of the variation of the quantity $\lambda$ in a given metal, for example, under the influence of pressure [50]. In this case the valency of the metal does not vary. In addition the variation of the compaction parameter, $r_s$, of the electron gas exerts a stronger influence on the quantity $\lambda$ through the variation of the phonon frequencies (on account of the variation of the electron contribution to them) than directly.

Among the simple metals, metallic hudrogen can be expected to possess the highest superconducting transition temperature for the following reasons.

First, in hydrogen the radius, $R_c$, of the ion core is equal to zero; therefore, the quantity $\overline{V^2(q)}$ will have a much greater value in it than in the remaining

TABLE 4.2

| $r_s$ | $\Delta_0$, Ry | $T_c$,K |
|---|---|---|
| 0.1 | 0 | 0 |
| 0.5 | $3 \cdot 10^{-5}$ | 1 |
| 6.9 | $5 \cdot 10^{-4}$ | 30 |
| 1.3 | $8 \cdot 10^{-4}$ | 110 |
| 1.5 | $9 \cdot 10^{-4}$ | 140 |

simple metals. In fact, it is due to the absence in hydrogen of an ion core; as a result, the electron-ion pseudopotential coincides with the true Coulomb potential.

Second, because of its small atomic mass, hydrogen should possess a higher Debye temperature and, consequently, a larger pre-exponential function in the expression for the critical temperature. Specific calculations carried out in [51-53] show that this is indeed the case. Table 4.2, which was taken from [53], shows the results of calculations of the critical temperature $T_c$ and the energy gap $\Delta_0$ for metallic hydrogen for different values of the compaction parameter $r_s$.

Generally speaking, the specific $T_c$ values obtained in the various calculations [51-53] for the same value of $r_s$ differ significantly from each other; but the conclusion that it is possible to obtain in metallic hydrogen $T_c$ values as high as ~100 K remains general.†) Unfortunately, hydrogen is a dielectric at normal pressures and will go over into the metallic state only at pressures of several million atmospheres. Various sorts of approximate computations yield for the transition pressure $P_c$ values ranging from 1 to 10 Mbar [55-58]. Some experimental data have also been published recently that indicate the possibility of such a transition at pressures of several megabar [59-61].

Of course, it is hardly worth speaking about the practical use of the superconducting properties of metallic hydrogen at such pressures. But relatively recently there appeared the hope that metallic hydrogen can exist in a meta-

†We should, however, point out in this connection Gupta and Sinha's recent work [54], in which they call in question the results of previous calculations of the $T_c$ in metallic hydrogen. The essence of Gupta and Sinha's observations amounts essentially to the following. Allowance for the nonlinearity, neglected in previous computations, of the screening of the proton in metallic hydrogen leads to a situation in which the conduction electrons near the Fermi surface, which, strictly speaking, are the electrons that participate in superconductivity, do not feel the Coulomb potential of the proton but some other weaker potential. This potential effectively plays the same role as the pseudopotential in the other simple metals, and, according to Gupta and Sinha's [54] estimates, is sufficiently small and leads to a small electron-phonon coupling constant $\lambda = 0.261$. This, in turn, yields the value $T_c = 0.1$ K. Of course, the numerical values obtained in that paper for the quantities $\lambda$ and $T_c$ cannot be taken with absolute confidence; they do, however, indicate the need for further careful investigations of this problem.

stable state at $P = 0$. As a definite confirmation of this, we can cite the computations carried out by Brovman *et al.* [58]. They showed that metallic hydrogen can indeed exist in a structurally distinctive metastable phase at $P = 0$. But the lifetime of such a phase was essentially not computed in the cited papers.

Estimates made by Salpeter in [62] indicate that the metallic phase is very unstable against evaporation and that its lifetime is very short ($\tau \approx 10^{-3}$ sec). There is another possibility of decay of the metastable state of hydrogen besides evaporation, namely, decay as a result of a transition into the dielectric state with a doubling of the lattice constant. Since, according to the calculations carried out in [57, 58], the metallic phase at $P = 0$ consists of linear chains of hydrogen ions, the Peierls transition into the dielectric state is possible in such a system. Precisely such type of instability is pointed out in [63, 64]. The problem of metallic hydrogen has, unfortunately, not yet been definitely solved.

Let us now consider the properties of the transition metals and their compounds. In contrast to the simple metals, the critical temperature in the transition metals depends in a complicated manner on the number, $Z$, of valence electrons. This can be seen from Fig. 4.3, which was taken from [65]. For comparison, we show in Fig. 4.4 a plot of the density of electron states in the transition metals as obtained from measurements of the electron thermal capacity. Comparing the behavior of $T_c$ with that of the density of states $N(0)$,

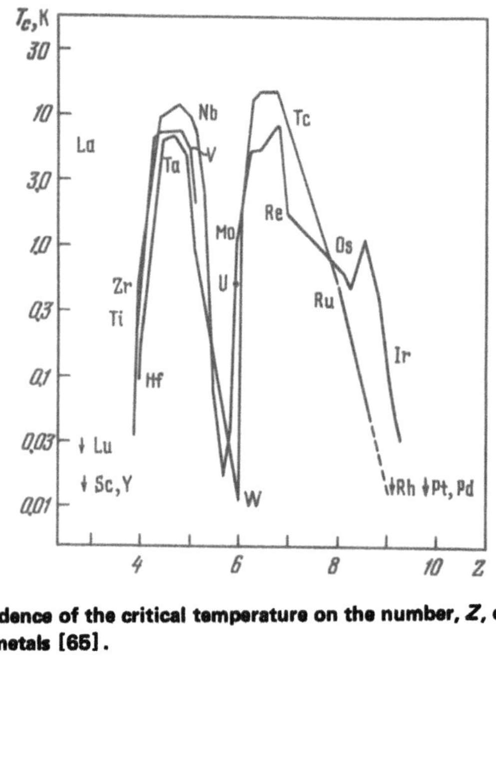


**Figure 4.3. Dependence of the critical temperature on the number, $Z$, of valence electrons for the transition metals [65].**

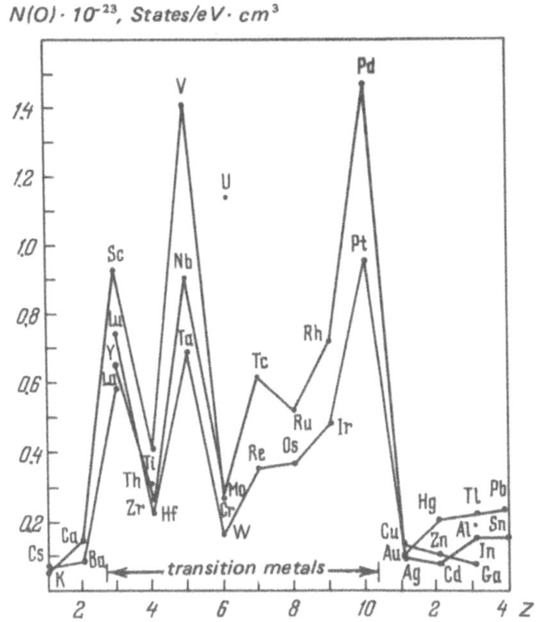


**Figure 4.4. Dependence of the density of electron states in the metals on the number, *Z*, of valence electrons [65].**

we see that at least in the range $4 \leqslant Z \leqslant 8$ the behavior of $T_c$ qualitatively corresponds to the behavior of $N(0)$.

The establishment of the existence of such a correlation is by itself still inadequate for the understanding of the physical nature of the electron-phonon interaction in the transition metals. First, the nature of such a complicated oscillatory dependence on $Z$ of the density of states itself is not entirely clear to us. Furthermore, as has been shown by McMillan [1] (see Section 4), the quantity $N(0)$ itself does not enter directly into the expression for the electron-phonon coupling constant $\lambda$. McMillan showed in his study of elements from group V (V, Na, Ta) and group VI (Cr, Mo, W) that, although the quantity $N(0)$ varies widely from element to element (its value for one element can be several times greater than its value for another), the product $N(0)\langle I^2 \rangle$, which enters into the expression, (4.70), for $\lambda$, is the same for all these elements to within 10%. The increase of the quantity $\lambda$ by a factor of nearly 2 in going from the group VI elements to the group V elements occurs largely as a result of a decrease in the mean phonon frequencies. In other words, the relation (4.93) is a very good approximation for the constant $\lambda$ in these elements. The correlation observed here between $T_c$ and the density of states arises largely as a result of the dependence of the phonon frequencies on $N(0)$ since the electron contribution to the lattice dynamics, $\omega_e^2$, is proportional to the electronic susceptibility $\chi \approx N(0)$.

Thus, as $N(0)$ increases, $\omega_e^2$ increases, and the mean square, $\langle\omega^2\rangle$, of the phonon frequencies decreases, with the result that $\lambda$ increases.

It has been experimentally observed that a dependence of the form

$$\lambda = \frac{\lambda_0}{1 - bN(0)} \tag{4.94}$$

is obeyed well in, for example, the compounds $V_3Ga_{1-x}Si_x$, $V_3Ga_{1-x}Sn_x$ [66]. Furthermore, in [67] the dependence of the phonon frequencies on the electronic characteristics is used to explain Matthias's rules [46], by which the existence of maximum $T_c$ values for $Z \approx 4.7$ and $Z \approx 6.8$ is predicted. The following expression is proposed in that paper for $\lambda$:

$$\lambda = \frac{\lambda_0}{1 - {}^4/_3 E_F N(0)\lambda_0}, \tag{4.95}$$

the quantity $\lambda_0$ being assumed to be a constant and equal to $\lambda_0 = (4.3)^{-1}$. As shown in that paper, the constant $\lambda$, as determined by the formula (4.95), in fact has maxima at $Z = 4.9$ and $Z = 6.8$, becoming infinite at these valencies. The corresponding phonon frequencies for these $Z$ values are equal to zero. It seems to us that the vanishing of the mean phonon frequencies is patently unphysical and is a consequence of extreme approximations.

Chui [68] has shown in detailed phonon-frequency computations, performed in conformity with experiment, that the phonon frequencies indeed have a maximum at $Z \approx 5.5$–$6$ and fall off toward the ends of the period, but the mean square of the frequencies $\langle\omega^2\rangle$ does not, of course, vanish. Furthermore, it is shown in [69] that the electron–ion interaction matrix element $\langle I^2\rangle$ is itself proportional to $E_F$, so that the quantity $N(0)\langle I^2\rangle$ behaves like $E_F N(0)$. Thus, clearly, not only the denominator in (4.95), but also the numerator, contributes to the dependence of the coupling constant $\lambda$ on the parameter $N(0)E_F$, a dependence that explains the Matthias rule.

It should be noted that, in general, it is doubtful that the dependence (shown in Fig. 4.3) of $T_c$ on the valency $Z$ can be explained by only the variation of the electron–phonon coupling constant. At least the absence of superconductivity in the transition metals with $Z > 8$ is clearly due not simply to the smallness of the constant $\lambda$ but first and foremost to the effect of the strong Coulomb correlations connected with the spin fluctuations [12]. The properties of such metals are considered in greater detail in [65] by Gladstone *et al.* Let us also note that, generally speaking, the Matthias rules are themselves not universal. In particular, they are not obeyed in amorphous transition metals and alloys [70].

The direct dependence of $\lambda$ on the electronic characteristics that arises on account of the quantity $N(0)\langle I^2\rangle$ has been observed in a number of transition metals. Weber, in particular, arrives at such a conclusion in [71], where he shows for the particular case of the carbides of the transition metals that the allowance for the variation of $l/m\langle\omega^2\rangle$ alone is not adequate for the correct description of the constant $\lambda$; the variation of the quantity $N(0)\langle I^2\rangle$ should also be taken into account. A similar conclusion was arrived at earlier by Ishikawa and Toth [72], who, to account for the experimental data, proposed an expression of the form

$$\lambda = \frac{a + bN(0)}{1 - b'N(0)},$$

which is reminiscent of (4.94) but makes additional allowance for the direct dependence of the quantity $N(0)\langle I^2\rangle$ on the electronic characteristics.

The inadequancy of the allowance for only the factor $\langle 1/\omega^2\rangle$ in the explanation of the high $T_c$ values in a number of materials, e.g., in Nb as compared to Mo or in NbC and TaC in comparison with HfC, is emphasized in [42], in which accurate microscopic calculations of the quantity $\lambda$ are carried out with the use of the expression (4.90) for the determination of $N(0)\langle I^2\rangle$ and the expression (4.80) for the constant $\lambda$ in the case of the compounds:

$$\lambda = \frac{N_{\mathrm{A}}(0)\langle I_{\mathrm{A}}^2\rangle}{M_{\mathrm{A}}\langle\omega_{\mathrm{A}}^2\rangle} + \frac{N_{\mathrm{B}}(0)\langle I_{\mathrm{B}}^2\rangle}{M_{\mathrm{B}}\langle\omega_{\mathrm{B}}^2\rangle}.$$

Below in Table 4.3 we give the results of these computations.

First, the surprisingly good agreement between the computed and the experimentally determined values of $\lambda$ and $T_c$ should be noted.

As can be seen from Table 4.3, the absence of superconductivity in HfC is due to the smallness of the value of $N(0)\langle I^2\rangle$ for it in comparison with the values in TaC and NbC and not be the difference in the phonon frequencies of these compounds. The following distinctive features can also be noted in a study of the table. First, in the case of the high-temperature superconductors the quan-

**TABLE 4.3**

| | | $N(0)\langle I^2\rangle$, eV/A$^2$ | $\langle I^2p\rangle$, (eV/A)$^2$ | $M_{\mathrm{A}}\langle\omega_{\mathrm{A}}^2\rangle$, eV/A$^2$ | $\lambda$ Exp. | $\lambda$ Computed | | $T_c$ Exp. | $T_c$ Computed |
|---|---|---|---|---|---|---|---|---|---|
| V | | 6.35 | 5.72 | 9.98 | 0.60 | 0.64 | | 5.30 | 6.5 |
| Nb | | 7.64 | 8.68 | 8.73 | 0.82 | 0.88 | | 9.22 | 10.6 |
| NbC | Nb | 5.87 | 20.24 | 12.87 | 0.61 | 0.45 | 0.63 | 11.1 | 10.8 |
| | C | 2.56 | 21.30 | 14.38 | | 0.48 | | | |
| TaC | Ta | 6.14 | 26.70 | 12.92 | 0.62 | 0.47 | 0.66 | 11.4 | 9.6 |
| | C | 3.07 | 27.91 | 16.22 | | 0.19 | | | |
| HfC | Hf | 1.08 | 56.57 | 15.66 | | 0.07 | 0.09 | 0 | 0 |
| | C | 0.15 | 5.27 | 12.45 | | 0.02 | | | |

tity $N(0)\langle I^2\rangle$ indeed remains more or less constant, although individually the quantities $N(0)$ and $\langle I^2\rangle$ vary greatly from material to material. The value of the matrix element $\langle I^2\rangle$ increases significantly in going from pure Nb and Ta to the compounds NbC and TaC. Still greater is the value of $\langle I^2\rangle$ in HfC. The increase in $\langle I^2\rangle$ is, according to Klein the Papaconstantopoulos [42], due to an increase in the fraction of ionic bonds in these elements and the fact that the conduction electrons interact more strongly with ions than with quasineutral atoms. Unfortunately, the number of conducting electrons decreases with the growth of the ionic bonding, which leads to the decrease of the density of states $N(0)$. It is of definite interest to analyze the possibility of optimizing this process and producing materials with partly ionic bonding, yet with a considerable number of conduction electrons or with a large value of $N(0)$.

Thus, although the softening of the phonons as a result of the electronic screening indeed plays an important role in the growth of the quantity $\lambda$ (the phonon spectra of Nb have been experimentally found by comparison with those of Mo to have anomalous features in the high-frequency region), this factor alone is often inadequate. It may be inferred that the growth of the quantity $N(0)\langle I^2\rangle$ plays an important role in the increase of $\lambda$ and $T_c$ in the presently known superconductors with the highest critical temperatures.

As another confirmation of the importance of the variation of the factor $N(0)\langle I^2\rangle$ in the transition metals, we can cite the data on the effect of pressure on $T_c$. As analysis of the experimental data shows, the variation of the critical temperatures with pressure cannot be explained on the basis only of the variation of $\langle 1/\omega^2\rangle$, it being necessary, in contrast to the case of the simple metals, to assume [30, 73] that the matrix element $N(0)\langle I^2\rangle$ also varies with pressure.

Using the McMillan formula for $T_c$ and the expression (4.70) for the coupling constant $\lambda$, we can show from the experimental data on $dT_c/dP$ that $N(0)\langle I^2\rangle$ increases with pressure (or with decreasing atomic volume $v$) in such a way that the parameter

$$\xi = -\frac{d \ln N(0) \langle I^2\rangle}{d \ln v}$$

is roughly a constant and lies in the range roughly from 2 to 4. Moreover, taking this quantity, for V for example, from the $dT_c/dP$ data, we can estimate the corresponding value of the coupling constant, $\lambda$, for $V_3Si$ (in which the volume per V atom is smaller than the corresponding volume in pure V by 15 to 20%) and then, using the McMillan formula, find $T_c$ [73]. It turns out that the $T_c$ value thus found for $V_3Si$ is close to the experimental value. We again arrive at the conclusion that a decisive role is played in the rise in $T_c$ in $V_3Si$, by comparison with the $T_c$ in V, by the increase in the factor $N(0)\langle I^2\rangle$ caused, possibly, by just the decrease in the volume per V atom in the A-15 lattice. In this case the softening of the lattice and the closeness of the structural transition may, of course, also play a definite, but not a decisive, role.

TABLE 4.4

| | Intermediate coupling | Strong coupling |
|---|---|---|
| *sp* Metals | Tl<br>$T_c = 2.4$ K<br>$\lambda = 0.795$<br>$\omega_{\log} = 52$ K<br>$N(0)\langle I^2 \rangle = 1.2$ eV/A$^2$ | Pb<br>$T_c = 7.2$ K<br>$\lambda = 1.55$<br>$\omega_{\log} = 56$ K<br>$N(0)\langle I^2 \rangle = 2.4$ eV/A$^2$ |
| *d* Metals | Nb<br>$T_c = 9.2$ K<br>$\lambda = 0.85$<br>$\omega_{\log} = 166$ K<br>$N(0)\langle I^2 \rangle = 4.7$ eV/A$^2$ | $Nb_3Sn$<br>$T_c = 18.1$ K<br>$\lambda = 1.67$<br>$\omega_{\log} = 125$ K<br>$N(0)\langle I^2 \rangle = 7.9$ eV/A$^2$ |

Summarizing, let us give for the purpose of illustration an extremely significant table (Table 4.4) [16] comparing the characteristics of superconductors with intermediate and strong coupling.

It can first and foremost be seen from Table 4.4 that in both the simple and transition metals the rise in $T_c$ in going over to strong coupling is caused not by the softening of the phonons, but primarily by the increase in the factor $N(0)\langle I^2 \rangle$. In the simple metals the values of $\omega_{\log}$ are quite small and, as a result, even for the maximum value of $\lambda$ (for the amorphous Pb-Bi alloy $\lambda = 2.13$), only moderately high $T_c$ values are attained (in PbBi, $T_c = 9$ K). Still higher values of $\lambda$ should, apparently, lead to lattice instability and the formation in the system of covalent bonds.

The situation is more favorable in the transition metals, in which, as can be seen from the table, the values of $\omega_{\log}$ are significantly higher. As a consequence of the partial localization of the *d* electrons, the system "resists" for a longer time the tendency toward instability as the quantity $N(0)\langle I^2 \rangle$ increases, and, as a result, we are able to achieve higher $T_c$ values in transition-metal-based compounds. But even here the lattice will sooner or later become unstable as $N(0) \langle I^2 \rangle$ increases. And to go over to even more highly localized electrons is impossible since in that case another danger appears: for highly localized electrons, instead of superconductivity, magnetic ordering arises in the system as a result of the Coulomb interaction. The problem of finding the optimum with respect to these factors also can by no means be considered to have been solved at the present time.

## 6. Superconductivity and Lattice Instability

Let us now specifically discuss the question of the interrelationship between superconductivity and lattice instability. This is, undoubtedly, one of the most important problems in the investigation of the possibility of raising $T_c$.

At present the opinion is fairly widely held that high critical temperatures can, under the conditions of the normal electron-phonon mechanism of superconductivity, be obtained precisely in structurally unstable compounds or in materials close to the lattice instability threshold [74]. Such a point of view is based primarily on the empirical fact that the highest critical temperatures that have now been attained have been by either compounds that are metastable, or compounds that undergo a structural transition at low temperatures, or at least those in which a softening of definite phonon modes occurs as the temperature is lowered. Such properties are possessed by materials with the A-15 structure ($V_3Si$, $Nb_3Sn$, $Nb_3Ge$), carbides and nitrides of the transition metals (NbN, NbC, TaC), and a number of other alloys and compounds. Moreover, guided by such considerations of the connection between lattice instability and high $T_c$, Testardi [75] was able to raise significantly the $T_c$ value by preparing alloys at temperatures close to the structural transition point and then rapidly quenching them, which allowed the preservation in the sample of the corresponding structure.

Following McMillan [1], authors usually relate the rise in $T_c$ in metals close to the structural transition point with a decrease in the quantity $\langle\omega^2\rangle$ and a corresponding increase in the constant $\lambda$. As we have already noted in the preceding section, however, both the analysis of the experimental data and the numberical computations that have been carried out for a number of metals and alloys show that allowance for the variation of only the quantity $\langle\omega^2\rangle$ does not allow us to account completely for the experimental data. The point is that in most cases the softening of the phonons embraces only a small part of the spectrum, and the total change in the quantity $\langle\omega^2\rangle$ turns out to be quite small. Let us note that the simple and transition metals behave very differently in this respect.

As is well known, in the simple metals an increase in the strength of the electronic coupling can lead to the transition of the metal into the covalent semiconductor or semimetallic state; Bi and gray tin are examples of this. It is noted in [49] that, in the Pb–Bi alloy, as the Bi concentration is raised and $\lambda$ increases, the phonon frequencies near the $X$ point of the Brillouin zone tend to decrease, which may be an indication of the closeness of the alloy to the covalent-instability threshold. Specific computations carried out in that paper show that the decrease of the phonon frequencies embraces only a small wave vector region, and the overall change in the quantity $1/M\langle 1/\omega^2\rangle$ because of this softening is not large. Moreover, it follows from [16, 49] that in the simple metals the variation of the quantity $N(0)\langle I^2\rangle$ as a result of the variation of the electronic characteristics is the most important factor determining the variation of the quantity $\lambda$. The fact that only a small fraction of the phonon modes in the simple metals soften even in the near-threshold region of the lattice instability is not surprising. The electrons of the simple metals are quasifree plane waves in the major part of phase space, significantly differing from such waves only in the vicinity of the Brillouin-zone boundaries. The softening of the phonon modes occurs on account of the contribution of the electrons from the small regions of phase space where the spectrum of the electron states differs greatly from $k^2/2m$. It is the smallness of these regions in the simple metals that leads to the smallness of the phase space of the softened phonons.

A completely different situation may obtain in the transition metals, in which the role of the factor $\langle\omega^2\rangle$ can be more important. The softening of the phonon modes often engulfs a substantial region of the Brillouin zone (see, for example [76]). In this case $T_c$ is found to correlate noticeably with the behavior of the phonon characteristics, in particular, with the structural transition temperature $T_L$. These correlations manifest themselves most clearly in recently obtained data on the variation of the superconducting- and structural-transition temperatures under pressure. In all the investigated systems ($Nb_3Sn$ [77], $V_3Si$ [78], $(Hf_{1-x}Zr_x)V_2$ [79], $V_xRu_{1-x}$ [80]), $T_c$ and $T_L$ varied in a converse manner: if $T_L$ fell, then $T_c$ rose, and vice versa. (Notice that in these materials $T_L > T_c$.) This, it would appear, indicates a direct (and positive) effect of the softening of the phonon modes on superconductivity.

One circumstance contradicting such a direct interpretation, however, has been experimentally observed: if we proceed only from the relation (4.94) or (4.95) and assume the quantity $N(0)\langle I^2\rangle$ to be a constant, then it is to be expected that $T_c \longrightarrow T_L$ will be the highest $T_c$ and that, subsequently, as the pressure is increased further, the lattice will become more rigid and $T_c$ will fall. Experimentally, however, it turned out that the VRu, $T_c$ continues to rise with pressure, although more slowly [80]. This fact becomes quite understandable if we take into consideration the result obtained in the preceding section that the controlling factor in the behavior of both $\lambda$ and the phonon frequencies $\langle\omega^2\rangle$ themselves is the electronic properties characterized by the parameter $N(0)\langle I^2\rangle$. It is precisely the growth of this quantity that gives rise to the increase in $\lambda$ and $T_c$ both directly and through the softening of the phonons. Upon the intersection of $T_c$ and $T_L$, the first factor remains, but the second ceases to exert any influence or even begins to have the opposite effect, and this leads to the observed behavior.

Thus, we see that in effect the proximity of the structural instability is propitious to the growth of $T_c$, although both of these factors are actually two consequences of the growth of the parameter $N(0)\langle I^2\rangle$. As has already been noted above, however, very strong attraction can lead to a real reconstruction of the lattice rather than to superconductivity; in other words, under the conditions of a strong interaction, instead of Cooper pairs (pairs of electrons having oppositely directed momenta and spins and "condensed" in momentum space), there can arise valence pairs (pairs of electrons having oppositely directed spins and condensed in coordinate space). In this case the material can, in general, cease to be a metal and go over into the dielectric state. The competition between superconductivity and covalent-bond formation has been qualitatively discussed in a number of papers (see, for example [36, 81]). The correlation of superconductivity with the formation of valence pairs and with the corresponding distortion of the lattice can be particularly clearly followed in the one-dimensional case. In this case if the effective phonon-induced attraction between the electrons is stronger than the Coulomb repulsion, then both superconductivity and a lattice instability of the Peierls type, e.g., the doubling of the lattice constant, are possible in the system. The connection between superconductivity and lattice reconstruction, the corresponding criteria, and the possible types of state that the system can go into have recently been investigated (see [82-85]). These questions will be discussed in greater detail in Chapter 6.

We shall also not discuss here in detail the specific theoretical models that have been proposed to explain the mechanism of structural transitions and its possible connection with superconductivity (such models as the Labbe-Friedel model [86] and the Gor'kov model [87] for compounds with the A-15 structure); these questions are considered in Chapters 5 and 6. To conclude the discussion of the problem of the correlation between superconductivity and lattice stability, let us try to formulate some results.

The increase of $T_c$ and the lattice instability are both consequences of a strong electron-ion interaction. In this case both the growth of the electron-ion interaction matrix element $N(0)\langle I^2\rangle$ and the accompanying phonon softening lead directly to the growth of $T_c$. But too much softening of the phonons can lead to instability of the system and to its reconstruction, which may limit the further growth of $T_c$. Therefore, as a rule, lattice instability plays a negative role in the problem of raising the superconducting-transition temperature (see, however, Chapter 5). Often, the highest $T_c$ values are indeed attained in systems that are close to the instability threshold; this instability is, however, usually not the cause of the rise in $T_c$ but rather an indication of a strong electron-ion coupling, which is, in fact, what secures the growth of the critical temperature. By itself, the instability of the system impedes the further enhancement of this coupling and, consequently, the growth of $T_c$. From this standpoint the search for new superconductors with record-breaking parameters precisely among materials that are close to the instability threshold is entirely justified since the proximity of the instability by itself indicates the existence of strong coupling. The most urgent problem, however, is to try to prevent, one way or another, the structural reconstruction and thus preserve the strong electron-ion coupling; then, possibly, we shall succeed in raising the $T_c$ values in metastable or structurally unstable systems.

## 7. Possibility of a Nonphonon Mechanism of Superconductivity in Three-Dimensional Systems

To conclude this chapter, we briefly discuss the possibility of obtaining high critical temperatures in three-dimensional systems by means of a nonphonon mechanism of superconductivity. Several nonphonon-mechanism models leading to interelectron attraction have been proposed. In particular, the possibility of the appearance of interelectron attraction in metals with different carrier groups (light $s$ and heavy $d$ electrons) [88-90] as a result of the $s$-electron transfer by the acoustic plasma oscillations that exist in a $d$-electron system has been considered.

By their nature, acoustic plasma oscillations are similar to normal sound in a system of ions and can exist only when there is a significant difference between the $s$- and $d$-electron masses, i.e., when $(m_d/m_s) >> 1$. The possibility of the appearance of such oscillations is easier to understand from an analysis of the permittivity of the system of $s$ and $d$ electrons [90, 91].

Let us represent $\varepsilon(q, \omega)$ in the form of a sum of contributions of the $s$- and $d$-electron polarizabilities:

$$\varepsilon(q, \omega) = 1 + 4\pi\alpha_s(q, \omega) + 4\pi\alpha_d(q, \omega), \tag{4.96}$$

where

$$\alpha_s(q, \omega) = -V(q)\Pi_{ss}(q, \omega),$$
$$\alpha_d(q, \omega) = -V(q)\Pi_{dd}(q, \omega). \tag{4.97}$$

The expression for the polarization operator $\Pi_{ii}(q, \omega)$ in the long-wave limit can be written in the form

$$\Pi_{ii}(q, \omega) = \frac{k_{F_i}^2 q}{2\pi^2} \int_{-1}^{1} x\, dx \frac{1}{\omega - v_i q x - i\delta}. \tag{4.98}$$

Here $v_i = k_{F_i}/m_i$ is the velocity of the $i$th type of electrons at the Fermi surface. When the conditions

$$\frac{qv_d}{\omega} < 1, \quad \frac{qv_s}{\omega} > 1 \tag{4.99}$$

are fulfilled, we obtain

$$\mathrm{Re}\,\Pi_{ss}(q, \omega) = -\frac{m_s k_{Fs}}{\pi^2}, \quad \mathrm{Im}\,\Pi_{ss}(q, \omega) = -\frac{m_s^2 \omega}{2\pi q}, \tag{4.100}$$

and, correspondingly, for $\Pi_{dd}(q, \omega)$ the expressions

$$\mathrm{Re}\,\Pi_{dd}(q, \omega) = \frac{k_{Fd}^3 q^2}{3\pi^2 m_d \omega^2}, \quad \mathrm{Im}\,\Pi_{dd}(q, \omega) = 0. \tag{4.101}$$

Using (4.97), (4.100), and (4.101), we find from the condition $\varepsilon(q, \omega) = 0$ that

$$\omega^2 = \frac{\omega_d^2}{1 + \left(\frac{k_s}{q}\right)^2 \left(1 + i\frac{\pi}{2}\frac{\omega}{qv_s}\right)}, \tag{4.102}$$

where

$$k_s^2 = \frac{4e^2 mk_{F_s}}{\pi}, \quad \omega_d = \frac{4e^2 k_{F_d}^3}{3\pi m_d}. \tag{4.103}$$

In the limit of small $q$, we finally obtain the dispersion law for the acoustic plasma vibrations in the form ($\Omega_q$ is the real part of the plasmon frequency and $\gamma_q$ is the decrement)

$$\begin{aligned} \Omega_q &= \frac{\omega_d}{k_s} q = \frac{1}{\sqrt{3}} \left(\frac{m_s n_d}{m_d n_s}\right)^{1/2} v_s q, \\ \gamma_q &= \frac{\pi}{4} \frac{\omega_d}{k_s^2} q = \frac{\pi}{12} \left(\frac{m_s n_d}{m_d n_s}\right) v_s q. \end{aligned} \tag{4.104}$$

Here $n_s$ is the total number of the $s$ electrons; $n_d$ that of the $d$ electrons. It is clear that we can speak of the existence of such plasmons as well-defined excitations only when the condition

$$\frac{\Omega_q}{\gamma_q} = \frac{\pi}{4\sqrt{3}} \left(\frac{m_s n_d}{m_d n_s}\right)^{1/2} \gg 1. \tag{4.105}$$

is fulfilled. Hopes for obtaining high $T_c$ values under conditions of acoustic-plasmon exchange are connected with the fact that the velocity of these plasmons and, consequently, the characteristic frequencies, $\omega_{\log}$ and $\overline{\omega}$, associated with them will be higher than the velocity of sound and the corresponding Debye frequencies. But, as can be seen from the expression (4.105), the condition for the existence of such excitation is itself quite a rigid one. In the majority of real transition metals $n_d > n_s$, while $m_d$ is, in its turn, really not so high in comparison with $m_s$. A careful analysis carried out in [91-93] shows that even within the framework of these simplest of approximations, the acoustic plasma vibrations exist only at very small values of the vector $q$. The statistical weight of such excitations is very small, and we cannot hope for the appearance of any significant coupling of the electrons to such vibrations.

On going outside the limits of applicability of the simplest approximation (i.e., on allowing for $s$- and $d$-electron transfer, the hybridization of the $s$ and $d$ bands [94], the interaction with the impurities [88], etc.), we clearly see that the existence of acoustic plasma vibrations will generally be impossible in real transition metals. An extensive analysis of the experimental data on the transition metals is performed in [92] with the object of discovering anomalies that indicate the existence of acoustic-plasma vibrations. In our opinion, practically all the effects found to exist in the transition metals and attributed in [92] to the influence of acoustic plasmons can be explained within the framework of the normal scheme with the electron–phonon interaction.

The consistent consideration of a system of ions and $s$ and $d$ electrons [95, 96] within the framework of the isotropic approximation already described by us leads to the interaction of the normal-sound and acoustic-plasma vibrations. In this case, at large values of the wave vector $q$ there occurs as a result of the additional screening of the ions by the $d$ electrons a decrease in the energy of the high-frequency phonon vibrations. Such a lowering of the frequencies, though it bears no direct relation to the nonphonon mechanism, may, in the opinion of Ganguly and Wood [95] and Fröhlich [96], lead to an increase in $T_c$ in the transition metals. Specifically, Ganguly and Wood try in [95] to relate the phonon anomalies in Nb and in the carbides of the transition metals to the manifestation of this mechanism. As already noted in the preceding sections, there can, indeed, be anomalies in the permittivity of the $d$ electrons, and they do affect the lattice dynamics, but they do not require for their explanation the introduction of such a concept as the concept of the acoustic plasmon.

We shall not discuss in detail the other models, considered in the literature, of possible nonphonon mechanisms of superconductivity in three-dimensional systems. We will only note that all these proposals (in particular, the proposition that exciton exchange is possible in alloys of metals with nonmetals [88] or in metals with injected nonmetallic materials [97, 98]) were considered in models so simplified that it is practically impossible to draw any correct conclusion about their effectiveness is real situations. These mechanisms are rather more likely to be manifested in inhomogeneous quasi-two-dimensional or quasi-one-dimensional systems. These questions will be considered in greater detail in the following chapters. A more comprehensive bibliography of papers pertaining to the subject of the present section can be found in Geilikman's review article [99].

The main shortcoming of all the nonphonon mechanisms proposed so far for superconductivity in three-dimensional systems, apart from their sketchiness, is that they all lead to positive values of the system's static permittivity [$\varepsilon_{tot}(q, 0) > 0$]. As shown in [15, 101], their effectivness is significantly lowered under these conditions. The reason why $\varepsilon_{tot}(q, 0)$ assumes only positive values is that only quasi-isotropic systems with a weak Coulomb interaction and no transverse exciton excitations are considered. As follows, in particular, from the results of the preceding chapter, a necessary condition for the existence of negative $\varepsilon_{tot}(q, 0)$ values even in a normal electron–phonon system is the presence of transverse phonons. This circumstance can be regarded as a consequence of local-field effects [102]. A consistent allowance for the local-field effects and the use in these computations of the complete permittivity matrix $\varepsilon(\mathbf{q} + \mathbf{G}, \mathbf{q} + \mathbf{G}', 0)$ are also necessary for a rigorous analysis of the effectiveness of the nonphonon mechanisms of superconductivity. This fact has also been emphasized by Bardeen [103] and Cohen and Louie [104]. Takada [105] and Uspenskii [106] have investigated the possibility of the occurrence of superconductivity in the one-component electron system when only the interelectron Coulomb interaction is taken into consideration. Takaga [105] has shown that if such a possibility in a three-dimensional system is realizable at all, then it is so only in the case of a strong Cooper interaction. Uspenskii [106] has studied the possibility of the existence in the one-component electron system of negative values of the static permittivity $[\varepsilon(q, 0) < 0]$ as a result of the presence in the system of exchange and correlation effects. He has shown that this is indeed possible in the paramagnetic phase in the vicinity of the transition into the magnetically ordered state. Since the results obtained in [105, 106] stem to a large extent from the exchage and correlation effects, it is clear that we should, in describing superconductivity, also go beyond the weak-coupling approximation and take the vertex corrections to the interelectron interaction into account [102].

# 5

# Possibility of an Increase in the Critical Temperature as a Result of a Structural-Transition-Induced Change in the Electron Spectrum†

## 1. Introduction

It is well known that the highest $T_c$ superconducting compounds undergo near the superconducting-transition temperature $T_c$ a structural transition at a temperature $T_p > T_c$ [1].

The view has been expressed [2] (see Chapter 4) that it is the closeness of $T_p$ and $T_c$ that secures the high values of the latter since a fairly marked softening of one of the optical phonon modes occurs in the vicinity of the former. In this case, because of this mode, the effective interelectron interaction constant $\lambda$, which is inversely proportional to the mean phonon-frequency value $\langle\omega^2\rangle^{1/2}$ [3], should increase.

On the other hand, the expression for $T_c$ contains the quantity $\langle\omega^2\rangle^{1/2}$ in the pre-exponential factor. Therefore, $T_c$ as a function of $\langle\omega^2\rangle^{1/2}$ should have a maximum. In [3] it is assumed that a state near this maximum is realized as a result of the structural transition.‡

The opposite point of view regarding the effect of a structural transition on $T_c$ has also been expressed.

Because of the large constant $\lambda$ in the high-temperature phase, this phase

†This chapter is based in large part on the material of [41].

‡It should be noted that for this maximum to appear, it is only necessary that $\langle\omega\rangle$ become low, which can occur in the absence of a structural transition.

becomes unstable at the temperature $T_p$. Th new phase below $T_p$, on the other hand, possesses a smaller electron–phonon coupling constant, which secures its stability. It is therefore asserted that the $T_c$ in the absence of the structural transition would be higher than the observed temperature.

So far, we have been discussing structural-transition-induced changes in the phonon characteristics and their effect on the $T_c$. Let us, however, note that $T_c$ depends quite strongly on the density of electron states near the Fermi level, $N(0)$, which can change during a structural transition. For example, the quantity $N(0)$ for the superconducting $V$–Ru alloys, which undergo low-temperature structural transitions, has been found [4] to change significantly when the temperature is lowered below $T_p$.

It follows from the dependence of the resistance and the magnetic susceptibility on temperature in the vicinty of $T_p$ that this change in $N(0)$ is due to the formation of a dielectric gap at some part of the Fermi surface (partial "dielectrization" of the conduction electrons). Although the total number of conduction electrons then decreases, the density, $N(0)$, of electron states may rise. Thus, partial dielectrization can lead to a rise in $T_c$. It is to the investigation of this possibility that the present chapter is devoted.

We shall consider the effect of the dielectric correlations on the superconducting pairing in a model in which the Fermi surface is assumed to have nearly congruent parts. This model allows a consistent description of such correlations since there exist in it, in the weak-coupling limit, a distinct class of diagrams, whose consideration corresponds to the generalized Hartree–Fock method. The formal description of the dielectric correlations in this limit is similar to the consideration of the superconducting correlations in the simplest BCS scheme.

Just as the BCS model turned out to be satisfactorily correct, not only qualitatively but also quantitatively even for materials with strong interactions, so, apparently, does the scheme in which the Fermi surface is assumed to have nearly congruent parts fairly well describe the dielectric correlations in many real systems.

This is corroborated, first, by the fact that, qualitatively, the character of the dielectric correlations is preserved in the opposite limiting case of strong interactions when the expansion in powers of the low-density parameter is valid [5].

Second, the model in question describes a great variety of properties of solids: metal–dielectric phase transitions [6], crystal-symmetry changes during phase transitions [7–9], antiferromagnetic [5, 10, 11], ferromagnetic [12, 13], and ferroelectric [14] ordering. Here a number of characteristics predicted in the model for the kinetic phenomena are observed in materials in which the interaction parameters are of the order of unity [15]. It is natural in the case of an interaction constant of the order of unity for the character of the dielectric correlations to be qualitatively preserved even when the condition for congruence of individual sections of the Fermi surface is fairly strongly violated.

Below we shall consider in detail the case in which, as a result of the dielectric correlations, a structural phase transition accompanied by partial dielectrization of the electron spectrum occurs at a temperature, $T_p$, higher than the superconducting transition temperature $T_c$.

The main results of the present chapter remain valid also for the case when the quantity $T_p$ is higher than the melting point $T_m$, i.e., when no structural phase transition occurs and the symmetric "protophase" with nearly congruent sections of the Fermi surface is unobservable (at least at normal pressure). Qualitatively, the nature of the effect that the dielectric and superconducting correlations have on each other is preserved in the opposite limiting case, when $T_p < T_c$, and even when the temperature $T_p$ is formally negative (provided $|T_p| < T_c$), i.e., when no structural transition occurs on account of insufficient congruence of the individual sections of the Fermi surface but the dielectric correlations are sufficiently strong.

Similar problems arise in the study of quasi-one-dimensional systems, for which Little has pointed out the possibility of the $T_c$'s rising as a result of the action of the exciton mechanism of superconductivity. Special interest has been shown in this problem in connection with the experiments performed by Heeger's group [16] on the quasi-one-dimensional compound TTF–TCNQ (see Chapter 7).

The electron spectrum, $E(p)$, of a one-dimensional metal possesses the property that $E(p) = -E(p + 2p_F)$, where $p_F$ is the Fermi momentum and the energy is measured from the Fermi level. This anomaly in the electron spectrum leads to the instability of the one-dimensional metal with regard to the appearance of a charge density wave with a period that is inversely proportional to $2p_F$. As a result, a structural transition into the dielectric state should occur at some temperature $T_p$. The expression for $T_p$ is similar to the expression for $T_c$, it being the same as the latter with the mean phonon energy replaced by the Fermi energy $E_F$ or by the halfwidth of the allowed band of the initial metallic state.

Strictly speaking, here the dielectric (electron–hole) and superconducting (electron–electron) pairings should be taken into account at the same time. This leads to a situation in which the dielectrization of the electron spectrum during the structural transition makes the superconducting pairing impossible in the case of the phonon mechanism of superconductivity when $\hbar\omega_D \ll E_F$ [17].

In the case of the exciton mechanism of superconductivity, when the cutoff energies in the equations for the dielectric and superconducting gaps can be of the same order of magnitude, the Hartree–Fock approximation (i.e., the ladder approximation) used in [17] is inadequate. But from the parquet approximation [18] follows the simultaneous existence of dielectric and superconducting pairings.

It was subsequently [19] elucidated that in the case in which there is one electron per unit cell, when the Umklapp processes are important in the one-dimensional case, besides the coexistence of the structural transition (dielectric pairing) and the superconducting transition, a transition into the antiferromagnetic state is also possible because of the possibility of the formation of spin density waves. Strictly speaking, because of fluctuations the dielectric or superconducting pairing, or both together, cannot occur in the one-dimensional case [20]. Therefore, we should, in considering quasi-one-dimensional systems, take account of the influence of the three-dimensionality effects, as is done in [21], also in the parquet approximation. Unfortunately, because of the logarithmic accuracy of this approximation, we can determine the possible states of a quasi-one-dimensional system but not the parameter values at which a particular state is realized, nor can be determine the critical temperature for the transition into the state. In particular, we are still not sure of the effect on the $T_c$ of the dielectrization that occurs in the structural transition, and the relation between the quantities $T_c$ and $T_p$ in the case when the superconducting and dielectric pairings coexist also remains unknown.

For the three-dimensional case the electron spectrum is not, generally speaking, such as to allow the occurrence of a structural transition with electron–hole pairing under conditions of arbitrarily weak interaction, as happens in the one-dimensional case. Nevertheless, it may turn out in the three-dimensional case that the electron spectrum does, even if for some distinct part of the Fermi surface, approximately satisfy the condition

$$E(\mathbf{p}) = -E(\mathbf{p}+\mathbf{q}) \tag{5.1}$$

for a discrete set of **q** vectors when the electron momenta **p** belong to this distinct part of the Fermi surface.

The condition (5.1) is satisfied, for example, over the entire Fermi surface for some crystal structures [22] when the one-electron spectrum is described in the tight-binding approximation for the electrons and allowance is made for the overlap of the electron wave functions for only nearest neighbors.

The second three-dimensional model in which the condition (5.1) is satisfied, and with which we shall begin our detailed exposition of the question of the effect on the quantity $T_c$ of the dielectrization that occurs during the structural transition, is the model of the semimetal with first-band-electron and second-band-hole Fermi surfaces of nearly the same shape and size, when $E_1(\mathbf{p}) = -E_2(\mathbf{p}+\mathbf{q})$. Here the vector **q** corresponds to the distance in momentum space between the extrema of the electron and hole bands.

In this case the system is unstable with regard to pairing of electrons and holes from different bands no matter how weak the interaction between them is [5, 23, 24].

In the case of singlet electron–hole pairing an electron charge-density wave is produced, while in the case of triplet pairing a spin-density wave leading to the formation of an antiferromagnetic state is produced [10], as happens in, for example, chromium and its alloys.

The presence in the Hamiltonian of interactions with a constant $g_2$ that corresponds to the transformation of particles of one band into particles of the other band leads to the fixing of the phase of the order parameter. Besides the purely real order parameter, which, as indicated above, gives rise to charge- or spin-density waves, a purely imaginary order parameter, which gives rise to charge- or spin-current density waves is also possible [24].

In the case when the band extrema coincide in momentum space, i.e., when $q = 0$, and the interband dipole transitions are allowed in the presence of a real order parameter, a ferroelectric ordering arises instead of a charge-density wave [14].

In the case of an imaginary order parameter, a state with a spontaneous current arises under these same conditions [26]. Allowance for the self-magnetic field of the spontaneous current leads to the formation of current domains [27]. An inhomogeneous current state arises when the self-magnetic field is neglected, but account is taken of the momentum dependence of the interaction constant $g_2$.

The diamagnetic susceptibility of such a system behaves according to the Curie–Weiss law as the critical temperature of the phase transition is approached [28], but the conductivity, in contrast to what obtains in superconductors, tends to zero as the temperature approaches absolute zero.

If the symmetry of the wave functions of the two bands in question is such that the matrix element of the momentum operator is equal to zero while that of the angular-momentum operator is nonzero, then ferromagnetic ordering of the electrons of the completely filled valence band of the semiconductor occurs in the case of an imaginary order parameter [29].

Fairly rigid conditions on the intensity of electron scattering by the impurities must be fulfilled in order for states with an imaginary order parameter to be realized [30].

Furthermore, as in the one-dimensional metal, a phonon instability obtains [31] that leads to a structural reconstruction by an ion charge-density wave. The character of this type of structural transition in compounds of the $A_4B_6$ group is investigated in [8, 9, 32].

The noncoincidence in shape of the electron and hole Fermi surfaces leads to the suppression of the dielectric pairing and the structural transition [33]. To be sure, the deviation of the screened electron–hole Coulomb interaction from a point interaction leads to a significant weakening of this suppression [34]. The situation is then possible when the dielectric gap is formed only at some part of the Fermi surface; in such a structural transition the semimetal transforms into the same semimetal but with lower electron and hole concentrations [35]. In the case of a single-band metal in which the condition (5.1) is fulfilled at some part of the Fermi surface, a transition into the semimetal state also occurs in a structural transformation.

The transition temperature $T_p$ decreases significantly when, as a result of doping, the electron and hole concentrations are not equal [11, 36]; in this case the doped semimetal goes over into the degenerate-semiconductor state at temperatures $T < T_p$.

Besides the pairing of electrons and holes from different bands and the structural transition into the semimetal state, superconducting (electron–electron) pairing is possible, as in any metal, within the limits of each band. Therefore, as in the case of the one-dimensional metal, we should consider the dielectric and superconducting pairings together. But because of the three-dimensional nature of the system, it does not become necessary to sum the parquet diagrams.

A similar assertion is valid for a single-band metal with a spectrum satisfying the condition (5.1). Naturally, there should exist a limiting process that takes quasi-one-dimensional systems into three-dimensional ones as the overlap integral for the electron wave functions of neighboring chains increases (see Chapter 7).

Below we shall find the conditions under which superconducting pairing can coexist with structural-transition-inducing electron–hole pairing. The effect of the latter on superconductivity manifests itself in a narrowing of the energy range for superconducting pairing as a result of the appearance of a dielectric gap, $\Sigma$, in the vicinity of the Fermi level. This should lead to the lowering of the superconducting transition temperature $T_c$.

On the other hand, because of the expulsion of the electron states from the region of the dielectric gap, the density of electron states above and below this gap increases significantly. Therefore, if the Fermi level turns out to be near the dielectric gap, then the elevated density of states facilitates the growth of $T_c$. Therefore, there should be optimal conditions for the effect of the dielectric pairing on the value of $T_c$.

Below we shall show that under these optimum conditions the quantity $T_c$ can significantly exceed its value in the absence of dielectric pairing (and, consequently, of a structural transformation).

Notice particularly that the electron states lying below the dielectric gap $\Sigma$ make an appreciable contribution to the quantity $T_c$ as a result of virtual transitions into states above the gap $\Sigma$ if the cutoff energy of the superconducting interaction is greater than the quantity $\Sigma$.

In the following sections we shall investigate in detail the case of the coexistence of superconductivity with a singlet, real dielectric order parameter. In this case there are two types of solutions, one of which corresponds to the coincidence of the phases of the superconducting order parameters from the different bands (the symmetric solution), while the second corresponds to a shift of these phases by $\pi$ (the antisymmetric solution). The dielectric correlations can lead to an increase in the critical temperature, $T_c$, of the superconducting transition only in the case of the symmetric solution.

For the case of the triplet dielectric pairing, the symmetric and antisymmetric solutions simply change places, and, thus, an increase in $T_c$ in the antiferromagnetic state is possible only for the antisymmetric solution.

## 2. The Ground State of a Semimetal in the Presence of Simultaneous Electron-Hole and Electron-Electron Pairings

Let us write the basic Hamiltonian (without allowance for the retardation of the electron-phonon interaction) in the form

$$H = \sum_i \int \psi_i^+(\mathbf{x})\, \varepsilon_i(\mathbf{p})\, \psi_i(\mathbf{x})\, d\mathbf{x} + \\ + \frac{1}{2}\sum_{ij} g_{ij} \int \left(\psi_i^+(\mathbf{x})\, \psi_i(\mathbf{x})\right)\left(\psi_j^+(\mathbf{x})\, \psi_j(\mathbf{x})\right) d\mathbf{x} + \\ + \left[g_{12}^{\mathrm{eph}} \int \psi_1^+(\mathbf{x})\, u(\mathbf{x})\, \psi_2(\mathbf{x})\, d\mathbf{x} + \mathrm{H.C.}\right], \qquad (5.2)$$

where $i, j$ are band indices ($i, j = 1, 2$). The last term in (5.2) describes the interband interaction, characterized by the coupling constant $g_{12}^{\mathrm{eph}}$, of the electrons with the static deformation, $u(\mathbf{x})$, of the crystal lattice (for the sake of brevity we have omitted the elastic lattice deformation energy). The corresponding term with the intraband electron-phonon interaction does not lead to a structural transition in the weak-interaction limit [31] and has, therefore, been dropped.

The electron spectrum in both bands is assumed, for the sake of simplicity, to be isotropic, although for electron-hole pairing to occur, it is only necessary that the electron and hole Fermi surfaces coincide in shape:

$$\xi_{1,2}(\mathbf{p}) = \mu \pm \xi, \quad \xi = \frac{p^2}{2m} - E_F,$$

where $\mu$ is the shift, due to doping, of the Fermi level in each of the bands. If we neglect the local impurity levels in the dielectric ($D$) phase [37], then we can assume (this is valid in the case when the impurity and valence bands overlap) that the difference, $\delta n$, between the electron and hole concentrations is fixed. This condition allows us to determine the quantity $\mu$ in the case when the dielectric and superconducting ($S$) pairings coexist.

Let us define the Green function of the system in the usual manner [38]:

$$G_{ij}^{\alpha\beta}(\mathbf{x} - \mathbf{x}') = \left\langle T\psi_{i\alpha}(\mathbf{x})\, \psi_{j\beta}^{+}(\mathbf{x}')\right\rangle, \\ F_{ij}^{\alpha\beta}(\mathbf{x} - \mathbf{x}') = \left\langle T\psi_{i\alpha}(\mathbf{x})\, \psi_{j\beta}(\mathbf{x}')\right\rangle. \qquad (5.3)$$

Let us investigate the case of the singlet $S$ and $D$ pairings, i.e., we shall assume that $G_{ij}^{\alpha\beta} = G_{ij}\delta_{\alpha\beta}$ and $F_{ij}^{\alpha\beta} = i\sigma_{\alpha\beta}^{y}F_{ij}$.

The function $G_{21}$ corresponds to $D$ pairing. It should be noted that although the electron-electron scattering amplitude does not have a logarithmic singularity when the electrons belong to different bands (in contrast to the case when the scattering occurs within the boundaries of each band or when the scattering involves an electron and a hole from different bands), the anomalous interband superconducting functions $F_{21}$ are, in the presence of dielectric pairing, not at all small compared to the corresponding intraband functions $F_{11}$ and $F_{22}$ [39].

After separating out the spin dependence of, and Fourier transforming, the functions (5.3), we can write the system of equations for them in the Matsubara representation for $T \neq 0$ in the form

$$\begin{pmatrix} i\omega_n - \xi_1 & -\Sigma & \Delta_{11} & \Delta_2 \\ \Sigma^* & i\omega_n - \xi_2 & \Delta_2 & \Delta_{22} \\ \Delta_{11}^* & \Delta_2^* & i\omega_n + \xi_1 & \Sigma^* \\ \Delta_2^* & \Delta_{22}^* & \Sigma & i\omega_n + \xi_2 \end{pmatrix} \begin{pmatrix} G_{11}(\omega_n, \mathbf{p}) \\ G_{21}(\omega_n, \mathbf{p}) \\ F_{11}^+(\omega_n, \mathbf{p}) \\ F_{21}^+(\omega_n, \mathbf{p}) \end{pmatrix} = \begin{pmatrix} 1 \\ 0 \\ 0 \\ 0 \end{pmatrix}, \tag{5.4}$$

where $\Delta_{ij} = g_{ij}F_{ij}(x, x)$, $\Sigma = \tilde{g}_{21}G_{21}(x, x)$, $\Delta_2 \equiv \Delta_{21}$, $\omega_n = \pi T(2n + 1)$ ($n$ is a whole number).

The interaction constants $g_{11}$ and $g_{22}$ (we shall, for the sake of simplicity, assume them to be equal) contain both the electron-phonon and the weakened—on account of the factor $\ln(E_F/\omega_D)$—intraband Coulomb interactions and are assumed to be negative, which corresponds to interelectron attraction. The corresponding interactions for the interband transitions are included in the constant $g_{21}$. Here the case when $g_{21} < 0$ is possible. Below we shall see that the $S$ pairing of electrons from different bands is possible for any sign of $g_{21}$, but that the $g_{21} < 0$ case is more favorable for the existence of superconductivity.

The constant $\tilde{g}_{21}$, which is responsible for the $D$ pairing, is equal to

$$\tilde{g}_{21} = g_{21} + \frac{2\left(g_{12}^{\text{eph}}\right)^2}{\omega_D}.$$

$D$ pairing occurs when $\tilde{g}_{21} > 0$. Below, it is assumed everywhere that this condition is satisfied. Let us emphasize that $g_{21}$ can be negative in this case. It should be pointed out that in the Hamiltonian (5.2), the interband interaction of the electrons is allowed for only in the form of a density-density correlation, the terms taking account of the mutual transformations of the particles of one band into the particles of the other being neglected. Allowance for these terms leads to the result that the transition into the $D$ phase can be a nearly second-order phase transition [40].

Let us write down the solution to the system of equations (5.4) [Det denotes the determinant of the system (5.4)]:

$$\begin{aligned} \mathrm{Det} \cdot G_{11}(\omega_n, \mathbf{p}) = {} & (i\omega_n + \xi_2)\,[(i\omega_n + \xi_1)(i\omega_n + \xi_2) - \\ & - |\Sigma|^2] - \Delta_{22}\left[\Delta_{22}^*(i\omega_n + \xi_1) - \Sigma \cdot \Delta_2^*\right] - \\ & - \Delta_2\left[\Delta_2^*(i\omega_n + \xi_2) - \Sigma^* \cdot \Delta_{22}^*\right], \end{aligned} \tag{5.5}$$

$$\text{Det} \cdot G_{21}(\omega_n, \mathbf{p}) = \Sigma^* [(i\omega_n + \xi_1)(i\omega_n + \xi_2) - |\Sigma|^2 - - |\Delta_2|^2] - \Sigma \overset{*}{\Delta}_{11} \cdot \Delta_{22} + \Delta_{22} \cdot \overset{*}{\Delta}_2 (i\omega_n + \xi_1) + + \overset{*}{\Delta}_{11} \cdot \Delta_2 (i\omega_n + \xi_2), \tag{5.6}$$

$$\text{Det} \cdot F^+_{11}(\omega_n, \mathbf{p}) = \overset{*}{\Delta}_{11} [\omega_n^2 + \xi_2^2 + |\Delta_{22}|^2] - - \overset{*}{\Delta}_2 (\overset{*}{\Delta}_2 \cdot \Delta_{22} + 2\Sigma^* \cdot \xi_2) + \overset{*}{\Delta}_{22} (\Sigma^*)^2, \tag{5.7}$$

$$\text{Det} \cdot F^+_{21}(\omega_n, \mathbf{p}) = \overset{*}{\Delta}_2 [-(i\omega_n + \xi_1)(i\omega_n - \xi_2) + + |\Sigma|^2 + |\Delta_2|^2] - \Sigma^* \cdot \overset{*}{\Delta}_{22} (i\omega_n + \xi_1) + + \overset{*}{\Delta}_{11} [\Sigma (i\omega_n - \xi_2) - \Delta_2 \cdot \overset{*}{\Delta}_{22}]. \tag{5.8}$$

Assuming that $g_{11} = g_{22}$, i.e., that $|\Delta_{11}| = |\Delta_{22}|$, we can easily show with the aid of the expressions (5.6)–(5.8) that the following two different cases exist ($\Sigma$ and $\Delta_2$ are real) [41]:

$$\Delta_{11} = -\Delta_{22} \equiv \Delta_a, \qquad \Delta_2 = 0, \tag{a}$$

$$\Delta_{11} = \Delta_{22} \equiv \Delta_s, \qquad \Delta_2 \neq 0. \tag{b}$$

Below we shall call the solution $\Delta_a(\Delta_s)$ the antisymmetric (symmetric) solution. The existence, besides the symmetric solution [39], of an antisymmetric one was pointed out by Aronov and Sonin [42].

We shall see below that in the weak-coupling limit under consideration (i.e., for $g_{21} m p_F \ll 1$), the parameter $\Delta_2$ in the case (b) is small ($\Delta_2 \ll \Delta_s$) in a broad range of values of the coupling constant. Therefore, the difference between the cases (a) and (b) essentially lies only in the fact that different phases have been chosen for the parameters $\Delta_{11}$ and $\Delta_{22}$. Let us point out that in both cases the anomalous function $F_{21}(x, x')$, (5.8), is by no means small and is, generally speaking, of the same order of magnitude as $F_{11}(x, x')$.

Let us write the determinant of the system (5.4) in the form

$$\text{Det} = (\omega_n^2 + \omega_+^2)(\omega_n^2 + \omega_-^2), \tag{5.9}$$

where $\omega_\pm$ is the energy of the elementary excitations (Fig. 5.1) in the *SD* phase (i.e., in the mixed phase):

$$\omega_\pm^2 = (E \pm \widetilde{\mu})^2 + \widetilde{\Delta}^2, \quad E^2 = \xi^2 + \widetilde{\Sigma}^2. \tag{5.10}$$

For the antisymmetric solution

$$\widetilde{\Sigma}_a^2 = \Sigma^2\left(1 + \frac{\Delta^2}{\mu^2}\right), \quad \widetilde{\Delta}_a^2 = \Delta^2\left(1 - \frac{\Sigma^2}{\mu^2}\right). \tag{5.11}$$

For the symmetric solution

$$\widetilde{\Sigma}_s = \frac{\mu\Sigma + \Delta \cdot \Delta_2}{\widetilde{\mu}}, \quad \widetilde{\Delta}_s = \frac{\Delta \cdot \mu - \Delta_2\Sigma}{\widetilde{\mu}}, \quad \widetilde{\mu}^2 = \mu^2 + \Delta_2^2. \tag{5.12}$$

In Fig. 5.1a, the curve $E_1(p)$ corresponds to the electron band of the semimetal, while $E_2(p)$ corresponds to the hole band, the extrema of these two bands being the vector **q** apart in momentum space. In Fig. 5.1b, the hatched region corresponds to the filling of the upper band by free carriers as a result of the doping.

In conclusion of this section, let us write down the equation for the determination of $\mu$ under the assumption that the difference, $\delta n$, between the electron and hole concentrations is given:

$$\delta n = 2 \int [1 - f_1(\mathbf{p}) - f_2(\mathbf{p})] \frac{d\mathbf{p}}{(2\pi)^3}, \tag{5.13}$$

where $f_i(\mathbf{p})$ is the distribution function for the electrons in the $i$th band, which distribution function can be determined with the aid of the expression (5.5) for $G_{11}$ and the analogous expression for $G_{22}$.

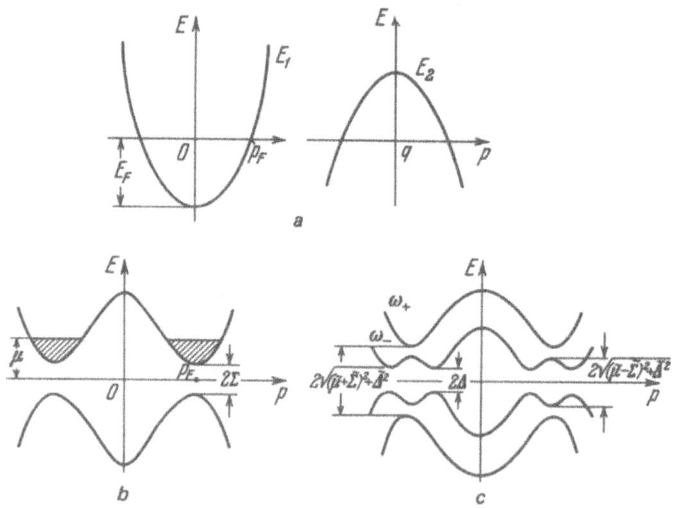


Figure 5.1. Electron spectra of the: (a) semimetal (*N* phase); (b) *D* phase; (c) *SD* phase.

For the subsequent discussion, it is convenient to go over from the variable $\delta n$ to the quantity, $\tilde{n}$, that has the dimension of energy:

$$\delta n = 4N(0)\tilde{n}. \tag{5.14}$$

In the case of a single-band metal with a spectrum satisfying the condition (5.1), the role of $\delta n$ is played by the deviation of the number of electrons per unit cell from unity.

## 3. Incompatibility of the Superconducting and Dielectric Pairings in the Case when the Electron and Hole Fermi Surfaces Coincide in Shape and Size

Let us first consider the antisymmetric case ($\Delta_2 \equiv 0$) at $T = 0$.

Because of the absence doping, the Fermi-level shift $\mu$ should be equal to zero. In this case, from Eqs. (5.6) and (5.7) we obtain for $\Sigma$ and $\Delta$ the equations

$$\beta_0 \equiv \ln z_0 = \frac{1-z^2}{2z}\ln\left|\frac{1+z}{1-z}\right|, \qquad z = \frac{\Sigma}{\Delta}. \tag{5.15}$$

$$\ln\left(\frac{\Delta_0}{\Delta}\right)^2 = \ln|1-z^2| + z\ln\left|\frac{1+z}{1-z}\right|. \tag{5.16}$$

The quantities $\Delta_0$ and $\Sigma_0$ introduced here correspond to the values of the order parameters in the pure $S$ and pure $D$ phases, respectively (in the absence of doping) at $T = 0$:

$$\Delta_0 = 2\omega_D \exp\left(-\frac{1}{|\lambda_{11}|}\right), \tag{5.17}$$

$$\Sigma_0 = 2E_F \exp\left(-\frac{1}{\tilde{\lambda}_{21}}\right). \tag{5.18}$$

$$\lambda_{11} = g_{11}N(0), \quad \tilde{\lambda}_{21} = \tilde{g}_{21}N(0).$$

The right-hand side of Eq. (5.15) decreases monotonically in the range from 1 to $-1$ as $z$ is varied in the range $(0, \infty)$. Consequently, the system (5.15), (5.16) possesses a nontrivial solution when the inequality

$$|\ln(\Sigma_0/\Delta_0)| < 1.$$

is satisfied. At the ends of the interval we have $\Delta = 0$ when $\Delta_0 = e\Sigma_0$ and $\Sigma = 0$ when $\Sigma_0 = e\Delta_0$. It is also easy to see that the parameter $\Delta$ decreases monotoni-

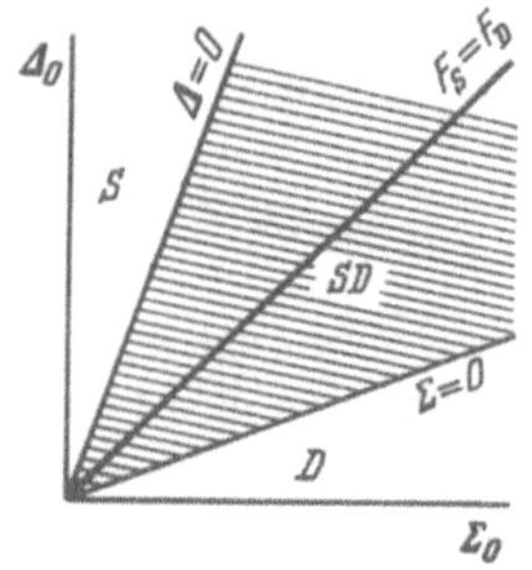


**Figure 5.2. Phase diagram for the undoped semimetal.**

cally as the coupling constant $|\lambda_{11}|$, i.e., $\Delta_0$, increases; $\Sigma$ as a function of $\Sigma_0$ behaves in much the same way.

Such an anomalous behavior of the solutions in the *SD* phase suggests that if the mixed *SD* phase is at all realizable in the region enclosed between the straight lines $\Delta_0 = e\Sigma_0$ and $\Sigma_0 = e\Delta_0$ in the $(\Sigma_0, \Delta_0)$ plane, then it is realized only in the metastable state (Fig. 5.2). This can easily be verified by computing the free energy corresponding to the solutions (5.15), (5.16).

In the general case (for $\Delta_2 = 0$) we have

$$\delta F_{SD} \equiv F_{SD} - F_N = -2\int\limits_C \left( \frac{|\Sigma|^2}{\tilde{\lambda}_{21}^2}\, d\tilde{\lambda}_{21} + \frac{|\Delta|^2}{\lambda_{11}}\, d\lambda_{11} \right), \tag{5.19}$$

where $F_N$ is the energy of the normal phase (when $\lambda_{11} = \tilde{\lambda}_{21} = 0$). The contour C joins the given point $(\Sigma_0, \Delta_0)$ with the coordinate origin in the $(\Sigma_0, \Delta_0)$ plane.

In the case under consideration the integration contour should lie entirely in the existence domain of the solutions, i.e., in the hatched region in Fig. 5.2. It is convenient to perform the integration in the expression (5.19) along the straight line $z = \text{const}$. Then with the aid of Eqs. (5.15) and (5.16) we find

$$\delta F_{SD} = f(z)\, \delta F_S = f\left(\frac{1}{z}\right) \delta F_D, \tag{5.20}$$

where

$$\ln f(z) \equiv \ln \left| \frac{1+z^2}{1-z^2} \right| - z \ln \left| \frac{1+z}{1-z} \right|, \tag{5.20a}$$

$$\delta F_S \equiv F_S - F_N = -N(0)\, \Delta_0^2, \tag{5.20b}$$

$$\delta F_D \equiv F_D - F_N = -N(0)\, \Sigma_0^2. \tag{5.20c}$$

The function $f(z)$ decreases monotonically as $z$ is varied in the range from 0 to $\infty$, with $f(0) = 1$ and $f(\infty) = e^{-2}$. Hence we can conclude that the $SD$ phase that has been found is energetically unfavorable as compared with the purely superconducting and purely dielectric phases. The dependence of the free energies $\delta F_{SD}$, $\delta F_S$, and $\delta F_D$ on $\beta_0 = \ln(\Sigma_0/\Delta_0)$ is shown in Fig. 5.3.

At first glance it may seem that the existence of the metastable $SD$ phase in the antisymmetric case indicates the possibility of the existence in the system of hysteresis phenomena, such as the "supercooling" of one of the phases (i.e., the $S$ or $D$ phase).

In fact, the transition from the $S$ phase into the $D$ phase (and vice versa) will occur on the line $\Sigma_0 = \Delta_0$. The point is that for $\delta n = 0$, the system of equations (5.6), (5.7) in the symmetric case also possesses a nontrivial solution when $\mu = \Delta_2 = 0$. This solution is realized on the equilibrium line for the pure $S$ and $D$ phases and has the form

$$\Sigma^2 + \Delta^2 = \Sigma_0^2.$$

The energy of this state obviously coincides with the energy of the pure $S$ and $D$ phases.

Thus, in the absence of doping, either the pure $S$ phase ($\Delta_0 > \Sigma_0$) or the pure $D$ phase ($\Sigma_0 > \Delta_0$) is realized in the system. Only on the line $\Sigma_0 = \Delta_0$ do we have a degenerate solution, corresponding to the coexistence of the $S$ and $D$ phases with an arbitrary relation between the parameters $\Delta$ and $\Sigma$. The same conclusion is arrived at in [42, 43]. A corresponding conclusion for narrow bands satisfying the condition (5.1) is drawn in [44, 45].

Because of the different cutoff energy values in the equations for $\Delta$ and $\Sigma$ ($\omega_D$ and $E_F$, respectively), we should, strictly speaking, take the presence of the parameter $\Delta$ in the electron spectrum into account when integrating from zero

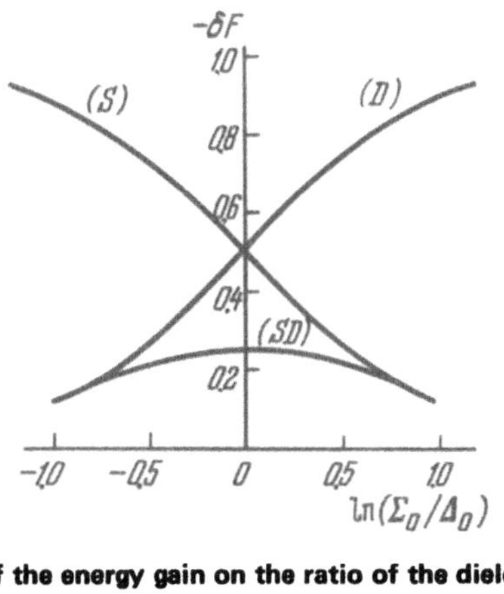


**Figure 5.3. Dependence of the energy gain on the ratio of the dielectric coupling constant to the superconducting constant for the *S*, *D*, and *SD* phases.**

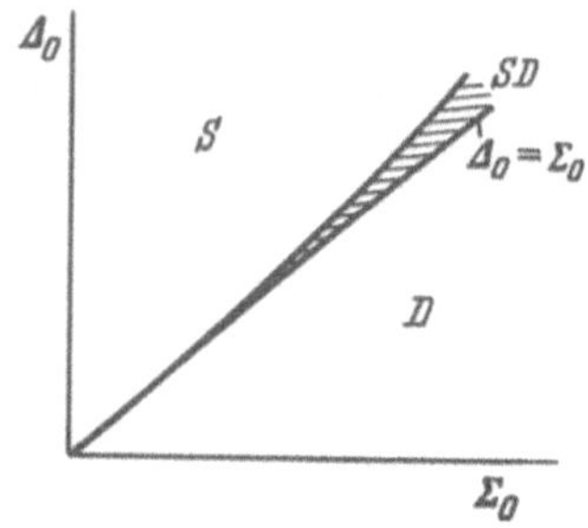


**Figure 5.4. Phase diagram for the undoped semimetal when allowance is made for the difference between the cutoff energies for the *S* and *D* pairings.**

to $\omega_D$ in the equation for $\Sigma$ and assume $\Delta$ to be equal to zero in the interval from $\omega_D$ to $E_F$. This leads, in the equation for $\Sigma$, to corrections proportional to $\Delta/\omega_D$ [46], and, as a result, the degeneracy in the *SD* phase on the bisectrix is removed and a finite existence region for the mixed *SD* phase appears in the vicinity of the bisectrix (Fig. 5.4).

A similar conclusion is drawn in [47] for an electron spectrum satisfying the condition (5.1), specifically, for the particular case of the one-dimensional metal. In this case, as the temperature is lowered, a structural transition occurs first at $T = T_p$. The magnitude of the dielectric gap $\Sigma$ increases with decreasing temperature. At a temperature, $T_c$, below the transition temperature for the pure *S* phase, the parameter $\Delta$ becomes nonzero, and increases with decreasing temperature, while the parameter $\Sigma$ begins to fall.

In [48] the possibility of the existence of a finite *SD*-phase region in the $(\Sigma_0, \Delta_0)$ diagram on account of the interelectron interaction terms that take the possibility of the transformation of particles of one band into particles of the other band into account is discussed. The phase diagram then has a shape similar to that of the diagram in Fig. 5.4.

In the following two sections we shall consider the possibility of the appearance of the *SD* phase when the condition $\xi_1(\mathbf{p}) = -\xi_2(\mathbf{p} + \mathbf{q})$ for a semimetal or the condition (5.1) for a metal is not exactly fulfilled. The violation of these conditions can be caused by doping (Section 4), or it may be because of the very nature of the one-electron spectrum, an effect that we shall conditionally call the electron-spectrum anisotropy effect (see Section 5).

For example, in the quasi-one-dimensional case the violation of the condition (5.1) arises as a result of the hopping of the electrons between neighboring chains. We shall show that in this case the $(\Sigma_0, \Delta_0)$ diagram contains an entire existence region for the mixed phase and that the critical temperature of the superconducting transition in the *SD* phase can significantly exceed the quantity $T_c$ for the *S* phase. Thus, we shall find the conditions under which the quantity $T_c$ rises as a result of the change that occurs in the electron spectrum during the structural transition.

## 4. Coexistence of the Dielectric and Superconducting Pairings in a Doped Semimetal

Let us first consider the antisymmetric solution ($\Delta_2 \equiv 0$). For $T = 0$ the basic system of equations (5.6), (5.7), (5.13) can be written in the form

$$\frac{1}{|\lambda_{11}|} = \mathcal{J}_1(\omega_D) + \left(\mu - \frac{\Sigma^2}{\mu}\right)\mathcal{J}_2, \tag{5.21}$$

$$\frac{1}{\widetilde{\lambda}_{21}} = \mathcal{J}_1(E_F) - \frac{\Delta^2}{\mu}\mathcal{J}_2, \tag{5.22}$$

$$\widetilde{n} = \mathcal{J}_3 + \frac{\Delta^2\Sigma^2}{\mu^2}\mathcal{J}_2, \tag{5.23}$$

where

$$\mathcal{J}_1(\widetilde{\omega}) = \frac{1}{2}\int_0^{\widetilde{\omega}} \frac{d\xi}{E}\left(\frac{E+\widetilde{\mu}}{\omega_+} + \frac{E-\widetilde{\mu}}{\omega_-}\right),$$

$$\mathcal{J}_2 = \frac{1}{2}\int_0^{\infty} \frac{d\xi}{E}\left(\frac{1}{\omega_-} - \frac{1}{\omega_+}\right),$$

$$\mathcal{J}_3 = \frac{1}{2}\int_0^{\infty} d\xi\left(\frac{E+\widetilde{\mu}}{\omega_+} - \frac{E-\widetilde{\mu}}{\omega_-}\right).$$

In the general case these quantities can be expressed in terms of elliptic integrals.

We shall solve the system (5.21)-(5.23) for $\Delta \longrightarrow 0$, i.e., in the limit of weak $S$ pairing.

For small $\Delta$ the integral $\mathcal{J}_2$ diverges logarithmically; therefore, Eq. (5.21) has no solutions in the limit $\Delta = 0$ ($\mu > \Sigma$). Physically, this means that in the presence of extraneous electrons (or holes) the $D$ phase is unstable with regard to the formation of Cooper pairs ($\lambda_{11} < 0$).

When $\Delta$ is small, corrections of the order of $\Delta \ln \Delta$ appear in the integrals $\mathcal{J}_1$ and $\mathcal{J}_3$; therefore, in the first approximation in $\Delta$, the influence of the $S$ pairing on the $D$ pairing and on the expression for $\widetilde{n}(\mu)$ can be neglected.

Then from Eqs. (5.22), (5.23) we obtain

$$\Sigma^2 = \Sigma_0(\Sigma_0 - 2\widetilde{n}), \quad \Sigma_0 \geqslant 2\widetilde{n}, \tag{5.24}$$

$$\tilde{n} = \sqrt{\mu^2 - \Sigma^2}. \tag{5.25}$$

Using the $\Delta \longrightarrow 0$ asymptotic form of the elliptic integral of the first kind for $\mathcal{J}_2$ and also the expressions (5.24) and (5.25), we find

$$\tilde{\Delta}_a = \frac{4\tilde{n}^2}{\Sigma_0} \exp\left(-\frac{\Sigma_0 - \tilde{n}}{\tilde{n}} \beta_0\right), \tag{5.26}$$

where $\Sigma_0 > 2\tilde{n}, \beta_0 \equiv \ln(\Sigma_0/\Delta_0) > 0$.

The obtained expression is valid when $\tilde{\Delta}_a \ll (4\tilde{n}^2/\Sigma_0)$ (weak coupling).

Since it is the quantity $\tilde{\Delta}_a$, and not $\Delta_a$, that figures in the expression (5.10) for the electron spectrum, all the properties of the *SD* phase are, apparently, expressible in terms of $\tilde{\Delta}_a$: At the end of this section we calculate $T_c$ and show that it is connected in the usual way with $\tilde{\Delta}_a$.

The excess-electron concentration $\tilde{n} = (\Sigma_0/2)$ is critical for *D* pairing, i.e., $\Sigma = 0$ for $\tilde{n} > \Sigma_0/2$ [see (5.24)]. It follows from the formula (5.26) that $\tilde{\Delta}_a = \Delta_0$ when $\Sigma_0 = 2\tilde{n}$, just as one would expect for the pure *S* phase. Below we present an expansion for $\tilde{\Delta}_a$ near the line $\Sigma = 0$, that coincides with the expression (5.26) for $\Delta_0 \ll \Sigma_0 < 2\tilde{n}$.

According to (5.26), $\tilde{\Delta}_a$ is a monotonically decreasing function of $\Sigma_0$ for $\Sigma_0 > 2\tilde{n}$, i.e., in the antisymmetric case *D* pairing prevents the onset of superconductivity in the system even in the presence of doping. Below we show that in the case of the symmetric solution $\tilde{\Delta}_s$ there exists in the $(\Sigma_0, \Delta_0)$ plane a region in which *D* pairing leads to the enhancement of superconductivity, i.e., to an increase in $T_c$. Using the expression (5.19), we obtain for the free energy of the *SD* phase in the antisymmetric case the expressions

$$F_{SD}^{(a)} = F_D - \frac{\tilde{n}\Delta^2}{\Sigma_0 - \tilde{n}} N(0), \tag{5.27}$$

$$F_D = -N(0)\left(\Delta_0 - 2\tilde{n}\right)^2. \tag{5.28}$$

Since the second term in (5.27) is of constant sign, $F_{SD}^{(a)} < F_D$, i.e., the formation of the *SD* phase is energetically more favorable than the formation of the purely dielectric phase.

Let us now proceed to consider the symmetric case. The system of equations for $\Delta, \Delta_2, \Sigma$, and $\mu$ at $T = 0$ has the following form:

$$\frac{\Delta}{|\lambda_{11}|} = \Delta \mathcal{J}_1(\omega_D) + \mu\tilde{\Delta} \cdot \mathcal{J}_2, \tag{5.29}$$

$$\frac{\Delta_2}{\lambda_{21}} = -\left(\frac{\Delta_2}{\mu}\right)\mathcal{J}_3 + \left(\frac{\widetilde{\mu}}{\mu}\right)\widetilde{\Delta}\cdot\widetilde{\Sigma}\cdot\mathcal{J}_2, \tag{5.30}$$

$$\frac{\Sigma}{\widetilde{\lambda}_{21}} = \Sigma\cdot\mathcal{J}_1(E_F) - \Delta_2\cdot\widetilde{\Delta}\cdot\mathcal{J}_2, \tag{5.31}$$

$$\widetilde{n} = \left(\frac{\mu}{\widetilde{\mu}}\right)\left[\mathcal{J}_3 + (\Delta^2 - \widetilde{\Delta}^2)\,\mathcal{J}_2\right]. \tag{5.32}$$

Only the integrals $\mathcal{J}_2$ and $\mathcal{J}_3$, which do not diverge at large $\xi$ figure in the equation for the interband superconducting gap $\Delta_2$. This means that the diagrams describing the scattering of a pair of electrons from different bands do not possess any singularities. It follows from Eq. (5.30) in the weak-coupling limit ($|\lambda_{21}| \ll 1$) that $\Delta_2 \sim \lambda_{21}$ in the general case. Therefore, it seems at first glance that we can set $\Delta_2 = 0$ in Eqs. (5.29), (5.31), and (5.32) (let us recall that $F_{21}$ does not vanish when we do this). This assertion is valid everywhere, with the exception of the case of extremely small values of $\Delta$. Indeed, setting $\Delta_2 = 0$ on the right-hand side of Eq. (5.30), we find in the limit when $\Delta \ll \Sigma \approx \mu$ that

$$\Delta_2/\Delta \sim \lambda_{21}\ln(\Sigma/\Delta),$$

i.e., the smallness of the coupling constant $\lambda_{21}$ can be compensated by the large value of $\ln(\Sigma/\Delta)$.

In the limit of weak $S$ pairing ($|\widetilde{\Delta}| \ll (4\widetilde{n}^2/\Sigma_0)$) we can set $\widetilde{\Sigma} = \Sigma$ and $\widetilde{\mu} = \mu$. It then follows from the expression (5.12) that, subject to the condition (5.24),

$$\widetilde{\Delta}_s \approx \Delta - \frac{\sqrt{\Sigma_0(\Sigma_0 - 2\widetilde{n})}}{\Sigma_0 - \widetilde{n}}\,\Delta_2.$$

After substituting into this formula the expressions obtained for $\Delta$ and $\Delta_2$ from Eqs. (5.29) and (5.30), respectively, we obtain

$$|\widetilde{\Delta}_s| = \frac{4\widetilde{n}^2}{\Sigma_0}\exp\left[-\frac{\widetilde{n}}{\Sigma_0 - \widetilde{n}}\beta^*\right],\quad |\widetilde{\Delta}_s| \ll \frac{4\widetilde{n}^2}{\Sigma_0}, \tag{5.33}$$

$$\beta^* = \beta_0\left[1 - \beta_0\lambda_{21}\frac{\Sigma_0(\Sigma_0 - 2\widetilde{n})}{(\Sigma_0 - \widetilde{n})^2}\right]^{-1}. \tag{5.34}$$

For $\Sigma_0 = 2\widetilde{n}$, the expression (5.33) for $\Delta_s$ goes over, as in the antisymmetric case, into the expression for $\Delta_0$, i.e., for the pure $S$ phase.

The contribution of $\Delta_2$ to $\widetilde{\Delta}_s$ amounts to the replacement of $\beta_0$ by $\beta^*$, which plays the role of an inverse "constant" for the renormalized interaction. Although, according to (5.34), the region in which it is essential to take $\Delta_2$ into

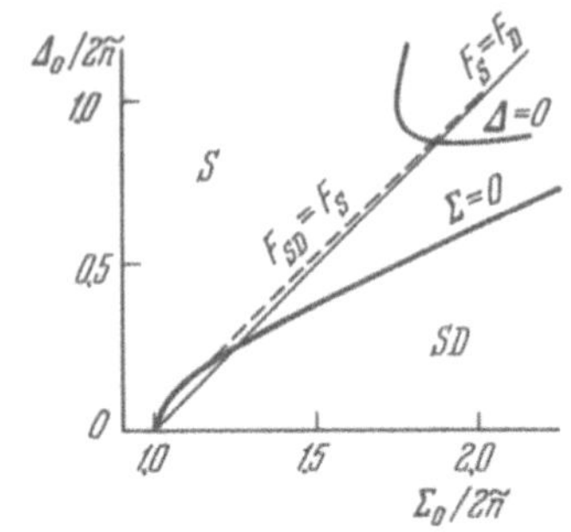


**Figure 5.5.** Phase diagram of a doped semimetal; the antisymmetric solution.

account is extremely narrow ($\beta_0^{-1} \ll \lambda \ll 1$), it is interesting to note that the behavior of $\tilde{\Delta}_s$ in this region depends on the sign of $\lambda_{21}$.

For the case $\lambda_{21} < 0$ (provided $\tilde{\lambda}_{21} > 0$), the expression (5.34) indicates that $\tilde{\Delta}_s$ increases with increasing $|\lambda_{21}|$. The opposite situation arises for $\lambda_{21} > 0$. In this case $\beta^*$ has a pole at a definite value of $\Delta_0$, i.e., here $\tilde{\Delta}_s = 0$. This corresponds in the $(\Sigma_0, \Delta_0)$ plane to the $\Delta = 0$ line shown in Fig. 5.6 (cf. Fig. 5.5). Below this line $\tilde{\Delta}_s = 0$ everywhere, i.e., the pure *D* phase should be realized. In fact, there exists above this line a first-order phase-transition line on which the mixed phase's free energies for the symmetric ($F_{SD}^{(s)}$) and antisymmetric ($F_{SD}^{(a)}$) cases coincide. The $\Delta = 0$ line is a metastability boundary for the symmetry *SD* phase.

It is easy to see from the expression (5.33) that, unlike $\tilde{\Delta}_a$, the quantity $\tilde{\Delta}_s$ as a function of $\Delta_0$ or $\tilde{n}$ possesses a maximum located within the *SD* phase, provided that $\beta_0 > 1$. As a function of $\tilde{n}$, the superconducting gap has a maximum at the point

$$\frac{2\widetilde{n}_{\max}}{\Sigma_0} = 2 + \frac{\beta_0}{2} - \left[\frac{\beta_0}{2} + \frac{\beta_0}{4}\right]^{1/2}. \tag{5.35}$$

It can be seen from this expression that the maximum falls in the physical region

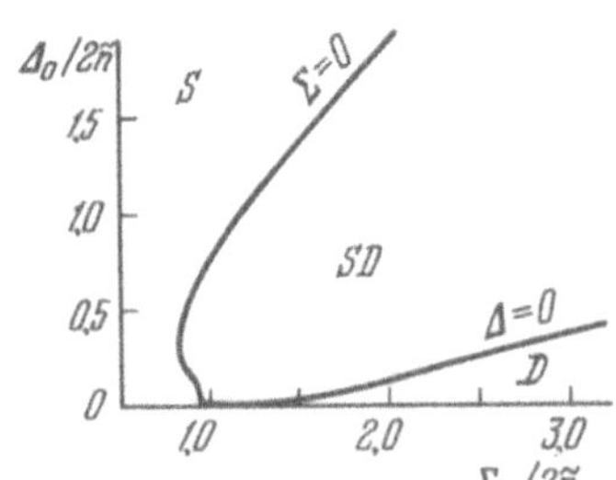


**Figure 5.6.** Phase diagram of a doped semimetal; the symmetric solution.

($2\tilde{n} < \Sigma_0$) when the condition $\beta_0 \geqslant 1$ is satisfied, i.e., when $\Delta_0 \leqslant \Sigma_0/e$. We can rigorously show with the aid of the expressions (5.33) and (5.35) that at its maximum the quantity $\tilde{\Delta}_s$ exceeds the value of the gap, $\Delta_0$, in the $S$ phase.

For the limiting case $\beta_0 >> 1$ we have ($e = 2.72$):

$$\frac{2\tilde{n}_{\max}}{\Sigma_0} \approx \frac{4}{\beta_0} \ll 1, \quad 8 \ll \beta_0 \ll \frac{1}{\lambda_{21}},$$
$$\left(\frac{\tilde{\Delta}_s}{\Delta_0}\right)_{\max} \approx \left(\frac{4}{e\beta_0}\right) e^{\beta_0} \gg 1. \tag{5.35a}$$

Thus, the increase in $\tilde{\Delta}_s$ (as well as in $T_c$) in the $SD$ phase relative to $\Delta_0$ can become arbitrarily large as $\Delta_0$ decreases. Allowance for the $\lambda_{21}$ dependence of $\beta^*$ (for $\lambda_{21} > 0$) leads to the saturation of this growth at $\lambda_{21} \approx \beta_0^{-1}$.

The $\Sigma_0$ dependence of $\tilde{\Delta}_s$ for a fixed value of $\tilde{n}$ has basically a similar character. Close to the critical—for the occurrence of the structural transition (i.e., of the $D$ phase)—concentration $\tilde{n} \lesssim \Sigma_0/2$, we find from the expression (5.33) that

$$\left(\frac{\partial\tilde{\Delta}_s}{\partial\Sigma_0^2}\right)_{\Sigma_0\approx 2\tilde{n}} = -\left(\frac{\Delta_0}{2\tilde{n}^2}\right)[1 - \beta_0 + 2\beta_0^2\lambda_{21}]. \tag{5.36}$$

This formula is a particular case of the more general expression (5.48) (for small $\phi$), which describes the change in $\tilde{\Delta}_s$ in the vicinity of $\Sigma = 0$.

The initial growth of $\tilde{\Delta}_s$ as a function of $\Sigma_0$ for $\lambda_{21} > 0$ occurs upon the fulfillment of the condition

$$2\beta_0\lambda_{21} < 1.$$

The maximum in the dependence $\tilde{\Delta}_s(\tilde{n})$ owes its origin to the competition between two factors. First of all, as a result of $D$ pairing, the density of states at the edges of the dielectric gap becomes infinite for the model without anisotropy being considered in the present section. In the presence of slight doping, the Fermi level will be located in the region of elevated—as a result of $D$ pairing—density of states, and this favors the growth of $\tilde{\Delta}_s$. On the other hand, as a result of the formation of the gap $\Sigma$, the energy interval in the vicinity of the Fermi level where there are electrons narrows. As the level of doping is raised, the number of such electrons increases, which is manifested in the pre-exponential factor in the expression (5.33).

It was assumed above that the cutoff energy, $\omega_D$, in the equations for the superconducting parameters is high compared to $\Sigma$ and $\tilde{n}$. In the opposite case (i.e., for $\omega_D << \tilde{n}, \Sigma$) the expression for $\tilde{\Delta}_s$ (and for $T_c$) has the form

$$\widetilde{\Delta}_s = 2\omega_D \exp\left(-\frac{\widetilde{n}}{\sqrt{\Sigma^2+\widetilde{n}^2}\,|\lambda_{11}|}\right),$$

i.e., the pre-exponential factor does not depend on $\widetilde{n}$, and the effect of doping on $\widetilde{\Delta}_s$ amounts to the decrease of the effective density of states with increasing $\widetilde{n}$.

Such a limit for a metal with a spectrum that satisfies the condition (5.1) is considered in [49]. This problem for an arbitrary relation connecting $\omega_D, \widetilde{n}$, and $\Sigma$ is solved in [50].

Let us compute the superconducting-transition temperature, $T_c$, of the *SD* phase under the assumption that it is low compared to the critical temperature, $T_p$, of the transition into the *D* phase. At $T = T_c$, we can limit ourselves in the expressions for $F_{ij}$ to the consideration of the terms linear in $\Delta$. Under the assumed condition that $T_c \ll T_p$ we can use the value of the parameter $\Sigma$ for $T = 0$. Let us limit ourselves to the consideration of the symmetric case, which is the most interesting case.

At $T = T_c$ the system of equations for $\Delta$ and $\Delta_2$ is similar to the system (5.29), (5.30) and has the form

$$\begin{aligned} \frac{\Delta}{|\lambda_{11}|} &= \Delta I_1(\omega_D) + \mu \cdot \widetilde{\Delta}_s I_2, \\ \frac{\Delta_2}{\lambda_{21}} &= -\left(\frac{\Delta_2}{\mu}\right) I_3 + \widetilde{\Delta}_s \cdot \Sigma \cdot I_2, \end{aligned} \tag{5.37}$$

where

$$I_1(\omega_D) = \int_0^{\omega_D} \frac{d\xi}{E}\left[\tanh\frac{E+\mu}{2T_c} + \tanh\frac{E-\mu}{2T_c}\right],$$

$$I_2 = \int_0^{\infty} \frac{d\xi}{E}\left[\frac{\tanh\frac{E-\mu}{2T_c}}{E-\mu} - \frac{\tanh\frac{E+\mu}{2T_c}}{E+\mu}\right],$$

$$I_3 = \frac{1}{2}\int_0^{\infty} d\xi\left(\tanh\frac{E+\mu}{2T_c} - \tanh\frac{E-\mu}{2T_c}\right).$$

In the limit $T_c \ll 4\widetilde{n}^2/\Sigma_0$ the integrals $I_1$ and $I_3$ reduce to those used above in the analysis for $T = 0$ ($\mathfrak{I}_1$ and $\mathfrak{I}_3$, respectively) since they do not have any singularities for $T_c \to 0$. In this limit the integral $I_2$ can easily be evaluated; as a result, the expression for $T_c$ coincides with (5.33) if we replace $\widetilde{\Delta}_s$ in it by $\pi T_c/\gamma$ ($\ln\gamma$ is the Euler constant).

Thus, we arrive at the conclusion that the quantity $\widetilde{\Delta}$, which determines the electron spectrum and whose value at $T = 0$ is connected with the quantity $T_c$ in the weak-coupling limit by the ordinary relation of the BCS theory, possesses a physical meaning in the *SD* phase.

Everywhere above the dielectric, $G_{21}$, and the superconducting, $F_{ij}$, Green functions were constructed with the wave functions of the metallic phase, all the characteristics of the superconducting phase being expressed in terms of the parameters of the symmetric metallic phase, which exists at $T > T_p$. In the case when $T_c << T_p$ we could have first solved the problem of the transition of the semimetal or metal with the spectrum $E(\mathbf{p}) = -E(\mathbf{p} + \mathbf{q})$ into the *D* phase (in the case of a semimetal with $\delta n \neq 0$, the *D* phase corresponds to the state of a doped semiconductor). After that, we should have solved the superconducting-transition problem, expressing the characteristics of the superconductor in terms of the parameters of the *D* phase. The problem is solved in precisely this way for $\delta n = 0$ in [43].

Two questions concerning the applicability of the ordinary BCS scheme arise here. First, because of the low concentration, $\tilde{n}$, of the free electrons in the *D* phase (in the case when $\tilde{n} < \omega_D << E_F$), the Fermi energy of these electrons turns out to be less than the phonon energy. Second, it looks as if this parameter is not involved in a logarithmic decrement of the Coulomb interaction in comparison with the electron–phonon interaction. Therefore, there arises the fear that the resulting interaction between the electrons will be a repulsive one.

These problems do not arise in the approach used above in the simultaneous description of the characteristics of the *D* and *SD* phases in terms of the parameters of the undistorted metallic phase since the renormalization of the interaction constants on account of the formation of the *D* phase is slight in the case of the approximation $(\Sigma/E_F) << 1$.

The physical essence of the weak renormalization lies in the fact that the above-used approach corresponds to that description of the superconducting pairing in the *D* phase in which, because of the small width, $\Sigma$ ($<< \omega_D$), of the dielectric gap, it is necessary to take into account the transition of the Cooper pairs from the valence band across the narrow forbidden band $2\Sigma$ into the conduction band.

For a similar reason ($\Sigma << \omega_D << E_F$), the logarithmic weakening of the Coulomb interaction is effected by all the electrons of the undistorted metallic phase, i.e., by the electrons of the valence band in the *D* phase and not only by the "extraneous" electrons $\delta n$, as is usually assumed in the description of the superconductivity of doped semiconductors [51].

It is the above-noted properties of semiconductors with narrow forbidden bands $\Sigma$ with respect to *S* pairing that distinguish the $A_4B_6$ semiconductors with band inversion, for which the applicability of the model under consideration is demonstrated in [8, 9, 32] and in which superconductivity occurs when they are doped [51]. Furthermore, the parameters $T_c$ and $T_p$ are found to be correlated in these semiconductors [52] in accord with the results of the analysis performed above.

In the case of semiconductors with wide forbidden bands, either the electrons of the conduction band or the holes of the valence band (depending on the type of doping) are responsible for superconductivity. In this case the weakening of the Coulomb repulsion is described by the factor $\ln(\tilde{n}/\omega_D)$. Since, moreover, such semiconductors have densities of states that are equal to zero at, and increase with distance from, the edges of the allowed bands, superconductivity can occur in them only when they are very heavily doped (i.e., when $\delta n \sim 10^{22}$ $cm^{-3}$), their parameters then being highly distorted.

In semiconductors in which the interaction between the electrons and the polar vibrations of the lattice is sufficiently strong, the effective superconducting coupling constant should increase with decreasing electron concentration. The opposite situation occurs in the high-electron-density limit (the BCS model with a delta-function interaction). In view of this, the situation could be realized [53] in which the Bose-condensation temperature for the Cooper pairs would be lower than the temperature at which they are formed.

This case corresponds to a weak interaction between the pairs, the radius of which is small compared to the mean distance between them, and was considered by Schafroth *et al.* [54] even before the construction of the BCS theory. In the low-electron-density limit, however, because of the absence of a large weakening logarithmic factor in the case of a wide forbidden band, the Coulomb repulsion makes superconductivity impossible [55].

We still have to consider another important question: How far in the ($\Sigma_0$, $\Delta_0$) plane does the above-found mixed-*SD*-phase region extend in the case of an arbitrary relation between $\Sigma_0$ and $\Delta_0$. (Let us recall that above we considered the case of weak superconductivity, i.e., the case when $\Delta \ll \Sigma$.)

For this purpose, first, let us determine the upper existence-domain boundary for the *SD* phase (i.e., the line $\Sigma = 0$) for increasing $\Delta_0$. It should be mentioned at once that the line $\Sigma = 0$ is not always an equilibrium line. This is so only in the case when the transition from the *SD* phase into the *S* phase is a second-order phase transition. If, on the other hand, the indicated transition is of first order (i.e., if it is accompanied by jumps in the parameters $\Sigma$ and $\Delta$), then the line $\Sigma = 0$ describes the lower boundary of the region of absolute instability of the *S* phase with regard to *D* pairing. The question of the nature of the transition can be resolved by investigating the solution for $\Sigma$ in the vicinity of the line $\Sigma = 0$.

Let us, first, investigate the antisymmetric case ($\Delta_2 \equiv 0$). It is convenient to carry out first the integration over $\xi$ in the equations (5.21)–(5.23) for $\Delta$, $\Sigma$, and $\tilde{n}$. As a result, we obtain (for $T \neq 0$)

$$\frac{1}{|\lambda_{11}|} = \pi T \sum_n \frac{1}{\tilde{\omega}_n} \operatorname{Re} \frac{1}{\Omega} \left[ \tilde{\omega}_n - i\mu + \frac{i\Sigma^2}{\mu} \right], \tag{5.38}$$

$$\frac{1}{\lambda_{21}} = \pi T \sum_n \frac{1}{\widetilde{\omega}_n} \operatorname{Re} \frac{1}{\Omega} \left[ \widetilde{\omega}_n + \frac{i\Delta^2}{\mu} \right], \tag{5.39}$$

$$\widetilde{n} = \mu + \pi T \sum_n \frac{1}{\widetilde{\omega}_n} \operatorname{Im} \frac{1}{\Omega} \left[ \widetilde{\omega}_n (\widetilde{\omega}_n - i\mu) + \frac{\Delta^2 \Sigma^2}{\mu^2} \right], \tag{5.40}$$

where

$$\widetilde{\omega}_n = \sqrt{\omega_n^2 + \widetilde{\Delta}^2}, \quad \Omega = \sqrt{\widetilde{\Sigma}^2 + (\widetilde{\omega}_n - i\mu)^2}, \quad \operatorname{Re} \Omega > 0.$$

In the first two equations it is assumed that the cutoff occurs at the frequencies $\omega_n \approx \omega_D$ and $\omega_n \approx E_F$, respectively. In the zeroth–in $\Sigma$–approximation, Eqs. (5.39) and (5.40) describe an $S$ phase with the parameters $\widetilde{\Delta} = \Delta_0$ and $\mu = \widetilde{n}$, while Eq. (5.38) describes the curve $\Sigma = 0$. At $T = 0$ the latter has the form

$$\beta_n = \cosh \varphi \ln\left(\coth \frac{\varphi}{2}\right), \tag{5.41}$$

where the parameter $\varphi$ is defined as follows:

$$\Delta_0 \equiv \widetilde{n} \sinh \varphi. \tag{5.42}$$

As can be seen from (5.41), the dependence $\Delta_0(\Sigma_0)$ is a monotonically increasing function on the line $\Sigma = 0$, as shown in Fig. 5.5, i.e., when the superconducting coupling constant ($\Delta_0$) is increased, the dielectric coupling constant ($\Sigma_0$) must be increased in order for $D$ pairing to occur in the $S$ phase. Such behavior indicates that the existence of superconductivity in the system impedes electron-hole pairing. Above it was shown for the case of weak superconductivity that $D$ pairing in its turn impedes the onset of superconductivity. Below we shall see that in the antisymmetric case this correlation between the $S$ and $D$ pairings obtains for any relation between the constants.

In the limiting cases the dependence $\Delta_0(\Sigma_0)$ for $\Sigma = 0$ assumes the form

$$\ln\left(\frac{\Sigma_0}{2\widetilde{n}}\right) = \frac{1}{2}\left(\frac{\Delta_0}{\widetilde{n}}\right)^2 \ln\left(\frac{2\widetilde{n}\sqrt{e}}{\Delta_0}\right), \quad \Delta_0 \ll \widetilde{n},$$

$$\ln\left(\frac{\Sigma_0}{\Delta_0}\right) = 1 + \frac{1}{3}\left(\frac{\widetilde{n}}{\Delta_0}\right)^2, \quad \Delta_0 \gg \widetilde{n}.$$

Thus, doping is unimportant in the region of large values of the parameters $\Delta_0$ and $\Sigma_0$ ($\widetilde{n}/\Delta_0 \ll 1$), and the line $\Sigma = 0$ coincides to within $(\widetilde{n}/\Delta_0)^2$ with the straight line $\Sigma_0 = e\Delta_0$ (see Fig. 5.2), which describes the lower metastability boundary for the $S$ phase. In this region the $\Sigma = 0$ line lies below the straight line $\Delta_0 = \Sigma_0 - 2\tilde{n}$, on which the $S$ and $D$ phases are in equilibrium, i.e., on which $\delta F_S = \delta F_D$ [see the formulas (5.20b) and (5.28)].

On the other hand, for small $\Delta_0$ the indicated curves are arranged in reverse order. Therefore, the initial part of the $\Sigma = 0$ line is a line of second-order phase transitions between the $S$ and $SD$ phases. To determine the $\Delta_0$ values up to which the curve $\Sigma = 0$ remains an equilibrium curve, let us consider the behavior of the solution for $\Sigma$, $\Delta$, and $\mu$ near this line.

From the expansion to second order in $\Sigma$ of the equation, (5.38), for $\Delta$ we obtain

$$\ln\left(\frac{\Delta_0}{\Delta}\right) = \frac{1}{2}\left(\frac{\Sigma_0}{\tilde{n}}\right)^2 \ln\left(\frac{\coth(\varphi/2)}{\cosh\varphi}\right) > 0, \tag{5.43}$$

i.e., in the antisymmetric case the superconducting parameter $\Delta$ always decreased ($\Delta < \Delta_0$) in the transition into the $SD$ phase.

From similar expansions of the equations, (5.39) and (5.40), for $\Sigma$, we can, using the expression (5.43) for $\Delta$, obtain [41] in the limiting cases of small and large $\Delta_0$ the following expressions:

$$\frac{\delta\Sigma_0}{\Sigma_0} = \begin{cases} (\Sigma/2\tilde{n})^2, & \Delta_0 \ll \tilde{n}, \\ -\frac{2}{3}(\Sigma/\Delta_0)^2, & \Delta_0 \gg \tilde{n}. \end{cases} \tag{5.44}$$

Here $\delta\Sigma_0$ denotes the change in $\Sigma_0$ on moving away from the $\Sigma = 0$ line under the condition that $\Delta_0 = \text{const}$.

The change in sign of the right member of (5.44) indicates a change in the character of the phase transition on the $\Sigma = 0$ line for $\Delta_0 \approx \tilde{n}$.

A numerical analysis of the point in the $(\Sigma_0, \Delta_0)$ plane at which $\delta\Sigma_0$ passes through zero yields [41]

$$(\Sigma_0/2\tilde{n}) = 1.104; \quad (\Delta_0/2\tilde{n}) = 0.144.$$

The $\Sigma = 0$ second-order phase-transition line terminates at this point. To the right, the $\Sigma = 0$ line should go over into a first-order $S$-to-$SD$ phase-transition line, determined by the condition $F_{SD} = F_S$. Since, according to (5.27), $F_{SD} < F_D$, the $F_{SD} = F_S$ line of first-order phase transitions (see Fig. 5.5) lies everywhere above the equilibrium line for the pure $S$ and pure $D$ phases.

We have already seen that in the presence of doping ($\delta n \neq 0$) the system of equations (5.21)–(5.23) does not admit of solutions corresponding to the nucleation of the $S$ phase in the $D$ phase (the line $\Delta = 0$), even in the limit as $\delta n \to 0$. The point is that the Fermi-level shift does not go to zero ($\mu \to \Sigma_0$) as $\delta n$ tends to zero (at $T = 0$), in contrast to the case in which it is assumed from the very beginning that the electron and hole concentrations are equal to each other

($\tilde{n} = 0, \mu = 0$). In the first case there are "excess" electrons at the edge of the allowed band; therefore, $S$ pairing is possible for any value of $\tilde{n}$ at $T = 0$. However, the quantity $\mu$ rapidly falls to zero at low temperatures $T \sim (\tilde{n}^2/\Sigma_0) \ll \Sigma_0$, and the system goes over into the pure $D$ phase ($\Delta = 0$).

Everywhere above we neglected the possibility of the formation of localized impurity states [37], assuming that the impurity band, on account of the finite concentration, overlaps the valence band. Since the binding energy at an impurity level increases with increasing $\Sigma_0$ [37], our analysis will not be valid at large $\Sigma_0$. The "excess" $\delta n$ electrons will with increasing $\Sigma_0$ become more and more localized on the impurities, the Fermi level will move into the region of the dielectric gap ($\mu < \Sigma$), and the superconductivity will disappear. Thus, there will appear in the region of large $\Sigma_0$ an $SD$-to-$D$ phase-transition boundary ($\Delta = 0$). Similar arguments apply in the symmetric case considered below.

The equation for $\Sigma$ in the symmetric case assumes, after integration over $\xi$, the form

$$\frac{\Sigma}{\lambda_{21}} = \pi T \sum_n \frac{1}{\tilde{\omega}_n} \operatorname{Re} \frac{1}{\Omega} [\tilde{\omega}_n \Sigma + i \tilde{\Delta} \Delta_2]. \tag{5.45}$$

Hence for the $\Sigma = 0$ line we shall have

$$\beta_0 \operatorname{ch} \varphi = \ln(\coth(\varphi/2)), \quad \Delta_0 \equiv \tilde{n} \cosh \varphi. \tag{5.46}$$

This $\Delta_0(\Sigma_0)$ function is shown for $\tilde{n} = \text{const}$ in Fig. 5.6 (the boundary between the $S$ and $SD$ phases). A characteristic feature of this dependence (in contrast to the antisymmetric case) is the existence of a region of double valuedness in the interval $0.9 \leqslant (\Sigma_0/2\tilde{n}) \leqslant 1$. The minimum value of $\Sigma_0 = 1.8\tilde{n}$ is attained on the curve $\Sigma = 0$ at $\Delta_0 = 0.66\tilde{n}$. The decrease of the function $\Sigma_0(\Delta_0)$ on the segment $0 < \Delta_0 < 0.66\tilde{n}$ indicates that intraband $S$ pairing favors interband $D$ pairing. We shall see below that in this same range of $\Delta_0$ values $D$ pairing in turn favors superconductivity.

It can be seen from (5.46) that for $\Sigma_0 \gg \tilde{n}$ the dependence $\Delta_0(\Sigma_0)$ assumes a linear character (i.e., $\Delta_0 \sim \Sigma_0$) on the $\Sigma = 0$ line. The equilibrium line, $F_S = F_D$, for the pure $S$ and $D$ phases ($\Delta_0 = \Sigma_0 - 2\tilde{n}$) everywhere passes below the line $\Sigma = 0$ (see Fig. 5.6). This indicates that a second-order phase transition from the $S$ phase into the $SD$ phase occurs on the $\Sigma = 0$ line. We can rigorously verify this by computing $\Sigma$ near the $\Sigma = 0$ line. We have up to terms of the order of $\Sigma^3$ the equation [41]

$$\frac{\delta\Sigma_0}{\Sigma_0} = \left(\frac{\Sigma}{2\tilde{n}}\right)^2 \frac{1}{\operatorname{ch}^4 \varphi} \left[1 + 5 \sinh \varphi \cdot \tanh^2 \varphi \cdot \ln\left(\coth \frac{\varphi}{2}\right) - 2(2 \tanh^2 \varphi - \sinh^2 \varphi) \ln^2 \left(\coth \frac{\varphi}{2}\right)\right]. \tag{5.47}$$

The right-hand side of Eq. (5.47) is positive for all $\varphi$ (in contrast to the antisymmetric case), i.e., a nontrivial ($\Sigma \neq 0$) solution exists in the ($\Sigma_0, \Delta_0$) plane only to the right of the $\Sigma = 0$ line. This proves the assertion made above that a second-order phase transition occurs at every point on the $\Sigma = 0$ line. Thus, the symmetric solution for the *SD* phase exists over a significantly wider range of $\Sigma_0$ and $\Delta_0$ values in comparison with the antisymmetric solution. The latter becomes energetically more favorable than the symmetric solution only when $\lambda_{21} > 0$, the $\Delta_0$ values are very small, and $\lambda_{21}\beta_0 \approx 1$. In the $\lambda_{21} < 0$ case the symmetric solution is energetically more advantageous everywhere.

The expression for $\widetilde{\Delta}_s$ near the $\Sigma = 0$ line in the *SD* phase has, for the symmetric case, the form [41]

$$\ln\left(\frac{\Delta_0}{|\widetilde{\Delta}_s|}\right) = \frac{1}{2}\left(\frac{\widetilde{\Delta}_s}{\widetilde{n}}\right)^2 \frac{1}{\cosh^2\varphi}\left[1 - \cosh^{-1}\varphi \cdot \ln\left(\coth\frac{\varphi}{2}\right) + \right.$$
$$\left. + 2\lambda_{21}\ln^2\left(\coth\frac{\varphi}{2}\right)\right]. \tag{5.48}$$

The last term on the right-hand side of (5.48) (i.e., the term containing $\lambda_{21}$) becomes important when $\varphi \ll 1$. For $\varphi \gtrsim 1$ the right-hand side of (5.48) changes sign at the point $\cosh\varphi \ln[\coth(\varphi/2)] = 1$. Returning to Eq. (5.46), we see that it is precisely at this point on the $\Sigma = 0$ curve that the quantity $\Sigma_0$ assumes its minimum value $\Sigma_0 = 1.8\tilde{n}$ at $\Delta_0 = 0.66\tilde{n}$. The increase of $\widetilde{\Delta}_s$ in the *SD* phase ($\widetilde{\Sigma} \neq 0$) corresponds to values of $\Delta_0 < 0.66\tilde{n}$.

For $\Delta_0 > 0.66\tilde{n}$, *D* pairing impedes the onset of superconductivity in the symmetric case as well.

It should be noted that under the weak-interaction assumption, the models considered by us lead to the result that the interaction-constant renormalization that occurs during the structural transition is insignificant. For example, the effect, discussed in Chapter 4 of the softening of the phonon modes during a structural transition on superconductivity turns out to be insignificant in comparison with the effect, exerted through the density of states, of the change that occurs in the electron spectrum during the structural transition. The point is that in the weak-interaction limit the softening of the phonon modes occurs for a narrow region in momentum space around the vector **q** determined by the condition (5.1). And as to the effective constants $\lambda_{ii}$, they contain integrated characteristics of the phonon spectrum.

## 5. Conclusion

In the preceding section we have shown that superconducting and dielectric pairings can coexist in semimetals with unequal electron and hole concentrations. The most important result is the conclusion that the superconducting-transition temperature $T_c$ can significantly exceed its value $T_{c_0}$, in the absence of dielectric pairing.

Let us give for the maximum value of the ratio $T_c/T_{c_0}$ an expression similar to the one (5.35a), for $(\tilde{\Delta}_s/\Delta_0)_{max}$:

$$\left(\frac{T_c}{T_{c_0}}\right)_{max} = \left(\frac{4}{e\beta_0}\right) e^{\beta_0}, \quad \beta_0 \equiv \ln\left(\frac{\Sigma_0}{\Delta_0}\right). \tag{5.49}$$

Thus, the quantity $T_c$ can significantly exceed the quantity $T_{c_0}$ in the case when $(\Sigma_0/\Delta_0) >> 1$.

Let us emphasize, however, that the realization of a $T_c$ that is of the order of $T_{c,\,max}$ is not a simple problem since the region in the $(\Sigma_0, \Delta_0)$ phase diagram (see Fig. 5.6) where dielectric pairing promotes superconducting pairing is fairly narrow: In the remaining *SD*-phase region the quantity $T_c$ turns out to be lower than the quantity $T_{c_0}$. Therefore, for the realization of sufficiently high $T_c$ values we should be in a position to choose purposely the parameters $\Sigma_0$ and $\Delta_0$ corresponding to the double-valuedness region of the $\Sigma = 0$ line in Fig. 5.6. At the same time, it is necessary that the "excess"-electron concentration approximately satisfy the condition (5.35).

Let us emphasize once more that the entire analysis presented above applies not only to semimetals with Fermi surfaces that roughly coincide in shape and size, but also to single-band metals whose spectrum, $\xi(\mathbf{p})$, approximately satisfies the condition $\xi(\mathbf{p}) = -\xi(\mathbf{p} + \mathbf{q})$. The requirement that parts of the Fermi surfaces should approximately coincide is not too rigid [34], especially if we take into account the fact that in real systems the interelectron-interaction constants, which determine the electron–hole pairing, are not small.

The main cause of the increase of $T_c$ from $T_{c_0}$ then lies in the increase in the density of states near the Fermi level or, rather, in the distinctive functional dependence of the density of states $N(\xi)$ on energy. In the range $\xi = \pm\Sigma$ the density of states is equal to zero, as a result of which the density of states at the edges of the dielectric gap rises formally to infinity. But it is well known [56] that in the weak-interaction limit in the two-dimensional case, when, for an ideal gas of particles, the density of states does not depend on the energy, a bound state whose binding energy is an exponential function of the interaction potential is always formed. The BCS solution is a solution of precisely this type since the density of states $N(\xi)$ can be assumed to be a constant in a narrow range in the vicinity of the Fermi level where the bound state arises.

For the one-dimensional case, when $N(\xi) \sim 1/\xi^{1/2}$, i.e., formally becomes infinite at $\xi = 0$, the bound-state energy is given by a power function of the interelectron attraction potential [56]. In our case, as a result of the electron–hole pairing, $N(\xi)$ significantly varies with energy in an interval of the order $\Sigma$ and becomes infinite at the edge of the dielectric gap, as for a one-dimensional ideal gas. In the region $\xi >> \Sigma$, $N(\xi)$ can be assumed to be a constant.

It is this quasi-one-dimensional behavior of $N(\xi)$ for $\xi \sim \Sigma$ that leads to the distinctive dependence of $\Delta$ on $\lambda$. The expression (5.33), for $\tilde{\Delta}_s$ for the case $\beta^* \approx \beta_0$ can be rewritten in the following form:

$$\tilde{\Delta}_s = \frac{4\widetilde{n^2}}{(\Sigma_0)^{1+(\tilde{n}/\Sigma_0)}}\left[\exp\left(-\frac{1}{\lambda}\right)\right]^{\tilde{n}/\Sigma_0}. \tag{5.50}$$

Thus, when the Fermi level turns out to be located near the dielectric gap [i.e., when $(\tilde{n}/\Sigma_0 \ll 1)$], the exponential function is taken with the very small power $\tilde{n}/\Sigma_0$, instead of one, in the case when $N(\xi) = \text{const}$. The smallness of the pre-exponential factor is because of the vanishing of the density of states in the interval $2\Sigma$. It is these two opposing tendencies that lead to the occurrence of a maximum in $T_c$ as a function of the level of doping or of the structural transition temperature $T_p$, which is proportional to $\Sigma_0$. Precisely such a dependence has been observed by Grassie and Benyon [52].

Thus, what is important for the quantity $T_c$ is not the density of states averaged over the range $2\omega_D$ around the Fermi level, a quantity that does not change as a result of the electron-hole pairing when $\omega_D > \Sigma$, but its change in the functional dependence of $N(\xi)$ [the fact of the $N(\xi)$'s becoming infinite].

Naturally, under real conditions $N(\xi)$ will not become infinite at the edges of the dielectric gap. For example, because of the scattering even by the nonmagnetic impurities, the singularity in $N(\xi)$ becomes smeared out over an interval of the order of $\hbar/\tau$ (where $\tau$ is the electron–impurity scattering time). But in the case when $\tilde{n} \gg \hbar/\tau$ the expression (5.50) for $\tilde{\Delta}_s$, or the expression (5.49) for the ratio $T_c/T_{c_0}$, remains practically unchanged.

The anisotropy or, rather, the noncongruence of the electron and hole parts of the Fermi surface that are responsible for the structural transition also leads to the disappearance of the singularity in $N(\xi)$ at the edges of the gap $\Sigma$.

The effect of this noncongruence on the possibility of the coexistence of the superconducting and dielectric pairings is similar to the effect, considered in Section 4, of the noncoincidence of the electron and hole concentrations as a result of impurity doping. Formally, common solutions of the equations for the parameters $\Delta$ and $\Sigma$ exist because the degree of noncongruence virtually does not enter explicitly into the expression for $\Delta$; on the other hand, it has a significant effect on the magnitude of $\Sigma$. There is a limit to the degree of noncongruence, above which the dielectric pairing and, consequently, the structural transition turn out to be impossible [33, 34].

For a slight noncongruence, the undoped ($\tilde{n} = 0$) system is an intrinsic semiconductor below the structural-transition temperature, i.e., the Fermi level lies in the middle of the forbidden band $E_g$. The extrema of the conduction bands and the valence band do not coincide in momentum space, and the minimum distance, $E_g$, in terms of energy between them is less than $2\Sigma$. When the quantity $E_g$ is less than some critical value, the superconducting pairing turns out to be posssible, i.e., there is a common solution for $\Delta$ and $\Sigma$. The condition under which the parameter $\Delta$ first appears with an infinitely small value has the form [57]

$$\frac{1}{|\lambda_{11}|} = \operatorname{Arc\,sinh} \frac{\omega_D}{\Sigma_0} + \frac{\pi}{2\sqrt{2}} \left(\frac{u}{\Sigma - u}\right)^{1/2}, \tag{5.51}$$

if, instead of the condition (5.1), the electron spectrum satisfies the condition $\xi(\mathbf{p}) = -\xi(\mathbf{p} + \mathbf{q}) + u$, where $2(\Sigma - u) \equiv E_g$, u being the degree of noncongruence. As can be seen from (5.51), superconducting pairing is, for $E_g \longrightarrow 0$, possible for any arbitrarily weak interelectron attraction.

The character of the onset of superconductivity in the semiconductor phase ($E_g > 0$) without doping ($\tilde{n} = 0$), when the Fermi level lies inside the dielectric gap, consists in the following: the excitation of a pair of electrons in the semiconductor (or dielectric) phase (*D* phase) requires the expenditure of energy of the order of $E_g$, which, for the noncongruent parts of the Fermi surface of the original metallic phase, is smaller than the quantity $2\Sigma$. If the energy, $\Delta_0$, required to bind them into a superconducting pair (in the absence of dielectric pairing) turns out to be larger than this quantity, then the *SD* phase turns out to be energetically more advantageous than the *D* phase. The *D* phase then turns out to be unstable against superconducting pairing. A similar relation between the width, $E_g$, of the forbidden band of a semiconductor and the energy required to bind an electron and a hole into an exciton should be satisfied in order for the exciton insulator phase to be realized [58].

It turns out that as the degree of noncongruence is increased, some of the states in the conduction band get to be occupied by electrons (let us recall that $\tilde{n} = 0$), whereas not all the states in the valence band are occupied, i.e., holes appear in it. The system in the *D* phase (at $T < T_p$) is then a semimetal [59] with an electron and hole concentration much lower than the carrier concentration in the corresponding state [of the semimetal or metal with a spectrum $\xi(\mathbf{p}) \simeq -\xi(\mathbf{p} + \mathbf{q})$] in the absence of dielectric pairing. Since in this case there are free carriers because of the slight (of the order of $\Sigma$) overlap of the bands in the *D* phase, the conditions for the appearance of the *SD* phase are similar in many respects to the conditions, considered in Section 4, for the case of unequal electron and hole concentrations. The explicit expressions for $T_c$, on the other hand, depend on the specific character of the noncongruence of the electron and hole parts of the Fermi surface.

The solution to one of the simplified variants of this problem is given in [57].

We must note one interesting distinctive feature of the solution for $T_c$ in the case when the bands in the *D* phase do not overlap and a dielectric–superconductor transition occurs at the point $T_c$. Since the quasiparticle excitation energy is determined simultaneously by both the superconducting, $\Delta$, and dielectric, $\Sigma$, gaps, the growth of the number of quasiparticles with increasing temperature will be slower than the growth when only the superconducting gap $\Delta_0$ is present. Therefore, although the magnitude of the superconducting gap in the *SD* phase can be smaller than the quantity $\Delta_0$, the superconducting-transition temperature $T_c$ may turn out to be higher than $T_{c_0}$. This is attested to by the fact that the ratio of $T_c$ to the superconducting gap $\Delta$ at $T = 0$ turns out to be higher [57] than the classical value of 0.57 obtained in the BCS theory.

A correlation, similar to the one considered in the present chapter, of the quantity $T_c$ with the structural-transformation temperature and the degree of doping has been observed in semiconductors of the $A_4B_6$ group [52] and the $NaWO_3$ bronze [60]. Apparently, suitable objects are the transition-metal compounds in which the spectrum of the conduction *d* electrons can approximately satisfy the condition (5.1). In particular, this apparently occurs in a number of layered superconductors (see Chapter 6) that are observed to undergo structural transformations.

The situation is possible in which the temperature $T_p$ is higher than the melting point $T_m$. In other words, the symmetric "protophase" that was the initial phase in our analysis is unobservable: for example, the cubic protophase of bismuth is unobservable under ordinary conditions, although it is convenient to choose it as an initial model [61].

Examples of systems with $T_p > T_m$, in which the scheme described in the present chapter is qualitatively realized are, apparently, the compounds with the A-15 lattice. The fact that a large fraction, $\alpha$, of their Fermi surface "goes out" to the Brillouin boundaries [62] can be interpreted as a partial dielectrization of an initial symmetric protophase in which the condition (5.1) is approximately fulfilled. The observed correlation $T_c(\alpha)$ [62, 63] corresponds to the correlation, considered in the present chapter, of $T_c$ with $\Sigma$ and $\tilde{n}$. On the other hand, the occurrence in a number of the A-15 compounds of a low-temperature structural transition is, possibly, a confirmation of the inference made in [35] of the existence of a sequence of structural transitions in the case of incomplete congruence of the individual parts of the Fermi surface of the initial phase.

In the compound VRu, a correlation of the quantity $T_c$ with the structural-transition temperature and the degree of dielectrization is observed when the composition is varied [4].

Suitable objects may be quasi-one-dimensional compounds (see Chapter 7) for which the condition (5.1) is approximately fulfilled. The situation is possible in which with the growth of the overlap integral of the electron wave functions on neighboring chains it becomes unnecessary to sum the parquet diagrams [18], but the structural instability still obtains. Therefore, the approximation used in the present chapter is sufficient for the description of the coexistence of the *S* and *D* pairings.

According to the above-presented results, quasi-one-dimensional materials with a feebly marked unidimensionality are required for the realization of superconductivity with high $T_c$ values. Otherwise the dielectrization of the electron spectrum during the Peierls transition will lead to the suppression of superconductivity.

To realize the optimum conditions for obtaining the highest $T_c$ value in the *SD* phase, we must be able to vary the overlap integral for neighboring chains or to dope the materials, thereby violating the condition (5.1).

Let us note a number of anomalies in the manifestation of the superconducting properties in the *SD* phase.

1. If in the *S* phase only the scattering by the magnetic impurities affect the quantities $\Delta$ and $T_c$, then in the *SD* phase, because the scattering by the nonmagnetic impurities does affect the quantities $\Sigma$ and $T_p$, this scattering will also affect the quantities $\Delta$ and $T_c$ since there is a strong correlation of the latter quantities with $\Sigma$ and $T_p$.

In particular, the situation is possible in which, because of electron scattering by the nonmagnetic impurities, the quantities $T_c$ and $\Delta$ will increase as the quantities $\Sigma$ and $T_p$ decrease, i.e., this is one of the methods of realizing the optimum conditions for the quantity $T_c$ in the *SD* phase. Then under the quantity $\Sigma_0$ we should understand the dielectric gap with allowance for the scattering by the impurities.

2. Since in the *SD* phase the electron spectrum contains both the superconducting and dielectric gaps (see Fig. 5.1c), this fact should be taken into account in the interpretation of experiments on threshold phenomena (infrared absorption, nuclear-spin relaxation, hypersound absorption, tunneling experiments) in the *SD* phase. For a number of compounds of the A-15 type that undergo a structural transition, the temperature dependence of the kinetic coefficients as obtained from nuclear magnetic resonance [64] and from ultrasound absorption [65, 66] is found to deviate significantly from the predictions of the BCS model [67]. Allowance for the effect of the dielectric correlations in the model considered by us on the corresponding coefficients [68] leads to a qualitative explanation of these experiments.

In [69] the effect of the dielectric correlations on the optical properties of superconductors is predicted.

3. Because of the coexistence of the *S* and *D* pairings, i.e., of the off-diagonal (ODLRO) and diagonal (DLRO) long-range orders, the superconducting current and the magnetic-field penetration depth are determined even at $T = 0$, not by all the electrons but by only a fraction proportional to the ratio of the quantity $\Delta$ to the sum $(\Delta + \Sigma)$ [18, 43]. The most important fact is the effect on each other of the superconducting, $\Delta$, and dielectric, $\Sigma$, order parameters. As a result, such a superconductor in a magnetic field is described by a system of three differential equations [70] (instead of two, as in the Ginzburg–Landau theory). This leads to a situation in which the damping of the magnetic field inside the bulk superconductor can be accompanied by oscillations.

4. It is well known that a superconductor in the *S* phase can go into an

inhomogeneous state upon the "parting," under the action of the internal magnetic field, of the Fermi surfaces of the electrons with opposite spin orientations [71]. As to the inhomogeneous state in the case of $D$ pairing (the deviation of the resultant momentum of the electron-hole pairs from $\mathbf{q}$), it is possible on account of doping or noncongruence [11, 25]. For this reason, because of the correlation between the parameters $\Delta$ and $\Sigma$ in the $SD$ phase, there can arise inhomogeneous $S$ pairing with all the pecularities of the electromagnetic properties [71] of such a state.

5. Besides the possibility, noted in (1) above, of controlling the quantity $T_c$ through the effect on the parameter $\Sigma$ of the scattering by the nonmagnetic impurities, we can vary $T_c$ in the $SD$ phase by means of any external influence that affects the magnitude of the dielectric gap. This can, for example, be pressure. The sign of the change in the quantity $T_c$ here will be determined by that part of the $(\Sigma_0, \Delta_0)$ diagram (see Fig. 5.6) that corresponds to the parameters of the material without the external influence.

Everywhere above, by $D$ pairing we meant the pairing of an electron with a hole into a singlet state. It is precisely this pairing that leads to the formation of electron charge density waves and to a structural transformation.

In the case of electron-hole pairing in the triplet state, a spin density wave arises, which leads to antiferromagnetic ordering [58].

The picture of the coexistence of the triplet electron-hole pairing with the superconducting pairing will be similar to the one considered in the present chapter: The onset of superconductivity is possible below the Néel temperature $T_N$, i.e., in the antiferromagnetic state, and the correlation between the quantities $T_c$ and $T_N$ will have the same character as the correlation between the quantities $T_c$ and $T_p$.

If, on the other hand, the singlet and triplet electron-hole pairings are realized simultaneously in the system, then this leads to ferromagnetic ordering [12].

The conditions for the onset of superconductivity are, of course, less favorable in that case considered in the present chapter because of the "parting" of the levels of electrons with oppositely directed spins.

Since such a ferromagnetic order significantly affects the density of electron states, however, the effect of the "parting" may be partially compensated by an increase in $N(\xi)$ near the Fermi level. This may ease the condition for the coexistence of superconductivity and ferromagnetism in such systems.

Let us note once again the class of materials in which electron-hole pairing and the related structural transformation apparently occur. These are the quasi-one-dimensional (see Chapter 7) and layered compounds of the type $NbSe_2$ and $TaS_2$, etc. (see Chapter 6), semiconducting compounds of the $A_4B_6$ group [32, 52], certain bronzes [60, 72], and the compounds VRu [4] and VS [73]. And similar structural transformations with the appearance of charge density waves appear on pure Ge and Si [74] and Pt [75] surfaces.

Qualitatively, the correlations, described in the present chapter, between the characteristics of the structural and superconducting transitions occur in a number of layered compounds, doped semiconductors of the $A_4B_6$ group [52], the compounds VRu [4], and sodium bronze [72].

By smoothly varying the parameters of these materials, we may hope to realize the conditions ensuring the closeness of the quantity $T_c$ to the optimum possibility, as revealed in the investigation carried out in the present chapter.

# 6

# Electronic Properties and Superconductivity of Layered Crystals

## 1. Introduction

In 1964 Little [1] put forward the hypothesis that high-temperature superconductivity can be realized in one-dimensional structures on the basis of the exciton mechanism. According to his hypothesis, the system should contain long linear conducting molecules (polyenes or polymers with metallic atoms in an organic matrix), and on either side of each of these molecules should be located groups of atoms with a high electronic polarizability. It is these side branches that should guarantee the mutual attraction of the conducting electrons as a result of the exchange of molecular excitations of the electronic type.

Also in 1964 Ginzburg [2] proposed the use for the realization of the electron machinism two-dimensional systems of the "sandwich" type. Such systems should contain a metallic film surrounded on both sides by dielectric or semiconducting films, which will secure the exciton mechanism of superconductivity for the electrons of the central metallic film. The advantage of both one- and two-dimensional systems in comparison with three-dimensional ones from the point of view of the problem of high-temperature superconductivity lies in the fact that it is easy to locate near the conducting filament (or plane) easily polarizable complexes for the realization of the exciton mechanism of superconductivity.

Little's and Ginzburg's ideas initiated work on the synthesis and investigation of systems with the required geometry and properties. It has become clear from the attempts to synthesize molecules of the one-dimensional metallic type that it is very difficult to obtain in linear molecules conductivities that even approach in magnitude the conductivity of a metal: all the presently known polymer molecules are semiconductors with a quite wide forbidden band. It has, however, turned out to be possible to produce three-dimensional crystals in which linear conducting filaments alternate with filaments of polarizable molecules. Because of the characteristics of the crystal structure, the electrons in such crystals can move largely along only the filaments, there being almost no motion in the other two directions. Such crystals have come to be called quasi-one-dimensional. Three classes of compounds of this type are known at present: complexes based on tetracyanoquinodimethane (TCNQ) [3], square-planar complexes of the mixed-valence transition elements of the platinum group [4], and crystals of the polymers $(SN)_x$ [5].

It has also turned out that Ginzburg's ideas about systems of the sandwich type can be realized at the molecular level. In 1971, Gamble and co-workers [6] succeeded in introducing organic molecules into layered compounds, specifically, into the dichalcogenides $TaS_2$ and $NbSe_2$. In the $TaS_2$ and $NbSe_2$ crystals the dichalcogenide layers are coupled to each other largely by weak van der Waals forces. Therefore, we can separate the layers, introducing between them molecules that can be coupled to the conducting dichalcogenide layers by forces somewhat stronger than the van der Waals forces. In the resulting, so-called intercalated compounds the conducting layers alternate with molecular layers that can play the role of a polarizable system for the exciton mechanism of superconductivity.

The crystal structure of intercalation compounds is such that the motion of the conduction electrons is almost two dimensional since the separation of the conducting layers is large (up to 60 Å), and the molecules hinder electron transitions between the conducting layers. Crystals of this sort are called quasi-two-dimensional. A striking characteristic of intercalation $TaS_2$ and $NbSe_2$ compounds is the preservation of superconductivity in them; in the case of $TaS_2$ the critical temperature $T_C$ even increases after the introduction of the molecules ($T_c = 0.8$ K in $TaS_2$, while $T_c = 3.5$ K in the case when the compound is intercalated by pyridine).

The first range of questions that were raised in connection with the problem of high-temperature superconductivity included the question of the possibility in principle of the existence of superconductivity in a system with a small dimension. From the point of view of the problem of high-temperature superconductivity, of not less importance is the question of the conditions under which the exciton mechanism would be effective. Both of these questions are discussed in the present chapter after a brief consideration of the structure and physical properties of the dichalcogenides of the transition metals and their intercalation compounds.

Furthermore, the vast field of investigations in the area of quasi-one-dimensional and quasi-two-dimensional crystals is connected with the fact that many physical properties of one- and two-dimensional electron systems significantly differ from the corresponding properties of ordinary isotropic three-dimensional crystals.

Thus, we obtain in the case of the quasi-two-dimensional, molecule-intercalated dichalcogenide compounds of the transition metals a new class of type-II superconductors with the Josephson interaction between the layers. The combination of a huge number of weakly interacting superconducting layers of atomic thickness leads to the appearance in such systems of a number of properties that are absent, or that practically cannot be observed, in either isotropic superconductors or at ordinary Josephson junctions. And although the critical temperature of intercalation compounds do not for the time being exceed 4 K, these systems are very interesting for the physics of superconductivity and are promising from the point of view of the practical application of their Josephson properties. This range of problems is, however, not directly connected with the problem of high-temperature superconductivity and will be considered only briefly in Section 6 of this chapter.

In the case of one-dimensional systems, the most pressing problem is the problem of the possibility of realizing the metallic and superconducting states in them (at low temperatures). Such effects as the Peierls instability, the localization of the electrons as a result of the disordered state of the lattice, and the Coulomb interaction between the electrons should facilitate the transition of one-dimensional electron systems into the dielectric ground state. Experimentally, the conductivity in almost all the quasi-one-dimensional crystals does decreases as the temperature decreases in the region $T \rightarrow 0$, and the determination of the cause of this effect is of interest both in itself and from the point of view of the problem of high-temperature superconductivity.

The questions connected with the specific character of the physical properties of electrons in one-dimensional systems are considered below in Chapter 7.

## 2. The Structure and Physical Properties of Layered Metals

**A. The Transition-Metal Dichalcogenides.** The general formula of the layered compound—the dichalcogenides—has the form $MX_2$, where M is a transition metal of the IV–VII groups and X = Se, S, or Te.

Crystals of these compounds consist of layers. each of which is a sandwich of two layers of X atoms with a layer of metallic M atoms between them. The bond between the M and X atoms in a sandwich is a strong (largely covalent) one, and these atoms form a two-dimensional hexagonal lattice. Therefore, the $MX_2$ layers are coupled in the crystal by weak van der Waals forces. The structure of the crystals is schematically shown in Fig. 6.1.

Two different two-dimensional crystal structures are formed, depending on the relative disposition of the two X-atoms planes inside the X–M–X sandwich. In one structure six X atoms surround one M atoms, forming an octahedron; in the other, a trigonal prism is formed.

The weak van der Waals interaction between the layers admits of various ways of packing for the $MX_2$ sandwiches in a crystal along the axis perpen-

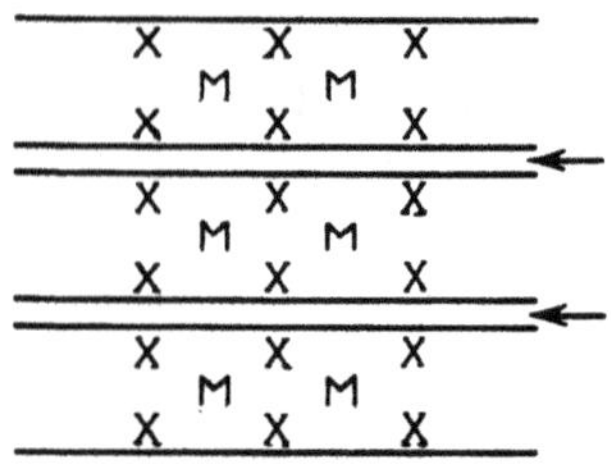


**Figure 6.1. Structure of the layers in the transition-metal dichalcogenides (view in the direction parallel to the layers). The arrow indicates the van der Waals gap.**

dicular to the layers (the *c* axis). Therefore, many layered compounds exist in several modifications. The most common types of layer packing are shown in Fig. 6.2.

The type of packing is characterized by a numeral, a capital letter, and additional symbols if the first two symbols do not uniquely define the structure (1T, 2H, 3R, 4Ha, 4Hb, and 6R). The numeral indicates the number of layers in the unit cell, while the capital letter characterizes the type of crystal symmetry (T stands for trigonal; H hexagonal; R rhombohedral). In the 1T phase a metal atom inside a sandwich has an octahedral surrounding; in the 2H, 3R, and 4Ha phases, a trigonal surrounding. In the 4Hb and 6R modifications the sandwiches alternate with octahedral and trigonal packings (more detailed information about the structure of the layered compounds can be found in [7, 8].

Since the interaction between the $MX_2$ layers is weak, it is not surprising that the physical properties of the layered compounds are determined largely by the two-dimensional structure of the lattice inside a sandwich. The three modifications of $TaS_2$ exhibit the strong dependence of the electronic properties of layered compounds on the type of crystal structure. The $1T$–$TaS_2$ crystals with trigonal packing of the Ta and S atoms in a $TaS_2$ layer are semiconductors, or, more precisely, semimetals whose conductivity increases slowly with increasing

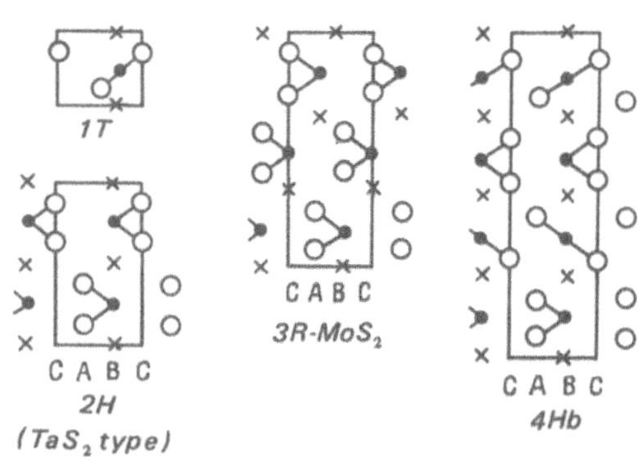


**Figure 6.2. Structure of various modifications of the hexagonal layered compounds [in the (1120) plane]. The points represent metal atoms; the circles, chalcogen atoms.**

temperature $(\sigma_{\|}^{-1}\,\partial\sigma_{\|}/\partial T = 4.0 \times 10^{-3}\ K^{-1}$ at $T = 300$ K for the conductivity $\sigma_{\|}$ along the layers). The 2H-$TaS_2$ crystals with octahedral packing are metals. In the 4Hb-$TaS_2$ compound, layers with trigonal and octahedral surroundings alternate, and in these crystals the conductivity is metallic along the layers, (because of the octahedral layers) and semimetallic across the layers [9]. This difference in the electronic properties of the crystals with trigonal and octahedral packings obtains for the other dichalcogenides of Ta and Nb.

The anisotropy in the elctronic properties of all the layered compounds is, first and foremost, because of the fact that the electrons can move freely inside the layers, while the overlap of the electron wave functions between the layers is small because of the van der Waals nature of the interlayer interaction.

In the most thoroughly investigated layered compounds, 2H-$TaS_2$ and 2H-$NbSe_3$, the anisotropy in the electron motion is not high. According to the data for the conductivities along and across the layers, the ratio $\sigma_{\|}/\sigma_{\perp}$ in these crystals lies in the range from 20 to 50 [10, 11], and roughly the same anisotropy is observed in the magnitude of the square of the upper critical magnetic field near $T_c$. Thus, the ratio $H_{c_2}^2(\|)/H_{c_2}^2(\perp) \approx 10 - 12$ in the compound $NbSe_2$ [11] and is roughly 45 in 2H-$TaS_2$ [12].

The anisotropy in the electron motion in the dichalcogenides is also manifested in the optical [13] and magnetic [9] properties (the electronic properties of the layered compounds are considered in greater detail in [14]). The anisotropy of the layered compounds is also exhibited in their mechanical properties [15]. In particular, the dichalcogenide crystals are easy to cleave along a plane parallel to the layers, and thin films can easily be obtained from these crystals. Thus, Frindt [16] obtained $NbSe_2$ crystals of a thickness of as low as two layers, and they remained superconducting, although the critical temperature decreased from 7 to 4.5 K as the number of layers was decreased from five to two. The ease with which films of small thickness can be obtained from layered crystals already allows them to be used for practical purposes, e.g., films of the superconductor $NbSe_2$ are used in devices like the quantum interferometer as a weak-coupling element [17]. Notice that films of such thickness can serve as a basis for the production of sandwich structures (see Chapter 8).

**B. Intercalation Compounds.** The weak van der Waals interlayer bonding in the dichalcogenides allows the introduction of extraneous atoms or molecules into the space between these layers. A large number of intercalation compounds of the dichalcogenides of Nb and Ta have thus far been produced with organic molecules [6, 18 - 20], ammonia [18], and metals [21].

The layered, molecule-intercalated compounds have the formula $MX_2$ (molecule)$_r$, where $r$ is a fraction close in the majority of cases to the form $1/n$ ($n$ is a whole number). Upon the introduction of the molecules, the distance between the $MX_2$ layers increases, attaining a value of 56 Å in the compound $TaS_2$ (octadecylamine)$_{1/3}$ [18] (in 2H-$TaS_2$ the layer spacing is approximately

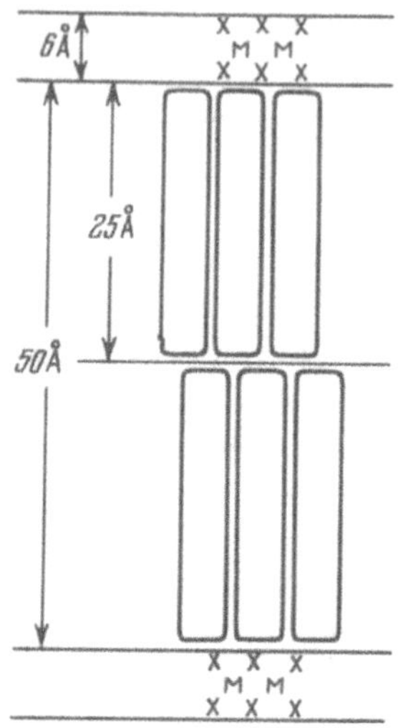


Figure 6.3. The arrangement of the octadecylamine molecules in the compound $TaS_2$ (octadecylamine)$_{1/3}$.

equal to 3 Å). The structure of this compound is schematically shown in Fig. 6.3.

Detailed x-ray analysis data on the arrangement of the layers and molecules are for the present available only for the crystals of $TaS_2$ with pyridine ($C_6H_5N$) [22]. In this intercalation compound the planar pyridine (Py) molecules settle at right angles to the layers, the mutual disposition of neighboring $MX_2$ layers being somewhat different from their arrangement in 2H–$TaS_2$ (see Fig. 1.7). Between the $TaS_2$ layers the pyridine molecules form a rectangular superlattice with a $2a\sqrt{3} \times 13a$ unit cell, where $a$ is the dimension of the hexagonal unit cell inside a $TaS_2$ layer. In this case the complete unit cell corresponds to the formula $TaS_2(Py)_{6/13}$, and not to $TaS_2(Py)_{1/2}$ (below for the sake of brevity we shall write in the formula of the compound $\frac{1}{2}$ and not $\frac{6}{13}$).

All the intercalated organic molecules contain nitrogen with two unbound electrons and possess, because of this, donor properties. It is suggested in [19] that the electrons from the nitrogen atom of the intercalated molecule partially go over into the conduction band of the dichalcogenide layers and that the lowering of the energy of the system upon intercalation is connected with this. Many experimental data corroborate the assumption that charge transfer occurs from the molecule to the $MX_2$ layers, and this assumption is the basis of the computation of the change in the electronic characteristics of the molecule and the conducting layer [23]. Indeed, some correlation is observed between the ability of the molecule to intercalate the dichalcogenide crystals and its ionization potential [19], although the intercalation energetics further depends on a number of other factors [20, 23].

The decrease in the electron density on the N atom of the molecule was found from data on photoelectric emission from $TaS_2(Py)_{1/2}$ under the action of x rays [19]. The changes in the NMR spectrum of $NbS_2(Py)_{1/2}$ crystals can also be interpreted with the aid of the assumption that charge transfer occurs from the Py molecule into the conduction band of the $NbS_2$ layers (according to these data, about 0.2 of an electron is transferred to the Nb atom [24]. The transfer of some electrons from the molecules to the dichalcogenide layers leads to a change in the electron density of the conducting layers and to the appearance near these layers of charged nitrogenous complexes at distances slightly smaller than the van der Waals distances.

These two factors, together with the electric dipole field of the intercalated molecules, alter the electron-phonon characteristics of the conducting layers. These changes are quite noticeable in a comparison of the magnetic susceptibilities [25] and the critical temperatures $T_c$ [18, 19] of the original and the layered intercalation compounds since the $T_c$ and the Van Vleck contribution to the paramagnetic susceptibility are highly sensitive to changes in the electronic structure. At the same time, quantities such as the density of states at the Fermi surface, the conductivity along the layers per conducting layer, and the Seebeck coefficient are practically unaltered by the intercalation [14].

The introduction of molecules between the conducting layers leads to a considerable reduction in the overlap of the electron wave functions of different layers if the molecule attaches itself to a single $MX_2$ layer [e.g., in the compound $TaS_2(Py)_{1/2}$]. According to the data of [10], the intercalation of Py into 2H-$TaS_2$ increases the conductivity anisotropy factor from 10 to a value exceeding $10^5$. In systems with such an anisotropy, the electron motion is essentially two dimensional. At the same time, intercalation of molecules into crystals of the 2H modification of $TaS_2$, $NbS_2$, and $NbSe_2$ does not lead to the disappearance of superconductivity, and in such systems superconductivity can become quasi-two-dimensional (see Section 5).

The crystals of the 2H and 3R modifications of $MoS_2$, $ZrS_2$, $HfS_2$, $WS_2$, and $WSe_2$, intercalated by atoms of the alkali metals [26–28] and atoms of the alkali-earth metals and Yb with ammonia [29], are also metals and superconductors. The parent-layered compounds in this case are semiconductors with spectra of the exciton type [7, 8]. The products of the intercalation of K, Rb, Cs, and Sr into $MoS_2$ are the stoichiometric compounds $K_{0.4}MoS_2$, $Rb_{0.3}MoS_2$, $Cs_{0.3}MoS_2$, and $Sr_{0.2}MoS_2$, which have critical temperatures around 6 K. The compounds with Li and Na are nonstoichiometric and have $T_c \approx 4$ K. In the compounds $A_x(NH_3)_yMoS_2$ with A = Ca, Ba, Yb, and Sr, the critical temperature does not exceed 5.7 K. According to x-ray structural analysis data [26], the intercalation leads to a 2 - 3.5 - Å increase in the $MoS_2$-layer spacing and a 1% increase in the two-dimensional-lattice constant, $a$, of a layer.

X-ray structural analysis [30] also shows that the Na atoms and the ammonia molecules in the compound $Na_x(NH_3)_yMoS_2$ occur between all the

$MoS_2$ layers, occupying vacancies inside the octahedron formed by the sulfur atoms of neighboring layers. The intercalated atoms form together with the impurities a superlattice of orthorhombic symmetry, intercalation by ammonia facilitating the ordering of the impurities. We do not for the present have more detailed data on the arrangement of the intercalated atoms. We know neither the electronic characteristics of these compounds nor the structure in which the conduction occurs (whether in the metal-atom layers or in the $MoS_2$ layers), although it may be inferred that it is the $MoS_2$ layers that become conducting because of the transfer of electrons from the metal atoms to them.

## 3. The Critical Temperature of Layered Superconductors

**A. Effect of Pressure on the Critical Temperature.** It was assumed at the beginning of the investigations of the superconductivity of layered compounds that their critical temperature, $T_c$, is largely determined by the interlayer interaction and that the value of $T_c$ should increase as this interaction becomes stronger because of the suppression of the two-dimensional fluctuations [31].

These arguments were based on the fact that, in the compound $Nb_{1+x}Se_2$, as $x$ is increased, the layer spacing increases while the $T_c$ falls [32]. But it became clear after the measurement of the $T_c$ in 2H-$NbSe_2$ and 3R-$NbS_2$ as a function of pressure that the variation of $T_c$ is correlated with the variation of the lattice constant, $a$, inside a layer rather than with the variation of the conducting-layer spacing [33]. Further, Dzyaloshinskii and Kats [34] have shown that the fluctuations in layered superconductors do not suppress the long-range superconducting order even in the case of very weak interlayer interaction (see Section 6). Sambongi [35], in a study of the effect of uniaxial compression along the $c$ axis on the $T_c$ in 2H-$NbSe_2$, found the $T_c$ to decrease with increase of such compression. Therefore, there is now no doubt whatsoever that the growth of the $T_c$ in $NbSe_2$ under pressure (to 8 K as the pressure is increased to 140 kbar [36 – 38]) is due to the modification of the lattice inside the layers.

**B. The Critical Temperature and Structural Transitions.** Structural transitions from one metallic polytype into another are observed in the 2H modifications of $TaSe_2$, $TaS_2$, and $NbSe_2$ at the temperatures $T_0 = 117, 75$, and 35 K, respectively. Apparently, these transitions are very nearly second-order transitions. Thus, in 2H-$TaS_2$ the enthalpy jump at the point $T_0 = 75$ K is roughly 1 cal/mol [39], and a 1.5% jump in the resistance is observed [40].

The structural transitions manifest themselves primarily in the variation of the kinetic and magnetic properties of the metals; these variations near $T_0$ are schematically depicted in Fig. 6.4. Structural investigation show that the transitions within the framework of one and the same modification of the layered compounds involve the appearance of a superlattice below the temperature $T_0$. Thus, in 2H-$TaS_2$ and 2H-$TaSe_2$, the superlattice constant in the plane of a

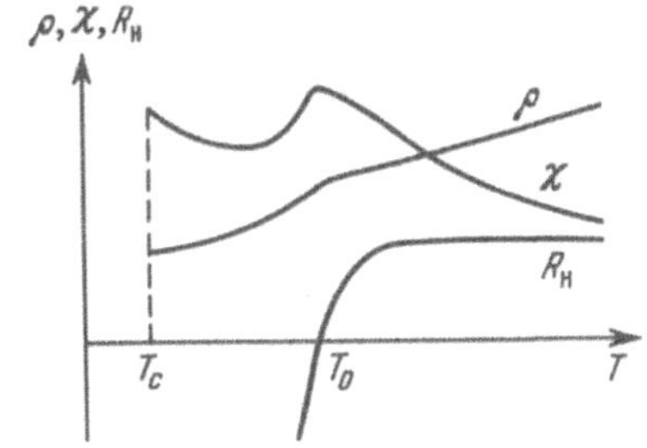


**Figure 6.4. Schematic representation of the behavior of the susceptibility χ, the resistance ρ, and the Hall coefficient $R_H$ during the structural transitions in the 2H modifications of the Nb and Ta dichalcogenides.**

layer is roughly equal to 3 times the lattice constant of the high-temperature polytype [40, 41].

According to the current ideas, the occurrence of these structural transitions is because sections of the Fermi surface can be superposed on each other after a definite displacement of them in momentum space. If the sections coincide after undergoing a relative displacement equal to the vector **Q**, then there appears below $T_0$ a sinusoidal displacement of the atoms from their equilibrium positions with wave vector **Q**. In this case there appears at the superposable sections of the Fermi surface below $T_0$ an energy gap corresponding to a Bragg reflection from the superlattice. But in the case of a two- or three-dimensional lattice, in which the Fermi surface also contains nonsuperposable sections, a transition involving the appearance of a superlattice (a charge density wave) turns out to be a metal-metal or a metal-semimetal transition, depending on the area of the superposable sections of the Fermi surface. In the one-dimensional case the analogous Peierls transition is always a metal-dielectric transition (see Chapter 7 and [42]). In the metal series $NbSe_2$, $TaS_2$, and $TaSe_2$ the self-covering portion of the Fermi surface increases since in this series $T_0$ and the quantity $\partial\rho/\partial T$ below $T_0$ increase.

Undoubtedly, the structural transition also affects the superconducting properties of the 2H metals. In fact, below the temperature $T_0$, part of the Fermi surface is covered by the dielectric gap, and this leads in the simplest case to the lowering of $T_c$ as a result of the decrease in the density of states at the Fermi surface. Thus, according to the data of [43], in 2H-$NbSe_2$, as the pressure is raised to 35 kbar, the transition temperature $T_0$ decreases from 33.5 to roughly 8 K, while $T_c$ increases from 7.1 to 8 K. Above 36 kbar the structural transition temperature $T_0$ becomes less than $T_c$, and the growth of $T_c$ with pressure practically stops.

Apparently, the decrease in $T_c$ as a result of the structural transition is even more drastic in $TaS_2$ and $TaSe_2$, and this is the cause of the low $T_c$ values for these crystals. This conjecture is confirmed in experiments on 2H-crystal structure disarrangement, which is accompanied by a rise in $T_c$ [44]. Defects were introduced into the crystals by bombarding them with argon ions or by intercalating molecules into, and subsequently deintercalating them from, the crystals [44].† After bombarding the compounds $TaS_2$ and $TaSe_2$, the quantity $T_c$ rose from 0.8 to 4.2 K and from 0.15 to 2.5 K, respectively. The result of the intercalation of molecules into, and the subsequent deintercalation of them from, 2H–$TaS_2$ yields roughly the same results.

The observed increase in $T_c$ can be related to the suppression of the structural transitions as a result of the disordering of the lattice. This conclusion follows from the theoretical ideas about the mechanism underlying the structural instability [42] and is corroborated by the fact that the anomaly in the resistance in the region around 75 K disappears after the bombardment of $TaS_2$ crystals.

**C. The Critical Temperature of Intercalation Compounds.** The most thoroughly investigated intercalation compounds are the 2H–$TaS_2$ ones, and they have higher $T_c$'s than the parent compound (from 2 to 4.5 K, whereas for 2H–$TaS_2$ the quantity $T_c$ = 0.8–2 K).

First, the intercalation compounds of 2H–$TaS_2$ provide proof of the nondependence of $T_c$ on the layer spacing. Thus, for $TaS_2$(octadecylamine)$_{1/3}$, with a layer spacing of 56 Å, the $T_c$ value is roughly the same as for the other intercalation compounds (≈3 K). Therefore, the change that occurs in $T_c$ as a result of intercalation is primarily because of the change in the electronic characteristics of the conducting layer (as a result of the transfer of some electrons from the molecules, the action of the potential of the charged molecular complexes, and the appearance of defects in the lattice after the intercalation).

Indeed, the tendency of the critical temperature $T_c$ to decrease as the number of molecules per Ta atom is increased has already been noted by Gamble *et al.* [19]. This tendency was confirmed after the preparation of *n*-alkylamine-intercalated 2H–$TaS_2$ compounds. The linear molecules of this series differ only in the number of carbonaceous radicals, which have little effect on the donating ability of the nitrogenous radical of the molecules [18]. It turns out that, as the length of the molecule increases, $T_c$ falls (from 4.2 K for $NH_3$ down to 1.8 K for octylamine) as long as the molecules set themselves parallel to the layers and the number of molecules per Ta atom decreases (the layer spacing remains almost constant in the process). Upon further increase of the molecule length, starting from *n*-nonylamine, the molecules set themselves at right angles to the layers, and the layer spacing increases with increasing molecule length, while the critical temperature $T_c$ remains almost a constant. The variation of $T_c$ undoubtedly depends on the type of molecules, but it is at present not clear precisely which of the characteristics of the molecule have a significant effect on the quantity $T_c$.

†The intercalated molecules get out of the crystals upon heating, and this process is called deintercalation.

There is reason to believe that the increase in $T_c$ is at least partially due to the suppression of the structural transition as a result of the intercalation. In fact, the molecules alter the parent-lattice constant in a layer (the condition for the superpossibility of sections of the Fermi surface may then be violated), and, what is more, intercalation impairs the crystal structure. The experimental data show that, when collidine and picoline molecules are intercalated into $2H–TaS_2$, the height of the peak in the dependence $\chi(T)$ at the point $T_0$ decreases and that this peak, as well as the kink in the dependence $\rho(T)$, practically completely disappears in $TaS_2(Py)_{1/2}$ crystals.

Undoubtedly, all the experimental data on the critical temperature of intercalation crystals show that the phonon mechanism of superconductivity is realized in these compounds. This is also attested by the low $T_c$ values for these superconductors and the possibility of explaining the rise in $T_c$ after intercalation within the framework of the conventional concepts expounded above. But why does the excitation mechanism turn out to be ineffective in the intercalation compounds? In which layered systems can we expect great success from the point of view of the problem of high-temperature superconductivity? We shall attempt to give answers to these questions in the following section.

## 4. Intercalation and the Problem of High-Temperature Superconductivity

**A. The Critical Temperature of Superconductors with the Exciton Mechanism.** Hopes of obtaining higher critical temperatures with the aid of the exciton mechanism are connected with the growth of $T_c$ as the mean frequency, $\Omega$, of the polarizable system increases in accordance with the formula

$$T_c = \Omega e^{-1/\lambda}, \tag{6.1}$$

where $\lambda$ is the effective constant of the interaction of the electrons via the polarizable system. However, the role of the Coulomb repulsion of the electrons, which is weak only when $\Omega \ll E_F$ because of the logarithmic cutoff of the Coulomb interaction in terms of frequency, begins to increase in importance with the growth of $\Omega$ [46]. With allowance for this effect, the formula for $T_c$ in the weak-coupling case has the form (see Chapters 1 and 2 and [47]):

$$T_c = \Omega \exp\left\{-\frac{1}{\lambda - \mu\,[1+\mu \ln(E_F/\Omega)]^{-1}}\right\}, \tag{6.2}$$

where $\mu$ is the dimensionless interelectron Coulomb interaction constant. The maximum value of $T_c$ for $\lambda = \mu$ is attained when $\Omega_{max} = E_F \exp(-1/\lambda)$, and

$$T_{c\,max} = E_F \exp(-3/\lambda). \tag{6.3}$$

If the exciton mechanism yields a critical temperature higher than the value given by the phonon mechanism, then $\omega_D < \Omega_{max} < E_F$, i.e., $\Omega_{max} \approx 0.1$–1 eV. However, a real increase in the critical temperature occurs on account of the exciton mechanism only in the case when the electrons interact strongly with each other via the polarizable system (i.e., when $\lambda \gtrsim \frac{1}{2}$). Similar estimates are made for the strong-coupling case in [47] (see p. xx). According to [47], $T_c$, as a function of $\Omega$, has a peak in this case too, but $\Omega_{max}$ and $T_{c\ max}$ are, respectively, $e^2$ and $e^4$ times smaller than the corresponding quantities determined above. Let us note, however, that the accuracy with which $T_c$ can be computed at present is hardly sufficient for the determination of the numerical factors in the formulas for $T_{c\ max}$ and $\Omega_{max}$.

Thus, a noticeable increase in the critical temperature in the case when the exciton mechanism obtains can be expected if the frequency of the polarizable system lies in the range from 0.1 to 1 eV. The energy, $\Omega$, of the lowest excited level in the molecules that have thus far been intercalated exceeds 5 eV, i.e., $\Omega \gtrsim E_F$ in these molecules. In this case a logarithmic cutoff of the Coulomb repulsion in terms of frequency does not occur, and we can hardly expect any manifestation of the exciton mechanism of superconductivity. But, we can, in principle, choose molecules with lower excitation energies (of the order of 1 eV). Let us consider the possibility of realizing the exciton mechanism for layered systems into which such molecules have been intercalated [48].

**B. The Effectiveness of the Exciton Mechanism in Layered Systems with Molecules.** The parameter $\lambda$ is determined by the exciton-mediated interaction, $V_{ex}(0.2k_F)$, between the conduction electrons in the case of zero energy transfer and large momentum transfers of the order of $2k_F$ (see Chapter 2) [49].

In order for the exciton mechanism to be effective, this interaction should not be weaker than the analogous phonon-mediated interaction $V_{ph}(0.2k_F)$. Only in the case when $\delta = V_{ex}(0.2k_F)/V_{ph}(0.2k_F) \gtrsim 1$ will be critical temperature, $T_c$, significantly depend on the parameters of the exciton system and can it be raised by choosing the requisite mean exciton frequency $\Omega$. Our problem is to estimate the ratio $\delta$ for quasi-two-dimensional systems with molecules as the polarizable medium.

In order that we can positively speak of the existence of the molecules and the conducting layers, it is necessary that they be separated from each other by a distance of the order of the van der Waals distance when the overlap of the wave functions of the conduction and molecular electrons is small. Let us consider the case of the van der Waals interaction between the conduction electrons and the electrons of the polarizable system in layered compounds in which the conducting layers alternate with molecular layers. Let us neglect the overlap of the wave functions of the conduction and molecular electrons, thus assuming that the interaction between the conduction electrons and the molecules has a purely electrodynamic character and that the motion of the conduction electrons is two dimensional. We shall assume the conducting layers to be infinitely thin and denote the distance between these layers and the molecules by $R$. Since, by assumption, this distance is of the order of the van der Waals distance, $R \approx 3$ Å.

Let us first consider a system in which the molecules are in form of squares (of side length $l$) and their planes, which we shall also assume to be infinitely thin, are parallel to the conducting planes. If we neglect the effects of the periodicity of the lattice, then the system can be assumed to be homogeneous in the plane $(x, y)$ of the layers. The coordinates of the planes are given by the expression $z_n = Rn$, where the $n$'s are whole numbers, the even $n$'s correspond, for example, to the conducting planes while the odd $n$'s correspond to the molecular planes.

In the random-phase approximation, the effective interaction between the electrons is determined by the equations

$$\begin{aligned} V(\omega, \mathbf{k}, n, n') &= V_0(k, n-n') - \\ &\quad - \sum_m \Pi_m(\omega, \mathbf{k}) V_0(\mathbf{k}, n-m) V(\omega, \mathbf{k}, m, n'), \\ V_0(\mathbf{k}, n-n') &= 2\pi k^{-1} \exp[-kR(n-n')], \end{aligned} \tag{6.4}$$

where $V(\omega, \mathbf{k}, n, n')$ is the Fourier transform of the effective interaction $V(\omega, \mathbf{r}, n, n')$ in the coordinates $\mathbf{r} = (x, y)$, $V_0(\mathbf{k}, n-n')$ is the Fourier transform of the Coulomb interaction between electrons located in the $n$th and $n'$th layers, and $\Pi_m(\omega, \mathbf{k})$ is the polarization operator for the $m$th plane:

$$\Pi_{2n}(\omega, \mathbf{k}) = \Pi_1(\omega, \mathbf{k}), \quad \Pi_{2n+1}(\omega, \mathbf{k}) = \Pi_2(\omega, \mathbf{k}). \tag{6.5}$$

For small $\omega$ ($\omega << E_F$), we have for the conducting layers

$$\Pi_1(0, \mathbf{k}) = \begin{cases} a_0^{-1}, & k \leqslant 2k_F, \\ a_0^{-1}\left[1 - \sqrt{1 - (2k_F k^{-1})^2}\right], & k > 2k_F, \end{cases} \tag{6.6}$$

where $a_0$ is the Bohr radius of the conduction electrons, $k_F$ is the Fermi momentum: $k_F = (2\pi n_1)^{1/2}$, $n_1$ being the surface density of the conductive electrons.

For the molecular planes

$$\begin{aligned} \Pi_2(\omega, \mathbf{k}) &= e^2 N_2 b(\mathbf{k}) \left[\frac{1}{\Omega - \omega - i\alpha} + \frac{1}{\Omega + \omega + i\alpha}\right], \\ b(\mathbf{k}) &= |\langle 0 | e^{i\mathbf{kr}} | f \rangle|^2, \quad \alpha \to 0, \end{aligned} \tag{6.7}$$

where $N_2$ is the surface density of the molecules and we have taken into account in $\Pi_2(\omega, \mathbf{k})$ the contribution of only the lowest excited state $f$, which lies at a distance equal to the frequency $\Omega$ from the ground state, and which has an intense transition from the ground state.

In (6.7) we have neglected the effects connected with the delocalization of the molecular excitations, assuming the interaction between the molecules in a layer to be a van der Waals interaction. The matrix element of the operator $\exp(i\mathbf{kr})$ is computed with the wave functions of the ground state and the excited state $f$ of the molecule. Let us note that the actual structure is not homogeneous in the $(x, y)$ plane and that the polarization operator in coordinate space depends on the coordinates $\mathbf{r}_1$ and $\mathbf{r}_2$ and not on their difference. The operator $\Pi_2(\omega, \mathbf{r}_1 - \mathbf{r}_2)$, whose Fourier transform with respect to $\mathbf{r}_1 - \mathbf{r}_2$ is given by the expression (6.7), is obtained from $\Pi_2(\omega, \mathbf{r}_1, \mathbf{r}_2)$ by averaging over $(\mathbf{r}_1 + \mathbf{r}_2)/2$, and thus it determines the molecules' polarization averaged over the plane.

The structure under consideration is periodic in the coordinate $z$, with period $2R$. Therefore, the quantities $V_{11}(\omega, \mathbf{k}, n, n') \equiv V(\omega, \mathbf{k}, 2n, 2n')$, $V_{12}(\omega, \mathbf{k}, n, n') \equiv V(\omega, \mathbf{k}, 2n, 2n' + 1)$ depend only on $n - n'$. Going over to the Fourier representations, $\widetilde{V}_{11}(\omega, \mathbf{k}, q)$ and $\widetilde{V}_{12}(\omega, \mathbf{k}, q)$, in terms of the discrete variable $n - n'$, we obtain

$$\widetilde{V}_{11}(\omega, \mathbf{k}, q) = \frac{2\pi}{k} \tanh(kR) \frac{P_1(\omega, k)}{P_1(\omega, k) P_2(\omega, k) - \cos^2 q \; \mathrm{ch}^{-2} kR},$$
$$P_{12}(\omega, \mathbf{k}) = 1 + 2\pi k^{-1} \tanh(kR)\, \Pi_{12}(\omega, \mathbf{k}). \tag{6.8}$$

The effective interaction between the electrons within a plane is given by the expression

$$V(\omega, \mathbf{k}, 0) = \int_0^{2\pi} \frac{dq}{2\pi} \widetilde{V}_{11}(\omega, \mathbf{k}, q) =$$
$$= \frac{2\pi}{k} \sinh(2kR) \frac{P_2(\omega, k)}{\{[2\,\mathrm{ch}^2(kR) P_1(\omega, k) P_2(\omega, k) - 1]^2 - 1\}^{1/2}}. \tag{6.9}$$

Let us now proceed to make comparative estimates for the exciton and phonon contributions to the effective interelectron interaction for $\omega = 0$. We obtain the contribution of the exciton mechanism to (6.9) by subtracting from $V(\omega, \mathbf{k}, 0)$ the Coulomb part of the effective interaction between the conductive electrons, $V_c(\omega, \mathbf{k}, 0)$, which is equal to $V(\omega, \mathbf{k}, 0)$ when $\Pi_2 = 0$. The contribution of the phonon mechanism at $\omega = 0$ coincides in order of magnitude with $V_c(0, \mathbf{k}, 0)$, and, thus,

$$\delta(k) = \frac{V_{ex}(0, k, 0)}{V_{ph}(0, k, 0)} = \frac{V(0, k, 0)}{V_c(0, k, 0)} - 1. \tag{6.10}$$

Substituting the expressions (6.6), (6.7), and (6.9) into (6.10), we obtain

$$\delta(k) < \frac{1}{2}\,\mathrm{ch}^{-2}\,kR\,\frac{\Pi_2'(0, k)}{\Pi_1(0, k)} = \frac{2\pi e^2 N_2 a_0}{\Omega}\,b(k)\cosh^{-2} kR. \tag{6.11}$$

Taking the estimates $\Omega \approx e^2/a_0N$ ($N$ is the number of $\pi$ electrons in the molecule) and $N_2 \approx 1/9a_0^2N$ into account, we can write (6.11) in the form

$$\delta(k) < b(k)\cosh^{-2}kR < 1. \tag{6.12}$$

The factor $\cosh^{-2}kR$ cuts off the exciton part of the effective interaction for momentum transfers $k > 1/R$. Its origin is clear: The conduction and molecular electrons are located at a distance $R$ from each other, and the transfer between them of momenta greater than $1/R$ is highly improbable. Thus, this factor stems from the van der Waals interaction between the layers and is organically connected with the structure of the layered compounds.

Let us now consider the factor $b(k)$. To estimate its order of magnitude, let us use for the molecule the model of a two-dimensional square potential well with infinite walls and width $l \approx 3a_0\sqrt{N}$, in which the $N$ electrons move. Then the single-electron wave functions and the energies of the electrons have the form

$$\begin{aligned} \psi_{k_1,k_2}(x, y) &= \frac{2}{l}\sin k_1 x \sin k_2 y, \qquad k_{1,2} = \frac{\pi n_{1,2}}{l}, \\ E_{k_1,k_2} &= \hbar^2(k_1^2 + k_2^2)/2m_e, \qquad 0 \leqslant x, \quad y \leqslant l, \end{aligned} \tag{6.13}$$

where $n_1$ and $n_2$ are whole numbers. The low frequency $\Omega \approx e^2/a_0N$ can be obtained in electron transitions accompanied by changes in both $n_1$ and $n_2$ (when the change in $n_1^2 + n_2^2$ is of the order of unity, near the values $n_1 \approx n_2 \approx \frac{1}{2}\sqrt{N}$). For such transitions

$$b(k) \approx \left(\frac{2}{l}\int_0^l \sin k_1x \sin k_1x \cos kx\, dx\right) \leqslant \begin{cases} (kl)^8 & kl \leqslant 1, \\ (kl)^{-4}, & kl \gg 1. \end{cases} \tag{6.14}$$

It can be seen from (6.14) that the factor $b(k)$ cuts off the effective molecule-mediated interaction between the electrons for momentum transfers of the order of the $1/l$. Its physical meaning is also clear: a molecule with linear dimensions $l$ cannot be effectively acted upon by an external field with wavelength less than $l$. For the structure under consideration the factor $b(k)$ is decisive since $l >> R$.

The contribution of the excitons to the pairing interaction, which determines $T_c$, is, in the three-dimensional isotropic case, given by the quantity

$$\overline{V}_{\mathrm{ex}} = \int_0^{2k_F} \frac{k\,dk}{2k_F} V_{\mathrm{ex}}(0, k), \tag{6.15}$$

where $\overline{V}_{\mathrm{ex}}$ is the value of $V_{\mathrm{ex}}(0, \mathbf{k})$ averaged over the angles and integrated over all the transferable momenta $k$.

Similarly, in a quasi-two-dimensional crystal, the contribution of the exciton mechanism is given by the quantity

$$\overline{V}_{\mathrm{ex}} = \frac{2}{\pi} \int_0^{2k_F} V_{\mathrm{ex}}(0, k) \frac{dk}{\sqrt{4k_F^2 - k^2}}. \tag{6.16}$$

Taking (6.10) and (6.12) into account, we obtain

$$\overline{V}_{\mathrm{ex}} = \frac{2}{\pi} \int_0^{2k_F} \delta(k) V_{\mathrm{ph}}(0, k) \frac{dk}{\sqrt{4k_F^2 - k^2}} \leqslant$$

$$\leqslant \frac{2}{\pi} \left[ \int_0^{2k_F} V_{\mathrm{ph}}(0, k) \frac{dk}{\sqrt{4k_F^2 - k^2}} \right] \frac{1}{2k_F} \int_0^{2k_F} \delta(k)\, dk = (2k_F l)^{-1}\, \overline{V}_{\mathrm{ph}} \tag{6.17}$$

The quantity $(2k_F l)^{-1}$ does not exceed 0.03 when $\Omega < 1$ eV, and the mean interelectron distance, $a$, in the conducting planes is of the order of 1.5 Å.

Similar computations for a structure with the molecules arranged such that their planes are perpendicular to the conducting planes lead to an expression for $b(k)$ that differs from (6.11) in having the factor $l/R$ since in this case the distance between the planes of the molecules is of the order of $R$ and $N_2 \approx 1/lR$. In such a system the difference between the exciton and phonon contributions is determined at least by the factor $(2k_F R)^{-1}$, and the exciton mechanism is ineffective in this case, too.

Thus, it is difficult to make the exciton mechanism effective in the case of layered compounds with molecular excitons because of the factor $b(k)$ in (6.7), a factor that cuts off the molecule-mediated interelectron interaction for momentum transfers $k \lesssim 1/l << 2k_F$. This result stems from the large linear dimensions of the molecules—an unavoidable price to pay for the low localized-molecular-excitation energies.

The estimates obtained above are approximate, and it is possible that numerical factors that will improve the situation will appear in more exact computations in which allowance is made for the specific structures of the molecules and the conducting layers. Little [50] has estimated the effectiveness of the exciton mechanism in layered compounds with certain specific molecules in the scheme used by us. He computed the factor $b(k)$ more accurately, using the method of molecular orbitals to compute the electronic wave functions of the molecule. The molecules had a prolate shape and were arranged at a distance $R$ from the conducting plane in such a way that their long axis was oriented at right angles to the metallic planes. It was assumed that $2k_F = 0.33 \times 10^8$ cm$^{-1}$ for the conduction electrons in the metallic planes. The results of the computations show that we can at best obtain $\delta \approx 0.5$ by implanting dicarbocyanic-dye molecules with a lowest-excitation energy of $\Omega = 1.25$ eV at a distance of 1.5 Å from the conducting plane. Let us note, however, that the value $R = 1.5$ Å is too small for the interaction to remain a van der Waals interaction. Even more important is the fact that, for the chosen value of $2k_F$, we obtain $E_F < 1$ eV, and the condition $\Omega << E_F$, which is necessary for the effective cutoff of the Coulomb interaction in terms of frequency, is not fulfilled. Thus, in essence, Little's concrete computations confirm the conclusion that molecule-mediated interelectron interaction is ineffective at large electron-momentum transfers.

**C. Systems with Alternating Metallic and Semiconducting Layers.** As has already been noted above, the ineffectiveness of systems with molecules as the polarizable medium is connected with the localized character of the molecular excitations. However, in principle, low exciton energies can be obtained through a drastic delocalization of the electrons by, for example, replacing the layers with the weakly interacting molecules with semiconducting layers.

In systems with delocalized electronic excitations a factor of the type $\cosh^{-2} kR$ in $V_{ex}(0, k)$ remains, but it is hoped that the related decrease in $\overline{V}_{ex}$ can be compensated by the increase in $V_{ex}(0, k)$ for $k \lesssim 1/R$. Therefore, layered compounds in which conducting and semiconducting layers alternate seem to be more promising systems for the realization of the exciton mechanism of superconductivity. Momentum transfers of the order of $2k_F$ could be realized in them in the case of excitation energies of the order of 0.1 eV if several valence bands $n$ and the conduction band of the semiconducting layers were arranged in momentum space at vector intervals of $\mathbf{g}_n$ with $|\mathbf{g}_n| \simeq 2k_F$ ($K_F$ is the Fermi momentum for the metallic layers).

The production of mixed crystals of $TaS_2$ with $WS_2$, and $NbSe_2$ with $MoS_2$ is reported in [8], but so far there is no information about their properties. The crystals $WS_2$ and $MoS_2$ are semiconductors, but they can be metallic in a system in which they alternate with metallic $TaS_2$ or $NbSe_2$ layers if some electrons are transferred to them from the $TaS_2$ or $NbSe_2$ layers. Therefore, it is as yet not clear whether the $WS_2$ and $MoS_2$ layers in crystals of the mixed type retain their "semiconducting" properties.

## 5. Fluctuations in Two-Dimensional and Quasi-Two-Dimensional Systems

**A. Absence of a Long-Range Superconducting Order Parameter in Two-Dimensional and One-Dimensional Systems.** The question arises in connection with Little's and Ginzburg's proposals for the realization of high-temperature superconductivity in one- and two-dimensional systems whether superconductivity can exist at all in such systems.

Doubts of this sort have arisen as a result of Rice's [51] and Hohenberg's [52] investigations, which have shown that fluctuations destroy long-range superconducting order in purely one- and two-dimensional systems. In this case the effect of the fluctuations in two-dimensional systems is much weaker, and therefore two-dimensional systems have, from this point of view, a clear advantage over one-dimensional systems [53]. Indeed, to destroy a long-range order in the one-dimensional case, it is sufficient to "break" the order at only one point, and, in the case of superconductivity, fluctuations of arbitrarily low energy are sufficient for the destruction of the phase coherence. In the two-dimensional case, on the other hand, a "rapture" along a line is necessary for the destruction of the order, while in the three-dimensional case a "rupture" over an entire surface is required for this purpose.

It is clear at present that the fluctuations in two-dimensional systems do not create any fundamental difficulties for the superconductivity effect. In [54, 55] it is demonstrated that superconducting properties can be manifested in two-dimensional systems without the realization of long-range order in them. And in the case of quasi-two-dimensional crystals the problem completely loses its acuteness since practically any slight overlap of the electron wave functions of neighboring layers suppresses the fluctuations and restores the long-range superconducting order.

**B. Fluctuations in Quasi-Two-Dimensional Systems.** Rice and Hohenberg's arguments are not applicable to quasi-two- and quasi-one-dimensional crystals, and, as will be shown below, long-range superconducting order is always realized in these crystals at sufficiently low temperatures. It is also significant that any overlap of the electron wave functions of neighboring layers or filaments results in the screening of the Coulomb interaction of the electrons of these systems [48, 57]. The presence of such screening facilitates the realization at low temperatures of the metallic state itself and leads to the suppression of the electron-density fluctuations, which destroy the long-range superconducting order [58].

It will be shown below that the fluctuations in quasi-two-dimensional systems are practically insignificant everywhere except in a very small neighborhood of $T_c$, and, from the point of view of the fluctuations, such systems are qualitatively not different from the normal isotropic three-dimensional crystals. We shall estimate the fluctuations, following Dzyaloshinskii and Katz [58], and compute the pair correlation function for the superconducting order parameter using the Ginzburg-Landau functional.

For a quasi-two-dimensional system the Ginzburg-Landau functional for the order parameter, $\psi_n(\mathbf{r})$, of the $n$th layer has the form

$$F\,[\psi_n(\mathbf{r})] = \sum_n \int d\mathbf{r} \Bigg\{ \Bigg[ \frac{\hbar^2}{2m_\parallel} \left( \frac{\partial \psi_n}{\partial \mathbf{r}} \right)^2 + \frac{1}{2\eta} |\psi_n|^2 \times \times \left( -\tau + \frac{1}{2N_e} |\psi_n|^2 \right) \Bigg] + \frac{t}{2\eta} \left[ 2|\psi_n|^2 - \psi_n \left( \psi^*_{n+1} + \psi^*_{n-1} \right) \right] \Bigg\}, \qquad (6.18)$$

where $m_\parallel$ is the effective electron mass for the motion along the layers, $\mathbf{r} = (x, y)$, and the $z$ axis is directed at right angles to the layers, $\tau = (T_c - T)/T_c$, and $N_e$ is the electrons density in a conducting layer.†

In the case when the electron motion between the layers can be described in the band scheme (i.e., when the mean-free path, $l_\perp$, for the interlayer electron motion is much greater than the interlayer distance $d$), the dependence of the electron energy on the quasimomentum, $\mathbf{k}$, has the form

$$E(\mathbf{k}) = \frac{\hbar^2}{2m_\parallel} (k_x^2 + k_y^2) + b \cos k_z d. \qquad (6.19)$$

The functional (6.18) is valid if $b \ll T_c$, and

$$\eta = \frac{7\zeta(3)\, E_F}{(4\pi T_c)^2}, \quad t = 7\zeta(3) \left( \frac{b}{8\pi T_c} \right)^2, \qquad (6.20)$$

where $\zeta(x)$ is the Riemann zeta function. For the diffusive interlayer electron motion ($l_\perp \ll d$), the expression (6.18) is valid when $\hbar/\tau_\perp \ll T_c$ ($\tau_\perp$ is the layer-to-layer jump time for an electron), and

$$\eta = \frac{7\zeta(3)\, E_F}{(2\pi T_c)^2}, \quad t = \frac{\pi\hbar}{8\tau_\perp T_c}. \qquad (6.21)$$

Let us find the $\tau$- and $t$-parameter regions in which the fluctuations remain weak. For this purpose, let us compute the correlation function for the Ginzburg-Landau order parameter,

†In [59] the parameter $r = 2t$ is introduced into the functional (6.18).

$$K_n(\mathbf{r}) = \langle \psi_n^*(\mathbf{r})\, \psi_0(0) \rangle, \tag{6.22}$$

using the quadratic–in $\psi_n(\mathbf{r})$–part of (6.18) and neglecting the fluctuations in the modulus of the order parameter, which are insignificant in the quasi-two-dimensional case [48, 60]:

$$K_n(\mathbf{r}) = Z^{-1} \int D\varphi_m e^{-F[\psi_m]/T} \psi_n^*(\mathbf{r})\, \psi_0(0),$$

(6.23)

$$F[\psi_m] = \psi_0^2 \int d\mathbf{r} \sum_m \left\{ \frac{\hbar^2}{2m_\parallel} \left( \frac{\partial \varphi_m}{\partial \mathbf{r}} \right)^2 + \frac{2t}{\eta} [1 - \cos(\varphi_{m+1} - \varphi_m)] \right\},$$

where

$$Z = \int D\varphi_m e^{-F[\psi_m]/T}, \quad \psi_m(\mathbf{r}) = \psi_0 \exp[i\varphi_m(\mathbf{r})], \quad \psi_0^2 = N_e\tau/2.$$

For the quantity

$$K(\infty) = \lim_{n, \mathbf{r} \to \infty} K_n(\mathbf{r}),$$

which characterizes the long-range superconducting order, we obtain from (6.23) the expression

$$K(\infty) = \psi_0^2 e^{-\mu}, \quad \mu \approx \frac{T}{\tau E_F} |\ln t|. \tag{6.24}$$

In the self-consistent field approximation $K(\infty) = \psi_0^2$, and, consequently, the contribution of the fluctuations is determined by the quantity $(1 - e^{-\mu})$. This quantity is small for practically any arbitrarily small value of the parameter $t$ because of the factor $T/E_F \ll 1$ in the expression for $\mu$. Therefore, allowance for the fluctuations does not alter the self-consistent-field solution in the entire temperature range, except in the immediate neighborhood of the point $T_c$ (in the critical region). In the three-dimensional case the region of intense fluctuations is even narrower since $\mu$ contains another factor $T_c/E_F$, which replaces the quantity ($|\ln t|$). In the quasi-one-dimensional case, as we shall see below (see Chapter 7, Section 5), $\mu$ does not contain small factors of the type $T/E_F$ or $T_c/E_F$ at all, and the fluctuations turn out to be intense everywhere when $t \ll 1$.

The fluctuations that have been taken into account by us with the aid of the Ginzburg–Landau functional are thermal fluctuations, and their contribution becomes small at $T \longrightarrow 0$. Besides them, there are, however, quantum fluctuations. They have, in the one-dimensional case, been considered by Efetov and Larkin [61], who found that their contribution in the region of not very low temperatures is small compared to the contribution of the thermal fluctuations. This result apparently remains valid in the case of quasi-two-dimensional systems.

## 6. Specific Character of the Superconducting Properties of Layered Crystals with the Josephson Interlayer Interaction

**A. Condition for the Realization of the Josephson Interlayer Interaction.** Although crystals with intercalated molecules do not seem to show much promise from the point of view of the realization of the exciton mechanism of superconductivity, they are of great interest for their physics and possible technical applications since the situation can be realized in them when the interlayer interaction is of the Josephson type.

In fact, the molecules can greatly reduce the overlap of the electron wave functions of neighboring layers, and in those cases when the energy, $E_\perp$, of the interlayer electron motion is small compared to the energy characteristic of Cooper pairing, the interlayer interaction is of the Josephson type (under these conditions the effect of the interlayer electron motion is weaker than that of the superconducting pairing of the electrons inside the layers).

In the case of layered crystals, this situation is realized when $t \ll 1$, and in the vicinity of $T_c$ the properties of superconductors with the Josephson interlayer interaction can be described with the aid of equations corresponding to the functional, (6.18), for the Ginzburg–Landau order parameter [62, 63, 14]. As in the case of ordinary Josephson junctions, the current, $j_{n,n+1}$, between the $n$th and $(n + 1)$th layers is given by the expression

$$j_{n,\,n+1} = j_c \sin(\varphi_n - \varphi_{n+1}), \tag{6.25}$$

where $\varphi_n$ is the phase of the order parameter in the $n$th layer and $j_c$ depends only on the modulus of the order parameter.

The electric field that is perpendicular and the magnetic field that is parallel to the layers cannot, when $t \ll \tau \ll 1$, change the modulus of the order parameter and, consequently, the quantity $j_c$. Such fields change only the phase difference $\varphi_n - \varphi_{n+1}$ in the expression, (6.25), for the current flowing between the layers, and effects qualitatively similar to those that are characteristic of Josephson junctions should be observed in quasi-two-dimensional superconductors when $t \ll \tau \ll 1$. Essentially, under these conditions, a layered compound consists of a set of parallel Josephson junctions with superconductors of atomic thickness (see Section 6.b).

At the same time, under the conditions when $t \ll \tau \ll 1$, we can assume the electron system to be two dimensional when considering those superconducting properties that are not directly connected with the interlayer electron motion. Thus, without allowance for the paramagnetic effect, a magnetic field that is parallel to the layers cannot at low temperatures destroy the superconductivity since the currents flowing between the layers can be entirely neglected when $t \ll \tau \ll 1$. We shall consider this effect and its consequences in greater detail in Section 6.c. For $(T - T_c)/T \gg t$ and $t \ll 1$, the fluctuations in layered superconductors have a two-dimensional character, and the corresponding effects will be considered in Section 6.d.

Let us estimate the quantity $t$ for layered and intercalation superconductors. The quantity $\hbar/\tau_\perp$ can be estimated from data for the conductivity, $\sigma_\perp$, in the direction perpendicular to the layers and from the value of the coefficient, $\gamma$, attached to the linear–in the temperature–term in the specific heat since

$$\sigma_\perp = e^2 D_\perp N(0), \quad D_\perp = d^2/\tau_\perp, \quad \gamma = {}^2/_3 \pi^2 k_B^2 N(0). \tag{6.26}$$

Using the experimentally obtained values for $\sigma_\perp$ and $\gamma$ [14], we find that in the case of $TaS_2(Py)_{1/2}$ the quantity $\hbar/\tau_\perp \approx 0.03$ K, i.e., the situation in this compound is close to the one in which the Josephson interlayer interaction is realized. In $TaS_2$ and $NbSe_2$ we obtain $\hbar/\tau_\perp \approx 10^2$–$10^3$ K, and what we have to deal with in these compounds is a simple anisotropy that does not lead to the appearance of any qualitatively new effects.

In the case of 4Hb–$TaS_2$ the quantity $\gamma$ is unknown, but if we take for $N(0)$ a value equal to half the value obtained in 2H–$TaS_2$ [the octahedral layers do not contribute to $N(0)$], and for the transverse conductivity we take the value, $\sigma_\perp = 100$ mho/cm, given in [25], then we obtain $t \approx 1$, and the compound 4Hb-$TaS_2$ occupies an intermediate position between three-dimensional and quasi-two-dimensional superconductors.

Let us note that the $\hbar/\tau_\perp$ estimate obtained from the transverse-conductivity data for the intercalation crystals is unreliable because of the fact that the measured $\sigma_\perp$ values do not characterize the internal conduction. The point is that the direct-current conductivity in the direction perpendicular to the layers in intercalation crystals is largely determined by the macroscopic defects in the packing of the layers. A more reliable estimate for $t$ is given by the data on the upper critical field of intercalation superconductors. From [64] it follows that

$$t = \frac{4c\hbar}{\pi e d^2 T_c} \left( \frac{dH_{c2}(\perp)}{dT_c} \right)_{T=T_c} \Bigg/ \left( \frac{dH_{c2}(\parallel)}{dT} \right)^2_{T=T_c}, \tag{6.27}$$

where the $H_{c2}(\perp, \parallel)$ are, respectively, the upper critical fields in the directions perpendicular and parallel to the layers.

From the data of [65] for $TaS_2(Py)_{1/2}$ we obtain $t \approx 1$, and in $TaS_2$ $(\text{aniline})_{3/4}$ we have $t \approx 0.25$. The intercalation of metals into the Ta, Mo, and Nb dichalcogenides yields compounds with an even smaller anisotropy [66, 67].

Thus, we cannot safely claim that the Josephson interlayer interaction is realized in any of the presently known compounds. But crystals with large intercalated molecules, which are the most interesting from this point of view, have, for the time being, not been investigated at all.

Let us note that new, carbon-based layered compounds with intercalated alkali metals, e.g., $C_8K$, are now being investigated. The compound $C_8K$ goes over into the superconducting state at temperatures below 0.55 K [68], and the lower and upper critical fields in it are highly anisotropic. Thus, according to magnetic-susceptibility measurements at $T = 0.097$ K on samples with $T_c \approx 0.14$ K, the fields $H_{c1}$ and $H_{c2}$ are, respectively, 5.9 and 77 Oe for **H** directed at right angles to the layers, and 15 Oe and I kOe for **H** oriented along the layers [69]. According to these data, the Josephson interlayer interaction is not realized in the investigated $C_8K$ crystals, but it is possible that further investigations performed on crystals with a more perfect structure will give more encouraging results.

**B. Josephson Effects in Quasi-Two-Dimensional Superconductors.** The nonstationary Josephson effect should be observed in layered superconductors upon the application of an electric field in the direction perpendicular to the layers. Thus, if a constant potential difference $U$ is applied between crystal surfaces that are parallel to the layers, then the current flowing between the layers should oscillate with frequency $\omega = 2eU/\hbar N$, where $N$ is the number of layers in the crystal [60, 63].

Natural oscillations corresponding to the Josephson plasma oscillations should also exist in quasi-one-dimensional superconductors [62]. In layered superconductors these effects are, however, complicated by the fact that the electric and magnetic fields cannot be screened by a conducting layer of atomic thickness alone. The effect of this circumstance on the nonstationary Josephson effects in a layered system located in an electric field has not yet been investigated. As to the behavior of layered quasi-two-dimensional superconductors in a weak magnetic field $H_{\|}$, it turns out, on account of this circumstance, to be quite distinctive, and, in particular, the vortex structure differs from both the Abrikosov-vortex structure for isotropic type-II superconductors and the vortex structure at ordinary Josephson junctions between two bulk superconductors [70].

The vortex structure for quasi-two-dimensional superconductors is investigated in [63], where the magnitude of the lower critical magnetic field for the direction parallel to the layers is found. A vortex in an $H_{\|}$ field is highly anisotropic. Its dimensions in the direction perpendicular to the layers are roughly equal to the London penetration depth for the transverse field, $\lambda_{\|} = (m_{\|}c^2/4\pi e^2 N_s)^{1/2}$, where $N_s$ is the superconducting-electron density at the given temperature [71].

The linear dimensions of a vortex in the direction along the layers are very large since they are roughly equal to the "Josephson" length $\lambda_{\perp} = (\Phi_0^2\eta/$

$4\pi^2 td^2 N_s)^{1/2}$, which determines the penetration depth of the field parallel to the layers ($\Phi_0$ is the magnetic-flux quantum). The vortex center is located between the layers, and nowhere inside the layers is the superconducting state destroyed (in contrast to the Abrikosov vortex with the normal state inside the vortex core). The field $H_{c1}$ turns out to be highly anisotropic (a vortex in an $H_\perp$ field is a normal vortex), and

$$H_{c1}(\parallel) = \frac{\Phi_0}{4\pi\lambda_\parallel\lambda_\perp} \ln\frac{\lambda_\parallel}{d}, \quad H_{c1}(\perp) = \frac{\Phi_0}{4\pi\lambda_\parallel^2} \ln\frac{\lambda_\parallel}{\xi_\parallel}, \tag{6.28}$$

where $\xi_\parallel$ is the correlation length along the layers.

The Meissner effect in a parallel field should be observed only in fields $H < H_{c1}(\parallel)$, but the corresponding penetration depth, $\lambda_\perp$, is very large. The penetration depth $\lambda_\perp$ and the field $H_{c1}(\parallel)$ have not yet been measured in intercalation compounds. Such measurements will give the most accurate information about the parameter $t$, which determines the interlayer interaction in layered superconductors.

**C. The Upper Critical Magnetic Field.** In layered superconductors with the Josephson interlayer interaction, a parallel magnetic field destroys the superconductivity on account of the orbital effects (the diamagnetic currents) only in a very small region around $T_c$, i.e., in the temperature region where $\tau \ll t$. When the opposite inequality is satisfied, the energy of the diagmagnetic currents is far less than the superconducting energy, and the orbital effect does not limit the magnitude of the upper critical magnetic field for the direction parallel to the layers.

The transition from the normal to quasi-two-dimensional behavior in a parallel magnetic field when $\tau \ll 1$ is described in [59, 64, 72] with the aid of differential-defference equations corresponding to the Ginzburg–Landau functional (6.18). The calculation shows that, for $\tau > 2t$, only the paramagnetic effect can lead to the suppression of the superconductivity in a parallel field. In such a situation the field $H_{c2}(\parallel)$ is bounded at low temperatures $T \ll T_c$ by the Chandrasekhar–Clogston paramagnetic limit $H_p = \Delta_0/\mu_B\sqrt{2}$ ($\Delta_0$ is the gap at $T = 0$ and $\mu_B$ is the Bohr magneton) if no mechanisms that suppress the paramagnetic effect exist in the crystals.

According to the experimental data given in [73], the upper critical magnetic field for the direction parallel to the layers in $TaS_2(Py)_{1/2}$ exceeds the paramagnetic effect by at least a factor of 2.5 at low temperatures. According to estimates made in [59, 63], this excess can hardly be explained by spin-orbit scattering, which leads to the suppression of the paramagnetic effect in ordinary superconductors.

It is shown in [74, 75] that the suppression of the paramagnetic effect in a parallel field in layered crystals can be related to the spin-orbit interaction of the electrons in a lattice without a center of symmetry. This is done under the assumption that the lowering of the symmetry of a crystal as a result of intercalation can be due to the action of the intercalated molecules, which, because of the incommensurability of their dimensions and those of the unit cell of the dichalcogenide layers, distort the lattice within the layers. There are for the present, however, no other experimental data that confirm the absence of a center of inversion in $TaS_2(Py)_{1/2}$, and the model proposed in [74, 75] remains purely hypothetical.

Efetov and Larkin [76] have proposed a model of triplet pairing of electrons from different layers to explain why the paramagnetic limit is exceeded in layered compounds. This pairing survives in arbitrarily strong magnetic fields (if the orbital effect is absent), but it is sensitive to ordinary impurities, which have different effects on electrons from different layers in the same way as the singlet pairing is sensitive to magnetic impurities. Therefore, this model can hardly be realized in such dirty superconductors as $TaS_2(Py)_{1/2}$. Thus, there is no satisfactory explanation at present for the effect whereby the paramagnetic limit is exceeded in $TaS_2(Py)_{1/2}$.

**D. Superconducting Fluctuations in Layered Systems.** Although the interlayer electron transitions suppress the order-parameter fluctuations, the fluctuations in quasi-two-dimensional systems turn out to be stronger than those in ordinary three-dimensional superconductors. In superconductors with the Josephson interlayer interaction, the fluctuations are essentially two dimensional in the temperature region where $\tau = (T - T_c)/T_c >> t$, and in this region the diamagnetic fluctuation susceptibility $\chi_\perp$ for a field perpendicular to the layers, the direct-current conductivity, and the specific heat increase with decreasing $\tau$ in proportion to $\tau^{-1}$ (in the three-dimensional case, like $\tau^{-1/2}$). In the region $\tau < t$, the fluctuations assume a three-dimensional character, and the broader the region of two-dimensional fluctuations is, the smaller the $t$ is [59, 77, 78].

Thus far, only $TaS_2(Py)_{1/2}$ among the intercalation compounds has had its diamagnetic fluctuations experimentally investigated [79, 80]. The data of [80] yield for the parameter $t$ roughly the same estimates that were obtained above from the measurements of the upper-critical-field anisotropy [77, 80]. In the investigated temperature range $\tau \lesssim 0.2$, the temperature dependence of the diamagnetic fluctuation susceptibility does not differ from the dependence characteristic of three-dimensional superconductors.

It follows from the theoretical calculations that, besides the thermal fluctuations, the zero-point vibrations of the Cooper pairs also make an appreciable contribution to the fluctuation susceptibility of quasi-two-dimensional superconductors at $T \gtrsim 5T_c$ (in the three-dimensional case this contribution is very small [78].† At low temperatures ($T << T_c$), as a result of these purely quantum fluctuations, the fluctuation diamagnetic moment is nonzero in a field

†A Cooper pair of electrons is a Bose-type formation, and as $T \longrightarrow 0$ the number of pairs in a given quantum state tends to $\frac{1}{2}$.

$H > H_{c2}$ and increases logarithmically as $H \to H_{c2}$ if $t << 1$. The fluctuations, however, have thus far not been experimentally investigated at low temperatures near $H_{c2}$, and the situation with the diamagnetic fluctuations in $TaS_2(Py)_{1/2}$ in the range $T > 2T_c$ is still far from being clear [25].

Thus, characteristic of layered superconductors with the Josephson interlayer interaction at temperatures not too close to $T_c$ are the following properties, which distinguish them from the ordinary isotropic or slightly anisotropic type-II superconductors:

1. The nonstationary Josephson effect should occur in them when they are located in an electric field directed at right angles to the layers.

2. The field $H_{c1}(\|)$ is anomalously low, and a vortex in this field is such that the superconducting state inside the layers is nowhere destroyed.

3. The upper critical magnetic field $H_{c2}(\|)$ is determined only by the paramagnetic effect and can be anomalously high if no mechanisms exist that suppress this effect.

4. In the not very close neighborhood of $T_c$ the fluctuations have a two-dimensional character, and the contribution of the zero-point vibrations of the Cooper pairs becomes appreciable at temperatures significantly higher than $T_c$ (at $T \gtrsim 5T_c$) and at $T \to 0$ in fields $H > H_{c2}$.

These characteristics of intercalation compounds are of great interest both from the standpoint of the physics of superconductors and for their possible practical applications. Apparently, the main obstacle to the observation of the Josephson effects is the imperfection of crystals intercalated by molecules. Therefore, the production of more perfect crystals is now one of the main problems in the field of layered compounds.

## 7. Conclusion

As has already been noted above, the ability of atoms or molecules to intercalate layered compounds was discovered under the influence of constructive ideas for the fabrication of systems with the exciton mechanism of superconductivity. Now, however, the attention being paid to layered compounds is, undoubtedly, largely connected not with the problem of high-temperature superconductivity, but with the problems of structural transitions [81], the Josephson interlayer interaction in crystals with molecules, and the anomalously high values of the critical field in the direction parallel to the layers in $TaS_2(Py)_{1/2}$. This change in interest is explained to a considerable extent by the low critical temperatures of the molecule-intercalated layered compounds. Theoretical estimates of the effectiveness of the exciton mechanism for this class of compounds also turn out to be unfavorable, and it is difficult to expect a superconductivity different from the phonon type to occur in crystals with molecules.

More promising from this standpoint are crystals with alternating metallic and semiconducting layers. From all appearances, crystals of this type will be produced in the near future, and their experimental investigation will show how promising the class of layered compounds is from the point of view of the fabrication of superconductors with the exciton mechanism. The main and basic obstacle to the realization of exciton-mediated strong interelectron interaction in this case is connected only with the relatively large van der Waals distance between the metallic and semiconducting layers. This obstacle can, apparently, be overcome only in sandwich structures in which the individuality of the metal and the semiconductor gets lost in the boundary layer, which largely determines the whole effect of conduction-electron interaction via other excitations of electronic character (see Chapter 8).

# 7

# Structural and Superconducting Properties of Systems with One-Dimensional Anisotropy

## 1. Introduction

The experimental and extensive theoretical investigation of one-dimensional and quasi-one-dimensional compounds began about 10 yr ago under the influence of the exciton-mechanism-of-superconductivity concept. And although it is now clear that the physics of such compounds is interesting in itself and that other practical applications of quasi-one-dimensional crystals are possible, the main purpose of synthesizing new crystals with one-dimensional anisotropy is to produce compounds with metallic conductivity and, ultimately, a superconductor with as high a critical temperature as possible. Three directions, which relate our hopes for obtaining high-temperature superconductors to systems possessing the one-dimensional type of anisotropy, have been marked out for this purpose.

The first direction derives from Little's work: It is assumed that the exciton mechanism of superconductivity can be realized in systems of the one-dimensional type.

The second direction is based on the possibility of cooperation between the superconducting and structural transitions in systems of the one-dimensional type (see Chapter 5).

Finally, the third direction sets as its object the realization of superconductivity in compounds without metallic atoms and, in particular, the fabrication of organic superconductors. The idea of this direction is based upon the following considerations. At present the main method of raising $T_c$ involves the fabrication of new compounds, some of which may turn out to have critical temperatures exceeding the presently known $T_c$'s.

In doing this, we should try to broaden the class of investigated systems as much as possible. In principle, organic synthesis affords us almost limitless possibilities in this. There are considerations that reinforce the hope that the work in this direction will meet with success. The experimental study of the existing organic crystals—the highly conducting TCNQ salts—shows that the Coulomb interaction between the conduction electrons in them is significantly weaker than the corresponding interaction in ordinary metal-atom compounds.† At the same time, the conduction electrons in TCNQ crystals are strongly coupled to the lattice vibrations, including the intramolecular ones.

Both of these factors favor superconductivity. In principle, the third direction does not have as its goal the synthesis of a compound with as large an anisotropy of the one-dimensional type as possible, and we shall see below that the problem of high-temperature superconductivity does not require that we strive to achieve this. But so far all of the existing methods of synthesizing nonmetal-atom compounds with metal-type conductivity yield crystals with one-dimensional anisotropy.

In fact, the basic structural elements for their synthesis are polymers [for poly(sulfurnitride) crystals] or planar molecules with conjugated bonds (TCNQ salts). In the first case the appearance of one-dimensional anisotropy is natural; in the second, it arises because of the fact that the $\pi$ electrons of the conjugated system are capable of producing a sufficiently strong overlap of the electron wave functions of neighboring molecules only in the direction perpendicular to the plane of the molecule. Therefore, the understanding of the specific nature of the physical properties of crystals with one-dimensional anisotropy is absolutely essential for the achievement of success in the field of synthesis of superconductors without metal atoms. Let us note that some success has already been achieved in the third direction: poly(sulfurnitride) crystals are superconductors with $T_c \approx 0.3$ K. Evidently, the next step in the development of this direction will be the synthesis of organic superconductors.

Quite a large number of compounds with a fairly high conductivity at room temperature (100 mho/cm and higher) and a strong anisotropy of the one-dimensional type (a room-temperature conductivity anisotrophy of the order of 20 and above) are now known. But the majority of these compounds become dielectrics at low temperatures. This circumstance is not accidental: Everything that we know at present about one-dimensional systems shows that it is difficult to make the undimensionality condition and the requirement that the ground state of the electrons should have a metallic character compatible with each other.‡

†For example, the mutual Coulomb repulsion of the conduction electrons in TTF–TCNQ does not have a significant effect on the properties of the system when the conduction bandwidth is 0.1–0.5 eV. In the same situation crystals with metal atoms, e.g., NiO, are Mott-Hubbard dielectrics with a gap of the order of 1 eV [1]. Apparently, the drastic weakening of the Coulomb interaction in molecular crystals is because of the high polarizability of the molecules as well as the fact that the electrons can "evade" each other even in the case when they belong to the same molecule.

‡The metallic state of an electron system is realized in the case when the single-electron excitations are not separated by an energy gap from the ground state and the wave functions of the single-electron states are delocalized.

In fact, the theoretical investigations of one-dimensional electron systems, which began long before quasi-one-dimensional-type crystals were experimentally obtained, show that the properties of these systems should significantly differ from those of crystals in which the electron motion is two or three dimensional. The following three assertions sufficiently characterize all the specific properties of one-dimensional systems.

1. A one-dimensional metallic system is, when the Coulomb interaction between the electrons is neglected, unstable against that change in the crystal-lattice constant that splits the partially filled band into a completely filled sub-band and an empty one.

In other words, as the temperature is lowered, a one-dimensional metal should undergo a lattice distortion with a wave number equal to twice the Fermi momentum. This assertion belongs to Peierls.

2. In a one-dimensional electron system with a half-filled band, the excited single-electron states are separated from the ground state by a gap when there is an arbitrarily weak repulsion between the electrons [2]. This assertion has been proved for the case when the interelectron interaction can be described within the framework of the Hubbard Hamiltonian, but there is no reason to believe that it is not valid for the actual interelectron Coulomb interaction. Thus, the interelectron Coulomb interaction also leads to the dielectric (Mott–Hubbard) transition when the temperature is lowered.

3. The single-electron states in a one-dimensional system are spatially localized in the presence of an arbitrarily weak random potential. Therefore, at low temperatures the conductivity of a one-dimensional system of electrons in a lattice with defects cannot be metallic [3, 4].

All these assertions show that, at least for three reasons, a one-dimensional electron system should be nonmetalic at low temperatures.

The question arises as to what extent these results are applicable to quasi-one-dimensional crystals in which the eledtron motion is not strictly one-dimensional. It is clear that we can obtain qualitatively new effects peculiar to one-dimensional systems in an analysis of some effect in quasi-one-dimensional crystals only when the kinetic energy, $E_\perp$, of the interfilament electron motion is much lower than the energy chracterizing the effect in question. [Thus, in the case of the Peierls or Mott–Hubbard transition the quantity $E_\perp$ should be much smaller than the corresponding (Peierls or Mott–Hubbard) gap.] Below by quasi-one-dimensional systems in the strict sense of this word, we shall mean only crystals that satisfy the above-formulated condition, for in the opposite case we have ordinary three-dimensional anisotropic systems to deal with.

Three classes of compounds with a fairly high (at room temperatures) and anisotropic conductivity are being extensively investigated at present. These are TCNQ-based charge-transfer salts [5–8], square-planar complexes of the mixed-valence transition metals (Pt, Ir), a typical representative of which is the salt

**TABLE 7.1**

| Compound | Re $\varepsilon(\omega)$ at 4.2 K and a frequency of $10^{10}$ Hz | $T = 300$ K | | $T = 20$ K | | $T_0$, K | References |
|---|---|---|---|---|---|---|---|
| | | $\sigma_{\parallel}$, mho/cm | $\sigma_{\perp}$, mho/cm | $\sigma_{\parallel}$, mho/cm | $\sigma_{\parallel}/\sigma_{\perp}$ | | |
| $K_2Pt(CN)_4Br_{0.3} \cdot 3H_2O$ | 1200 | 300 | 0.006 | $10^{-8}$ | $2.5 \cdot 10^3$ | $T_3 = 120$ | [5, 9] |
| TTF–TCNQ | 3000 | 500–1000 | 5 | $3 \cdot 10^{-7}$ | – | 55 | [7, 8, 19] |
| NMP–TCNQ | 350 | 180 | – | $10^{-5}$ | – | 7 | [5, 20] |
| $(SN)_x$ | – | 1000 | 20 | $1.4 \cdot 10^4$ | 500 | – | [11] |

$K_2Pt(CN)_4Br_{0.3} \cdot 3H_2O$ (KCP) [5, 9, 10], crystals of the polymers $(SN)_x$ [11, 12], and salts of TTF-type cations with the halogen I or Br [13-15].†

Table 7.1 gives an idea about the conductivity and anisotropy of these compounds at room and low temperatures. It can be seen from this table, as well as from other experimental data considered below, that KCP-type and TCNQ-based salts are highly anisotropic, and the ground state of their electron system is dielectric. Only poly(sulfurnitride) $(SN)_x$ crystals remain metallic at low temperatures (and are superconducting at $T < T_c \approx 0.26$ K [12]), but the anisotropy in them is not very high, and they are, apparently, ordinary three-dimensional crystals with a high anisotropy. The experimental data show that the properties of the platinum complexes at low temperatures are due to the Peierls metal-dielectric transition. We shall briefly consider the theoretical ideas about, and the experimental data on, this transition in Section 2.

The TCNQ salts with asymmetric cations (phenazine NMP, quinolinium Qn, acridinium Ad) also undergo metal-dielectric phase transitions in the temperature range from 10 to 20 K, but these transitions clearly have a Mott-Hubbard character.‡ Section 3 is devoted to this class of quasi-one-dimensional compounds.

In the TCNQ salts with symmetric cations the metal-dielectric transitions occur in the 40-60 K range, and, according to the x-ray data, the transition in TTF-TCNQ is a Peierls transition. The properties of these salts are considered in Section 4. Finally, Section 5 is devoted to the crystals of the polymers $(SN)_x$. In Section 6 we discuss the possibility of realizing superconductivity in quasi-one-dimensional and organic compounds.

## 2. The Peierls Transition in Square-Planar Complexes

**A. The Structure and Anisotropy of KCP.** In the platinum complexes a Pt atom and four CN radicals form the planar structure shown in Fig. 7.1. In the crystals these planar atomic complexes, $Pt(CN)_4$, form chains or columns, two of which are shown in Fig. 7.1 (the bottom part). Of the 10 platinum-atom electrons, two participate in the bonding with the CN radicals, and 0.3 of the remaining eight $d$ electrons are transferred to the halogens. Formally, the platinum ions have charges of +2 and +4; therefore, complexes of the type $K_2Pt(CN)_4Br_{0.3} \cdot 3H_2O$ (abreviated to KCP) are called mixed-valence complexes. Five $d$ orbitals of the platinum atom in the crystal form bands, and the top, partially filled band in KCP is of the $d_{z^2}$ type (it is $\frac{5}{6}$ filled).

†The most recent review article on the properties of quasi-one-dimensional compounds is Toomb's paper [16], and new results in the field of quasi-one-dimensional compounds can be found in the proceedings of the International Conferences in Siofok (Hungary 1976) and Dubrovnik (Jugoslavia 1978) [17, 18].

‡The structure of the molecules, the full names, and the abbreviated designations are given in Table 7.2.

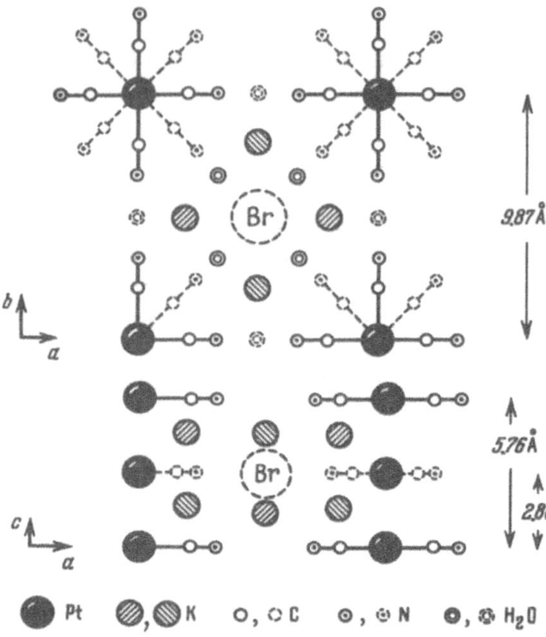


**Figure 7.1. Spatial arrangement of the Pt atom, the CN radicals, and the chain of these complexes in a KCP crystal [9]. Top: view along the chains. Bottom: view in the direction perpendicular to the alignment of the chains.**

The quasi-unidimensionality of the motion of the $d_{z^2}$ band's holes is determined by two factors. First, as can be seen from Fig. 7.1, the $d_{z^2}$ orbitals are aligned along the Pt-ion chains, and they ensure a good overlap of the neighboring ions in a chain but not of the nearest Pt ions of neighboring columns. Second, the Pt-ion spacing, $c$ ($c = 2.89$ Å), in a chain is only slightly greater than the distance (2.77 Å) between nearest ions in metallic platinum. The large platinum-chain spacing, $a$ ($a = 9.87$ Å), leads to a very slight overlap of the electron wave functions of neighboring conducting chains.

On the basis of these simple band concepts, it can be expected that KCP will exhibit all the properties of a quasi-one-dimensional metal. It is evident from light-reflection measurements that, at frequencies $\hbar\omega > 0.4$ eV, the electrons of KCP behave like a one-dimensional electron gas with a plasma frequency of 2.88 ev [9]. The electrical conductivity of KCP at room temperature also turns out to be high (up to 300 mho/cm). At temperatures above 200 K the conductivity falls with increasing temperature [5], and in this temperature region the Seebeck coefficient† for the direction along the chains is roughly constant, small

†The Seebeck coefficient is the coefficient of proportionality between the thermoelectromotive force and the temperature gradient in a crystal.

in magnitude ($\approx 1\ \mu W/K$), and positive, which indicates a hole conduction, in accord with the band concepts [21].

At low temperatures and low frequencies, however, KCP is a dielectric. In fact, below 200 K the conductivity decreases with decreasing temperature, while the Seebeck coefficient along the chains changes its sign and rapidly increases in absolute value with decreasing temperature. The behavior of the permittivity at low temperatures and at low frequencies is also not characteristic of a metal since as the frequency is decreased, the real part of the permittivity along the chains, Re $\varepsilon_{\parallel}(\omega)$, tends to a constant positive value and not to a negative infinity, as it should in the case of free electrons. (Measurements performed at 4.2 K at a frequency of $10^{10}$ Hz gave the value Re $\varepsilon_{\parallel} \approx 10^3$ and showed that $\varepsilon_{\parallel}$ increases with increasing temperature [22].)

Thus, the electrical and optical measurements show that, at high temperatures and frequencies, the electrons in KCP indeed constitute a quasi-one-dimensional metallic system. At the same time, the ground state of the crystal is undoubtedly dielectric, and a metal-dielectric transition occurs in it when the temperature is lowered.

It became clear after experiments on diffuse x-ray scattering that the main properties of KCP at low temperatures should be related with the Peierls transition [23]. These experiments showed that the Pt chains exhibit the superstructure that is predicted by the theory of the Peierls instability. Thus, the superlattice constant turns out to be greater than the Pt-atom spacing in a chain by a factor of roughly 6.7 in the case of the $K_2Pt(CN)_4Br_{0.3} \cdot 3H_2O$ salt and 8.0 in the case of $Rb_2Pt(CN)_4Br_{0.3} \cdot 1.3H_2O$. These constants correspond exactly to the quantity $2k_F$ ($k_F$ is the electron Fermi momentum) since the $d_{z^2}$ electronic band is roughly $\frac{5}{6}$ filled in KCP and $\frac{7}{8}$ filled in the salt with Rb.

**B. The Peierls Transition in a One-Dimensional System in the Molecular-Field Approximation.** In the free-electron model, the inevitability of the Peierls transition in a one-dimensional system is due to the fact that the Fermi surface in such a system is a point, and a periodic potential with period $2k_F$ always leads to the appearance of a gap at the Fermi surface and to the lowering of the energy of the electron gas (Fig. 7.2). Since a periodic potential can always be produced by means of a suitable distortion of the lattice, the ground state in the one-dimensional case should be a dielectric, and the lattice should exhibit a superstructure with period $2k_F$ at low temperatures. It is precisely such a superstructure that is observed in KCP at temperatures below room temperature.

An investigation of the Peierls instability of a one-dimensional system of noninteracting electrons within the framework of the Fröhlich Hamiltonian† in the molecular-field approximation yields the following picture for the second-order structural transition:

†The Peierls instability within the framework of the plasma model and the role of the mutual Coulomb repulsion of the electrons are discussed in Chapter 3.

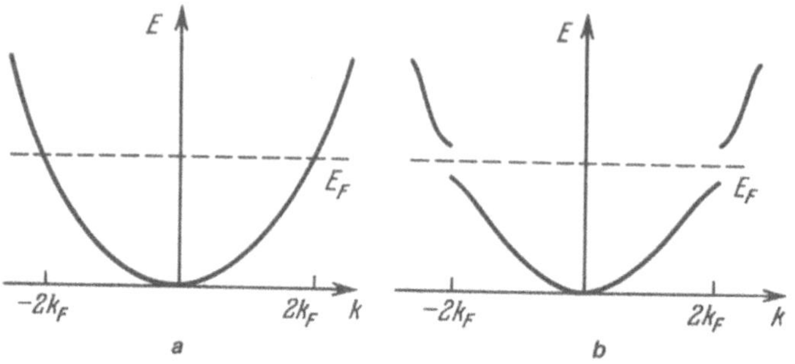


**Figure 7.2. Electron spectra: (a) in a one-dimensional homogeneous lattice without a Peierls distortion; (b) in the periodic potential, $V(2k_F)$, produced by an ion-displacement wave of wavelength $(2k_F)^{-1}$.**

1. The initial, high-temperature lattice turns out to be unstable at the Peierls-transition temperature

$$T_p = T_F \exp(-1/\lambda), \tag{7.1}$$

where $\lambda$ is the dimensionless electron-phonon interaction constant in the Fröhlich model.

2. Below the temperature $T_p$ static-lattice distortions with wavelength $2k_F$ appear. If the superlattice constant is highly incommensurable with the initial-lattice constant, then we can, in the first approximation, use the continuous-medium model. The Peierls displacement wave then has the form

$$u(x) = u_0 \cos(2k_F x + \varphi); \tag{7.2}$$

here the phase, $\varphi$, of the wave is indeterminate, i.e., the ground state of the Peierls dielectric in the continuous-medium model is degerate with respect to the phase (as in the case of a superconductor).

3. Below the temperature $T_p$ there appears in the electron spectrum a gap $\Delta(T)$ that increases with decreasing temperature, attaining the value $\Delta(0) = 1.76k_B T_p$ at $T = 0$.

4. In the Fröhlich model for a continuous medium (without allowance for the mutual Coulomb repulsion of the electrons), the phase degeneracy of the ground state leads to the appearance of a Fröhlich collective mode, $\omega_-$, corresponding to the oscillations of the phase, $\varphi$, of the displacement wave. The oscillations of the amplitude, $u_0$, of the superstructure give rise to an amplitude mode $\omega_+$ [24]. The frequencies of these modes at $T = 0$ are given by the expressions

$$\omega_-(q) = \sqrt{\frac{m}{m^*}}\, v_F q, \quad \frac{m^*}{m} = 1 + \frac{4\Delta^2(0)}{\lambda\omega_0^2(2k_F)\hbar^2},$$

$$\omega_+^2(q) = \lambda\omega_0^2(2k_F) + \frac{3}{4}\frac{m}{m^*} v_F^2 q^2, \tag{7.3}$$

where $\omega_0(2k_F)$ is the initial (bare) frequency of the phonons with momentum $2k_F$. For $q \to 0$ the Fröhlich collective mode corresponds to the motion of the displacement wave and the electrons along the crystal. Such a motion leads to the transfer of electron charge, and therefore the Fröhlich collective mode is optically active. Its contribution to the permittivity along the chains is determined in [25, 26], and, with allowance for electron transfer across the gap and the contribution of the Fröhlich mode, the expression for Re $\varepsilon_{\parallel}(\omega)$ for $T = 0$, $q \to 0$, and $\omega << \Delta$ has the form

$$\mathrm{Re}\,\varepsilon_{\parallel}(\omega) = 1 + \frac{\hbar^2\omega_p^2}{6\Delta^2} - \frac{\omega_p^2 m/m^*}{\omega^2}, \quad \omega_p^2 = \frac{4\pi e^2 N_e}{m}, \tag{7.4}$$

where $N_e$ and $m$ are the electron density and mass, respectively. Such a form of Re $\varepsilon_{\parallel}(\omega)$ corresponds to infinite conductivity $[\sigma(\omega)]$ for $\omega \to 0$.

If we also take account of the fact that at low temperatures the single-electron scattering in a current-carrying Peierls dielectric is negligible because of the energy gap in the single-particle excitation spectrum, then we arrive at the conclusion that the Peierls dielectric possesses properties similar to those of a superconductor. But, in fact, the effect of the commensurability of the initial-lattice and superlattice constants leads to a situation in which, as $q \to 0$, the frequency of the collective mode tends to a quantity $\omega_T > 0$ [accordingly, in the formula (7.4) for $\varepsilon_{\parallel}(\omega)$ the quantity $\omega^2$ is replaced by $(\omega^2 - \omega_T^2)$]. This effect is called the pinning of the Fröhlich collective mode at the initial-lattice sites.†

†In a real lattice there is no degeneracy with respect to the phase $\varphi$ [see (7.2)], and $\varphi$ is fixed in the ground state. But the higher the degree of incommensurability of the quantity $2k_F$ and the initial-lattice constant $c$, the lower the potential barrier for the $u_n$-displacement wave motion involving $n$ ions along the chain, and the smaller the quantity $\omega_T$. If the ratio $2k_F/c$ is the rational fraction of the form $n/m$, then at $T = 0$ the pinning frequency

$$\omega_T \approx \frac{m\Delta}{\hbar}\left(\frac{\Delta}{E_F}\right)^{(m/2)-1}$$

[25].

The scattering of the Fröhlich collective mode by impurities and the interaction of the Peierls displacements on different chains apparently lead to a similar result. But the superconducting properties of the Peierls dielectric do not disappear without leaving a trace. The Fröhlich collective mode leads to a high polarizability in the ground state of the Peierls dielectric and to the absorption of electromagnetic waves at low frequencies of the order of $\omega_T$.†

Furthermore, in the fluctuation regime above $T_p$ the Fröhlich collective mode can, in principle, make a fairly large contribution to the dc conductivity [27–30]. But the question of the existence of such a contribution cannot be regarded as having been theoretically elucidated completely since this contribution in the fluctuation regime has so far been computed only semiphenomenologically. Experiment also does not, for the present, give a unique answer to the question of the existence of such a contribution: This contribution is certainly absent in KCP, while its existence in compounds of the TTF–TCNQ type cannot be regarded as having been sufficiently conclusively demonstrated (see Section 4 below).

5. A significant reconstruction of the phonon spectrum also occurs during the Peierls transition. As the temperature approaches $T_p$ from above, the phonons with quasimomenta $q \approx 2k_F$ "soften," and at the transition point itself the frequency of the phonon with $q = 2k_F$ vanishes. Below the transition point one of the optical branches remains a low-frequency branch, and it corresponds to the Fröhlich collective mode. The difference between the Peierls dielectric and any other presently known dielectric is connected with the Fröhlich collective mode. Although there is a gap in the electron spectrum in the Peierls-dielectric state, and this gap can be wide, the response of such a system to a low-frequency electric field is strong precisely because of the presence of the collective mode.

6. In a chain of atoms of the same kind, a periodic potential with wave number $2k_F$ can be produced only by a sinusoidal displacement of the atoms along the chain. In this case the superlattice corresponds to the condensation of acoustic phonons with quasimomenta equal to $2k_F$. In more complex crystals a periodic potential can also be produced by the atomic displacements corresponding to optical vibrations with quasimomentum equal to $2k_F$. In such a situation the Peierls transition temperature is determined by the dimensionless constant $\lambda$, to which all the optical phonons with quasimomenta equal to $2k_F$, besides the acoustic phonons, contribute [31, 32].

If we denote the constant of the interaction between the electrons and a given vibration $s$ in the Fröhlich Hamiltonian by $g_s$, then

†Let us note that the longitudinal normal modes of the system are determined from the condition $\varepsilon_{\parallel}(\omega) = 0$, i.e., the normal-mode frequency for $q \longrightarrow 0$ becomes, when allowance is made for the mutual Coulomb repulsion of the electrons, equal to $\frac{3}{2}\lambda\omega_0(2k_F)$. The absorption determined by the quantity Im $\varepsilon_{\parallel}(\omega)$, however, has a peak in the region $m_T << \lambda\omega_0(2k_F)$.

$$\lambda = \sum_s \lambda_s, \quad \lambda_s = \frac{g_s^2}{\hbar\omega_s(2k_F)} N(0), \tag{7.5}$$

where $N(0)$ is the density of electron states at the Fermi surface. In such a complex crystal, as the temperature $T$ approaches $T_p$ from above, the mode that is a superposition of the acoustic vibration and the optical vibrations softens, the contribution of the given vibration $s$ to this superposition being determined by the corresponding constant $\lambda_s$ (the greater $\lambda_s$ is, the more strongly the vibration in question will be represented in the soft mode). In the case when the crystal is made up of molecules, the intramolecular vibrations also contribute to the Peierls instability (see Section 4c).

**C. The Role of Fluctuations and the Influence of the Three-Dimensional Effects.** In a one-dimentional system, allowance for the fluctuations (i.e., the consideration of the problem outside the limits of applicability of the molecular-field approximation) significantly alters the form of the transition. The crystals that we now have to deal with, however, are three dimensional, i.e., the filaments in them always interact with each other. The Peierls displacements on different chains interact with each other as a result of the tunneling of electrons between the chains [33], the Coulomb interaction between the CDW connected with the Peierls displacements, and the direct interaction between the Peierls displacements of different chains. Estimates of these contributions to the three-dimensional interaction of the CDW can be found in [34]. Either the picture of an almost "one-dimensional" transition in which the fluctuations are important in the entire temperature range below $T_p$ or the ordinary three-dimensional transition with a narrow temperature region in which the fluctuations are intense in realized, depending on the total energy of interaction between the CDW on different chains as compared to the energy of the CDW on one chain. The intensity of the fluctuations in one-dimensional and quasi-one-dimensional systems significantly depends on the number of independent functions describing the small deviations from the order parameter of the transition. Lee *et al.* [35] have suggested that the fluctuations associated with the Peierls transition are determined by one independent real function, which describes the variation of the amplitude of the ion-displacement wave during the structural transition. Let us first consider what role the fluctuations play within the framework of such an assumption.

If we denote the energy of interaction between the CDW on different chains by $E_\perp$, then the fluctuations are important only in a very small neighborhood of $T_p$ when $E_\perp >> k_B^2 T_p^2 / E_F$, and in this case the transition is, from the standpoint of the fluctuations, a normal "three-dimensional" transition, well described by the self-consistent field approximation.

When $E_\perp << k_B^2 T_p^2 / E_F$, a transition of the "one-dimensional" type is realized [10], and the picture of the transition differs greatly from the one that is described in Section 2b.

All the characteristics of the transition can be computed in the approximation that is adiabatic for the electrons, i.e., in the case when $\hbar\omega \ll k_B T_p$, and the picture of the transition is as follows [10, 35, 36].

The most important changes in the electron system occur not at the temperature $T_p$, but at the temperature $T_3 \approx T_p/4$. There appears in the density of states at the Fermi surface in the $T_3$ region a sharp dip corresponding to the appearance of a gap $\Delta = 1.76 k_B T_p$ on further lowering the temperature $T \to 0$.

Correlations between the ion displacements along a chain appear below the temperature $T_p$, but the correlation length, $\xi_\parallel$, along a chain becomes infinitely large only at the temperature $T_3$. Correlations between the displacements of the ions of neighboring chains as a result of the interchain coupling appear in the region around $T_3$ (in this region the correlation length, $\xi_\perp$, in the direction perpendicular to the chains increases), but the usual three-dimensional long-range order with small fluctuations is not established at any temperature (the fluctuations destroy the correlations between the displacements on different chains).

The most important change in the thermodynamic quantities (e.g., the anomaly in the specific heat) is observed in the vicinity of the temperature $T_3$ [36].

The qualitative features of this picture of the transition are connected with the fact that in the case of a phase transition occurring in a one-dimensional system, and described by one real function, $u(x)$, varying slowly in space, the absence of long-range order [$\langle u(x)\rangle = 0$] is because of the possible vanishing of this function at some point [the minimum of the free energy of the system is attained when the order parameter is constant in space, i.e., when $u(x) = u_0$]. The absence of a nonzero mean value of the order parameter $u(x)$ is possible only in a situation in which the function $u(x)$ changes its sign at some points. Since the vanishing of the order parameter at a point requires a finite activation energy, the probability for the destruction of the long-range order is exponentially small at low temperatures. Exact numerical computations [36] show that this probability becomes negligible at $T < T_p/4$.

The question, however, arises whether one slowly varying real function is sufficient for the description of all the small deviations from the function, (7.2), characterizing the long-range order that arises in the Peierls transition. At least in the continuous-medium ("jellium") model, i.e., in the case when the superlattice and the parent lattice are highly incommensurable, the answer is clearly in the negative: Small deviations from the function (7.2) are described not by one, but by two slowly varying independent functions. One of these functions corresponds to the fluctuations in the amplitude, $u_0$, of the order parameter, while the other corresponds to the fluctuations in the phase, $\varphi$, of the displacement wave. The independence of these functions is a consequence of the fact that the function (7.2) oscillates rapidly in space.† Therefore, in the incommensurable case the fluctuations associated with the Peierls transition are virtually entirely similar to the fluctuations in a system with a complex order parameter (e.g., the fluctuations associated with the superconducting transition).‡

Calculations [36] show that the fluctuations in an uncharged system with a complex order parameter are significantly more intense than those in a system with a real order parameter since the energy of the system is degenerate with respect to the continuous quantity–the phase of the order parameter–and there is, in general, no potential barrier for the phase fluctuations. The commensurability effects lead to a finite potential-barrier height (as do any other mechanisms that cause the pinning of the Fröhlich collective mode), but in the extreme-incommensurability case the potential barrier for the phase fluctuations will remain much smaller than the barrier for the amplitude fluctuations. According to [36], in the situation with a complex order parameter there does not exist in a one-dimensional system any temperature of the order of $T_p$ below which the correlation length becomes very long; it is long only at $T \ll T_p$. In a quasi-one-dimensional system with a complex order parameter, a "three-dimensional" transition is realized when $E_\perp \gg k_B T_p$; in the opposite limiting case long-range order is established below the temperature $(T_p E_\perp / k_B)^{1/2}$ [37]. Thus, the fluctuation theory used in [35] to describe the Peierls transition is valid only in the case of strong pinning of the Peierls superlattice, and it can, for example, be used to describe the Peierls transition with a doubling of the lattice constant.

For the incommensurable Peierls transition, it is mainly the phase changes that are important for the computation of the fluctuations. The corresponding computations of the correlation functions with allowance for the mutual Coulomb repulsion of the electrons show that the charges that appear during the phase oscillations do not lead to a significant change in the role of the fluctuations in the electron system as compared to their role in an uncharged system [38].

**D. Experimental Data on the Peierls Transition in Crystals of the Platinum Complexes**

*The Structural Data.* An investigation of diffuse x-ray scattering in KCP and $Rb_2Pt(CN)_4Br_{0.23} \cdot 1.3H_2O$ crystals at room temperature [23] has revealed that

†This situation is similar to the amplitude and phase modulations of a high-frequency wave in radio engineering.

‡This observation is valid for uncharged systems. The Coulomb repulsion makes the fluctuations associated with the Peierls and superconducting transitions different. Allowance for the mutual Coulomb repulsion of the electrons in the same chain leads to the appearance of an energy gap in the spectrum of the phase oscillations associated with the Peierls transition.

either elastic or almost-elastic x-ray scattering occurs in these crystals with a change in momentum corresponding to the quantity $2k_F$. The scattering indicates that the atomic displacements have a sinusoidal character with periodicity $6.7c$ in KCP and $8c$ in the Rb salt. The diffuseness of the scattering indicates the absence of correlations between the ion displacements in different chains.

Diffuse x-ray scattering does not allow us to distinguish between elastic and almost-elastic scattering; therefore, we cannot, on the basis of this experiment, say whether the displacement wave with momentum $2k_F$ is a static wave corresponding to long-range order within a chain or whether it corresponds to low-frequency phonons with momenta approximately equal to $2k_F$.

Experiments on inelastic neutron scattering [39] unambiguously indicate the existence at room temperature of a giant Kohn anomaly in the phonon spectrum (the softening of the phonons with momenta around $2k_F$).† At the same time a peak can be observed at this temperature in the almost elastic neutron scattering involving a $2k_F$ momentum change along a chain [40], in accord with the diffuse x-ray scattering results.

At temperatures below 120 K (measurements have been performed down to 77 K) the diffuseness of the x-ray scattering decreases, and the nature of this change indicates the appearance of those three-dimensional correlations for which the ion displacements in neighboring chains are antiphased [41]. These measurements did not yield the temperature dependence of the correlation length in the direction perpendicular to the chains; this information was provided by an almost-elastic-neutron-scattering investigation performed by Renker *et al.* [42] and Lynn *et al.* [43] in the temperature range from 300 down to 6 K.

According to the results presented in [42, 43], the scattering involving a $2k_F$-momentum transfer along a chain is the most intense at room temperature. As the temperature is lowered, this scattering weakens, but a peak appears at the momentum transfer $\mathbf{Q} = (\pi/a, \pi/a, 2k_F)$ ($a$ is the spacing of the Pt-atom chains and the first two coordinates $Q_x$ and $Q_y$ correspond to the direction perpendicular to the chains). But the $\mathbf{Q} = (0, 0, 2k_F)$ scattering remains fairly strong even at 6 K, i.e., long-range three-dimensional order does not set in right down to 6 K.

A comparison of the $\mathbf{Q} = (0, 0, 2k_F)$ and $\mathbf{Q} = (\pi/a, \pi/a, 2k_F)$ scattering intensities shows that, as the temperature is decreased, the correlation length for the displacements on different chains increases sharply in the vicinity of 120 K (up to $3a$) and then remains practically constant down to 6 K.

†In the three-dimensional isotropic case the phonon spectrum is singular at the quasimomentum $q = 2k_F$ at low temperatures. This singularity arises as a result of the Fermi step in the momentum distribution of the electrons. It was discovered by Kohn and is called the Kohn anomaly in the phonon spectrum of metals. This is essentially what gives rise in the one-dimensional case to the Peierls transition at $T < T_p$ and to the dip in the phonon spectrum near the quasimomentum $2k_F$ at temperatures $T > T_p$.

The correlation length along the chain, $\xi_{\parallel}$, is greater than $60c$ at room temperature [41], and greater than $100c$ at $T < 160$ K [43]. The magnitude of the amplitude $u_0$ for the displacements of the Pt atoms (together with the CN radicals) along the chains at low temperatures constitutes about 0.5% of the platinum-atom spacing [44], and, according to the data of [43], $u_0$ decreases monotonically with increasing temperature above 80 K.

*Variation of the Density of Electron States.* The appearance of a gap during the Peierls transition can be detected from the temperature dependence of the optical properties of the crystals.

According to Brüesch *et al.* [45], there is an energy gap $2\Delta(0) \approx 0.18$ eV in KCP at low temperatures. Since $\Delta(0) = 1.76 k_B T_p$, we obtain for the temperature $T_p$ a value of about 500 K.

The temperature dependence of the electrical conductivity $\sigma_{\parallel}$ in the range from 40 to 100 K can be described by the exponential law $\sigma_{\parallel} \sim \exp(-\Delta/T)$, and the activation energy $\Delta$ is roughly equal to 0.08–0.09 eV. This value is also in good agreement with the value obtained above for $\Delta(0)$.

Measurements of the paramagnetic susceptibility $\chi_p$ yield complete information about the temperature dependence of the value of the density of states, $N(0)$, at the Fermi surface since

$$\chi_P = \frac{1}{2} g^2 \mu_B^2 N(0). \tag{7.6}$$

There is observed in KCP at room temperature a small shift in the NMR frequency for the $^{195}$Pt nuclei with respect to the NMR frequency for the nonconducting salt $K_2Pt(CN)_4 \cdot 3H_2O$; this shift can be interpreted as a Knight shift.† The shift disappears in the 100–200-K range, and the NMR spectrum no longer depends on the temperature in the 1.5–110-K range [46]. EPR measurements on KCP crystals yield roughly the same temperature dependence for $\chi_p$ [47].

Measurements of the EPR frequency show that the $g$-factor anisotropy is characteristic of the $d_{z^2}$ band. The EPR-signal intensity, which is proportional to the paramagnetic susceptibility, decreases with decreasing temperature in the range from 130 to 65 K (measurements at higher and lower temperatures are impossible because of the broadening of the resonance line). From these data we can determine the dependence of the density of states $N(0)$ on temperature in the measurement range (at 85 K the signal intensity corresponds to $10^{20}$ spins per cm$^3$, i.e., to a paramagnetic susceptibility $\chi_p \approx 1.2 \times 10^{-4}$ mol$^{-1}$). The conversion of the signal intensity to the quantity $N(0)$ $(E_F)$ yields a reduction in the density of states from $1.3 \times 10^{36}$ erg$^{-1}$ $\cdot$ mol$^{-1}$ at 130 K to $0.4 \times 10^{36}$

†The effective magnetic field acting on the nuclei depends on the spin state of the electron system. The deviation of the effective field from the external field is called the Knight shift. For nonmagnetic compounds this shift is proportional to the paramagnetic susceptibility of the electrons.

$erg^{-1} \cdot mol^{-1}$ at 65 K. In the tight-binding model the density of states in the original band can be expressed in terms of the total bandwidth $W$ with the aid of the relation

$$N(0) = 4N_{at}/\pi W \sin(\pi p/2), \tag{7.7}$$

where $N_{at}$ is the density of the atoms, $p$ is the degree of occupation of the band, and $W/4$ is the resonance spacing of the molecules. For $N \approx 1.3 \times 10^{36}$ $erg^{-1} \cdot mol^{-1}$, we obtain $W \approx 0.6$ eV. Since the dip in the density of states is already noticeable at 130 K, this value is only a lower bound, and the quantity $W$ may, in fact, turn out to be 1.5–2 times greater. The total bandwidth has a value of 5 eV in metallic platinum, and values of the order of 1–2 eV might be expected for it in KCP. The EPR data are in good agreement with this estimate.

Specific heat measurements on KCP at low temperatures (from 0.06 to 1.2 K) [48] show that the linear—in the temperature—contribution to the specific heat is very small, and if we interpret it as the electron contribution, then it corresponds to a density of states at the Fermi surface of $N(0) = 0.8 \times 10^{35}$ $erg^{-1} \cdot mol^{-1}$. Thus, a comparison of the magnetic and specific-heat data reveals that the density of states at the Fermi surface decreases by roughly a factor of 20 as the temperature $T$ is lowered from 130 to 1 K. The presence of electron states inside the Peierls gap at the Fermi surface at low temperatures is due, apparently, to the presence of KCP of an inherent lattice disorder caused by the random distribution of the Br ions [10, 49].

*The $\omega_-$ Soft Optical Mode and the Phonons.* According to Brüesch *et al.* [45], the absorption spectrum of KCP at low temperatures has in the 0.002-eV energy region a peak that can be interpreted as a Fröhlich-collective-mode excitation. If this interpretation is correct, then the frequency of the collective mode with zero quasimomentum is $\omega_T \approx 0.002$ eV, and so low an excitation-energy value for this optically active mode allows us to explain the high permittivity value at low temperatures. In fact, according to the formulas (7.3) and (7.4), for $\omega_0(2k_F) \approx 0.006$ eV [39], $\lambda \approx 0.5$ and $\varepsilon_{\parallel}(0) \approx 1200$ [5], we obtain the value $\omega_T \approx 0.002$ eV, which agrees with the data given in [45]. The Raman-scattering spectrum of KCP has been found to contain a strongly temperature-dependent mode that can be interpreted as an $\omega_+$ amplitude mode [50]. The frequency of this mode for $T \longrightarrow 0$ is 0.005 eV, and this value is in good agreement with the estimates obtained for $\omega_+(0)$ from the formulas (7.3) when we set $\lambda \sim 0.5$ and $\omega_0(2k_F) = 0.006$ eV.

The experimental investigation of the inelastic scattering of neutrons in KCP [43, 51, 52] confirms the existence at temperatures below 160 K of two longitudinal modes with momentum $\mathbf{Q} = (0, 0, 2k_F)$ in the parent lattice and energies 0.0025 and 0.006 eV [when $\omega_0(2k_F) = 0.008$ eV]. These modes can be interpreted as phase and amplitude modes, respectively.

In [52] an attempt is made to estimate the ratio $m^*/m$ for these two modes from the experimental data with the aid of the relations (7.3). The expressions (7.3) are, however, valid only for low momenta $k \ll \xi_{\parallel}^{-1} = \Delta_0/\hbar v_F$. This limitation makes it virtually impossible to estimate the quantity $m^*/m$ from the experimental frequency- and amplitude-mode dispersion data since the changes that occur in the frequencies $\omega_+$ and $\omega_-$ as $k$ is varied from zero to a value much less than $\xi^{-1}$ are too small, and it is hardly possible to measure them experimentally by the method of inelastic neutron scattering.

*Interpretation of the Experimental Data on KCP.* The first attempts to explain the results of the measurements on KCP were based on the theoretical "one-dimensional"-Peierls-transition model with the Ginzburg–Landau order parameter represented by a single real function [10, 35].

In the region near 120 K a rapid decrease in the density of electron states (according to the magnetic data) and the growth of the correlation length in the direction perpendicular to the chains are observed. Therefore, the temperature 120 K can be identified with the quantity $T_3 \approx T_p/4$. This estimate corresponds to the roughly 0.09-eV gap value obtained from the optical data and the temperature dependence of the electrical conductivity. But for the correlation length $\xi_{\parallel}$, the theoretical model used yields a value approximately equal to $7c$ at room temperature, while the neutron and x-ray scattering data show that the experimental $\xi_{\parallel}$ value exceeds this estimate by at least an order of magnitude.

This experimental contradiction casts doubt on the applicability to KCP of the "one-dimensional"-transition model with a single real function. We have noted above that there are theoretical objections to the application of such a description to KCP crystals, in which the superlattice is fairly highly incommensurable with the parent lattice. But at present there is essentially no satisfactory model that can explain the strong temperature dependence near 120 K of the density of electron states, the specific heat, and the transverse correlation length with the simultaneous existence in KCP of a very long longitudinal correlation length even at 300 K.

It has also been noted above that the theory is at variance with the experimental data for the dispersion of the phase mode. We can only suggest that the internal disorder characteristic of KCP crystals is important for the understanding of the situation. It is clear from the results of [38] that the internal disorder of the crystals leads not only to the lowering of the Peierls transition temperature [49, 53, 54], but also to the absence of long-range order for the Peierls displacements at all temperatures. It is also shown in [38, 55-58] that the lattice disorder leads to a finite contribution of the Fröhlich collective mode to the permittivity of the crystal and the dc conductivity at $T = 0$ in the case when the Peierls displacements are incommensurable with the parent crystal lattice.

Thus, the superlattice predicted by the theory of the Peierls instability is undoubtedly observed in KCP. But the computation of the temperature dependences of the physical characteristics of the crystal in the case of a Peierls transition occurring in a disordered crystal lattice is very complicated, and we have for the present only particular results in this direction. At present these results cannot quantitatively explain all those temperature dependences of the transition parameters that are observed in KCP.

## 3. TCNQ Salts with Asymmetric Cations

**A. The Structure and Electronic Properties of the Highly Conducting TCNQ Salts.** The TCNQ molecule is a planar molecule with saturated bonds. It possesses strong acceptor properties and can form charge-transfer salts with other atoms or molecules. The charged TCNQ molecule in such salts are in the anion-radical state (Fig. 7.3).

The change that occurs in the chemical bonds when the neutral TCNQ molecule goes over into the $TCNQ^-$ anion leads to a change in the intramolecular distances, and for this reason x-ray structural analysis enables us to distinguish the neutral molecule from the charged molecule. We now know quite a large number of molecules that can form charge-transfer salts with TCNQ. Some of them are given in Table 7.2. In the simple $K^+(TCNQ)^-$ salts there is one $TCNQ^-$ molecule for each $K^+$ cation. In the complex salts there are two or more TCNQ molecules for each cation [e.g., $K^+(TCNQ)_2^-$].

In the charge-transfer crystals the planar TCNQ molecules are packed in stacks, these stacks forming parallel columns in the crystal. The cations occupy positions between the TCNQ columns. This type of structure is shown in Fig. 7.4 for the simple NMP–TCNQ salt and in Fig. 7.5 for the TTF–TCNQ salt.†

The unit cell in this case contains one molecule of the complex, with the cation at the center of the cell and TCNQ molecules at the corners. The columns consisting of coplanar TCNQ molecules are parallel to the *b* axis, with a spacing of 3.24 Å between the planes of the molecules (this spacing is slightly less than the

†The abbreviated disignations of the cation molecules are given in the text. Their structures and full names are given in Table 7.2.

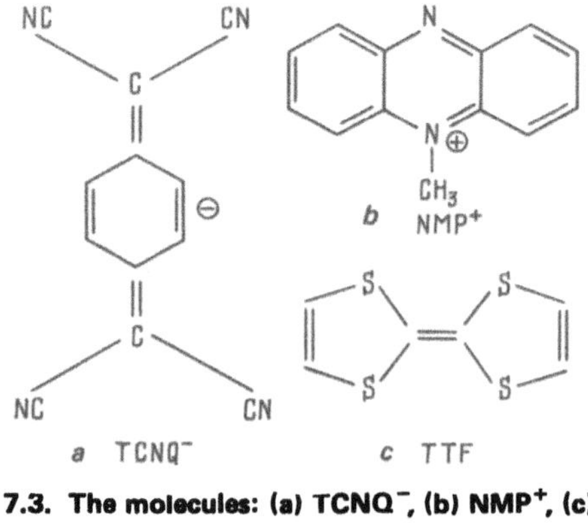


Figure 7.3. The molecules: (a) $TCNQ^-$, (b) $NMP^+$, (c) TTF.

**TABLE 7.2**

| No. | Cation | Structure | Designation |
|---|---|---|---|
| 1 | *N*-Methylphenazine | 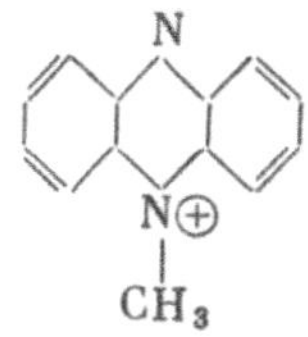 | NMP |
| 2 | Quinolinium |  | Qn |
| 3 | Acridine | 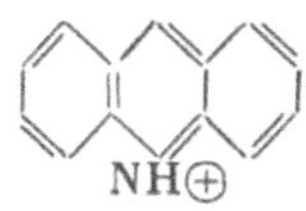 | Ad |
| 4 | Tetrathiatetracene | 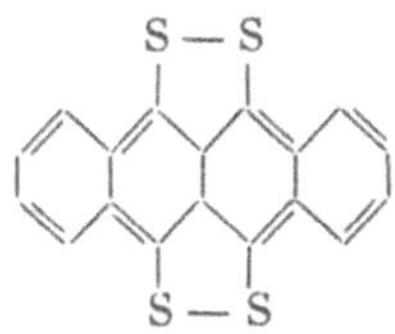 | TTT |
| 5 | Tetrathiafulvalene | 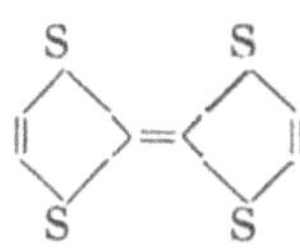 | TTF |
| 6 | Tetraselenafulvalene | 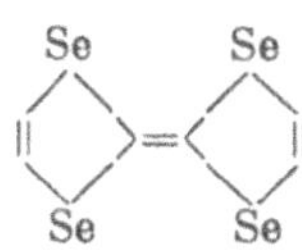 | TSeF |
| 7 | *cis*-Diselenadithiafulvalene | 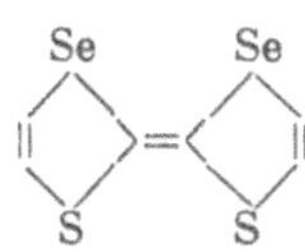 | *cis*-DSeDTF |
| 8 | *trans*-Diselenathiafulvalene |  | *trans*-DSeDTF |
| 9 | *cis*-Dimethyltetrathiafulvalene | 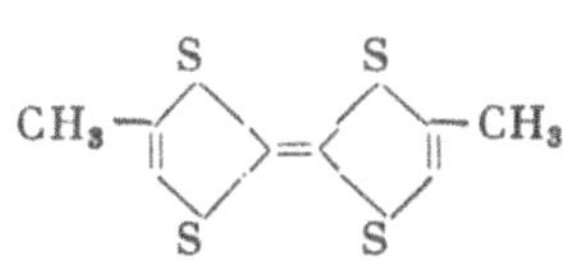 | *cis*-ATTF |
| 10 | Hexamethylene-tetraselenafulvalene | 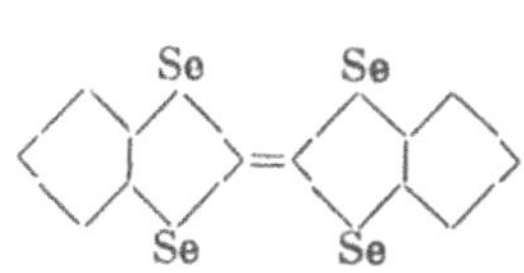 | HMTSF |

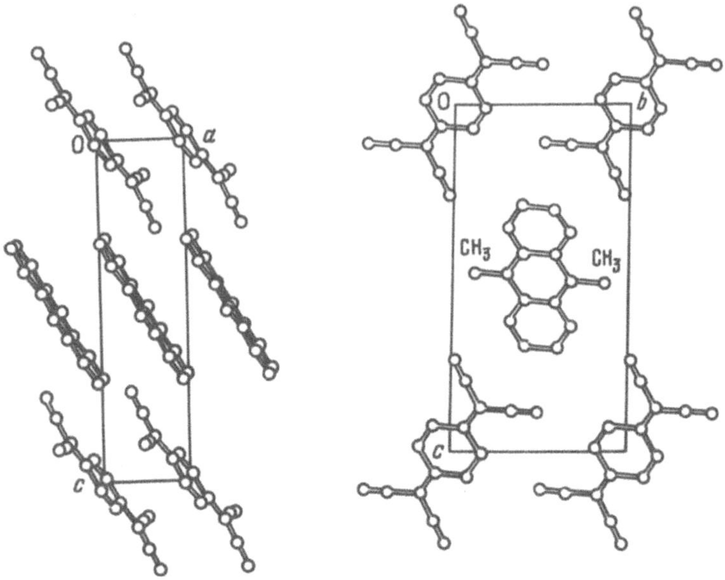


Figure 7.4. The crystal structure of the TCNQ complex with phenazine [59].

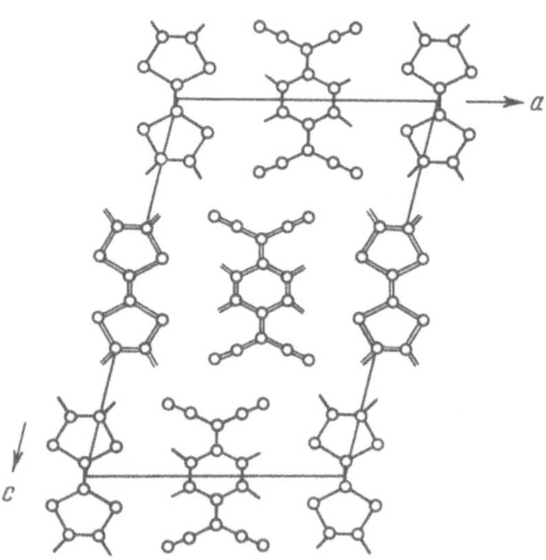


Figure 7.5. The arrangement, according to the x-ray data [40], of the TTF and TCNQ molecules in the TTF-TCNQ crystal.

van der Waals distance). The column spacing is much greater, ranging from 7 to 15 Å. The outer electrons of the TCNQ molecules are in $\pi$ orbitals elongated in direction perpendicular to the planes of the molecules. Therefore, the overlap of the electron wave functions of neighboring TCNQ molecules is much stronger for molecules within a column than for molecules in different columns.

From the foregoing it is clear why the conductivity of the TCNQ salts are highly anisotropic (the conductivity along the columns attains at room temperature a value lying in the range from $10^2$ to $10^3$ mho/cm, which is from $10^2$ to $10^3$ times higher than the conductivity in the transverse direction). Therefore, in the case of charge-transfer TCNQ salts we must deal with linear conducting chains regularly arranged in a three-dimensional crystal.

The TCNQ salts can be grouped into several classes differing in their physical properties [5]. These properties are determined mainly by the structure of the columns. Salts whose columns are already highly Peierls distorted at room temperature belong to the class of salts with moderate conductivity. On the basis of band considerations, these salts should be simple dielectrics with completely filled and completely empty electron bands. A comparison of the magnetic and electric properties reveals, however, that they are actually Mott-Hubbard dielectrics (the conduction activation energy in them greatly exceeds the energy gap for magnetic excitations).

The so-called highly conducting salts form a more interesting class of salts. It follows from the x-ray data that all the TCNQ molecules in a column in these salts are equivalent at room temperature, i.e., they are arranged such that all the intermolecular distances and angles along a column are the same. Furthermore, in the complex salts of this class with one electron for two TCNQ molecules, all the intramolecular distances in the TCNQ molecules are identical at room temperature, and these distances are intermediate between the distances in the neutral molecule and the anion (TCNQ) molecule. In other words, the electron density is in this case distributed evenly among all the molecules of the TCNQ column.

All the TCNQ complexes produced thus far (except the HMTSF-TCNQ complex; see below) undergo the metal-dielectric transition as the temperature is lowered. But the type of this transition significantly depends on the properties of the cations.

The cations NMP, Qn, Ad, and others similar to them (see [5]) are charged molecules with filled electron shells, and there are, apparently, no conduction electrons in a chain of such cations.†) All of them are asymmetric with respect

†For the time being the experimental data do not allow us to estimate sufficiently reliably the extent of electron transfer from the cation chains to the TCNQ ones. From the data for the second moment of the NMR line in NMP-TCNQ the extent of transfer is indirectly estimated to be 0.94 (i.e., about 6% of the electrons remain on the NMP chains) [60]. But this estimate was obtained on the basis of certain assumptions, and there are as yet no other experiments that corroborate these assumptions, in particular, the assumption that the electron transfer is a partial one.

to the position of the charged complex (see Table 7.2), and the randomness in the orientational disposition of the cation in the crystal leads to an inherent disorder in the crystal structure of the complexes. Thus, Fig. 7.4 shows the structure of the NMP–TCNQ complex with the asymmetric NMP cation. The NMP molecule is shown in this figure with methyl ($CH_3$) radicals. Actually, the cation contains only one $CH_3$ radical, but it can take up either of two alternative orientations in the crystal (as shown in the figure) with equal probability. The positive charge of the cation resides on the nitrogen atom with which the methyl radical is bonded. Therefore, the random orientation of the cation leads to a random variation of the potential along a chain. In accordance with the results presented in [49, 53, 54], the Peierls instability can be absent in this case. According to the experimental data, the paramagnetic susceptibility in a highly conducting salt with an asymmetric cation increases with decreasing temperature [61], whereas in the presence of the Peierls transition the paramagnetic susceptibility should, under these same conditions, decrease. Below we shall see that this growth gives grounds for believing that the metal–dielectric transition in salts of this type can have a Mott–Hubbard character, although we cannot entirely exclude the possibility of a Peierls transition in the TCNQ chain in the case of partial charge transfer from the cation chains.

The molecules TTT, TTF, and their analogs have a filled electron shell, and they become cationic radicals in charge-transfer crystals [7, 60, 62, 63]. Therefore, in crystals with such cations conduction electrons can be present in both the cation and the anion chains (see Fig. 7.5). Furthermore, the cations of this class are symmetric or slightly asymmetric. The properties of this class will be considered in greater detail in Section 4.

**B. The Dielectric Character of the Ground State and the Metal–Dielectric Transition**

*The Electrical Properties [5, 64, 65].* At low temperatures the conductivity of the highly conducting salts is nonmetallic in respect to its temperature dependence. In TCNQ salts with NMP, Qn, and Ad the conductivity decreases with decreasing temperature when $T < 200$, 230, and 150 K, respectively. At very low temperatures the decrease of the conductivity is found to be slower than an exponential decrease (Fig. 7.6). The temperature dependence of the Seebeck coefficient is also not characteristic of a metal. In all these salts the permittivity in the SHF region is positive at all temperatures and attains large values ($10^2$ - $10^3$) at liquid-helium temperatures (in the moderately conducting crystals the permittivity does not, under these same conditions, exceed 10).

All these data show that the ground state of the crystals at low temperatures is dielectric. But at the same time, the fact that the electron system is highly polarizable indicates that the highly conducting TCNQ crystals with asymmetric cations are close to the metallic state, i.e., the gap for the single-electron excitations in them is not wide, and these one-electron states are highly delocalized.

*The Magnetic Properties.* In all the highly conducting TCNQ salts with

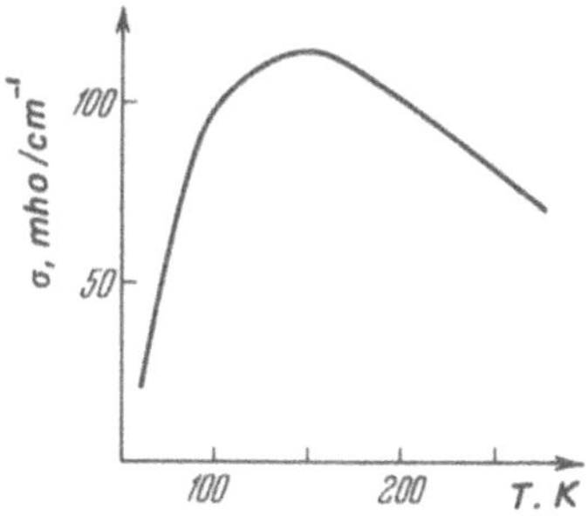


**Figure 7.6. The temperature dependence of the electrical conductivity of the complex Ad/TCNQ)$_2$ along the chains [64].**

asymmetric cations the paramagnetic susceptibility is found to be roughly constant at temperatures above 100 - 200 K, but to increase according to the law $\chi_p \sim T^{-\alpha}$, with $0 < \alpha < 1$, at temperatures below 30 K (Fig. 7.7).

In high fields ($g\mu_B H >> k_B T$), the paramagnetic moment, $M$ in these same crystals increases with increasing field intensity without reaching saturation, and $M \sim H^{1-\alpha}$. Arguments are adduced in [66] to show that these magnetic proper-

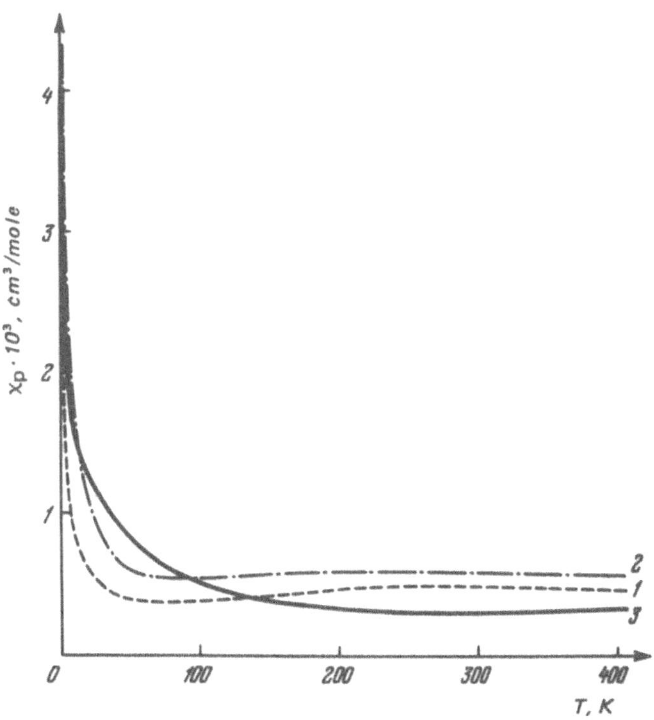


**Figure 7.7. The temperature dependences of the susceptibilities of the complexes (1) Qn(TCNQ)$_2$, (2) Ad(TCNQ)$_2$, and (3) NMP-TCNQ [61].**

ties are not connected with the paramagnetic impurities but reflect the properties of the electron system of the salts with asymmetric cations. To account for these low-temperature magnetic properties, Bulaevskii and Shchegolev [66] have proposed the model of localized spins with a one-dimensional Heisenberg antiferromagnetic random interaction. This model describes both the magnetic data and the dependence of the electronic specific heat of the salts on temperature and magnetic field at extremely low temperatures, down to 0.01 K [67, 68]. It also gives a qualitatively correct picture of the temperature dependence of the Knight shift in $Qn(TCNQ)_2$. According to the experimental data of Shchegolev *et al.* [69], the Knight shift in this compound is proportional to the susceptibility at $T > 7$ K, but it vanishes when the temperature is lowered below 7 K, whereas the paramagnetic susceptibility continues to increase according to a power law. It is shown in [70] that precisely such a behavior of the Knight shift is to be expected in a one-dimensional spin chain with a random exchange interaction in those chases when the paramagnetic susceptibility increases according to a power law as $T \to 0$.

The model of localized spins with a random one-dimensional interaction is a semiphenomenological model and cannot answer the question of the origin, and the positions in the crystal, of the localized spins, nor can it answer the question of the connection between the disorder in the disposition of the cations and the random character of the exchange interaction between the localized spins. Theodorou and Cohen [71] have shown that a one-dimensional electron system with a strong on-site Coulomb repulsion and random electron-energy values at the sites is, from the point of view of the low-temperature magnetic properties, equivalent to a spin model with a random-exchange interaction. But we do not at present have sufficiently reliable information on the basis of which we can determine the chains in which these localized spins are sited. In the case of total transfer, the localized electrons can reside only in the TCNQ chains, and the growth of the susceptibility with decreasing temperature indicates that the electrons of the TCNQ chains are in the Mott-Hubbard dielectric state. In the case of partial transfer, the electrons of the cation chains may turn out to be localized (the model proposed in [60] for the interpretation of the NMR linewidth in NMP-TCNQ; see the footnote on p. 000). In this situation we cannot exclude the possibility of a Peierls transition in the TCNQ chains, and superlattices with tripled and sextupled periodicities have now been discovered in NMP-TCNQ [72]. A phase transition has been observed in this compound in the 7-K region. A change in the temperature dependence of the paramagnetic susceptibility [73] and a jump in the specific heat occurs here [20].

We can estimate from the data for the paramagnetic susceptibility at room temperature the density of electron states at the Fermi surface [see (7.5)] and from the density of states the specific heat, $c_e$, of the electron system at the transition point $T_0$ since

$$c_e(T_0) = \gamma T_0, \quad \gamma = \frac{2\pi^2}{3} k_B^2 N(0). \tag{7.8}$$

The corresponding data for the specific-heat jump and the results of the $c_e(T_0)$ computations for the NMP-TCNQ salt are presented in Table 7.3 together with the data for KCP and TTF-TCNQ. It can be seen from the table that in all the cases the specific-heat jump, $\Delta_c$, at the transition point is, in order of magnitude, equal to $c_e(T_0)$, i.e., the transitions have an electronic character (for NMP the susceptibility increases as the temperature is decreased from 200 to 7 K; therefore, for this salt the specific heat at the transition point is actually higher than the value found from the data for the paramagnetic susceptibility at 300 K).

The appearance of a dielectric gap at the transition point $T_0$ in NMP–TCNQ has also been detected from the optical absorption in this salt as a function of the frequency (in the 4 - 65-cm$^{-1}$ range) and the temperature (in the 4.2 - 35-K range) [20]. At low temperatures $T \lesssim T_0$, the frequency dependence of the absorption in this salt has a peak, the position of which shifts toward the region of lower frequencies as the temperature increases.

At high temperatures the paramagnetic susceptibilities of the TCNQ salts with NMP, Qn, and Ad can be described within the framework of the model of noninteracting electrons [73]. But the conductivity of these crystals is, on the basis of its temperature dependence, not metallic since it increases with increasing temperature in the 100 - 200-K temperature range. Therefore, it may be inferred that at high temperatures the interaction of the electrons is not essential for the description of the system (there is no gap in the single-electron spectrum) but the electrons are spatially localized as a result of the internal disorder of the system, and their mobility has an activation character [74].

Gogolin *et al.* [76] have constructed a theory that allows the description of the temperature dependence of the complex conductivity in a disordered metal. The effects of the disorder and the electron–phonon scattering lead to that temperature dependence of the complex conductivity that is experimentally observed in the metallic phase of the TCNQ salts with asymmetric cations [77] It is, however, possible that the Coulomb interaction between the electrons in the salts with asymmetric cations is not weak and that it affects the properties of the crystals even at high temperatures [78]. Indeed, the mutual Coulomb repulsion of the electrons can be strong in comparison with the other parameters of the system (the electron-bandwidth and the temperature), but the electron system will exhibit metallic properties if the number of electrons is not equal to the number of sites (i.e., if the band is not half-filled). In this situation, there is no gap for the single-electron excitations [79], and the kinetic properties are largely determined by their interaction with the lattice, as observed in [74, 76, 77]. But in this case the electrons do not encounter each other at one and the same site, so that their paramagnetic susceptibility can be described in

TABLE 7.3

| Compound | $T_0$ | $x_p$, cm$^3$/mol ($T$ = 300 K) | $\frac{\Delta c}{R}$ | $\frac{\Delta c}{\nu T_0}$ | Total specific heat $c(T_0)/R$ | $\Delta(0)$, K | References |
|---|---|---|---|---|---|---|---|
| NMP-TCNQ | 7.2 | $6 \cdot 10^{-4}$ | 0.2 | 5.15 | 3 | 25 | [5, 46] |
| TTF-TCNQ | $55 \div 60$ | $6 \cdot 10^{-4}$ | 0.3 | 1.1 | 14 | 120 | [7, 8, 31, 52, 57] |
| KCP | $T_p$ = 500 K | $4 \cdot 10^{-4}$ | – | – | – | 800 | [25, 27] |
| $(SN)_x$ | – | $0.1 \cdot 10^{-4}$ | – | – | – | – | [12, 72] |

the model of localized spins with the Heisenberg exchange interaction [80]. Let us also note that the temperature dependence of the thermo-emf of salts of the $Qn(TCNQ)_2$ type can be explained within the framework of the strong-Coulomb-interaction idea [81], whereas the noninteracting-electron model does not explain these experimental data. Thus, we now have theories that satisfactorily describe the thermodynamic properties of the low-temperature dielectric phase and the kinetic properties of the disordered-metal phase. The quantitative description of the dielectric and metallic phases is based on the notion of disorder in the system. Therefore, we can conclude that the properties of the salts with asymmetric cations are largely determined by the mutual Coulomb repulsion of the electrons and the disorder connected with the asymmetry of the cations. The correctness of the idea that the disorder in the salts with asymmetric cations plays an important role is also confirmed by the fact that the behavior of the TCNQ salts with symmetric cations above the metal–dielectric transition point does indeed turn out to be entirely different: Their conductivity increases monotonically as the temperature is decreased from room temperature to the point $T_0$, below which they go over into the Peierls-dielectric state.

## 4. The Peierls Transition in TCNQ Salts with Symmetric Cations

**A. Characteristics of the Structure.** In this section we consider those TCNQ salts in which metal–dielectric transitions of the Peierls type have been observed. Among them are the salts with the cations numbered 4 – 9 in Table 7.2. In contrast to the salts considered in the preceding section, the salts with the cations 4 – 9 go over into the diamagnetic-dielectric state as the temperature is lowered.

It is now an established fact that the metal–dielectric transition that occurs in TTF–TCNQ crystals is connected with the Peierls instability. Apparently, the Peierls instability is also characteristic of all the other salts with cations 5 – 9: Their magnetic and electric properties are qualitatively similar to the properties of the TTF–TCNQ salts.

In the compounds $K^+(TCNQ)_n^-$, where $K$ = TTT, TTF, and TSeF, there are two electrons for a cation molecule and $n$ TCNQ molecules (in the salts with the asymmetric cations NMP, Qn, and Ad, the corresponding number of electrons is one). These two electrons are distributed over the anion and cation chains. Thus, the transfer of electrons from the TTF to the TCNQ molecules in TTF–TCNQ can be seen from the photoemission spectrum of the electrons [82, 83], and the transfer ratio lies in the range 0.56 – 0.57 [84]. According to the x-ray structural-analysis data for room temperature, all the cation and anion molecules in $TTT(TCNQ)_2$ and TTF–TCNQ are equivalent [53, 60, 85, 86].

The cations TTF and TSeF are completely symmetric, and the TTF–TCNQ and TSeF–TCNQ crystals possess no internal disorder. The cations *cis*-ATTF and *cis*-DSeDTF are asymmetric, but the magnetic and electric properties of their salts are similar to the properties of the TCNQ salts with the symmetric cations TTF and TSeF [63, 87], and they will be considered in this section.†

**B. The Electronic Properties.** The anisotropy in the electron motion in the salts with symmetric cations is manifested in the electric and optical properties of the crystals and is quite strong. Thus, in TTF–TCNQ the conductivity for the direction along the chains (i.e., along the *b* axis) is several orders of magnitude higher than the conductivity for the transverse direction (i.e., along the *a* axis) ($\sigma_b/\sigma_a \approx 500$ at 300 K and $\sigma_b/\sigma_a \approx 10^4$ near the transition point [8, 88]). The permittivity has roughly the same degree of anisotropy in the SHF band (at 4.2 K, $\varepsilon_b \approx 3200$ and $\varepsilon_a \approx 6$ [19]). The anisotropy in the electron spectrum in TTF–TCNQ can be estimated from the data on the magnetic field dependence of the nuclear relaxation time $T_1$ [89, 90]. These estimates yield for the transverse resonance integrals values of the order of 10 - 30 K, and the quantum-mechanical calculations [91] yield roughly the same values for the integrals.

According to the optical data [92, 93], in the case of an electric-field vector polarized along the chains, the electrons of TTF–TCNQ behave like free electrons at frequencies, $\hbar\omega$, above 0.4 eV, and a plasma edge corresponding to a plasmon energy of 1.4 eV can clearly be seen in the reflection of light of this polarization. At the same time, the reflection for other directions of polarization indicates that the electron motion along the *a* and *c* axes has a dielectric character at frequencies, $\hbar\omega$, of up to at least 2 eV.

The conductivity of the TCNQ salts with TTT, TTF, TSeF, ATTF, and DSeDTF increases as the temperature is decreased from room temperature to the critical temperature, $T_0 \approx 55$ K, of the transition into the dielectric state (Fig. 7.8). For the TTF, TSeF, and DSeDTF the decrease of the resistivity $\rho$ as the temperature is decreased to the point $T_0$ can be described by the dependence

$$\rho(T) = \rho_0 + \rho_2 T^{\nu}, \quad \nu = 2.3 - 2.4. \tag{7.9}$$

the parameter $\rho_2$ depending little on the mode of preparation of the samples [63].‡ The dependence (7.10) leads to a peak conductivity that is higher than the conductivity at room temperature by roughly a factor of 30 [$\rho_2 \approx (1.8 - 2.9) \times 10^{-8}\ \Omega \cdot \mathrm{cm/K}\nu$].

According to the Pennsylvania group [8, 62, 63], some TTF–TCNQ crystals exhibit just before the point $T_0$ a conductivity growth that is more rapid than

†Undoubtedly, the random potential introduced by the cations *cis*-ATTF and *cis*-DSeDTF is small compared to the one produced by the asymmetric cations NMP, Qn, and Ad.

‡The dependence (7.9) with $\nu = 2$ describes the experimental conductivity data just as well.

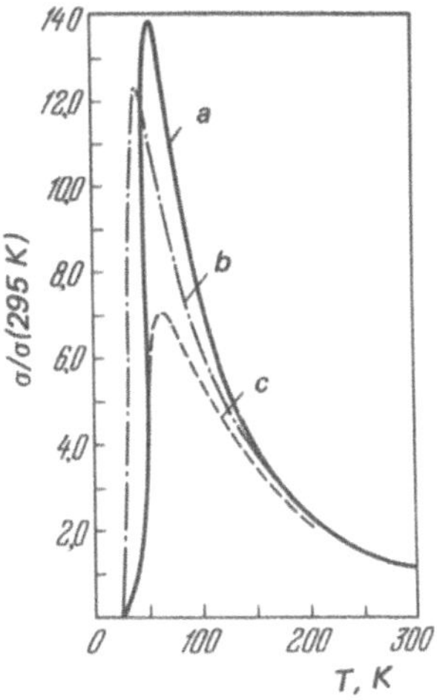


**Figure 7.8.** The temperature dependence of the relative conductivity $\sigma(T)/\sigma(295\ K)$ for (a) TTF-TCNQ; (b) TSeF-TCNQ; (c) DSeDTF-TCNQ [63].

the growth predicted by (7.9), and for some of them the ratio $\sigma(60\ K)/\sigma(300\ K)$ is as large as 150. For the time being, however, no other research group has been able to confirm this result, and the question of the existence of a giant conductivity peak remains open (see [94, 95]).

As the temperature is lowered below $T_0$, the conductivity falls off sharply, and the crystals become dielectrics with a gap for the single-electron and magnetic excitations. In TTF-TCNQ, according to the paramagentic-susceptibility measurements [87, 96] and the data for the infrared photoconductivity [97] and the dc electrical conductivity [8] in the SHF band [19], this gap is approximately 120 K wide.†

The temperature dependence of the electronic specific heat for typical TTF-TCNQ crystals is the same as that of the electrical conductivity: it is proportional to $T^{-2}$ above, and exponentially small below, $T_0$ [99]. In the metallic phase the Seebeck coefficient, $S_b$, for the direction along the chains in TTF-TCNQ is negative and proportional to the temperature. In the region around 60 K the Seebeck coefficient $S_b$ changes sign, in accord with the notion that the transition has a dielectric character. The Seebeck coefficient, $S_a$, for the direction along the $a$ axis also changes its sign at 60 K, but its temperature dependence indicates that the conductivity along the $a$ axis is not metallic at all temperatures below 300 K [100]. The paramagnetic susceptibility of $TTT(TCNQ)_2$ [60] and TTF-TCNQ [7, 86] decreases monotonically as the temperature is decreased below room temperature. In TTF-TCNQ this decrease is quite slow in the temperature range from 300 to 70 K but becomes more rapid below 60 K (Fig. 7.9).

†According to the optical data [98], the dielectric gap is much wider—about 0.14 eV.

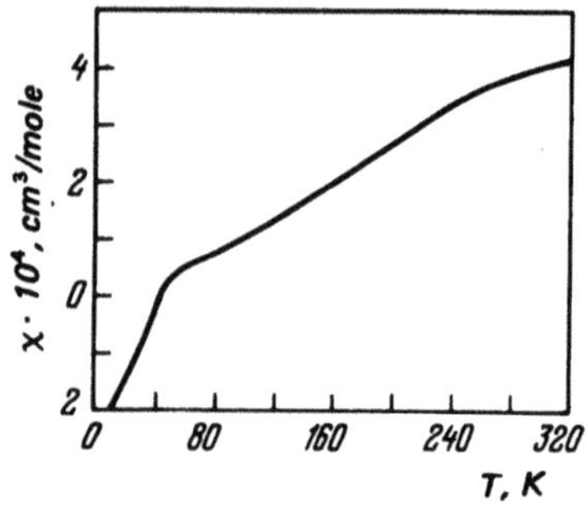


Figure 7.9. The temperature dependence of the magnetic susceptibility of TTF-TCNQ [86].

In TTF-TCNQ the temperature of the electrical-conductivity peak and the temperature at which the susceptibility begins to decrease rapidly are roughly the same; also equal are the electrical-conductivity and paramagnetic-susceptibility activation energies.

In $TTT(TCNQ)_2$ the conductivity peak occurs at 90 K (the peak conductivity exceeds the room-temperature value by roughly a factor of 3.5), while the susceptibility begins to decrease rapidly at about 55 K. It should be noted, however, that $TTT(TCNQ)_2$ crystals contain solvent molecules, and that these molecules are randomly distributed in the crystals. This circumstance can lead to the lowering of the conductivity peak and to the shifting of the peak temperature toward the region of higher temperatures (as in the salts with asymmetric cations).

**C. The Peierls Transition.** The decrease of the paramagnetic susceptibility below $T_0$ in salts of the type under consideration excludes the possibility of the transition's having a Mott–Hubbard character and shows that the low-temperature phase of the crystals corresponds to a dielectric with a completely filled conduction band.

There is no internal disorder in TCNQ salts with symmetric cations (in the *cis*-ATTF and *cis*-DSeDTF salts it is slight), and it is suggested in [62] that a metal-dielectric transition of the Peierls type occurs in the quasi-one-dimensional crystals of these compounds at the temperature $T_0$. The behavior of the magnetic susceptibility is consistent with this conjecture. The magnitude of the specific-heat jump at the point $T_0$ in TTF-TCNQ also indicates an electronic origin for the metal-dielectric transition in this compound [101] (see Table 7.3).

The hypothesis, put forward in [62], that the transition in TTF-TCNQ has a Peierls character has now been confirmed by direct x-ray studies of the superlattice at temperatures below 60 K [102, 103].

The most complete data on the lattice distortions that arise during the Peierls transition in TTF-TCNQ have now been obtained from the neutron-scattering data [104, 105]. Even at room temperature there is a very sharp dip in the dependence of the phonon energy on the quasimomentum $q$ at $q = 2k_F$. At 54 K a superlattice structure of periodicity 3.4$b$, where $b$ is the parent-lattice constant of the crystal in the direction of the chains, appears along the chains. Starting from 49 K, the lattice constant in the direction, $a$, perpendicular to the chains varies monotonically, and, as the temperature is lowered from 49 to 38 K, this lattice constant increases from $2a$ to $4a$, the final increase to $4a$ occurring discontinuously at 38 K. The lattice constant along the $c$ axis remains constant. A long-range order in the superlattice structure along the $b$ and $a$ axes is established below 45 K (below this temperature the width of the scattering peak for the superlattice does not exceed the resolving power of the apparatus). These transitions are described within the framework of the Landau theory in [90].

All these data apparently indicate that distortions occur in the TCNQ chain at 54 K, while the period of the TTF chains begins to change at 49 K. The interaction between the displacements in these chains leads to the variation of the lattice constant along the $a$ direction (planes of TCNQ chains alternate with those of TTF chains in this direction). Such a picture of the transition is consistent with the changes that have been found to occur in the position and width of the EPR line by Tomkiewicz *et al.* [106]. According to the data presented by these authors in [106], the magnitude of the $g$ factor changes, as the temperature is lowered, from a value characteristic of the TCNQ molecules to a value equal to the $g$ factor of the TTF molecule. Let us note that the 3.4$b$ superlattice constant corresponds to the transfer of 0.56 electrons from a TTF to a TCNQ molecule. According to the data of [107], a superlattice structure of periodicity 1.7$b$ has also been observed in TTF–TCNQ. The causes of the appearance of this superlattice are discussed in [107, 108].

Let us now discuss the question of the contribution of the various vibrations to the development of the Peierls instability in TTF–TCNQ crystals. An estimate is given in [31] for the contribution of the acoustic and intramolecular vibrations to the total dimensionless constant, $\lambda$, that determines the Peierls transition temperature [see (7.5)]. For the TCNQ molecule seven intramolecular molecule-symmetry oscillations with frequencies ranging from 334 to 1602 $cm^{-1}$ make a contribution to $\lambda$ roughly equal to 0.48 [the density of states was determined from the formula (7.7) with $p = 1$ and the value $b = 0.125$ eV, corresponding to a total band width of 0.5 eV]. At the same time, according to the estimates made in [31], the contribution of the acoustic vibrations is roughly equal to 0.08, i.e., the Peierls instability in TTF–TCNQ is largely due to the interaction of the conduction electrons with the intramolecular vibrations.†

The experimental verification of the conjecture that intramolecular distor-

†Let us note that, in principle, the Peierls transition in quasi-one-dimensional molecular crystals could also be connected with librational-displacement waves [109].

tions occur in TTF–TCNQ below the Peierls transition point is of great interest not only for the theory of the Peierls transition in molecular crystals. It is also very necessary for the prediction of the superconducting-transition temperature for crystals made up of organic molecules and possessing metallic conductivity (see Section 6).

Let us note that, because of their low $T_p$'s, crystals of the TTF–TCNQ type allow us to investigate experimentally the influence of crystal-structure defects on the properties of the Peierls transition. According to the theoretical ideas [49, 53, 54], in the case of a structural transition of the Peierls type, the introduction of disorder along the chains should lead to the lowering of the metal-dielectric transition temperature (as measured from the anomalies in the static, but not the kinetic, characteristics of the system). The accuracy of the published measurements of the magnetic susceptibility of ATTF–TCNQ and TTF–TCNQ is not sufficient for the determination of the transition-temperature shift that occurs under the influence of the disorder [101].

**D. On the Fröhlich Collective Mode in TTF–TCNQ.** The approximately thirtyfold increase that occurs in the conductivity as the temperature is lowered from room temperature to 60 K in the TTF–TCNQ salt can, in principle, be explained within the framework of the conventional one-electron transfer theory (although it is at present not entirely clear precisely what electron-scattering mechanism leads to the temperature dependence (7.9).†

On the basis of the conventional ideas, the growth of the conductivity according to the law (7.9) is in no way connected with the approach to the Peierls-transition point along the temperature axis: This growth is characteristic of the metallic state; it stops near the critical point for the structural transition into the dielectric state, and the conductivity falls off sharply upon further lowering of the temperature.

But the observed [52, 62, 110] hundredfold (and more) increase in the ratio $\sigma(T)/\sigma(300\ \mathrm{K})$ as $T \longrightarrow T_0$ cannot be explained by the one-electron transfer mechanism: The mean freepath of an electron near the conductivity-peak temperature becomes unrealistically long (of the order of $10^{-4}$ cm). Therefore, it is suggested in [27] that the growth of the conductivity of TTF–TCNQ crystals as the transition point is approached is because of the contribution of the conductivity of the Fröhlich collective mode in the fluctuation regime. According to estimates, this contribution for TTF–TCNQ crystals can attain values ranging from $10^5$ to $10^6$ mho/cm (the peak conductivity of the samples investigated in [62, 110] attained precisely such values). The possibility of such an interpretation of the peak in the conductivity makes the question of the existence of the Fröhlich collective mode in the Peierls-dielectric phase of TTF–TCNQ crystals especially interesting. It is shown in [34, 111] that the frequency of the op-

†In ordinary, slightly anisotropic systems, electron scattering by phonons leads to the dependence (7.9) with $\nu = 5$.

tically active Fröhlich mode in TTF–TCNQ is determined by the interaction between the CDW on the donor and acceptor chains, and this mode should be observed in the far-infrared region at temperatures below $T_0$.

Jacobsen *et al.* [98] have observed in TTF–TCNQ at 4.2 K in the 80-cm$^{-1}$ frequency region an absorption peak that disappears at 65 K. These authors interpret this peak as a collective-mode excitation. But even at 4.2 K the peak in the 80-cm$^{-1}$ region is, in contrast to the corresponding peak near 30 cm$^{-1}$ in KCP, very broad, and, in principle, another interpretation of its origin is possible (the frequencies of the intramolecular vibrations of TTF and TCNQ lie in this region). It is possible that the Fröhlich collective mode in TTF–TCNQ has now been detected by Eldridge and Bates in experiments on absorption at energies of 10 cm$^{-1}$ [112]. Let us also note that in many investigations the nonlinearities in the current-voltage characteristics of TTF–TCNQ [113] and a number of other quasi-one-dimensional and layered compounds with CDW [114, 115] are connected with the existence of nonlinear excitations for the CDW phase [116, 117]. But so far nonlinear solutions for the excitations of the CDW phase have been obtained only in the purely one-dimensional models, and in [116] the phase excitations in a disordered incommensurable lattice are considered, while in [117] soliton excitation by an electric field in a commensurable superlattice is investigated. None of the considered models describes the real situation for the incommensurable superlattice in TTF–TCNQ, where the pinning of the Fröhlich collective mode is due to the interaction between the CDW on the donor and acceptor chains [111]. Therefore, the interpretation of the nonlinear electric properties of the compounds with CDW within the framework of nonlinear phase excitations can hardly be considered to be justified until these excitations have been investigated in the many-chain model. Thus, for the time being the experimental proof of the existence of the Fröhlich collective mode in TTF–TCNQ both below the transition point and in the fluctuation regime can hardly be considered to be sufficiently reliable. Meanwhile, the question of the Fröhlich collective mode is, perhaps, one of the most interesting in the Peierls-transition phenomenon since it is precisely the existence of this collective mode that is unique for crystals with a superlattice structure.

## 5. Metallic Systems without Metal Atoms

The attempts of experimenters to produce compounds of the one-dimensional type with metallic conductivity in the ground state has so far been successful only in a few cases. Crystals of the polysulfur-nitride polymer $(SN)_x$, the charge-transfer HMTSF–TCNQ crystals [118], and the crystals of TSeT with bromine and chlorine $(TSeT)_2Cl$ [119] indeed remain metallic in their conductivity at very low temperatures. [The ground state of the $(SN)_x$ crystal is, as has already been noted above, superconducting with $T_c \approx 0.3$ K, and the conductivity in HMTSF–TCNQ below 4.2 K remains constant and roughly equal to 4000 mho/cm right down to 0.045 K.] These new metallic systems are being investigated with a view of obtaining answers to the following fundamental questions.

1. What is the specific nature of the electronic characteristics of metallic-type compounds without metal atoms?

2. To what extent are these compounds one dimensional with respect to their electronic properties, and why is it that they do not undergo a transition into the dielectric state?

3. Do the superconductors of this class possess some new superconducting properties, and on which characteristics does their critical temperature depend?

The HMTSF–TCNQ crystal is now the most thoroughly investigated of the compounds. Because of the large size of the cation, the HMTSF–TCNQ crystal is structurally different from salts of the TTF–TCNQ type, its lattice being characterized by a disorder in the direction perpendicular to the conducting axis. This crystal has been found to possess, even at room temperature, a one-dimensional-type superlattice structure of periodicity $2.7b$; a three-dimensional superlattice ordering has not, for the time being, been detected.

Above 50 K the temperature dependence of the resistivity is similar to (7.9) with $\rho_2 = 8.8 \times 10^{-7}\ \Omega \cdot \text{cm/K}\nu$ and $\rho^{-1} = (1.3\text{–}2.2) \times 10^3$ mho/cm at 300 K; at lower temperatures the conductivity decreases somewhat with decreasing temperature and then remains constant below 4.2 K. The Seebeck coefficient is constant and small in magnitude above 150 K, changes its sign at 150 K, and rapidly increases in absolute value when the temperature is lowered further.

Measurements of the resistivity anisotropy [120] and the magnetic-field dependence of the nuclear-relaxation time $T_1$ [89] show that the electronic structure of HMTSF–TCNQ is such that the conduction band is much wider along the $a$ direction than along the $c$ direction (the chains are stretched out along the $b$ axis). Thus, according to the data presented in [120], the ratio $\sigma_a/\sigma_c \geqslant 10$ and $\sigma_b/\sigma_c \approx 33$. Thus, in the case of HMTSF–TCNQ we deal with quasi-two-dimensional, rather than quasi-one-dimensional, crystals. It is precisely in this case that there occurs a structural transition with the formation of CDW and the electron system remains metallic as $T \longrightarrow 0$. This occurs because, as a rule, the CDW transition in two-dimensional and three-dimensional crystals closes only part of the Fermi surface, leading only to the worsening of the metallic properties and, in particular, to the lowering of the critical temperature of the superconducting transition or to the complete suppression of the transition (see Chapter 6). Apparently, it is precisely such a situation that is realized in $(TSeT)_2Cl$ [119] and the transition-metal trichalcogenides of the type $TaSe_3$ [121].

Let us now consider the properties of poly (sulfurnitride) crystals. The structure of the polymeric chain of $(SN)_x$ is shown in Fig. 7.10.

These chains are arranged parallel to each other in the crystal. The motion of the electrons along the $(SN)_x$ chain is metallic at all temperatures and at all

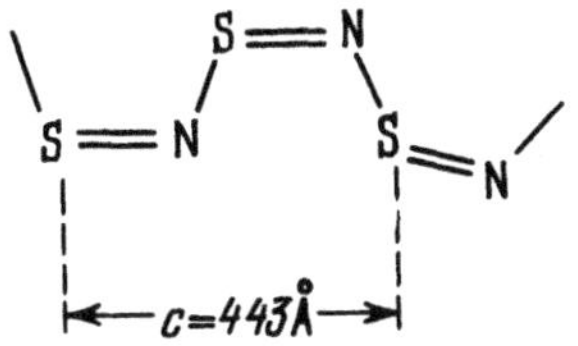


Figure 7.10. The $(SN)_x$ polymer chain [121].

investigated frequencies. A plasma reflection edge is observable in measurements of light reflection from the crystals at room temperature [126], and the plasma frequency in $(SN)_x$ is roughly equal to 2.4 eV. Right down to 1.5 K the specific heat contains a linear - in the temperature - term with coefficient $\gamma = 18 \times 10^{-6}$ J/g · K$^2$ [127], which corresponds to a density-of-states value at the Fermi surface of $N(0) = 0.35 \times 10^{34}$ erg$^{-1}$ · cm$^{-3}$ and a conduction-bandwidth of the order of 1 eV. According to DiSalvo's measurements [127], the paramagnetic susceptibility of the crystals is almost constant in the range from 300 to 4 K, and it roughly corresponds to the same $N(0)$ value as the coefficient $\gamma$ in the specific heat.

Crystals of $(SN)_x$ have now been produced that have a longitudinal conductivity of the order of $3 \times 10^3$ mho/cm at room temperature and $5 \times 10^5$ mho/cm at 5 K [125, 128, 129].

As in HMTSF-TCNQ, the conductivity of $(SN)_x$ crystals along the chains increases with decreasing temperature, attains a maximum, and remains constant (below 1 K), the ratio $\sigma(1\ K)/\sigma(300\ K)$ ranging from 6 to 1000 for different samples. The Seebeck coefficient in $(SN)_x$ crystals also duplicates qualitatively the temperature dependence that is observed in HMTSF-TCNQ [118]. Thus, the kinetic properties of $(SN)_x$ and HMTSF-TCNQ crystals are qualitatively similar and differ from the properties of ordinary metals without magnetic impurities. In the case of $(SN)_x$ the entire set of kinetic properties indicates that we are dealing with a semimetal, and the band-structure calculations confirm the correctness of this conjecture [129].

Let us now consider the data on the degree of anisotropy in the electronic properties of $(SN)_x$ crystals. The x-ray structural analysis data [122] show that the spacing between sulfur atoms of neighboring chains is 3.1 Å, and not 3.5 Å, as it should have been if the interchain interaction were a van der Waals interaction. Therefore, the overlap of the electron wave functions of neighboring chains in $(SN)_x$ crystals can be considerable. The band-structure calculations show that the anisotropy in the electronic spectrum is actually not very great [130 - 132]. Thus, according to the data given in [131], the change in the electron energy when the quasimomentum is varied in the direction perpendicular to the chains is about 0.5 eV for a total bandwidth of about 3 eV.

The optical experiments indicate that the plasma reflection edge is observable when the light is polarized in such a way that the electric vector is perpendicular to the chains, although it is not as sharp as when the polarization is in the parallel direction [133, 134]. The degree of anisotropy in the conductivity is roughly equal to 50 at room temperature, and it increases to 500 at 20 K. Apparently, this anisotropy in the conductivity of $(SN)_x$ samples is largely because of the filamentary structure of the crystals. According to the electron-microscope data, the diameter of a filament may be very small (as small as 200 Å) [128]. Poor contacts between the filaments lead to a reduction in the observed conductivity and an increase in its anisotropy. Thus, of the compounds considered above, $(SN)_x$ crystals have the lowest degree of anisotropy in their electrical and optical properties.

Apparently, it is precisely the three-dimensional effects that suppress the Peierls instability in $(SN)_x$ crystals [135]. The $(SN)_x$ crystal is essentially an ordinary three-dimensional crystal with a one-dimensional-type anisotropy: specifically, three-dimensional effects do not occur in it.

Let us now consider the superconducting properties of poly (sulfurnitride) crystals. As has already been noted above, the critical temperature of these crystals is roughly equal to 0.3 K at atmospheric pressure [12]; at a pressure of 9 kbar it is equal to 0.54 K [125]. This rapid growth with pressure is quite unusual: In the case of the nontransition metals the critical temperature decreases with increasing pressure. According to the estimates of Greene *et al.* [12], the 0.3-K critical temperature value is consistent with the Debye-temperature value (170 K) determined from the low-temperature specific-heat data and with the electron-phonon interaction constant ($\lambda \approx 0.3$) obtained from a comparison of the specific-heat coefficient $\gamma$ and the paramagnetic susceptibility ot the crystal. The Coulomb-interaction constant, $\mu^*$, in the expression for $T_c$ (see Chapter 2) was assumed to be roughly equal to 0.1, i.e., to the value characteristic of ordinary metals.

The Debye temperature changes insignificantly as the pressure is increased, and almost the entire increase in the quantity $T_c$ should be attributed to the increase in the electron-phonon interaction constant [125].

Measurements of the upper critical field, $H_{c2}$, of $(SN)_x$ crystals (they are type-II superconductors) yield a value of about 50 for the degree of anisotropy in the correlation length [128, 136]. For $T \ll T_c$, the upper critical field for the direction parallel to the filaments is anomalously high, and at $T = 0$, $H_{c2}(\parallel) = 8.1$ KOe. It is higher than the paramagnetic limit, $H_p$, for this compound, if we assume that the BCS relation between the gap $\Delta(0)$ and the quantity $T_c$ applies here ($T_c = 0.3$ K yields $H_p \approx 5$ KOe).†

†The Cooper pair consists of electrons with oppositely directed spins. In a magnetic field, because of the Zeeman splitting, these electrons have different energies, and this effect leads to the breakup of the pair. The corresponding field at $T = 0$ is $H_p = \Delta(0)/\mu_B\sqrt{2}$, where $\Delta(0)$ is the superconducting gap at $T = 0$. The paramagnetic effect limits the quantity $H_{c2}$ in the case when the orbital effect (the twisting of an electron in a magnetic field) turns out for some reason to be weakened. In $(SN)_x$ crystals this weakening of the orbital effect is connected with the anisotropy: The electron motion in the plane perpendicular to the chains is hindered.

The conventional suppression mechanism for the paramagnetic effect is connected with spin-orbit scattering on impurities. But $(SN)_x$ does not contain heavy atoms; therefore, the spin-orbit scattering evidently cannot lead to the strong suppression of the paramagnetic effect. The riddle of the excess of $H_{c2}$ over the paramagnetic limit is $(SN)_x$ can, apparently, be regarded as solved, following experiments on the direct measurement of the gap $\Delta(0)$ in $(SN)_x$ [137]. It turned out that $2\Delta(0)/T_c \approx 6$ in $(SN)_x$, and the paramagnetic limit is, in fact, not attained. But the gap measurements raise the problem of the explanation of so high a $\Delta(0)/T_c$ value in $(SN)_x$. In principle, the ratio $\Delta(0)/T_c$ should be large in the case of highly anisotropic quasi-one-dimensional superconductors since the fluctuations in the superconducting order parameter decrease the quantity $T_c$ to a much greater extent than they decrease $\Delta(0)$ (see below). But as has been noted above, there are no other experimental data confirming the presence in $(SN)_x$ crystals of that high degree of anisotropy, which is necessary for such an explanation.

According to the data given in [129], the temperature dependence of the fluctuation conductivity, $\Delta\sigma$, above $T_c$ in $(SN)_x$ crystals has a one-dimensional character $[\Delta\sigma \sim (T - T_c)^{-3/2}]$, which is apparently due to the filamentary structure of the crystals.

In conclusion of this section, let us note that the fabrication of the $(SN)_x$ superconductor demonstrated for the first time that superconductivity is possible in systems without metal atoms. Undoubtedly, this is the first success in the synthesis and investigation of compounds possessing one-dimensional anisotropy and superconducting properties. In the following section we shall discuss further prospects in this direction of search for high-temperature superconductors.

## 6. Superconductivity in Quasi-One-Dimensional and Organic Crystals

In Section 1 we noted three directions of investigation, which for some reasons or other related the hope for obtaining high-temperature superconductors to the fabrication of systems of the quasi-one-dimensional type. Below we consider these directions in greater detail, focusing our main attention on the correlation between the specific nature of the electronic properties of one-dimensional systems and superconductivity.

It has already been noted in Section 1 that, as a result of the Peierls instability, the interelectron Coulomb interaction, and the lattice disorder, the ground state of the electrons in quasi-one-dimensional crystals has s dielectric character. Of greatest interest from the point of view of superconductivity are crystals in which the effective interelectron Coulomb interaction is weak but the electron-phonon interaction is strong. In crystals with such properties the ground state of the electron system is determined by the competition between the Peierls and superconducting instabilities.

Generally speaking, the Peierls instability is an effect that is stronger than the Cooper instability. In fact, the Peierls transition is activated by the entire conduction band, while the superconducting transition is produced by only an energy range of the order of $\hbar\omega_0$ ($\hbar\omega_0 << E_F$, where $\omega_0$ is the frequency of the polarizable system and is roughly equal to $\omega_D$ for the electron-phonon mechanism of superconductivity.† Therefore, systems with a sufficiently strong anisotropy of the one-dimensional type should, as the temperature is lowered, undergo a Peierls transition, below the critical temperature of which the crystals become dielectrics, with the result that superconductivity is precluded [138]. This trend is broken when the Peierls instability turns out to be suppressed or when it can coexist with the superconducting instability.

Two cases in which such coexistence turns out to be possible are now being investigated. The first is realized in systems with a very strong anisotropy when $\hbar\omega_D > k_B T_p >> E_\perp$ (in the case of the exciton mechanism, $\omega_D$ should be replaced by the frequency, $\omega_0$, of the polarizable medium). Then the two instabilities turn out to be coupled to each other because of the specific properties peculiar to systems with one-dimensional electron motion [139-141]. The problem of the determination of the ground state in such a situation is mathematically very complex, and we do not as yet have a definite answer concerning the nature of the ground state.

The superconducting and Peierls states can also coexist in systems in which the interchain electron-transition process suppresses the Peierls instability, decreasing the Peierls-transition temperature and the magnitude of the dielectric gap to such an extent that Cooper pairing becomes energetically advantageous. This situation is considered in detail in Chapter 5, and we only note here that the electron motion in the systems in question should not be highly anisotropic, i.e., the conduction bandwidth, $E_\perp$, for the interchain electron motion should be of the order of the Peierls transition temperature. It is difficult now to say whether the conditions under which the Peierls and superconducting instabilities can coexist will ever be practially realized. It appears that greater success in the achievement of superconductivity can be expected in systems without the Peierls transition. As noted in the introduction, high critical temperatures could be provided in this case by the exciton mechanism or the normal phonon mechanism in compounds of large organic molecules.

The suppression of the Peierls instability is brought about by the interchain electron transitions, which destroy the condition for the superposability of parts of the Fermi surface [10, 135], the lattice disorder [49], and the presence at

†This argument is applicable not only to systems with the electron-phonon interaction, but also to systems in which the conduction electrons interact with electron-type excitations (e.g., with those that are considered in the treatment of the exciton mechanism of superconductivity). If the interaction with these excitations is stronger than the interaction with the phonons, then a static Peierls electronic-excitation wave of wave vector $2k_F$ develops.

the Fermi surface of many bands with different $k_F$ values [37]. In the first case we should try to raise the probability for the interchain electron transitions. This way seems at present to be the most realistic because it only requires the energy of the interchain electron motion to be bounded from below. But in point of fact this method leads to ordinary crystals with one-dimensional anisotropy.

A higher degree of anisotropy in the electronic properties is admissible in the many-band and lattice-disorder cases. Thus, to suppress the Peierls instability through the introduction of lattice disorder, it is sufficient to take a random potential $V_r \gtrsim T_{p_l}$, where $T_{p_l}$ is the transition temperature in the perfect lattice (with $V_r = 0$). In this case, in order for the delocalized character of the electron wave function to be preserved, the energy, $E_\perp$, of the interchain electron motion should be greater than the quantity $V_r^2/E_F$ obtained by Schante [118], and in such a situation the case when $E_\perp << T_{p_l}$ is possible. But according to calculations carried out by Larkin and Mel'nikov [142], superconductivity is apparently impossible in this case. In the case of several bands, it is necessary that the interaction between the Peierls displacements in different chains be sufficiently strong, otherwise the displacements in each chain will occur independently.†

Thus, the suppression of the Peierls instability (to a level where the Cooper pairing of the electrons can become energetically advantageous) occurs only in systems in which the quantity $E_\perp$ is not very small.

There is another factor that is hindering the fabrication of superconductors with very small values of the parameter $E_\perp$. This factor is the superconducting fluctuations. In Chapter 6, Section 5, we noted that the superconducting-transition temperature $T_c$ in quasi-two-dimensional systems practically does not depend on the degree of overlap of the electron wave functions of neighboring layers. We encounter a different situation in quasi-one-dimensional superconductors. Calculations that take account of the effect of the fluctuations on the temperature at which long-range superconducting order is established (see [37]) and calculations entirely analogous to the quasi-two-dimensional calculations [50] lead to the estimate $T_c \approx (E_\perp T_{c0}/k_B)^{1/2}$ for $E_\perp << k_B T_{c0}$ ($T_{c0}$ is the superconducting-transition temperature computed without allowance for the fluctuations, i.e., in the molecular-field approximation).

Thus, in synthesizing systems that are of interest from the point of view of superconductivity, we should not, apparently, aim at obtaining crystals with a very strong anisotropy of the one-dimensional type. Our aim should rather be the opposite if the problem is to raise the critical temperature of the superconducting transition by raising the frequency of the polarizable system.

Let us now briefly discuss the question of the prospects of the exciton mechanism in compounds of the quasi-one-dimensional type. Here we en-

†The situation with different $k_F$ values in the chains is realized in the complex salt TTT $(TCNQ)_2$, but the Peierls transition apparently still occurs in crystals of this salt.

counter essentially the same difficulty that arises in the case of layered compounds with intercalated molecules (see Chapter 6, Section 3), and that is that we do not as yet really know how to make the interaction of the electrons with the molecular electronic excitations sufficiently strong. In [144] Davis *et al.* carry out calculations that show that we can expect the exciton mechanism to be effective only in structures in which the conducting system is covalently bonded to the polarizable complexes and each atom of the conducting system is surrounded by more than three polarizable complexes. According to the computations for some specific hypothetical system, the exciton mechanism can give critical temperature values of the order of 100 K if the electron-phonon interaction turns out to be weaker than the electron-exciton interaction. But, apparently, the system considered in [144] is unstable since for small momentum transfers the exciton-mediated mutual attraction of the electrons is stronger than their mutual Coulomb repulsion (see Chapter 2). Therefore, more promising for the production of high-temperature superconductors is, apparently, the approach involving the synthesis of crystals with metallic conductivity from organic molecules or polymers in which the electrons interact as strongly as possible with the intramolecular vibrations.†

The frequencies of the intramolecular vibrations can be fairly high, and, from this point of view, they can very well replace those electronic excitations that are considered in the exciton mechanism of superconductivity. Let us estimate the values of $T_c$ that can be obtained on account of the intramolecular vibrations. The formula for the critical temperature with allowance for the acoustic and intramolecular vibrations has the form (see Chapter 2)

$$T_c = \frac{\hbar\overline{\omega}}{k_B} e^{-\frac{1+\lambda}{\lambda-\mu^*}}, \qquad \ln\overline{\omega} = \frac{1}{\lambda}\sum_s \lambda_s \ln\omega_s,$$

$$\mu^* = \frac{\mu}{1+\ln(E_0/\overline{\omega})}, \qquad \lambda = \sum_s \lambda_s, \tag{7.10}$$

where $\omega_s$ is the vibration frequency, $\lambda_s$ is the dimensionless vibration-electron interaction constant, and $E_0$ is the electron energy, which lies between the Fermi and plasma energies. Let us note that the formulas (7.10) were derived under assumptions that clearly are not valid in molecular crystals. Therefore, they can yield only rough estimates

For a chain of TCNQ molecules‡ with $E_0 \approx 1$ eV, we have on the basis of the data given in [31] that $\lambda \approx 0.56$ and $\overline{\omega} \approx 600$ K. For $\mu = \lambda$ the quantity

†Let us note that the use of intramolecular vibrations to raise the critical temperatures of superconductors was suggested by Kresin [145]. In this case ordinary metals with intercalated molecules are what Kresin had in mind.

‡We assume that the Peierls transition is suppressed by the three-dimensional effects, which do not very drastically change the conduction bandwidth of the TCNQ chain (of the order of 0.5 eV).

$T_c$ for the above-given values of the parameters is roughly equal to 6 K. Higher $T_c$ values can be obtained by increasing $\lambda$ or decreasing $\mu$. In principle, the situation when $\mu < \lambda$ is possible in conducting molecular crystals (see Chapter 2). However, a detailed quantitative analysis of the electron-phonon and Coulomb interactions in crystals of this type has not yet been published.

It follows from what has been said in this section that, in order to obtain organic superconductors, we must synthesize from organic molecules compounds in which the electron wave functions overlap sufficiently strongly in at least two crystallographic directions (i.e., crystals with as small an anisotropy of the one-dimensional type as possible). If superconductivity will be realized in such systems, then to raise the critical temperature, we must choose molecules in which the electrons interact as strongly as possible with the intramolecular vibrations and which are such that the effective Coulomb interaction between the conduction electrons will be as weak as possible in the crystals.

## Notes Added in Proof

The organic superconductors $(TMTSF)_2X$. $X = ClO_4$, $PF_6$, $AsF_6$, $SbF_6$, and $TaF_6$ are now synthesied [146]. For $X = ClO_4$ the critical temperature $T_c \approx$ 1.3 K at ambient pressure. The upper magnetic critical fields perpendicular to the chains are not very large in these compounds, so the overlapping of the electron wave function of neighboring chains TMTSF is large enough. Apparently, the usual three-dimensional anisotropic superconductivity is realized in this family [147], their anisotropy seeming to be close to quasi-two-dimensional superconductors (see Chapter 6).

# 8

# Superconducting Systems of the "Sandwich" Type

## 1. Introduction

Occupying a special place among the systems that are attracting a great deal of attention to themselves in connection with the possible manifestation of the exciton mechanism of superconductivity are systems of the sandwich type, i.e., thin metallic films surrounded by dielectric or semiconducting coatings.

In contrast to truly quasi-two-dimensional layered structures (considered in Chapter 6), in which the alternating conducting and nonconducting layers have thicknesses of the order of atomic (or molecular) dimensions, in systems of the sandwich type these layers are presumed to have macroscopic thicknesses. In such systems the conduction regions are spatially separated from the dielectric or semiconducting layers, with which the exciton excitations are associated. In this case the conditions for the existence of the excitons are slackened; in particular, they are more weakly damped in this case than in the bulk case (see Chapter 1).

Let us note that, below, under the term "exciton," we shall mean not only the collective electronic excitations of the molecules of the dielectric, or the electron-hole exictations in the semiconductor, but, generally, polarization excitations of all kinds and vibrational modes of both the bulk and surface types that arise in the electron subsystem near the metal–coating interface and have energies that are significantly less than the Fermi energy $E_F$.†

The exchange of excitations of this sort can, by assumption, lead to the strengthening of the effective attraction between the electrons of the superconductor and to a corresponding increase in the transition temperature [1–4]. Here the dielectric or semiconducting coatings can easily be changed, selecting materials with the optimum exciton characteristics. The last circumstance is extremely attractive from the experimental point of view, and therefore the

†We shall not consider the special case of systems having negative permittivity $\varepsilon < 0$ and characterized by excitation energies $\omega \gtrsim E_F$ (see Chapter 2).

investigation of such systems is of indubitable interest in the search for the exciton mechanism of superconductivity.

An obvious difficulty peculiar to systems with plane geometry is that the coupling between the conductive electrons and the excitons (or, in another language, the polarizations) existing in the dielectric or semiconducting layers turns out to be weaker than the coupling in three-dimensional systems or for thin one-dimensional filaments imbedded in a nonconducting matrix. If the thickness of a layer in a sandwich is sigificantly greater than the distance over which the screening of an external electromagnetic influence in a metal occurs, then most of the superconducting electrons will remain insensitive to the electrical polarizations that occur in the dielectric or semiconducting layer, as a result of which the effectiveness of the exciton mechanism will be significantly reduced. This can be seen, in particular, from the estimate [5, 6] of the effective-interaction constant $\lambda_{eff}$ in the case of an inhomogeneous metallic film. According to Cooper [5] and Kirzhnits and Maksimov [6], the constant $\lambda_{eff}$ is equal to the average value of the constants characterizing the various parts of the film:

$$\lambda_{eff} = (\lambda_1 L_1 + \lambda_2 L_2)/(L_1 + L_2), \tag{8.1}$$

where $\lambda_1$ is the constant for the interaction in the surface layer; $L_1$ the thickness of this layer; and $\lambda_2$ and $L_2$ are the corresponding quantities pertaining to the remaining part of the film. Therefore, in order for the effect of the surface layer on the effective interaction constant to be appreciable, it is necessary that the thickness of the whole metallic film be of the same order of magnitude as the thickness of the surface layer, i.e., that it be at any rate not greater than 10–20 Å.

The presence of exciton modes in the coatings leads not only to the possibility of varying the effective-interaction constant for the surface layer of the metal, but also to the variation of the characteristic frequency, $\overline{\omega}$, determining the critical temperature: $T_c \sim \overline{\omega} \exp(-1/\lambda_{eff})$ (see Chapters 1 and 2). Thus, in order to determine the effect that the dielectric or semiconducing coating has on the critical temperature of the metallic film, it is necessary, first, to estimate the value of the effective-interaction constant $\lambda_{eff}$ in the presence of exciton modes near the surface of the film and, second, to estimate the effective frequency, $\overline{\omega}$, that characterizes a system of the sandwich type.

The specific character of the present problem is the fact that the detailed study of the properties of the system near the metal–coating interface is of paramount importance. Unfortunately, the consistent microscopic theory of surface phenomena in systems with finite dimensions is extremely complex and not yet sufficiently developed. This makes it difficult in a number of cases to obtain reliable quantitative estimates from "first principles." In the majority of the theoretical papers devoted to the present problem, the authors use the phenomenological approach and various model arguments to one or another degree, making it possible to obtain only a qualitative picture.

We shall make the theoretical estimates on the basis of the formula for the critical temperature (cf. Chapter 3, as well as [7-9]):

$$T_c = \bar{\omega} \exp\left\{-\frac{1+\lambda}{\lambda-\mu^*}\right\}; \tag{8.2}$$

here

$$\ln \bar{\omega} = \frac{1}{\lambda} \frac{2}{\pi\Sigma} \sum_{\mathbf{k}_1, \mathbf{k}_2} \int_0^\infty d\omega \, \mathrm{Im}\, V_\omega(\mathbf{k}_1, \mathbf{k}_2) \frac{\ln \omega}{\omega}, \tag{8.3}$$

$$\lambda = -\frac{2}{\pi\Sigma} \sum_{\mathbf{k}_1, \mathbf{k}_2} \int_0^\infty \frac{d\omega}{\omega} \mathrm{Im}\, V_\omega(\mathbf{k}_1, \mathbf{k}_2), \tag{8.4}$$

and $\mu^*$ is the Coulomb "pseudopotential:"

$$\mu^* = \frac{\mu}{1 + \mu \ln\left(E_F/\bar{\omega}\right)}, \tag{8.5}$$

where $\mu$ is a parameter characterizing the Coulomb repulsion, $\Sigma = v_F S_F/(2\pi)^3$, $v_F$ is the velocity at the Fermi surface, $S_F$ is the Fermi surface area;

$$V_\omega(\mathbf{k}_1, \mathbf{k}_2) = \int\int d\mathbf{r}\, d\mathbf{r}' \psi^*_{\mathbf{k}_1}(\mathbf{r})\, \psi^*_{-\mathbf{k}_1}(\mathbf{r}') \times \\ \times V_\omega(\mathbf{r}, \mathbf{r}')\, \psi_{\mathbf{k}_2}(\mathbf{r}')\, \psi_{-\mathbf{k}_2}(\mathbf{r}), \tag{8.6}$$

$V_\omega(\mathbf{r}, \mathbf{r}')$ is the Fourier transform of the retarded interaction between two electrons located at the points $\mathbf{r}$ and $\mathbf{r}'$:

$$V(\mathbf{r}, \mathbf{r}', t) = \frac{1}{2\pi} \int_{-\infty}^{\infty} V_\omega(\mathbf{r}, \mathbf{r}')\, e^{i\omega t} d\omega, \tag{8.7}$$

the $\psi_{\mathbf{k}}(\mathbf{r})$ are the electron wave functions (the spin indices have been dropped).

We shall consider layered systems in which the major role is played by the two types of interactions responsible for superconductivity: the electron-phonon and electron–exciton interactions.† Below we shall assume that the energies of the phonon and exciton excitations in the system are much lower than the Fermi energy $E_F$.‡ Therefore, in considering the interaction of the electrons through the exchange of phonons and excitons, we can assume that the momenta of the interacting particles lie on the Fermi surface. In view of this, the summation in (8.3), (8.4) is performed over the states lying on this surface.

The expression (8.4) corresponds to the averaged–over the Fermi surface–coupling constant characterizing the interaction of the electrons with the elementary excitations. It can easily be seen that if the integration in the formula (8.4) is performed over the phonon frequency region only, then we obtain the normal expression for the electron-phonon coupling constant [10]:

$$\lambda_{\mathrm{ph}} = -\frac{2}{\pi \Sigma} \sum_{\mathbf{k}_1, \mathbf{k}_2} \int_0^{\tilde{\omega}_{\mathrm{ph}}} \frac{d\omega}{\omega} \operatorname{Im} V_\omega(\mathbf{k}_1, \mathbf{k}_2), \tag{8.8}$$

where $\tilde{\omega}_{\mathrm{ph}}$ is the maximum frequency of the phonon spectrum.

The constants characterizing the coupling of the electrons to the other excitations are expressed by similar formulas if the integration over $\omega$ is performed over the frequency regions characteristic of these excitations (we assume that these regions do not overlap). In particular, for the electron–exciton interaction constant we have

$$\lambda_e = -\frac{2}{\pi \Sigma} \sum_{\mathbf{k}_1, \mathbf{k}_2} \int_{\tilde{\omega}_{\mathrm{ph}}}^{\tilde{\omega}_e} \frac{d\omega}{\omega} \operatorname{Im} V_\omega(\mathbf{k}_1, \mathbf{k}_2), \tag{8.9}$$

where $\tilde{\omega}_e$ is the maximum frequency of the exciton spectrum, the total coupling constant $\lambda$ being the sum of the corresponding quantities: $\lambda = \lambda_{\mathrm{ph}} + \lambda_e$.

The interaction energy $V(\mathbf{r}, \mathbf{r}', t)$ in the case of an inhomogeneous system of the sandwich type depends, generally speaking, on the coordinates of the two points, $\mathbf{r}$ and $\mathbf{r}'$, in contrast to the case of an isotropic homogeneous medium,

†We shall not explicitly consider the volume plasma oscillations in the electron subsystem. These oscillations play an important role, leading as they do to the screening of the Coulomb interaction and the logarithmic suppression of the repulsion [see (8.5)], but plasmon exchange cannot by itself change the sign of the interaction and secure superconducting pairing (for greater details, see Chapters 1 and 2).

‡Otherwise the Coulomb repulsion will not be suppressed (see Chapter 1 and 2 for greater details); as typical values we can assume: $\omega_{\mathrm{ph}} \sim (1\text{–}3) \times 10^{-2}$ eV, $\omega_e \sim 0.4\text{–}0.5$ eV, and $E_F \sim 5\text{–}10$ eV.

when the energy is a function of the difference $\mathbf{r} - \mathbf{r}'$. This circumstance is also reflected in the constitutive equations for the electromagnetic field, which relate the induction $\mathbf{D}$ to the electric-field intensity $\mathbf{E}$ [11]:

$$D_i(\mathbf{r}, t) = \int\limits_{t'<t} dt' \int d\mathbf{r}' \varepsilon_{ij}(\mathbf{r}, \mathbf{r}', t - t') E_j(\mathbf{r}', t'),$$
$$D_i(\mathbf{r}, t) = E_i(\mathbf{r}, t) + 4\pi \int\limits_{t'<t} dt' \, j_i(\mathbf{r}, t'), \tag{8.10}$$

where $\mathbf{j}$ is the current in the system.

The quantity $\varepsilon_{ij}$ in (8.10) would depend only on the difference, $\mathbf{r} - \mathbf{r}'$, of the coordinates if the medium were homogeneous, but in the presence of a boundary the effective permittivity tensor $\varepsilon_{ij}$ will also depend in the general case on the two coordinates: $\mathbf{r}$ and $\mathbf{r}'$. The problem of the microscopic treatment is the determination of the form of the interaction potential (8.7) with allowance for the boundary, or, equivalently, the derivation of an expression for either $\varepsilon_{ij}(\mathbf{r}, \mathbf{r}', t)$, or the conductivity tensor, $\sigma_{ij}$, of the medium (see [12, 13]):

$$\varepsilon_{ij}(\mathbf{r}, \mathbf{r}', t) = \delta(t)\, \delta(\mathbf{r} - \mathbf{r}')\, \delta_{ij} + 4\pi \int\limits_{t'<t} dt' \sigma_{ij}(\mathbf{r}, \mathbf{r}', t).$$

Let us note that in some cases the computation of $\varepsilon_{ij}$ and $\sigma_{ij}$ with the aid of the kinetic equation for a bounded system leads to expressions that depend essentially on only one space variable. Thus, for example, in the problem of the anomalous skin effect in a metal [14], in which the conductor is assumed to occupy the half-space $z > 0$, the response function depends only on $|z + z'|$ or $|z - z'|$, depending on the character of the reflection from the surface. In the general case, however, we cannot expect such a simplification, and the problem of finding the interaction in a bounded system is an extremely complex one.

In the phenomenological description, the properties of the medium can be specified by choosing one or another concrete form of the functions $\varepsilon_{ij}(\mathbf{r}, \mathbf{r}', t)$. If we write the matrix element of the interaction (8.6) with the aid of an effective permittivity, $\varepsilon(\mathbf{k}, \mathbf{k}', \omega)$ (for the sake of simplicity, we limit ourselves to the isotropic case, for which $\varepsilon_{\mu\nu} = \varepsilon\delta_{\mu\nu}$).

$$V_\omega(\mathbf{k}_1, \mathbf{k}_2) = \frac{4\pi e^2}{(\mathbf{k}_1 - \mathbf{k}_2)^2 \, \varepsilon(\mathbf{k}_1 - \mathbf{k}_2, \mathbf{k}_1 + \mathbf{k}_2, \omega)},$$

then the formulas (8.3) and (8.4) can be expressed directly in terms of the permittivity:

$$\ln \bar{\omega} = \frac{1}{\lambda}\frac{2}{\pi}\int \frac{dS_{\mathbf{k}}}{v_F}\int d\omega \frac{\ln \omega}{\omega}\frac{4\pi e^2}{k^2}\,\mathrm{Im}\,\frac{1}{\bar{\varepsilon}(\omega, \mathbf{k})}, \tag{8.11}$$

$$\lambda = -\frac{2}{\pi}\int \frac{dS_{\mathbf{k}}}{v_F}\int \frac{d\omega}{\omega}\frac{4\pi e^2}{k^2}\,\mathrm{Im}\,\frac{1}{\bar{\varepsilon}(\omega, \mathbf{k})}, \tag{8.12}$$

where $dS_{\mathbf{k}}$ is an element of area of the Fermi surface of the metal, $\mathbf{k} = \mathbf{k}_1 - \mathbf{k}_2$ is the momentum transferred during the interaction, $\bar{\varepsilon}(\omega, \mathbf{k})$ is the permittivity averaged over the resultant momentum, $\mathbf{k}_1 + \mathbf{k}_2$, on the Fermi surface [such an averaging occurs automatically in the summation over one of the variables, $\mathbf{k}_1$ or $\mathbf{k}_2$, in (8.2)-(8.9)]:

$$\bar{V}_\omega(\mathbf{k}) = \left\langle \frac{4\pi e^2}{k^2 \varepsilon(\mathbf{k}, \mathbf{k}_1 + \mathbf{k}_2, \omega)} \right\rangle_{\mathbf{k}_1+\mathbf{k}_2} = \frac{4\pi e^2}{k^2 \bar{\varepsilon}(\omega, \mathbf{k})}. \tag{8.13}$$

It can be seen from this that, knowing the effective permittivity $\bar{\varepsilon}(\omega, \mathbf{k})$, we can compute the parameters determining the critical temperature, (8.2), of the system. In this case the zeros of $\bar{\varepsilon}(\omega, \mathbf{k})$ determine the possible excitation modes of the system, while the peaks of the imaginary part of $1/\bar{\varepsilon}(\omega, \mathbf{k})$ determine the contribution of these excitations to the magnitude of the coupling constant $\lambda = \lambda_{\mathrm{ph}} + \lambda_e$, where $\lambda_{\mathrm{ph}}$ and $\lambda_e$ are given by the formulas (8.8) and (8.9).

The effectiveness of the exciton mechanism of superconductivity in a sandwich can be judged from the change in the value of the critical temperature of the metallic film upon the deposition of the coating. In order to simplify the estimates, we shall assume that $\omega_{\mathrm{ph}}$ and $\lambda_{\mathrm{ph}}$ do not change as a result of the deposition of the coating,† and that the entire change in $T_c$ is only because of the switching on of the electron–exciton interaction. In the specific computations carried out below, we shall assume everywhere that the metallic film is characterized by a typical—for superconducting materials—electron-phonon coupling constant value $\lambda_{\mathrm{ph}} \sim 0.5$-$1.0$. At the same time, in the simplified models considered below the electron–exciton coupling in sandwiches turns out to be quite weak, with $\lambda_e \lesssim 0.1$-$0.2$, i.e., $\lambda_e \ll \lambda_{\mathrm{ph}}$; we shall assume that the latter condition is satisfied. In view of this, the mean logarithmic frequency [7, 8] computed from the formula (8.11) turns out to be equal to

$$\bar{\omega} = \omega_{\mathrm{ph}}\left(\frac{\omega_e}{\omega_{\mathrm{ph}}}\right)^{\lambda_e/(\lambda_e+\lambda_{\mathrm{ph}})} \sim \omega_{\mathrm{ph}} \ll \omega_e, \tag{8.14}$$

†Let us note that in the practical preparation of thin films, the deposition of a coating is, generally speaking, accompanied by a change in the crystal structure of the film with a corresponding change in $\omega_{\mathrm{ph}}$. In the theoretical estimates presented below, it is assumed that no such change occurs in the crystal structure and the phonon spectrum of the film.

i.e., to be much lower than the optimum frequency $\overline{\omega}_{opt}$ (see Chapter 2). In this case we can show that the increase in $T_c$ as a result of the deposition of the coating is largely because of the increase in $\overline{\omega}$; and the change in the index of the exponential function in (8.2) can be neglected. Therefore, in the numerical computations carried out below we shall use the simple formula [8]

$$\frac{(T_c)_{\text{coat}}}{(T_c)_{\text{without coat}}} \approx \left(\frac{\omega_e}{\omega_{\text{ph}}}\right)^{\lambda_e/(\lambda_e+\lambda_{\text{ph}})}, \tag{8.15}$$

which allows us to estimate the effect of the coating on the critical temperature of the film.

The results of some specific computations of the interaction energy $\overline{V}_\omega(\mathbf{k})$ in films, results that allow us to find $\lambda_e$ and roughly estimate the change that can occur in the critical temperature of a sandwich as a result of the presence of the exciton mechanism, are presented in Section 2. As will be seen from the exposition, such estimates are tied with a number of simplifying assumptions, and cannot, therefore, be regarded as entirely reliable. Furthermore, in Section 3 we indicate a number of additional effects, the comprehensive allowance for which may turn out to be important in the construction of a more complete theory of superconducting layered structures.

But despite their sketchiness, the presently available theoretical computations are useful since they give the necessary orientation in a discussion of the properties of such systems.

## 2. Estimates of the $T_c$ for "Sandwiches" in Certain Models

Recently a number of authors have considered two somewhat differing mechanisms that can, generally speaking, lead to an increase in the critical temperature of sandwiches.

The first, originally proposed by Ginzburg [1] and analyzed in greater detail in [8, 15, 16], is connected with the penetration of polarization waves from the dielectric or semiconducting coating into the surface layer of the metal (the polarization mechanism). Because of polarization-wave (exciton) exchange, the conduction electrons in the metallic film experience additional mutual attraction (in comparison with the normal attraction because of phonon exchange), which facilitates the formation of Cooper pairs and the increase of $T_c$.

According to the other variant, first considered by Allender *et al.* [17], an additional increase in the binding energy of the Cooper pairs occurs as a result of the interaction of the conduction electrons (which tunnel from the metallic film over some distance into the material of the coating) with the electronic excitations existing in the coating (the tunneling mechanism).

Below we shall discuss in detail the possibility of raising $T_c$ in sandwiches with the aid of each of these mechanisms.†

**A. The Polarization Mechanism.** It has been noted above that the problem of computing the $T_c$ of thin films amounts to that of finding the interaction energy of two electrons with allowance for the film boundary and the properties of the material of the coating.

Because of the complexity of the problem, it is expedient to use a simple electrodynamic approach that provides a correct qualitative and semiquantitative description of many of the properties of the system of interest to us. In this approach, the polarizability of each of the media is described right up to the very boundary with the aid of the corresponding volume permittivity $\varepsilon(\omega, \mathbf{k})$.‡

In the case of an isotropic unbounded medium, the Fourier transform of the interaction energy (8.7) can be written in the form

$$V_\omega(\mathbf{r}-\mathbf{r}') = e\varphi_\omega^0(\mathbf{r}-\mathbf{r}') = \int \frac{4\pi e^2}{k^2} \frac{e^{i\mathbf{k}(\mathbf{r}-\mathbf{r}')}}{\varepsilon(\omega, \mathbf{k})} \frac{d\mathbf{k}}{(2\pi)^3}, \tag{8.16}$$

where $\varphi_\omega^0(\mathbf{r}-\mathbf{r}')$ is the electric-field potential.

In the case when a plane interface is present (at $z = 0$), it is convenient to consider the Fourier expansion of the interaction energy in terms of the coordinates whose axes are directed along the $z = 0$ plane. It is not difficult to see that the corresponding Fourier component of the interaction can be written in the form

$$V_\omega(\varkappa, z, z') = \int \frac{4\pi e^2}{q^2+\varkappa^2} \frac{e^{iq(z-z')}}{\varepsilon(\omega, \varkappa, q)} \frac{dq}{2\pi} + \sum_i C_i(\omega, \varkappa, z') e^{iq_i z} \equiv$$

$$\equiv e\varphi_\omega^0(\varkappa, z-z') + e\delta\varphi_\omega(\varkappa, z, z'), \tag{8.17}$$

where the wave vector $\mathbf{k} \equiv \{\varkappa, q\}$ has a component, $\varkappa$, lying parallel to the $z = 0$

†Let us note that a more general type of electron interaction, to which the two indicated mechanisms contribute, should undoubtedly occur in real superconducting films.

‡It should be noted that the band structure of the material near the surface can differ significantly from the band structure of bulk samples. Thus, for example, it is known that surface (Tamm) electron states exist at a (111) surface of silicon in a layer of thickness of the order of 5 Å and that the ions are slightly shifted from the crystal-lattice sites [18, 19]. Such electron-spectrum reconstruction and lattice distortion lead, of course, to an appreciable change in the permittivity of the surface layer.

Since we are interested solely in the polarization properties of the material of the coating in the immediate neighborhood of the surface, in such cases we shall by the term "permittivity of the coating" always mean the effective permittivity of the surface layer of the coating; it is only this quantity that will enter into all the formulas.

plane and a component, $q$, oriented in the direction perpendicular to this plane, while the quantities $q_i = q_i(\omega, \varkappa)$ satisfy the equation

$$[q_i^2(\omega, \varkappa) + \varkappa^2]\,\varepsilon(\omega, \varkappa, q(\omega, \varkappa)) = 0. \quad (8.18)$$

The electric-field potential $\varphi^0_\omega$ in (8.17) corresponds to an unbounded medium, while the term $\delta\varphi_\omega$ reflects the effect of the surface on the interaction and decreases rapidly with distance from the $z = 0$ plane. From the mathematical point of view, $\delta\varphi_\omega$ is an arbitrary linear combination of the solutions to the homogeneous differential equation for the potential that must be added to the solution to the inhomogeneous linear equation in order to obtain the general solution. The arbitrary coefficients $C_i$ entering into (8.17) should be determined from the boundary conditions.

Following the conventional electrodynamic approach,† we naturally require that the well-known continuity conditions for the potential and normal component of the electric induction be satisfied at the boundary:

$$\varphi|_{z\to+0} = \varphi|_{z\to-0}, \quad D_z|_{z\to+0} = D_z|_{z\to-0}. \quad (8.19)$$

In the case when Eq. (8.17) has more than one solution, we need for the determination of the coefficients $C_i$ more boundary conditions in addition to (8.19) [13].‡ Below we shall limit ourselves to relatively simple functions for the permittivity, the use of which makes the introduction of additional boundary conditions unnecessary.

The above-described method can be used to determine the interaction between two electrons located in the metal at the points $\mathbf{r}$ and $\mathbf{r}'$ near a boundary with a dielectric (we assume that the metal fills up the half-space $z < 0$). Since the frequencies figuring in the present problem satisfy the condition $\omega \ll \omega_p$ ($\omega_p \sim 20$ eV is the plasma frequency of the metal), we can neglect the frequency dependence in the expression for the permittivity of the metal and use the simple expression that takes account only of the screening properties of the electron gas:

$$\varepsilon_m(\omega, \mathbf{k}) = 1 + k_{Dm}^2/k^2, \quad (8.20)$$

where $k_{Dm}$ is the reciprocal screening length in the metal.

†Let us note that the application of the concepts of macroscopic electrodynamics to objects with characteristic dimensions equal to a few atomic distances is far from being an obvious procedure, and the description of an inhomogeneous system in terms of an effective permittivity can lead to a corresponding error in the estimates. The use in this case of boundary conditions of the type (8.19) also requires additional justification.

‡Such additional conditions are introduced in an implicit form in, for example, [15]. Let us also note that boundary conditions somewhat different from (8.19) are used in [8, 15], but that these differences are unimportant for the problems considered below.

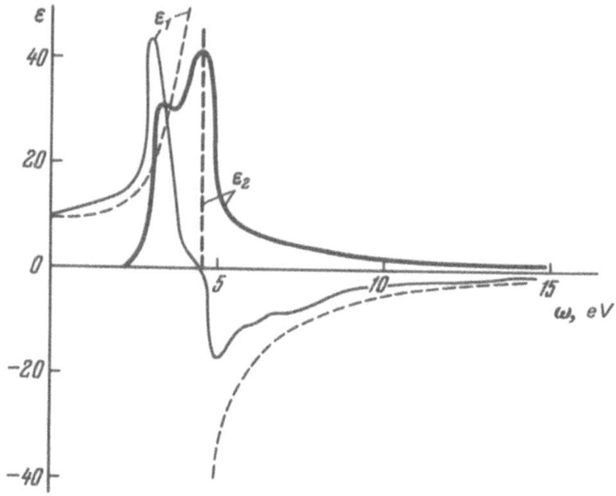


**Figure 8.1. Dependence of the permittivity $\varepsilon = \varepsilon_1 + i\varepsilon_2$ on the frequency $\omega$ for Si. The continuous curve is the experimental curve [18, 19]; the dashed curve is the plot of the model expression (8.21) with $\varepsilon_\infty = 1$, $\varepsilon_0 = 12$, $\omega_1 = 5.1$ eV.**

As to the material of the coating, below we shall, for the sake of simplicity, neglect the dependence of the permittivity of the coating on **k** (i.e., we shall neglect spatial dispersion), and assume that $\varepsilon_{co}(\omega, \mathbf{k}) = \varepsilon_{co}(\omega, 0)$. Optical measurements allow us to determine directly the frequency dependence of the permittivity $\varepsilon_{co}(\omega, 0)$. For many materials $\varepsilon_{co}(\omega, 0)$ is well known (see, for example, [20]). For the purpose of carrying out the computations in analytic form, however, instead of the experimental dependence, it is convenient [21] to use the model expression

$$\varepsilon_{co}(\omega, 0) = \varepsilon_\infty + \frac{(\varepsilon_0 - \varepsilon_\infty)\,\omega_1^2}{\omega_1^2 - \omega^2 - i\delta} \tag{8.21}$$

[here $\varepsilon_\infty$ is the value of the permittivity $\varepsilon_{co}(\omega, 0)$ at frequencies higher than the frequency of the exciton excitation; $\varepsilon_0 \equiv \varepsilon_{co}(0, 0)$]. The expression (8.21) satisfactorily reproduces the distinctive features of the experimental curve $\varepsilon_{co}(\omega, 0) = \varepsilon_1 + i\varepsilon_2$ in the resonance region (see Fig. 8.1, where we show the data for Si [18, 19]).

Using (8.20) and (8.21), we obtain the following expression for the interaction energy [8, 16]:

$$V_\omega(\varkappa, z, z') = e\varphi_\omega(\varkappa, z, z') =$$

$$= \int \frac{4\pi e^2 e^{iq(z-z')}}{k_D^2 + q^2 + \varkappa^2}\frac{dq}{2\pi} + \frac{4\pi e^2}{2\sqrt{\varkappa^2 + k_D^2}}\,\frac{\varepsilon(\omega)\varkappa - \sqrt{\varkappa^2 + k_D^2}}{\varepsilon(\omega)\varkappa + \sqrt{\varkappa^2 + k_D^2}} \times$$

$$\times \exp\left\{(z + z')\sqrt{\varkappa^2 + k_D^2}\right\}, \tag{8.22}$$

where we have introduced the notation: $\varepsilon(\omega) \equiv \varepsilon_{co}(\omega, 0)$ and $k_D \equiv k_{Dm}$.

The first term in (8.22) represents the normal screened Coulomb repulsion, while the second describes the penetration of the polarization waves from the dielectric into the metal. It can be seen from the expression (8.22) that the effect of the coating is to make possible the appearance in the metallic film's surface layer of excitations with frequency, $\omega_e(\varkappa)$, determined by the equation

$$\varepsilon(\omega)\,\varkappa + \sqrt{\varkappa^2 + k_D^2} = 0. \tag{8.23}$$

These excitations penetrate into the metal to a depth of $1/k_D$ ($\lesssim 1$ Å for typical metals).

If $\varepsilon(\omega)$ is given by the formula (8.21), then the frequency of such exciton excitations can be expressed in terms of the parameters $\varepsilon_0$, $\varepsilon_\infty$, and $\omega_1$ as follows:

$$\omega_e(\varkappa) = \omega_1 \left[ \frac{\varepsilon_0 \varkappa + \sqrt{\varkappa^2 + k_D^2}}{\varepsilon_\infty \varkappa + \sqrt{\varkappa^2 + k_D^2}} \right]^{1/2}. \tag{8.24}$$

When the parameters are chosen as follows: $\omega_1 \sim 0.2$–$0.3$ eV, $\varepsilon_\infty = 2$–$5$, and $\varepsilon_0 = 5$–$15$, the frequency of the exciton excitations for $\varkappa = k_D$ falls within the favorable–from the point of view of supercondictivity–frequency range from 0.4 to 0.8 eV (cf. the footnote on p. 286).

It is not difficult now to estimate with the aid of the formulas (8.2)–(8.15) the effect that the coating can have on the critical temperature of the metallic film. As the interaction energy $V_\omega(\mathbf{r}, \mathbf{r}')$ entering into these formulas, we can use the expression (8.22); and, in computing the matrix element (8.6), we can allow for the finiteness of the film thickness by performing the integration over space in (8.6) within the limits $0 \leqslant z \leqslant L$, where $L >> k_D^{-1}$ is the film thickness. Here the coating is assumed to be on only one side of the metallic film. Let us, for simplicity, first assume that, in spite of the presence of the boundary, the electron wave functions are plane waves and that, as usual, electrons with oppositely directed momenta and spins participate in the superconducting pairing. Under such simplifying assumptions the interaction matrix element (8.6) depends only on the momentum transfer $\mathbf{k} = \mathbf{k}_1 - \mathbf{k}_2$:

$$V_\omega(\mathbf{k}) = \frac{4\pi e^2}{k^2 + k_D^2} \frac{1}{\mathcal{V} L} \left[ L + \frac{\varepsilon(\omega)\,\varkappa - \sqrt{\varkappa^2 + k_D^2}}{\varepsilon(\omega)\,\varkappa + \sqrt{\varkappa^2 + k_D^2}} \frac{\sqrt{\varkappa^2 + k_D^2}}{2\left(k^2 + k_D^2\right)} \right], \tag{8.25}$$

where $\mathcal{V}$ is the volume of the film, $L$ its thickness.

For the purpose of estimating the average–over the Fermi surface–interaction strength, (8.13), entering into the formulas (8.11) and (8.12), it is convenient to use the mean-value theorem:

$$\int \frac{dS_{\mathbf{k}}}{v_F} f(\mathbf{k}) = f(\bar{k}) \frac{S}{v_F} = f(\bar{k}) N(E_F).$$

Below we shall assume that $\bar{k} = k_F$ and $\varkappa = (2k_F/3)^{1/2}$, which yields the correct order of magnitude of the integrals to be computed (this can be verified by exact calculations). Furthermore, we shall assume that the value of the parameter $\mu^*$ in (8.5) in the film differs little from the value in the bulk metal.

In the free-electron gas approximation† we have for the density of electron states at the Fermi surface the expression

$$N(E_F) = \frac{3}{4} \frac{Z}{E_F} \frac{1}{\Omega_a} = \frac{k_D^2}{8\pi e^2}, \qquad k_D^2 = \frac{4}{\pi} \frac{k_F}{a_B}, \tag{8.26}$$

where $Z$ is the valency of the atom, $\Omega_a$ is the atomic volume, and $a_B$ is the Bohr radius.

As a reuslt, for the electron–exciton coupling constant $\lambda_e$ we obtain the formula

$$\lambda_e = \frac{1}{2L} \frac{k_D^2 (\bar{\varkappa}^2 + k_D^2)}{(\bar{k}^2 + \bar{k}_D^2)^2} \frac{(\varepsilon_0 - \varepsilon_\infty)\bar{\varkappa}}{\left(\varepsilon_\infty \bar{\varkappa} + \sqrt{\bar{\varkappa}^2 + k_D^2}\right)\left(\varepsilon_0 \bar{\varkappa} + \sqrt{\bar{\varkappa}^2 + k_D^2}\right)}. \tag{8.27}$$

Given the typical parameter values: $k_F = 1.5$ Å$^{-1}$, $L = 10$ Å, $\varepsilon_0 = 23$, $\varepsilon_\infty = 3$, we obtain $\lambda_e \approx 5 \times 10^{-3}$, i.e., a very small value. If we take $\omega_e = 0.8$ eV and $\omega_{ph} = 0.02$ eV, $\lambda_{ph} = 0.5$, then the increase in $T_c$ resulting from the deposition of the coating is, according to the formula (8.15), given by the relation

$$(T_c)_{\text{coat}} / (T_c)_{\text{without coat}} = 1.04. \tag{8.28}$$

Even if we double the above-obtained $\lambda_e$ value, considering that the coating is deposited on both sides of the film, we still find that the effect of the coating is weak ($\lesssim 10\%$).‡

More consistent (but also more tedious) calculations within the framework

†This approximation describes the nontransition metals well.

‡According to [16], in which a similar problem is considered, the effect of the coating is even weaker. This is, however, because the interaction matrix element is computed in [16] under the extremely crude asumption that $V_\omega(\varkappa, z, z') \approx V_\omega(\varkappa, z = z' = L/2)$ (i.e., the interaction strengths near the boundaries of, and deep inside, the film are assumed to be the same). As a result, the quantity $\lambda_e$ turns out in [16] to be exponentially dependent on the film thickness ($\lambda_e \sim e^{-Lk_D}$), and the effect of the coating is negligible. On the other hand, according to the formula (8.27), the quantity $\lambda_e$ is inversely proportional to $L$ [cf. (8.1)], and the effect of the coating is no longer so weak.

of the above-expounded model [8], with account taken from the very beginning of the finiteness of the film thickness $L$, as well as the effect of the boundary conditions† on the form of the electron wave functions, and with the use of the values of the integrals as calculated on a computer, lead to a similar result: the rise in the $T_c$ of sandwiches due to the mechanism of polarization-wave penetration into the metal can hardly be more that 10-20%, even for optimally chosen parameters for the dielectric. The main reason for this rather weak effect of the coating on the $T_c$ of the film is the smallness of the penetration depth ($\sim 1/k_D$) of the polarization waves into the film, as a result of which the exciton mechanism is realized only in the portion $1/k_D << 1$ of the film.

A computation similar in many respects to the one described above is carried out in [22]. The main differences are that the spatial dispersion of the permittivity of the substrate and the dimensional quantization of the electron states in the film are taken into consideration in [22]. The first factor reduces $\lambda_e$ by roughly a factor of 1.5 in comparison with the above-presented results. Allowance for the second factor and its effect on the permittivity of the film leads to the increase of the Coulomb repulsion $\mu$ with decreasing film thickness. Thus, the parameter $\mu$ increases by almost a factor of 3 when $d = 4$ Å, but it is already close to its bulk value when $d = 10$ Å. The found increase of $\mu$ is, in all probability, due to the boundary conditions used in [22] for the electron wave function ($\psi|_{z=0} = 0$), as a result of which the electrons in the case of small thicknesses concentrate largely at the center of the film. The increase of $\mu$ should be much slower under more realistic conditions. As a result, the estimate $\lambda_e \approx 0.1$ is obtained in [22] for a two-sided coating of a film with thickness $d = 3$ Å, $\varepsilon_0 = 30$, and $\varepsilon_\infty = 1$. A recalculation with the formula (8.27) and the above-used parameter values $L = 10$ Å, $\varepsilon_0 = 23$, and $\varepsilon_\infty = 3$ yields for a one-sided coating a value practically equal to the value found above: $\lambda_e \approx 5 \times 10^{-3}$.

In [23] an investigation is carried out on the effect of the tunneling of electrons from the film into the substrate material on the interelectron interaction in the metallic film. It turns out that the effect is to change $T_{c\ \mathrm{coat}}/T_{c0}$ by only a few percent, which justifies the use of the cruder model employed above for the electron-density distribution. The estimate $\lambda_e \sim 10^{-2}$ found in [23] with comparable parameter values is in good agreement with the estimate obtained from the formula (8.28).

In [24] the case of a small metallic ball imbedded in a dielectric medium is considered. Such a variant should, for purely geometrical reasons, be more advantageous than the case of the metallic film: The ratio of the volume of the

†Let us note that the quantity $\lambda_e$ depends quite strongly on the boundary conditions, i.e., on the form of the electron density distribution at the film boundaries. Thus, the $\lambda_e$ values are obtained in [8] in two limiting cases: (a) $\psi = 0$ at the boundaries, i.e., $\psi \sim \sin kz$; (b) $\partial\psi/\partial z = 0$, i.e., $\psi \sim \cos kz$. The numerical computer calculation with $\varepsilon_\infty = 1$, $\varepsilon_0 = 30$, $L = 10$ Å, and $k_F = k_D$ yielded the values; $\lambda_e^{(a)} = 2.4 \times 10^{-3}$; $\lambda_e^{(b)} = 2.1 \times 10^{-2}$.

surface layer of a ball of diameter $d$ ($>>r_D$) to the volume of the ball is 3 times greater than the corresponding ratio for a film of thickness $d$, surrounded on both sides by a dielectric. The computation carried out in [24] confirms this qualitative observation.

We might have hoped to obtain a much stronger effect for the superconducting semiconducting (semimetallic) film. Indeed, the density of electron states is significantly lower in this case than in a metal, and therefore the parameter $1/k_D$ can be substantially increased and the screening accordingly weakened.

The question of the superconductivity of three-layered structures of the type metal-semiconducting film-metal or dielectric-semimetalic film-dielectric has recently been considered in greater detail in a number of theoretical papers [25-30]. Special attention is given in these papers to the possible effect on the semiconducting or semimetallic film's superconducting properties of a specific plasmon superconductivity mechanism resulting from the interaction of degenerate free carriers with the surface plasma oscillations of the electrons and holes at the boundary between the layers.†

According to [25-30], the exchange of such plasmons in a semiconducting film leads to an additional increase in the transition temperature $T_c$ and can, under certain conditions, be more effective than the exchange of excitons.‡ According to [30], however, the increase in $T_c$ due to this mechanism is also not large. (Thus, the estimate $T_c \lesssim$ 10-15 K is obtained in [30] for the optimum case of the semiconducting film of thickness $\lesssim$30 Å, coated on both sides with a semimetal.

Thus, the above-presented estimates for the effectiveness of the purely "polarization" superconductivity mechanism, stemming from the penetration of polarization waves from the external medium into the superconducting film, allow us to conclude that such a mechanism is quite capable of appreciably raising the film's $T_c$, although we can hardly expect to obtain on its basis critical temperatures significantly higher than those dictated by the conventional

†As has already been noted, plasmon exchange in a homogeneous isotropic metal amounts essentially to the screening of the Coulomb repulsion, i.e., to the replacement of the Coulomb potential $V_c = 4\pi e^2/k^2$ by the screened potential $V = 4\pi e^2/k^2 \varepsilon$(see Chapter 2). For this reason, the presence of volume plasma oscillations in the electron subsystem does not lead directly to a supplementary (to the phonon or exciton) mechanism of attraction between the electrons. The topic of discussion in [25-30] is the additional oscillation branch that arises as a result of the presence of the boundary and the external medium. (For this reason, the "plasmon" mechanism can be regarded as a variant of the more general exciton mechanism.) The mechanism considered in [25-30] is a two-dimensional analog of the mechanism of superconductivity in multivalley (multiband) semiconductors and semimetals [31-33] or in $s$-$d$ transition metals [34-36].

‡Let us note that the theoretical estimates in [30] correspond to the approximation in which the surface plasmons are actually not distinguishable from the volume plasmons. This approximation is a rather rough one.

phonon mechanism. Nevertheless, the search for real systems in which the indicated exciton mechanism manifests itself is of definite interest.

Below we shall show that the estimates for the purely "tunneling" mechanism of superconductivity are somewhat more favorable than the estimates for the purely "polarization" mechanism.

**B. The Tunneling Mechanism.** Let us now proceed to consider the other variant of the exciton mechanism of superconductivity in sandwiches that is connected with the Cooper pairing of electrons tunneling from the metallic film into the material of the coating.†

An important parameter of the system, which determines to a large extent the effectiveness of the superconductivity mechanism under consideration, is the depth of penetration of the electrons with $E \sim E_F$ tunneling from the metal into the coating. The penetration depth can easily be estimated if we use the method developed by Heine [37] to determine the electron wave functions. In the method in question it is assumed that the boundary between the metal and the coating is the $z = 0$ plane, and that the electron wave functions on the left or right side of the plane satisfy the same Schrödinger equation that obtains in the interior of the material (metal or coating) on that side. In such an approach, if the band structures of the bulk materials are known, then the problem of finding the wave functions for the metal-coating system reduces to that of matching the known wave functions of the parent materials on the $z = 0$ plane.

In the metal-semiconductor (or metal-dielectric) system, the Fermi level for the entire system falls within the forbidden energy region of the semiconductor (dielectric). Electron states with such energies do not occur in the interior of the coating; they do occur near the surface, with corresponding electron wave functions that fall off exponentially with distance into the material (i.e., that are proportional to $e^{-\alpha z}$), and describe the process (of interest to us here) of electron tunneling from the metal into the coating. The electron density corresponding to these states decreases with distance into the coating like $e^{-2\alpha z}$ and is characterized by the penetration depth $D = \frac{1}{2}\alpha$.

If, for the sake of simplicity, we assume the energy gap of the semiconductor (dielectric) to be isotropic, then the penetration depth of the electrons is given by the expression [17]

$$D(E) = \frac{1}{2\alpha} = \frac{\hbar^2 \left(k_F^2 - k_\parallel^2\right)^{\frac{1}{2}}}{2m\left(\Delta^2 - E^2\right)^{\frac{1}{2}}}, \tag{8.29}$$

where $2\Delta$ is the width of the energy gap, $E$ is the electron energy, as measured

†Here we assume that there is a direct chemical bond between the metallic film and the material of the coating and that no separating interlayers of any kind (e.g., oxide layers) exist. We also neglect the effect of band deformation near the film boundary (cf. [17]).

from the middle of the energy gap, and $k_{\|}$ is the magnitude of the electron wave vector in the plane parallel to the interface.

If we average the penetration depth $D(E)$ over the entire energy region lying within the gap, then for the mean penetration depth we find

$$D = \frac{1}{2\Delta} \int_{-\Delta}^{\Delta} D(E)\, dE = \frac{\pi}{3} \frac{\hbar^2 k_F}{2m\Delta}. \tag{8.30}$$

Substituting into the last formula the values $\Delta = 1$ eV, $k_F = 1.5$ Å$^{-1}$, which are typical values for narrow-band semiconductors, we easily find that the mean penetration depth in this case is quite large: $D \approx 5$ Å. This means that if the thickness, $L$, of the metallic film is sufficiently small, the electrons with energy $E \sim E_F$ will be in the coating for a considerable part of the time $b$:

$$b = \frac{\gamma D}{L + \gamma D} \tag{8.31}$$

(the film is assumed to be coated on only one side). In the formula (8.31) we have introduced the coefficient, $\gamma$, that characterizes the decrease in the value of the electron density directly at the interface in comparison with the value in the interior of the film. For coatings with a large penetration depth (i.e., for $Dk_F >> 1$), the coefficient $\gamma$ is close to unity. In this case, which is the case of greatest interest to us, taking for a numerical estimate the typical values $D = 5$ Å and $L = 10$ Å, we have $b \approx D/(L + D) \sim \frac{1}{3}$.

How will two electrons, 1 and 2, that have penetrated into the coating interact with each other? In analogy with the analysis presented in the preceding subsection, we can conclude that the interaction of interest to us will have the form†

$$V_\omega(\varkappa, z, z') = V_\omega^0(\varkappa, z - z') + \delta V_\omega(\varkappa, z + z') + \delta V'_\omega(\varkappa, z, z')\cdot \tag{8.32}$$

The first term on the right-hand side of the exprfession (8.32) represents the screened interelectron Coulomb interaction (the same interaction that occurs in the interior of the coating), which can be expressed in the usual manner in terms of the permittivity, $\varepsilon_{co}(\omega, \mathbf{k})$, of the material of the coating:

$$V_\omega^0(\varkappa, z - z') = \int \frac{4\pi e^2}{k^2} \frac{e^{iq(z-z')}}{\varepsilon_{co}(\omega, \mathbf{k})} \frac{dq}{2\pi} \qquad (\mathbf{k} \equiv \{\varkappa, q\}). \tag{8.33}$$

†Of cource, here, as above, reservations need to be made concerning the problematical nature of the application of the concepts of macroscopic electrodynamics to objects with spatial dimensions of the order of several angstroms.

The second term in (8.32) reflects the influence of the boundary. If we neglect the penetration of the electric field into the metallic film, then, according to the classical-electrodynamic treatment (see, for example, [38]), this term amounts to the interaction of the electron 1 with the image of the electron 2, i.e., to the interaction of the charge $-|e|$ with a charge $+|e|$ whose position and that of the electron 2 are symmetrically situated with respect to the $z = 0$ plane:

$$\delta V_\omega (\varkappa, z + z') = -\int \frac{4\pi e^2}{k^2} \frac{e^{iq(z+z')}}{\varepsilon_{co}(\omega, \mathbf{k})} \frac{dq}{2\pi}. \tag{8.34}$$

It can be seen from the expression (8.34) that the same elementary excitations [i.e., zeros of $\varepsilon_{co}(\omega, \mathbf{k})$] that correspond to the interaction $V^0_\omega$ correspond to the interaction $\delta V_\omega$. It is clear that the absolute value of $V^0_\omega$ should be greater than that of $\delta V_\omega$ since the distance between the electrons 1 and 2 is always less than the distance between the electron 1 and the image of the electron 2. Allowance for the finiteness of the penetration depth of the field into the metal leads to the smearing of the image of the electron 2 over an area $\sim 1/k_D$ without significantly changing the form of $\delta V_\omega$. Thus, the term $\delta V_\omega$ only slightly reduces the direct interaction $V^0_\omega$; it does not introduce new types of excitations and does not change the order of magnitude of the interaction strength. Below we shall, for the sake of simplificity, drop the term $\delta V_\omega$, which should not significantly affect the final results.

Let us, finally, discuss the third term $\delta V'_\omega$, which represents the change in the interaction energy of the electrons under the influence of the electrons that have tunneled into the coating. Since the density, $n_t$, of such electrons is significantly lower than the total density, $n_v$, of the valence electrons in the coating ($n_t/n_v \sim \Delta/E_F \sim \frac{1}{5}$), this will not lead to an appreciable enhancement of the screening of the interelectron interaction for $k \sim k_F$. (The change in the electron density will probably be even smaller since the electrons located near the bottom of the valence band will tunnel in the opposite direction from the coating into the metal [37].) The term $\delta V'_\omega$ can have the greatest effect on the form of the interaction in the case of relatively low-momentum transfers $k \lesssim k_{D\,co}/[1 + \varepsilon_{co}(0, 0)]^{1/2}$, leading to additional screening of the long-range Coulomb interaction. Since it is the region of momenta $k \sim k_F \sim k_D$ that is important for superconductivity, this additional screening turns out to be unimportant.

The penetration of conduction electrons into the coating also leads to the appearance of a nonzero imaginary part of $V_\omega(\mathbf{k})$ in the entire energy region within the dielectric gap and to a finite elementary-excitation lifetime. Qualitatively, this effect can be described by adding an imaginary part, $i\Gamma(\omega, \mathbf{k})$, to $\varepsilon_{co}(\omega, \mathbf{k})$ in the first term of the expression (8.32). We can infer from the random-phase-approximation formulas for the permittivity of an electron gas (see, for example, [20]) that, for momentum transfers $k \sim k_F$ and frequencies $\omega \sim \omega_e << E_F$, the damping constant $\Gamma(\omega, \mathbf{k})$ will be small since it is of the order of $\omega_e/E_F \lesssim 0.1$. Such weak damping, like the above-indicated change in the screening, has little effect on the parameters of the system and the value of the critical temperature. Therefore, in the subsequent computations, which have the nature of estimates, we shall also drop the term $\delta V'_\omega$ and use the expression (8.33) as the interaction $V_\omega(\varkappa, z, z')$.

In order to simplify the subsequent computations, let us assume that the wave functions of the electrons that have tunneled into the coating are ordinary plane waves and that the density of states of these electrons has the form $N_t(E) = bN_0(E) = DN_0/(L + D)$, where $N_0(E)$ is the density of electron states with energy $E$ in the metallic film. As a result of these simplifications, the problem of computing the electron–exciton coupling constant in the spatially inhomogeneous film-coating system reduces to the corresponding problem for a spatially homogeneous system with permittivity $\varepsilon_{co}(\omega, \mathbf{k})$.

Since the specific form of the function $\varepsilon_{co}(\omega, \mathbf{k})$ is important for the subsequent computation, let us discuss this question in greater detail. The expression for the permittivity, $\varepsilon_{co}(\omega, \mathbf{k})$, of the material of the coating is normally not expressible in a simple and computationally convenient form. Therefore, we shall use for this purpose the following approximate formula:

$$\varepsilon_{co}(\omega, \mathbf{k}) = 1 + \left[\frac{1}{\varepsilon_{co}(\omega, 0) - 1} + \frac{k^2}{k_{Dco}^2}\right]^{-1}, \tag{8.35}$$

where $k_{Dco}$ is the reciprocal screening length for the material of the coating, which can be computed from the number of valence electrons. It is easy to see that the formula (8.35) for $k = 0$ is exact, and, for $\omega = 0$, it very well describes the wavelength dependence of $\varepsilon_{co}(\omega = 0, \mathbf{k})$ [21] (Fig. 8.2).

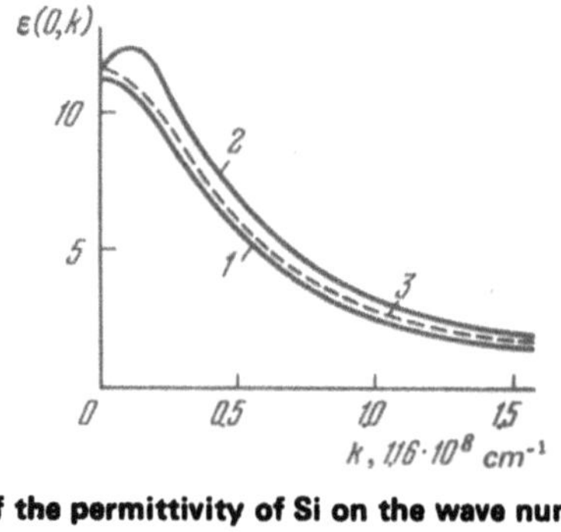


**Figure 8.2. Dependence of the permittivity of Si on the wave number $k$: (1) pseudopotential calculation [39]; (2) calculation carried out in [40]; (3) the model (8.35) with the parameter values $\varepsilon(0, 0) = 12$, $k_{Dco} = 2.1$ Å$^{-1}$.**

For $k \neq 0$ and $\omega \neq 0$, the expression (8.35) turns out to be quantitatively correct in the long-wave region and reproduces qualitatively correctly the character of the variation of the permittivity in the short-wave region $k \sim k_{D\text{co}}$ (Fig. 8.3). Below, as $\varepsilon_{\text{co}}(\omega, 0)$, we shall use the model expression (8.21).

Let us note that when we use the expressions (8.21) and (8.35), the exciton-excitation frequency, $\omega_e(k)$, corresponding to the condition $\varepsilon_{\text{co}}(\omega, \mathbf{k}) = 0$ increases by a factor of $2 - \varepsilon_\infty^2/\varepsilon_0$ when $k$ increases from zero to $k_D$, i.e., by a factor of less than 2. Such a relatively weak spatial dispersion does not always emerge† and is a direct consequence of the use of the approximate expression (8.35); the latter circumstance may introduce some error into the results of the calculation, but it does not alter the qualitative conclusions. We shall assume that $\varepsilon_{\text{co}}(\omega, \mathbf{k})$ is given by the formulas (8.21) and (8.35). Then applying the mean value theorem and carrying out the simple computations, we obtain

$$\lambda_e = -\frac{2}{\pi} b \int \frac{dS_{\mathbf{k}}}{v_F} \int \frac{d\omega}{\omega} \operatorname{Im} \frac{4\pi e^2}{k^2 \varepsilon_{\text{co}}(\omega, \mathbf{k})} = \frac{4\pi e^2}{\bar{k}^2} b N(E_F) \times$$

$$\times \frac{(\varepsilon_0 - \varepsilon_\infty) k_{D\text{II}}^4}{\left[(\varepsilon_\infty - 1)\bar{k}^2 + \varepsilon_\infty k_{D\text{co}}^2\right]\left[\varepsilon_0 k_{D\text{co}}^2 + (\varepsilon_0 - 1)\bar{k}^2\right]}. \qquad (8.36)$$

Using for the density of states $N(E_F)$ the formulas (8.26), which describe the nontransition metals well, we obtain from (8.36), for $\bar{k} = k_F$, $b = \frac{1}{3}$, $k_F a_B = 0.8$, $k_{D\text{co}} = 1.4 k_F$, $\varepsilon_0 = 23$, $\varepsilon_\infty = 3$, and $\lambda_{\text{ph}} = 0.5$, the magnitude of the coupling constant $\lambda_e = 0.05$ and the value

$$(T_c)_{\text{coat}} / (T_c)_{\text{without coat}} \sim 1.4.$$

Since the density of states at the Fermi level in a number of transition metals (niobium, vanadium, etc.) is 3 to 4 times higher than the corresponding density in the nontransition metals, we could, according to (8.36), expect for films of such metals values of $\lambda_e \sim 0.1$–$0.2$ and $(T_c)_{\text{coat}}/(T_c)_{\text{without coat}} \sim 2$–$3$. It should, however, be mentioned that, when we use a formula of the type (8.36) to make such estimates, we, in fact, presuppose that the *s*-, *p*-, and *d*-electron penetration depths into the semiconductor are equal.

It is, however, well known that the *d* electrons, which make the dominant contribution to $N(E_F)$ in niobium and vanadium, are localized near the transition-metal ions in a region of radius $\sim 1$ Å, and for them the penetration depth clearly will be considerably less than for the *s* and *p* electrons. For this reason,

†For example, the surface plasmons are characterized by a stronger spatial dispersion: for $k = 0$ we have $\omega_{p\,\text{sur}}^2(k = 0) = \omega_p^2/(1 + \varepsilon_0)$, whereas for $k \sim k_F$ we have $\omega_{p\,\text{sur}}(k \sim k_F) \sim \omega_p$, i.e., for $\varepsilon_0 \equiv \varepsilon_\infty(0, 0) \gtrsim 15$ the frequencies will, as a result of the dispersion, differ by a factor of more than 4.

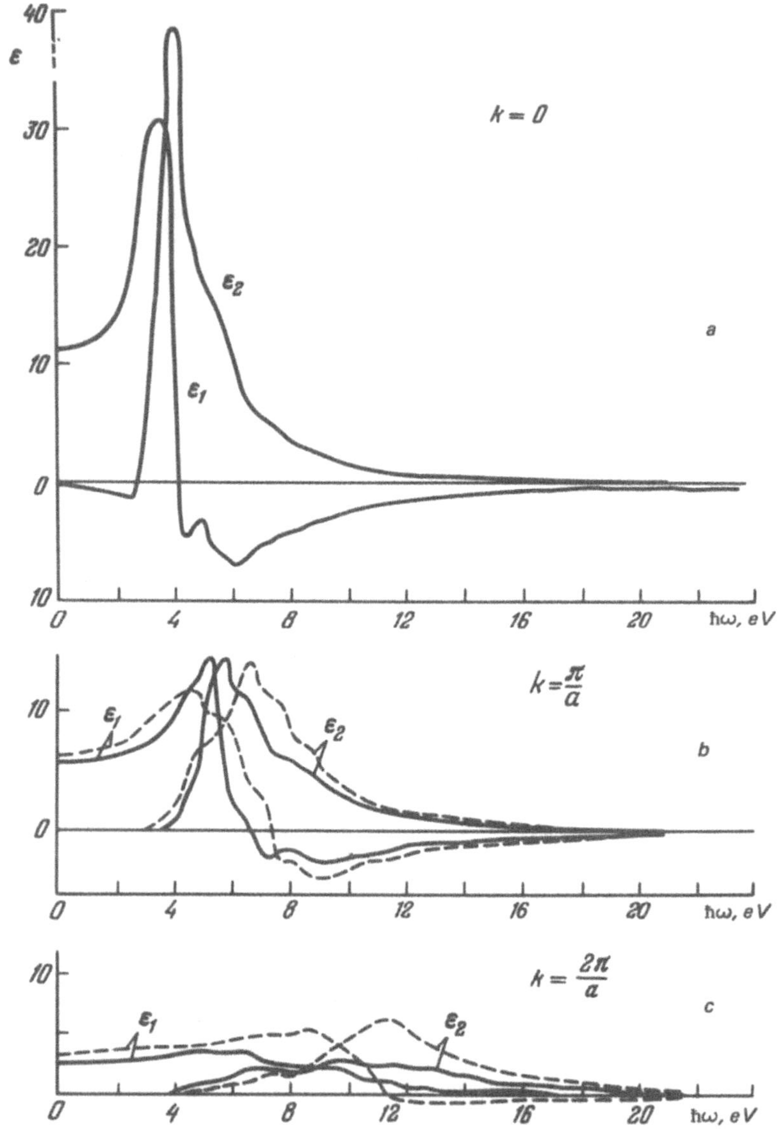


**Figure 8.3.** Dependence of the permittivity $\varepsilon = \varepsilon_1 + i\varepsilon_2$ on frequency. The continuous curves are the result of the exact calculation [41]; the dashed curves depict the dependence as given by the approximate formula (8.35) with the parameters: (a) $k = 0$ [the approximation (8.35) coincides in this case with the exact formula]; (b) $k = \pi/a$ ($a$ is the lattice parameter); (c) $k = 2\pi/a$.

it is the $s$ and $p$ electrons, which have a relatively small $N(E_F)$ value, that will, in the main, interact with the "exciton" excitations of the semiconductor, and, as a result, the electron–exciton coupling constant for niobium and vanadium films may turn out to be roughly the same as for nontransition-metal films.

The above-performed qualitative analysis allows us to formulate for the coating material a number of requirements, without the satisfaction of which it is difficult to expect an appreciable increase in the critical temperature of sandwiches: (a) the electron-penetration depth into the coating should be sufficiently large, or, more precisely, it should not be less than 4–5 Å, i.e., the forbidden band inside which the Fermi level is located should not be wider than 2–3 eV [cf. (8.30)]; (b) the material of the coating should have at the "exciton" frequency $\hbar\omega_e \sim 0.5$–$1$ eV for $k \sim k_F$ a sufficiently high-power peak in $\mathrm{Im}\,\varepsilon_{co}^{-1}(\omega, \mathbf{k})$, so that the integral over this frequency range will be sufficiently large:

$$-\frac{2}{\pi}\int \frac{d\omega}{\omega}\,\mathrm{Im}\,\frac{1}{\varepsilon_{co}(\omega, \mathbf{k})} \gtrsim 0.3 - 0.4$$

(in this case $\lambda_e \gtrsim 0.05$, and, for $\lambda_{ph} = 0.5$, we have $(T_c)_{\text{coat}}/(T_c)_{\text{without coat}} \gtrsim 1.4$).

If to satisfy the requirement (a) is relatively easy (for this purpose it is sufficient to use narrow-band semiconductors as the coating material), to verify whether the coating material satisfies the requirement (b) is significantly more complex. Indeed, although the permittivity $\varepsilon_{co}(\omega, \mathbf{k})$ for $k \sim k_F$ can, in principle, be determined in, for example, slow-neutron-scattering experiments, such data are not available for the overwhelming majority of materials. We therefore have to estimate $\varepsilon_{co}(\omega, \mathbf{k})$ with the aid of the approximate formula (8.35), which contains the permittivity $\varepsilon_{co}(\omega, \mathbf{k} = 0)$, a quantity that can be determined from optical measurements. If for $\varepsilon_{co}(\omega, \mathbf{k} = 0)$ we use the model formula (8.21), then the above-presented analysis shows that the parameters of the coating should satisfy the requirements:

$$\varepsilon_0 - \varepsilon_\infty \gtrsim 20 \text{ and } \varepsilon_0 \gg \varepsilon_\infty;$$

in this case the dominant contribution to the static permittivity, $\varepsilon_{co}(0, 0)$, should be made by the frequency region below $\omega_e$.

At present we do not know any materials that satisfy the above-listed requirements. And what is more, the question arises whether the existence of such materials is, in principle, possible. Thus, in particular, the opinion is expressed in [42] that the limitations imposed by the so-called "$f$-sum rules" drastically reduce the effectiveness of the exciton mechanism since the electron–exciton interaction can be strengthened only by decreasing the intensity of the plasmon exchange and correspondingly weakening the screening of the mutual

Coulomb repulsion of the electrons. The latter should lead to the growth of the Coulomb pseudopotential $\mu^*$, (8.5), and, according to [42], the possible increase in $\lambda_e$ will always be canceled out by the increase in $\mu^*$ [see the formula (8.2)]. Thus, according to [42], the exciton mechanism turns out to be always suppressed. Let us discuss this question in greater detail.

According to the "$f$-sum rule," which remains valid in the vicinity of a surface [43, 44] and which relates the number of elementary excitations in the electron subsystem to the density of the electron gas, the permittivity $\varepsilon_e(\omega, \mathbf{k})$ should satisfy the equation†

$$\int_0^\infty \omega\, d\omega\, \mathrm{Im}\, \frac{-1}{\varepsilon_e(\omega, \mathbf{k})} = \frac{\pi}{2}\omega_p^2, \qquad \omega_p^2 = \frac{4\pi n e^2}{m}, \tag{8.37}$$

where $n$ is the density of the valence-electron gas.

In the systems being considered by us, there are, besides the phonons, at least two other types of elementary excitations: (1) exciton excitations with frequency $\omega_e(k) \sim 0.5$-$1$ eV and (2) plasmons of frequency $\omega_p(k) \sim 20$ eV. (Above we were interested only in the frequency region $\omega \sim \omega_e$; therefore, the plasma excitations with frequencies $\omega_p >> \omega_e$ were not explicitly introduced.) As is well known, the inverse dielectric function $\varepsilon^{-1}(\omega, \mathbf{k})$ should have a pole at an elementary-excitation frequency. Therefore, in the phenomenological description of a material with such excitations, as the quantity $\varepsilon^{-1}(\omega, \mathbf{k})$, we can use the following model expression, which satisfies the "$f$-sum rule" (8.37):

$$\varepsilon^{-1}(\omega, \mathbf{k}) = 1 + \frac{\alpha(k)\,\omega_p^2}{\omega_e^2(k) - \omega^2} + \frac{(1 - \alpha(k))\,\omega_p^2}{\omega_p^2(k) - \omega^2}. \tag{8.38}$$

Computing, in accordance with (8.36) and (8.38), the contribution of the exciton pole to the coupling constant, we find

$$\lambda_e = -\frac{4\pi e^2}{k^2} N(E_F)\, b \frac{2}{\pi} \int \frac{d\omega}{\omega}\, \mathrm{Im}\, \frac{1}{\varepsilon(\omega, \bar{k})} = \\ = \frac{4\pi e^2}{\bar{k}^2} N(E_F)\, b \frac{\alpha(\bar{k})\,\omega_p^2}{\omega_e^2(\bar{k})} \sim b \frac{\alpha(\bar{k})\,\omega_p^2}{\omega_e^2(\bar{k})}. \tag{8.39}$$

If we substitute the typical values $\omega_p \sim 20$ eV, $\omega_e \sim 1$ eV into (8.39), then, to obtain, for example, the value $\lambda_e \sim 0.1$, it is sufficient to take a very small

†The quantity $\varepsilon_e(\omega, \mathbf{k})$ in (8.37) characterizes only the electron subsystem for a fixed ion subsystem; therefore, the phonon contribution to $\varepsilon_\varepsilon(\omega, \mathbf{k})$ should not be taken into consideration here.

value for the quantity, $\alpha$, determining the intensity of the exciton peak in (8.38): $\alpha \sim 2 \times 10^{-3}$. This means that a sufficiently strong electron–exciton interaction can be obtained practically without decreasing the intensity of the plasmon exchange, i.e., without increasing $\mu^*$. Thus, in contrast to [42], we arrive at the conclusion that, in principle, we can, not contradicting the "$f$-sum rule," obtain a sufficiently large value of $\lambda_e$ with the screening of the Coulomb interaction maintained.†

Allender *et al.* [17] have suggested that, as a coating material, we can use such typical semiconductors as Si, Ge, and InSb. According to the calculations carried out in [17], systems of this sort can have quite a high critical temperature (~ 100 K). It is, however, shown in [9, 45] that such a high $T_c$ value is obtained in [17] on account of the direct use of the random-phase approximation and the "$f$-sum rule," as a result of which the entire residue at the plasma pole turns out to be shifted into the exciton frequency region, and the contribution of the exciton pole becomes anomalously large. The value $\alpha \sim 1$ corresponds to this situation in the formula (8.39).

A more consistent computation performed in [9] for an InSb coating with the use of the experimentally found permittivity $\varepsilon_{co}(\omega, 0)$ and the formula (8.35) shows that in this case the $\lambda_e$ value is small ($\lambda_e < 0.01$), i.e., the effect of the coating is negligible.

This result is easy to understand. In semiconductors of the indicated type there occur very intense electron transitions from the valence band into the conduction band with frequency $\omega \sim 2\Delta$ ($2\Delta$ is the forbidden bandwidth), which result in $\mathrm{Im}\,\varepsilon(\omega, \mathbf{k})$'s having a large magnitude when $\omega \sim \Delta$. As a result, the quantity $\mathrm{Im}\,\varepsilon^{-1}(\omega, \mathbf{k}) = \varepsilon_2/(\varepsilon_1^2 + \varepsilon_2^2)$ turns out to be very small (of the order of $\frac{1}{20}$), which is responsible for the small value of the electron–exciton coupling constant.‡

In [42] an exciton superconductivity mechanism is considered that consists of the exchange of surface plasmons existing near the metal–semiconductor interface. The frequency, $\omega_{ps}$, of such surface plasmons is relatively low for $k = 0$: $\omega_{ps}(k = 0) = \omega_p^2/(1 + \varepsilon_0)$ (where $\varepsilon_0 >> 1$ is the static permittivity of the coating), but increases rapidly with increasing wave-vector transfer, so that when $k \sim k_F$ we have $\omega_{ps}^2(k \sim k_F) \sim \omega_p^2$.

The estimates made in [42] show that the coupling constant characterizing the exchange of such excitations is severely limited because of the "$f$-sum rule" and is all in magnitude. In the light of the foregoing, such a result is quite natural since the relatively low-frequency surface plasmons with $k = 0$ have a

†Let us emphasize that a large value of $\lambda_e$ is obtained only when we use the inequality $\omega_p >> \omega_e$, i.e., when the exciton and plasma frequency regions are far from each other.

‡The discrepancy between the results of [17] and [9, 45] is, in our opinion (see [9]), precisely because of the fact that the coupling constant $\lambda_e$ is actually determined by the quantity $\mathrm{Im}\,\varepsilon^{-1}$ [9, 45], and not by $\mathrm{Im}\,\varepsilon$ [17].

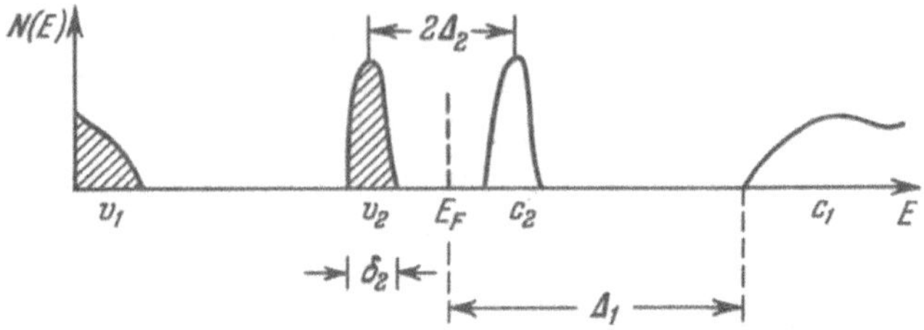


**Figure 8.4. Schematic view of the density of states for the model semiconductor.**

very small statistical weight, while the plasmons with $k \sim k_F$, which have a large statistical weight, overlap in frequency with the ordinary volume plasmons, and their role essentially amounts to the screening of the Coulomb repulsion (see the footnote on p. 305).

The above-presented estimates, which were made with the use of experimental data obtained for different materials, show that none of the coatings thus far investigated have been found to be materials that could guarantee a sufficiently high effectiveness of the exciton mechanism in superconducting sandwiches. This, however, does not mean that such materials cannot be found. As an example, below we propose a model of some hypothetical material whose properties are favorable from the point of view of its being used as a coating for the purpose of raising the $T_c$ of the system.

Let us consider some hypothetical material (a semiconductor) possessing a density of electron of states of the type depicted in Fig. 8.4 (we have, for the sake of simplicity, adopted a model that is symmetric with respect to the Fermi energy $E_F$). Its main characteristic is the presence in the density of states of two narrow peaks corresponding to the valence band $v_2$ and the conduction band $c_2$. Let the width of these bands be $\hbar\delta_2 \sim 0.1$–$0.2$ eV and the mean width of the energy gap between them be $2\hbar\Delta_2 \sim 0.2$–$0.4$ eV. Let us suppose that, apart from these two bands, there exist another valence band, $v_1$, and conduction band, $c_1$. Let us assume the width of these bands to be several electron volts and the distance between them to be $2\hbar\Delta_1 \sim 2$–$5$ eV.

Let us compute the permittivity function corresponding to the band structure in question. In the random-phase approximation $\varepsilon(\omega, \mathbf{q})$ is given by the expression [20]

$$\varepsilon(\omega, \mathbf{q}) = \\ = 1 + \frac{8\pi e^2}{\hbar q^2} \sum_{\mathbf{k}} \frac{|I_{v_2c_2}(\mathbf{k}, \mathbf{q})|^2 [\varepsilon_{v_2}(\mathbf{k}) - \varepsilon_{c_2}(\mathbf{k}+\mathbf{q})]}{[\varepsilon_{v_2}(\mathbf{k}) - \varepsilon_{c_2}(\mathbf{k}+\mathbf{q})]^2 - \omega^2 - i\delta} + \widetilde{P}(\omega, \mathbf{q}), \tag{8.40}$$

where

$$I_{v_2c_2}(\mathbf{k}, \mathbf{q}) \equiv \langle c_2, \mathbf{k}+\mathbf{q} | e^{i\mathbf{q}\mathbf{r}} | v_2, \mathbf{k} \rangle$$

is the matrix element of the interband transitions $v_2 \leftrightarrow c_2$ and $\widetilde{P}(\omega, \mathbf{q})$ denotes the contribution to $\varepsilon(\omega, \mathbf{q})$ from the $v_2 \leftrightarrow c_1$, $v_1 \leftrightarrow c_2$, and $v_1 \leftrightarrow c_1$ interband transitions.

Since the $v_2$ and $c_2$ bands are assumed to be very narrow, we can set

$$\varepsilon_{v_2}(\mathbf{k}) - \varepsilon_{c_2}(\mathbf{k}+\mathbf{q}) \approx 2\Delta_2$$

and

$$|I_{v_2c_2}(\mathbf{k}, \mathbf{q})|^2 \approx |I_{v_2c_2}(\mathbf{q})|^2 \equiv I_{\mathbf{q}}^2$$

in the expression (8.40), which can then be rewritten in the form

$$\varepsilon(\omega, \mathbf{q}) = 1 + \frac{16\pi^2}{3} \frac{I_{\mathbf{q}}^2 (2\Delta_1)^2}{(2\Delta_1)^2 - \omega^2} \frac{\mathrm{Ry}}{\hbar\Delta_1} \frac{V_{\mathbf{k}}}{{}^4/{}_3\pi k_F^3} \left(\frac{k_F}{q}\right)^2 k_F a_{\mathrm{B}} + $$

$$+ \widetilde{P}(\omega, \mathbf{q}) = 1 + A(\mathbf{q}) \frac{(2\Delta_1)^2}{(2\Delta_1)^2 - \omega^2} + \widetilde{P}(\omega, \mathbf{q}). \qquad (8.41)$$

Here $V_{\mathbf{k}}$ is the volume of the $v_2$ band in $\mathbf{k}$ space and $k_F$ is determined from the total number of valence electrons of the coating material. It is natural to assume that $V_{\mathbf{k}}$ is equal to the volume of the Brillouin zone. Then $V_{\mathbf{k}}/\frac{4}{3}\pi k_F^3 = 1/z$, where $z$ is the total number of valence electrons per unit cell.

The band structure assumed above can probably be realized only in relatively complex compounds with large $z \gtrsim 20$.† Taking for numerical estimates the typical values $z = 20, k_F a_B \sim 1, \hbar\Delta_1 = 2 \times 10^{-2}$ Ry, we obtain

$$A(q = k_F) \approx \frac{8\pi^2}{3} \cdot 0.5 \cdot 10^2 \cdot 10^{-1} I_{k_F}^2 \approx 10^2 I_{k_F}^2.$$

The magnitude of the square of the matrix element $I_{\mathbf{q}}$ depends on the form of the wave functions describing the electron states in the $v_2$ and $c_2$ bands. For example, if the wave functions are ordinary plane waves (i.e., if $\psi_{\mathbf{k}} \sim e^{ikz}$, then $I_{\mathbf{q}}^2 = 1$. In the present case, because of the small width of the bands, the wave functions should be highly localized, and the square of the matrix element will be significantly smaller than the value for the case of plane waves. Taking, by way of illustration, the value $I_{k_F}^2 \sim \frac{1}{10}$, we find that the coefficient $A(k_F)$ is quite large, being of the order of 10.

Let us not analyze the structure of the term $\widetilde{P}(\omega, \mathbf{q})$. Since the electron transitions determining its magnitude have energies greater than $\Delta_1 - \Delta_2$,

†The band structure of the pure elements and a considerable number of compounds with two to three atoms per unit cell has been studied relatively successfully. The absence in them of anything like the above-described density-of-electron-states picture compels us to direct our attention to more complex compounds.

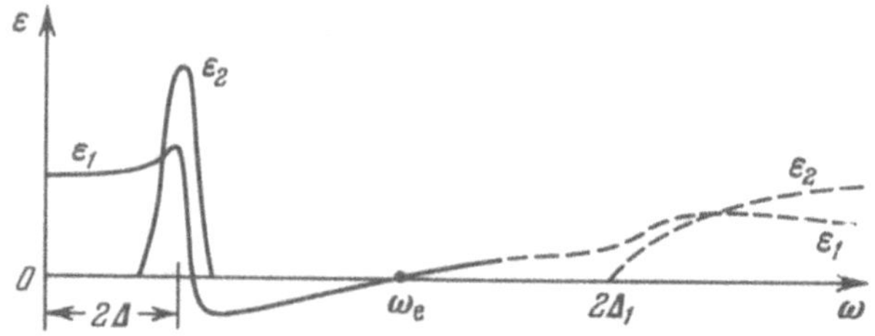


**Figure 8.5.** Diagrammatic form of the frequency dependence of $\varepsilon = \varepsilon_1 + i\varepsilon_2$ for the model semiconductor with a band structure of the type depicted in Fig. 8.4 for $\delta_2 = \Delta_2 = \Delta$, $\Delta_1 >> \Delta$. The dependence on the wave-number transfer $q$ is important only in the $v_1 \leftrightarrow c_2$, $v_2 \leftrightarrow c_1$, and $v_1 \leftrightarrow c_1$ transition region, which is indicated in the figure by dashed curves.

$\mathrm{Im}\, P(\omega, \mathbf{q}) = 0$ for $\omega < \Delta_1 - \Delta_2$ (Fig. 8.5 shows the schematic form of the function $\varepsilon(\omega, \mathbf{q})$; notice that the dependence on $q$ is important only in the region $\omega > 2\Delta_1$). It is clear that the dominant contribution to $P(\omega, \mathbf{q})$ will be made by the $v_1 \leftrightarrow c_1$ transitions, which are analogous to the interband transitions in semiconductors of the type Si, Ge, etc. Therefore, the quantity $\tilde{\varepsilon}(\omega, \mathbf{q}) = 1 + P(\omega, \mathbf{q})$ can be assumed to be similar to the permittivity function of these semiconductors [cf. Fig. 8.3 and the formulas (8.21) and (8.35)].

It is easy to see that the permittivity, $\varepsilon(\omega, \mathbf{q})$, given by the expression (8.41) has a zero corresponding to the "exciton" excitations at the frequency, $\omega_e(\mathbf{q})$, given by the equation

$$\omega_e^2(\mathbf{q}) = (2\Delta_1)^2 [A(\mathbf{q}) + \tilde{\varepsilon}(\omega_e(\mathbf{q}), \mathbf{q})]/\tilde{\varepsilon}(\omega_e(\mathbf{q}), \mathbf{q}).$$

If $\omega_e(\mathbf{q})$ lies in the region where $\mathrm{Im}\,\varepsilon(\omega, \mathbf{q}) = 0$ (Fig. 8.5), then the damping of the excitations will be weak. For $q \sim k_F$, we find with the typical values $\tilde{\varepsilon}(\omega_e, k_F) \sim 5$ and $A(k_F) \sim 10$ that $\omega_e(k_F) \sim 6\Delta_1 \sim 0.6$–$1.2$ eV.

Using the formula (8.36), we easily find that, when the coating described here is used, the electron–exciton coupling constant will be equal to

$$\lambda_e = bN(E_F)\frac{4\pi e^2}{\bar{k}^2}\frac{A(\bar{k})}{A(\bar{k}) + \tilde{\varepsilon}(\omega_e(\bar{k}), \bar{k})} \approx \frac{1}{3} b \sim 0{,}1,$$

with $(T_c)_{\mathrm{coat}}/(T_c)_{\mathrm{without\ coat}} \sim 2$. Thus, a coating of this type could, in principle, guarantee a significant increase in the critical temperature of a metallic film.

How real is the possibility of the existence of a material with the above-described band structure? In the first place, how real are the narrow $v_2$ and $c_2$ bands? It is well known that the density-of-electron-states curves of compounds with quasi-one-dimensional and quasi-two-dimensional crystal structures have narrow peaks similar to those in question. Thus, in particular, the $N(E)$ curves of the compounds with the A-15 structure have been found to have a peakwidth of the order of 0.05–0.1 eV and a value $N(E) \approx 12$ states per electron volt per unit cell [46].

A band structure of this type is realized in, for example, the $(2 \times 1)$-structure Si(111) surface. In this case the $v_1$ and $c_1$ bands differ significantly from the volume valence and conduction bands of silicon. The $v_2$ and $c_2$ bands arise during the formation of the $2 \times 1$ structure as a result of the splitting of the surface electron state $s_1$ (often called the dangling bond band). The characteristic widths of the $v_2$ and $c_2$ peaks, as well as the spacing between them, are of the order of several tenths of an electron volt. Such a band structure follows not only from the results of calculations (see, for example, [47, 48]) but is also observed in optical-absorption experiments [49]. Unfortunately, for the Si(111) surface, the matrix element of the transition between $v_2$ and $c_2$ is very small (this is because of the localization of the $v_2$ and $c_2$ states on different atoms), which makes the use of this material as a backing unpromising. It may, however, be hoped that other materials will be found that will possess similar band structures but for which the transition matrix elements will be much greater.†

Thus, the rough estimate of the effectiveness of the "tunneling" mechanism of superconductivity in films turns out to be somewhat more favorable than the estimate for the case of the purely "polarization" mechanism. If we ignore the possibility of the existence of other neglected effects that mask or suppress the growth of the $T_c$ of films, then, according to the above-presented estimates, we can, with a fortunate choice of the coating, expect to raise the $T_c$ of the film by a factor of 2–3.

## 3. Surface Effects in Layered Structures

As has already been noted, the above-presented theoretical estimates of the effect of coatings on the $T_c$ of thin films cannot be regarded as entirely reliable because of the use of a number of simplifying assumptions. Some of these assumptions are mentioned in the text; other circumstances, which we shall touch upon in the present section, can also play an important role. Thus, for example, the above-employed description of the surface of a metal as an ideal, defect- and impurity-free plane boundary is undoubtedly an abstraction that is far from reality.

Allowance for the states localized at the interface can drastically change the result and can in some cases lead not to an increase, but to a decrease in the $T_c$ of the system [50]. The consistent allowance for the Umklapp processes [51]

†It should, incidentally, be noted that the presence of the $v_2$ and $c_2$ peaks in the density of states should lead to strong screening of the electron–ion interaction and, as a consequence, to a relatively low stability of the crystal lattice of the material in question.

in bounded systems can also be of fundamental importance.† And other often insufficiently investigated phenomena occurring at the boundary of a thin film can play a special role.

It should be said that an important feature of the investigations that have been going on in recent years is precisely the comprehensive allowance for the surface phenomena in solids. This line of research has been organized into a big independent division of solid-state physics, and we cannot set ourselves the task of giving any detailed review of the great number of theoretical and experimental papers in this field. Let us note that the majority of such papers are not devoted directly to the study of the influence of surfaces on superconductivity. But they contain discussions on a number of surface effects (such as the real lattice structure of a crystal and the electron-density distribution near a boundary, the computation of the density of electron states with allowance for the surface, the appearance of additional vibration branches at a layer boundary and their dispersion properties, and the computation of the permittivity of the substances in a surface layer), without allowance for which the existing estimates of the effect of a surface on the superconducting-transition temperature of layered structures are not quite convincing. Below we make reference to only some of the papers that have a bearing on the problem of interest to us.

The electron theory of metals has always been devoted mainly to the investigation of the various volume properties of conductors, and here it has scored great successes (see, for example, [54, 55]). On the other hand, the theory of metallic surfaces is relatively poorly developed because of a number of additional difficulties connected, in particular, the necessity to take the inhomogeneity and the destruction of translational invariance into consideration.

Recently, however, quite a number of papers have been published in which the theory of inhomogeneous systems and, in particular, the theory of metal surfaces, are developed (see, for example, [56, 61]; a review of the earlier investigations of the surface properties of metals can be found in [62, 63]). Here the electronic structure of metal surfaces is investigated with the use of different approximations (the "jellium" model [58, 59], the hydrodynamic model of a metal [64], the pseudoatomic approximation [65], and more realistic approaches that take account of the ion-lattice structure [58-60, 63, 66, 67]). The effects connected with the spatial quantization of the transverse electron

†It is, for example, well known that allowance for the Umklapp processes [51] leads to the deviation of the local field acting on an ion from the mean macroscopic field, which alters the condition for the stability of the lattice [52, 53]. The fact that in real superconductors [10] the magnitude of the electron–phonon coupling constant is sometimes several times greater than the value that follows from the stability condition [20] is related by authors precisely to the Umklapp processes. Unfortunatley, this important question, as applied to the electron–exciton interaction, has practically not been investigated.

motion in films are studied in [68-72], while the effect of the sharp decrease in the electron density near the boundary of a metal on the surface vibration modes is considered in [60, 64, 73-81].

Besides the study of pure metal surfaces, much attention has been given in recent years to the investigation of the surface properties of semiconductors and dielectrics and the phenomena that occur at the boundary between metals and nonmetallic coatings.

One of the interesting phenomena of this sort is the experimentally observed dependence of the $T_c$ of thin films on the coating thickness [82-84]. If for dielectric coatings the $T_c$ of the film monotonically approaches its bulk value, then for semiconducting, semimetallic, and metallic coatings its dependence on the coating thickness exhibits characteristic oscillations. According to [82], this effect is connected with the specific character of the behavior of the conduction-electron wave function at the film–coating and coating–vacuum interfaces and bear no direct relation to the exciton mechanism of superconductivity.

Let us also mention the series of papers [85-87] devoted to the study of the effect of the surface of a metal on the spectrum of the excited states of both the collective electrons and the electrons localized on the individual molecules of the dielectric coating (see also [88]). The interaction of an excited molecule with the surface plasmons (i.e., with the charge-density oscillations) is studied in, for example [89]. The excited states of a metal–dielectric junction that are connected with the possible charge transfer from the dielectric to the metal are considered in [90], while the spectrum of the long-wave surface states in a dielectric bordering on a metal is studied in [91]. A number of general questions of exciton theory, including those that involve the consideration of the boundary effects, are elucidated in [13, 92].

In many semiconductors the surface effects are complicated by the presence of a region of surface charge, which spreads for many lattice constants (see, for example, [93]) and can in a number of cases exert a strong influence on the energy levels near the surface. It has been experimentally demonstrated, for example, that in GaAs the electric field in the surface region is so strong that it is impossible for excitons to exist there (cf. [94, 95]; apropos of the field effects at the boundaries of semiconductors, see also [96-99]).

It should also be noted that the deposition of a metallic coating significantly alters the electronic structure of the surface layer in comparison with that of the bulk material. This effect can exert an appreciable influence on the exciton mechanism of superconductivity and should be taken into consideration when selecting the coating material with the best polarization properties. Thus far, however, in doing this we have been guided only by the properties of large samples of the material in question. At the same time, it is well known that the electronic structure of a material near its free surface is substantially reconstructed as a result of both the appearance of electron surface (Tamm) states and the distortion of the crystal lattice.† The deposition of a coating in turn modifies significantly the electronic structure of this layer. Thus, under the influence of a coating, the dangling bond band of width 0.5–1 eV, that exists at the free surface of a semiconductor [47, 48] becomes greatly broadened as a result of the formation of a chemical bond between the atoms of the semiconductor and the metal. Such a change in the band structure is observed in photoemission experiments [102] and also follows from pseudopotential calculations [103, 104].

In the case of metals with impurities on their surfaces, there also arises effects connected with the appearance of electric fields at an interface. In this case, depending on the electron–metal and electron–impurity binding energies, electrons cross either from the metal into the impurity or from the impurity into the metal. In both cases an ionic bond is established between the metal and the impurity at the surface, a bond that underlies the phenomenon of chemisorption (see, for example [105–108]).

Let us also mention papers that deal with the plasma surface states in semiconductors [109], as well as with other surface excitations [110–125]. Thus, for example, a theory of the vibration modes in a semi-infinite crystal is developed in the electrostatic approximation in [110]; the retardation is taken into account in [111], and the spatial dispersion is allowed for in [112] (see also [113–118]). Other geometries (the sphere and the cylinder) are considered in [119, 120]. The results of the cited papers are often used to explain the energy losses that occur during the scattering of electrons and ions [121–123], to explain infrared-radiation absorption [124, 125], and in other methods of determining the spectrum of the excited states at the surface of a solid. The excitations occurring at a metal–dielectric (semiconductor) interface are also considered in [126–130]. Potential and surface-band calculations with allowance for the boundary of the solid can be found in, for example [131–144], where the calculations are carried out both with the use of model potentials [131–137] and in the self-consistent approximation [138–141]. Devoted to the determination of the form of the permittivity function of various media are, for example [145–147].

We have listed these papers with the purpose of demonstrating that the theoretical description of a metallic film in contact with a medium is a difficult problem, complicated by a host of circumstances whose roles are often insufficiently understood and which can significantly affect the conclusions that can

†As a rule, a surface layer of atoms shifts deep into the material through a distance of the order of several tenths of an angstrom. Furthermore, the so-called surface crystal structures, which are sometimes extremely complex, are usually formed on the surface. One and the same surface can, depending on its purity and temperature, have several surface structures (see, for example [101]).

be drawn. It is clear from the foregoing that the estimates presented in Sections 1 and 2 require a more comprehensive quantitative investigation.

In recent years much attention has been given to the experimental study of the superconductivity of thin films and, in particular, the possibility of raising their critical temperatures $T_c$ through their interaction with substrates. It was noted long ago [148–152] that the $T_c$'s of films of many nontransition metals, deposited at cryogenic temperatures, can exceed the critical temperatures of the bulk metals. Since the presence of a free surface should lower the mean-phonon frequency in thin films and increase their critical temperatures $T_c$, as compared to those of the bulk metals, the observed increases, $\Delta T_c$, are often related by authors precisely to the "softening" of the phonon spectrum in films [10, 153].

As a rule, thin films have a granular structure, and the resulting microdisorder in the crystal structure plays an important role in the explanation of the possible shifts $\Delta T_c$ [154]. A substrate, on the other hand, often plays only an indirect role, facilitating the stabilization of the disorder [155, 156]. Strains develop in a cold substrate during the deposition of a film on it, and this may lead to a shift $\Delta T_c$ [157].

The oxidation of films [158], as well as the absorption of inert-gas molecules on their surface [157, 159], leads to a strongly pronounced shift $\Delta T_c$, but the sign of the effect can be different for different metals.

The deposition of dielectric coatings also has an irregular effect on the observed shift $\Delta T_c$. For example, a SnS, $Tl_2Se$, $SiO_2$, or Ge coating on Sn decreases the $T_c$, while a TlCl, $Tl_2Se$, or Ge coating on Tl increases the $T_c$ [160–162]. Let us also mention the possible role of the surface phonons [159], which contribute to the shift $\Delta T_c$, the "charge-transfer" effect [163], and the role of the fluctuations [164–166], which lead to a shift, $\Delta T_c$, in the $T_c$ of granular films [167]. The observed change $\Delta T_c$ due to the above-indicated causes is usually a fraction of a degree but can attain several degrees in particular cases. And for the low-$T_c$ soft metals the $T_c$ is sometimes found to increase several times over.

As to the exciton mechanism, which should also contribute to the shift $\Delta T_c$, we do not at present have any reliable data on its manifestation in thin films. Experiments with ultrathin films [168, 169] with thickness $\lesssim 20$ Å, epitaxially grown on different substrates, normally can also be interpreted with the aid of the conventional (not exciton) mechanism.

Much attention has been given in recent years to the theoretical and experimental studies of the effect of organic-dye coatings on the superconducting properties of metallic films (see, in particular, the reports published in [170–172]). A specific characteristic of the metal–organic dye systems is the fact that their constituent molecules are easily polarized and interact with a broad continuum of conduction electrons. This property can manifest itself in the onset of electron exchange between the molecules and the metal. As a result of the exchange, the individual molecules can no longer be regarded as a localized system of charges (in contrast to ordinary dielectrics); in other words, the "hybridization" of the electron states in such systems is important. Furthermore, the strong electrostatic interaction of the molecules with the metal can affect the spectrum of the excited states of the coating molecules. Finally, the electronic structure of the metal in the junction region can also undergo a change as a result of the interaction with the coating. These and other aspects of the problem of the interaction between the molecules of a dielectric and a metal in the presence of a chemical bond are discussed in [85-87, 173-177]. Although conclusive evidence of the important role of organic coatings has not yet been obtained, the study of such systems is of great interest.

Thus, the question of the manifestation of the exciton mechanism in films and the possibility of distinguishing it in a background of other masking effects, taking all the complicating circumstances into account, has yet to be solved, theoretically or experimentally. We can only state that we do not at present know of any convincing arguments that negate the alluring possibility of significantly raising the $T_c$ of films with the aid of the exciton mechanism of superconductivity. Therefore, the continuation of the comprehensive investigation of this problem is absolutely necessary.

# 9

# Superconductivity under Nonequilibrium Conditions

## 1. Introduction

Normally, by the state of a superconductor without a current is meant the equilibrium state, investigated in the preceding chapters. At the same time, such characteristics of the material as the electron spectrum or the density of electron states, the electron wave functions or the interelectron interaction matrix elements, and the distribution functions of the quasiparticles (the electrons and phonons), on which the parameters of the superconducting state essentially depend, can vary within fairly wide limits when the system is under the influence of external sources (electromagnetic or acoustic fields, electric fields in tunnel structures, etc.).

For example, under equilibrium conditions the quasiparticle distribution function is determined by only one parameter: the temperature $T$. As the temperature increases, the number of quasiparticles increases, and, because of the collective character of the Cooper pairing, the binding energy of the pairs or the magnitude of the superconducting gap decreases, which leads to the existence of a critical temperature $T_c$. Since not just the number of quasiparticles, but also the form of the quasiparticle distribution function can vary within wide limits under the action of an external source, there arises the problem of finding that form of this function for which the quantity $T_c$ can exceed its equilibrium value. The object of the present chapter is to investigate the possibility of raising the superconducting-transition temperature of a system in a state of partial thermodynamic equilibrium.

To describe the nonequilibrium state of a superconductor in terms of the time-dependent variables of the external fields, we must use the system of temporal equations for the order parameter $\Delta$ and the excitation-distribution function $f(\xi)$ [1]. Depending on the conditions of the problem, we can derive different relations between the times necessary for the establishment of the equilibrium state with respect to the parameter $\Delta$ and the function $f(\xi)$ separately. In the case of sufficiently fast quasiparticle relaxation in terms of energy, the quasiequilibrium (stationary) state, which we shall be concerned with below, can be realized. In this state the magnitude of the superconducting gap, $\Delta$, is determined by the standard BCS-theory equation [2] but with a nonequilibrium quasiparticle distribution function $f(\xi)$. The nonequilibrium character of the phonon distribution function can be taken into account in the Éliashberg model [3], which is more rigorous than the BCS model.

In Section 2 we discuss the question of the possibility of raising the $T_c$ values in systems that are superconducting and in the equilibrium state by changing, with the aid of an external field, the electron spectrum, the energy distribution functions of the electrons and phonons, and the effective interelectron interaction constant. We consider the optimum relations between the parameters of the material and the external source.

In Section 3 we investigate the possibility of the appearance of the superconducting state in systems in which superconductivity is impossible under equilibrium conditions. We show that for both the model of a layered semiconductor with a narrow forbidden band and for the conventional model of a metal the superconducting state is possible in the case of interelectron repulsive interaction and an inverted population $[f(\xi) > \frac{1}{2}]$. In the semiconductor model the existence of the forbidden band $E_g$ makes possible the buildup of electrons in the conduction band and holes in the valence band up to the corresponding Fermi quasilevels $\pm\mu$ [inverted population: $f(\xi) = 1$ for $\xi < \mu$ and $f(\xi) = 0$ for $\xi > \mu$]. From this point of view the greater the quantity $E_g$ the better. Since, however, the superconducting gap in the presence of interelectron repulsion occurs not at the Fermi quasilevels but near the conduction- and valence-band extrema, the quantity $E_g$ should not exceed some critical value [see the formula (9.16)], and the density of states at the conduction- and valence-band edges should be nonzero. To fulfill the latter condition, we need either quasi-two-dimensional (layered) or quasi-one-dimensional (filamentary) semiconductors.

In the case of the metal model with an inverted excitation distribution, the condition imposed on the constant density of states is fulfilled, as in the equilibrium case. It is, however, difficult to obtain an inverted population (in a layer of width $\mu \ll E_F$ near the Fermi level). The latter turns out to be possible if the superconducting gap arising as a result of the interelectron repulsion is greater than the Debye phonon energy (the self-sustaining solution). Since the gap $\Delta$ is proportional to the pump intensity [through the quantity $\mu$ in the formula (9.28)], for superconductivity to occur in this case it is necessary that some pump-intensity threshold be exceeded.

In Section 3 we consider one possible way of transferring a system into the self-sustaining superconducting state.

In Section 4 we investigate some properties of superconductors with an inverted excitation distribution in the presence of interelectron repulsion. It is shown that such a state possesses infinite conductivity (i.e., is an ideal conductor). But there are fundamental differences between the magnetic properties of such systems and those of equilibrium superconductors. First, instead of being ideal diamagnets (in accordance with the Meissner effect), superconductors with an inverted distribution should be "ideal paramagnets." The magnetic field should penetrate into them with spatial oscillations. Thus, they are similar to ordinary type-II superconductors, the only difference being that in their case the penetration should occur even when the magnetic field is infinitely weak. Second, their superconducting-transition temperature is determined not by the magnitude of the superconducting gap $\Delta$, but by the magnitude of the Fermi-degeneracy energy, $\mu$, of the quasiparticles, which is determined by the pump intensity. Therefore, the quantity $T_c$ can be of the order of room temperature even for moderate pump intensities.

Thus, a number of qualitatively new phenomena are possible under nonequilibrium conditions. To corroborate this, we can indicate a number of interesting properties of a superconductor that manifest themselves in the nonequilibrium state, e.g., the variable Josephson effect [4], the appearance on the superconductor of an electric voltage proportional to the difference between the chemical potentials of the Cooper pairs and the quasiparticles [5], the appearance of a thermo-emf in anisotropic or inhomogeneous superconductors [6].

## 2. The Possibility of the $T_c$'s Rising under the Action of an External Field

As has already been indicated in the first section, we shall consider quasi-equilibrium states. In this case we must solve the ordinary BCS equation [2] for the gap $\Delta$ (see, for example, Chapter 2)

$$1 = -g \int_0^{\tilde{\omega}} N(\xi) \frac{1 - 2f(\xi)}{\sqrt{\xi^2 + \Delta^2}} d\xi, \tag{9.1}$$

in which in place of the equilibrium quasiparticle distribution function

$$f(\xi) = \left[\exp\left(\frac{\sqrt{\xi^2 + \Delta^2}}{T}\right) + 1\right]^{-1} \tag{9.2}$$

we should substitute the solution to the kinetic equation for $f(\xi)$ [1].

In the expression (9.1) the effective matrix element, $g$, of the interelectron interaction is assumed to be a constant in the region of energies $\pm\tilde{\omega}$ near the Fermi level and zero outside this region. The quantity $g$ can vary as a result of the action of the external source.

The density-of-states function $N(\xi)$ in the BCS theory is assumed to be a constant, equal to $N(0)$ in the band $\pm\tilde{\omega}$. Under the action of a high-power monochromatic field, the electron spectrum near the Fermi level can change significantly. If the energy scale for the region where this change is appreciable turns out to be of the order of, or greater than, the quantity $\tilde{\omega}$, then there simply occurs a renormalization of $N(0)$, which is entirely analogous to the renormalization of the matrix element $g$. If, on the other hand, this scale is less than the quantity $\tilde{\omega}$, then it is necessary to take the functional dependence $N(\xi)$ into account in the integration over $\xi$ in Eq. (9.1).

The effect on the quantity $T_c$ of the change that occurs in the matrix element $g$ as a result of the change induced in the electron wave functions by a strong electromagnetic field is considered in [7], although the changes in the electron spectrum and the quasiparticle distribution function are neglected in these papers.

The possibility of raising the $T_c$ with the aid of an electromagnetic field of frequency lying in the exciton-absorption region of the system is investigated in [8]. In this case the interelectron attraction constant $g$ can be increased as a result of the effect of the exciton mechanism of superconductivity, in contrast to the usually considered exciton mechanism (see Chapter 1), in which the role of $\tilde{\omega}$ is played by the excitation energy of the system in the excitonic state (in a semiconductor this energy is of the order of the forbidden bandwidth $E_g$). In the present case, because of the presence of real excitons in the external field, this role is played by the ground $\leftrightarrow$ excited exciton level transition energy (which is of the order of the exciton binding energy $E_{ex}$).

The latter energy can be much greater than $\hbar\omega_D$ but much smaller than the quantity $E_g$. Therefore, it may be hoped that the weakening of the Coulomb repulsion as a result of the presence of the well-known factor $\ln(E_F/\tilde{\omega})$ ($\tilde{\omega} = E_{ex}$) will be sufficiently effective, that the cutoff energy $\tilde{\omega}$ will be closer to the optimum value (see Chapter 1) than the quantity $\hbar\omega_D$ in the phonon mechanism, and that $\tilde{\omega} \approx E_g$ in the exciton mechanism in zero field.

Such a mechanism will be effective only when the excitation (exciton) concentration is fairly high. But then the interexciton interaction will lead to their disintegration. Furthermore, the breakup of the excitons is facilitated by the screening of the Coulomb potential by the free electrons responsible for the superconductivity. The role of the two exciton-disintegration effects can be insignificant only in quasi-one-dimensional or layered systems [9] in which the Coulomb interaction is virtually unscreened.

A similar mechanism for raising $T_c$ is investigated in [10] in which it is shown that allowance for the two-phonon term in the electron–phonon interaction Hamiltonian leads to the occurrence of effective attraction between the electrons in an energy band of width of the order of $4\hbar\omega_D$ near the Fermi level, instead of the $2\hbar\omega_D$ band found when only the single-phonon term is taken into consideration. True, the effective attraction that arises as a result of the presence of such a term is proportional to the number of really existing (and not virtual, as obtains in the conventional phonon mechanism) phonons. In order for this interaction to be comparable to the normal interaction, it is necessary that the phonons be generated under the action on the system of a high-power, high-frequency sound of frequency of the order of $10^{12}$ Hz. The assertion is made in [10] that the power of the presently available sound generators in this frequency range is sufficient for the raising of a $T_c$ several times over.

The effect of an electromagnetic-field-induced change in the electron spectrum on the superconducting properties of a semiconductor is studied in [11].

It is well known [12] that a strong electromagnetic field of frequency, $\Omega$, greater than the forbidden bandwidth $E_g$ leads to the creation of a nonequilibrium steady state. The conduction and valence bands then, respectively, contain electrons and holes with field-frequency-dependent Fermi quasilevels $\mu_0$ ($2\mu_0 = \hbar\Omega - E_g$). Furthermore, there appears in the energy spectrum of the electrons and holes a gap, $\Lambda$, of a dielectric nature for the momenta $|\mathbf{p}| = p_F = [m(\hbar\Omega - E_g)]^{1/2}$ (for simplicity, we assume the electron and hole masses to be equal and isotropic).

The quantity $\Lambda$ contains two terms. The first (the field gap) is proportional to the product of the dipole moment, **d**, of the interband transition and the field amplitude **E**. The second term is determined by the Coulomb interaction between the electrons in the conduction band and the holes in the valence band [11].

The problem of determining the effect of the dielectric gap $\Lambda$ in this case on the superconductivity turns out to be similar in many respects to the problem, considered in Chapter 5, Section 3, in which the presence of electrons and holes and the instability of the system against electron–hole pairing and reconstruction of the lattice in the equilibrium state were because of the anomalies in the single-electron spectrum. The case, considered in [11, 12], of a semiconductor with a wide forbidden band $E_g$ does not exhibit corresponding anomalies. But these anomalies appear in the nonequilibrium state when electrons and holes appear in the system as a result of the action of the powerful source. There are two important differences between this case and the equilibrium case with electron-hole pairing. First, because the electrons and holes are widely separated in energy (the difference between their energies being of the order of $E_g$), only the density–density type of terms in their interaction Hamiltonian are important (see Chapter 5). As to the terms corresponding to the interband transitions of the particles, they do not lead to any instabilities. Therefore, in particular, a reconstruction of the lattice should not occur in such a nonequilibrium state, although, as a result of the density–density type of interaction, dielectric gaps will appear at, or in the neighborhood of, the Fermi quasilevels of the electrons and holes.

Second, as has already been indicated above, there appears in the expression for the dielectric gap a term of a noncollective nature—the field gap. As a result, for the undoped semiconductor, which in the pumping regime corresponds to the case considered in Chapter 5, Section 2 of an undoped semimetal, there appears in the $(\Sigma_0, \Delta_0)$ plane an entire SD-phase region instead of the single SD-phase line (or rather the very narrow region around the bisectrix) that occurs in the equilibrium case of an undoped semimetal.

Because of the appearance of the dielectric gap $\Lambda$ near the Fermi quasilevels, the magnitude of the superconducting gap $\Delta$ decreases:

$$\Delta = \sqrt{\Delta_0^2 - \Lambda^2} \tag{9.3}$$

Here $\Delta_0$ corresponds to the magnitude of the superconducting gap that would obtain in the absence of the gap $\Lambda$ at the electron and the hole Fermi quasilevels. As can be seen from the formula (9.3), there exists for the wave's electric-field intensity **E** a critical value determined by the condition $\Lambda \equiv e\mathbf{E}\mathbf{d} = \Delta_0$ (**d** is the dipole moment of the interband transition). Superconductivity turns out to be impossible at any temperature in fields of higher intensity.

Thus, in the case of an undoped semiconductor, a strong field, at the same time as it produces carriers (electrons and holes), leads together with the Coulomb interaction to the appearance of a dielectric gap at the Fermi quasilevels, and this has a destructive effect on the superconductivity.

It is of interest from the point of view of the possibility of raising $T_c$'s to investigate the case of a pumped, doped semicondcutor [13]. In this case, similarly to those of a doped semimetal (see Chapter 5, Section 4), the Fermi quasilevels of the electrons and holes in the $D$ phase are located outside the dielectric gap $\Lambda$. The electron-spectrum renormalization that occurs as a result of the appearance of $\Lambda$ leads to an increase in the density of states at the Fermi quasilevels, which facilitates the increase of $T_c$.

But the vanishing of the density of states inside the gap $\Lambda$ has the opposite effect on $T_c$.

When the recombination of the nonequilibrium carriers is neglected, which is possible provided

$$\frac{\Lambda^4}{\hbar^3 \omega_D^3 \mu_0} \gg \frac{\tau_{ph}}{\tau_R},$$

the investigation of the $SD$ phase in this case is similar to the case investigated in detail in Chapter 5, Section 4. Here $\mu_0 = p_F^2/2m$, $\tau_{ph}$ is the characteristic time of the relaxation in energy terms as a result of the interaction between the electrons and the phonons, and $\tau_R$ is the characteristic time of the recombination via the dielectric gap, $E_g$, of the semiconductor.

The expression for $T_c$ has the following form:

$$T_c = 2.28 \frac{\tilde{n}^2}{\Lambda} \exp\left(-\frac{\tilde{n}}{\Lambda}\beta_0\right), \quad \beta_0 = \ln \frac{0,57\,\Lambda}{T_{c_0}}, \quad \tilde{n} = \frac{\mu_0 N_d}{3n_0}, \tag{9.4}$$

where $N_d$ is the concentration of the impurity under the assumption that it is fully ionized, $n_0$ is the carrier concentration due to the pump, and $T_{c_0}$ is the superconducting-transition temperature for $\Lambda \to 0$.

As can be seen from the expression (9.4), $T_c$ as a function of $\Lambda$ for a given doping level $\tilde{n}$ has a maximum at the value $\Lambda = \Lambda_0$, determined by the condition

$$\frac{\Lambda_0}{\tilde{n}} + 1 = \ln \frac{0.57\,\Lambda_0}{T_{c_0}}. \tag{9.5}$$

The peak $T_c$ value is equal to

$$T_{c\max} \approx 2.28\,\Lambda_0 \left(\ln \frac{1.54\,\Lambda_0}{T_{c_0}}\right)^{-1}. \tag{9.6}$$

Thus, the $T_c$ increases as $\Lambda$ (and, consequently, $\tilde{n}$) increases along the line (9.5). If $\tilde{n} \gg \hbar\omega_D$, then the pre-exponential factor $4\tilde{n}^2/\Lambda$ in (9.4) should be replaced by $2\hbar\omega_D$. It is likely that the satisfaction of the optimum conditions (9.6) will turn out to be a simpler problem under nonequilibrium conditions than in the equilibrium case of semimetals or metals with the spectrum $\varepsilon(\mathbf{p}) = -\varepsilon(\mathbf{p} + \mathbf{q})$ (see Chapter 5, Section 4) since it will be easier to vary the parameter $\Lambda$ by varying the field than to select an equilibrium system with the appropriate value for the dielectric parameter $\Sigma_0$.

Above we assumed that the system remains homogeneous during the electron and hole pumping. If, on the other hand, electron–hole drops form in the system [14], then the above-considered *SD* phase can be realized inside a drop. It is also assumed that the interelectron interaction has an attractive character. In the opposite case ferromagnetic ordering can occur during the pumping a doped semiconductor with a wide forbidden band $E_g$ [15].

The optimum objects for the realization of the condition (9.6) may be semiconductors of the $A_4B_6$ group, in which the extrema of the conduction and valence bands coincide (which is favorable to the formation of the gap $\Lambda$ because the electron and hole Fermi surfaces are then closer in shape), and which become superconducting even under equilibrium conditions (i.e., the interelectron interaction corresponds to attraction).

The nonequilibrium-carrier concentration $n_0$ should significantly exceed the doping-impurity concentration $N_d$, which can be achieved by suitably choosing the field frequency $\Omega(h\Omega - E_g \gg \tilde{n})$ and the field intensity ($\Lambda\tau_R > 1$). Let us note that, as a pump source for the $A_4B_6$ semiconducting compounds and their alloys, it is convenient to use $CO_2$ lasers. Then by varying the composition of the alloy, we can choose the optimum forbidden bandwidth $E_g$. For the case when $E_g \approx 0.1$ eV, $m \approx 10^{-2}m_0$, $N_d \approx 10^{19}$ cm$^{-3}$, and $T_{c_0} \approx 1$ K, the value $\Lambda \approx 10^{13}$ sec$^{-1}$, which is necessary for a value of $T_c \approx 10^2$ K, is attained at a laser intensity of the order of $10^5$ W/cm$^2$.

Let us now consider the possibility that the variation of the excitation distribution function $f(\xi)$ under the action of an external field will lead to the variation of the quantity $T_c$.

The effect of an electromagnetic field of frequency $\hbar\omega < 2\Delta(T)$ on $T_c$ is considered in [16]. In this case new excitations, i.e., excitations other than those that arise under thermodynamic equilibrium conditions, are not produced. The effect of the field amounts to the modification of the energy distribution, $f(\xi)$, of the already-existing excitations. To wit, the number of low-energy excitations decreases on account of transitions into states with higher energy. But, as can be seen from the formula (9.1), because of the presence of the term $(\xi^2 + \Delta^2)^{1/2}$ in the denominator, the states with low energy $\xi$ enter into the expression for $\Delta$ with a greater weight. Therefore, the decrease of $f(\xi)$ for low $\xi$ should lead to the increase of $\Delta$ and $T_c$ and, consequently, to the growth of the critical values for the current and magnetic field [17]. These effects have been observed in experiments on superconducting bridges [18]. The quantity $T_c$ increased by several percent, although, according to the theoretical estimates [16], a roughly 50% increase could have been expected.

A similar effect should be produced by the action of hypersound of frequency $\hbar\omega < 2\Delta(T)$ on a superconductor. Tredwell and Jacobsen [19] have reported the observation of a severalfold increase in the quantity $\Delta$ during the action of hypersound of frequency $10^{10}$ Hz on a thin aluminum film in a Josephson junction. In this case the experimental data are in good agreement with the theory.

Tredwell and Jacobsen [19] relate this significant difference in the effects of SHF fields and hypersound on the superconducting characteristics to the fact that, besides the effect on the excitation distribution function, a number of other effects occur during the action of SHF oscillations on a Josephson junction. This makes the observation of the effect suggested in [16] difficult. Gray [20] has observed an increase in the quantity $\Delta$ in the course of tunnel injections.

Aronov and Gurevich [21] have proposed a method of decreasing $f(\xi)$ below the thermodynamic-equilibrium distribution through the use of resonance transitions occurring under the action of an electromagnetic field of frequency $\hbar\omega \gg 2\Delta$. It is assumed that, as a result of the action of the monochromatic source, electrons are excited from a low-lying band into partially vacant—at $T \neq 0$—states below the superconducting gap. Thus, the proposal is to decrease the number of quasiholes in the superconductor. The number of electrons above the gap $\Delta$ (i.e., of quasielectrons) can be decreased with the aid of another monochromatic source that will also effect the resonance transition of these electrons into a higher-lying empty band.

The superconducting pairs will, of course, break up under these conditions ($\hbar\omega \gg 2\Delta$), but the breakup rate is inversely proportional to $\omega^2$. Then for a source of nonmonochromaticity less than $2\Delta$, the electron temperature $T_e$, which enters into the expression (9.1), can be lowered below the lattice temperature. The reduction of the number of excitations to a value smaller than the equilibrium value is described by Aronov and Gurevich in [21] with the introduction of a Fermi function, which, however, has a nonzero negative chemical potential $\mu$. If we assume that, as $\Delta \longrightarrow 0$, the chemical potential $\mu$ remains finite and negative, then the expression (9.1) shows that trivial solutions for $\Delta$ are impossible at a finite temperature ($T \neq \infty$). From this Aronov and Gurevich draw the conclusion that the quantity $T_c$ becomes infinite. But it follows from the kinetic equation for $f(\xi)$ that the quantity $\mu$ should vanish together with $\Delta$. Thus, the quantity $T_c$, which is determined by the condition $\Delta = 0$, should not change, even though the value of $\Delta$ may exceed the equilibrium value at any electron temperature $T_e < T_c$.

A similar idea of decreasing the electron temperature by extracting excitations from a superconductor in a tunnel structure has been put forward by the same authors [22].

In the absence of resonance interband transitions, a field with frequency $\hbar\omega > 2\Delta$ leads largely to the breakup of the superconducting pairs, i.e., $f(\xi)$ becomes greater than the equilibrium value, and therefore the quantities $\Delta$ and $T_c$ fall. This phenomenon has been observed in laser-irradiated lead [23]. Owen and Scalapino [24] have suggested that the relaxation of the excitations above (below) the gap $\Delta$ occurs at a much faster rate than the processes of excitation recombination across the superconducting gap $\Delta$, i.e., the processes of binding them into Cooper pairs. Then the action of radiation would lead to the establishment of a stationary excitation distribution function with a nonzero positive chemical potential $\mu$:

$$f(\xi) = \left[\exp\left(\frac{\sqrt{\xi^2+\Delta^2}-\mu}{T}\right)+1\right]^{-1}, \quad \mu > \Delta \tag{9.7}$$

[cf (9.2)].

Then an inverted excitation distribution $[f(\xi) > \frac{1}{2}]$ obtains in the interval $0 < \xi < (\mu^2 - \Delta^2)^{1/2}$, and the contribution of this region in the integral (9.1) is negative. When the value of $\mu$, which is proportional to the pump intensity, exceeds some critical value [24], the superconductivity should disappear even at $T = 0$. The gap $\Delta$ has been observed [25] to decrease at low pump intensities in accordance with the predictions of the theory [24].

It, however, follows from the analysis of the kinetic equation for $f(\xi)$ [1] that the assumption made by Owen and Scalapino [24] that the processes of recombination across the gap $\Delta$ are slow processes is valid only when $f(\xi) \ll 1$ [26] and $2\Delta < \hbar\omega_D$, the latter condition being valid for all superconductors (see, for example [27]). Consequently, it is impossible to obtain an inverted distribution in superconductors with $2\Delta < \hbar\omega_D$, i.e., a quantitative comparison of the experimental data presented in [25] with the theoretical results obtained in [24] does not make sense.

The form of the function $f(\xi)$ may turn out to be very important not only for the quantities $\Delta$ and $T_c$, but also for a number of properties of superconductors under the action of a pump source. For example, the Meissner effect may not occur in superconductors described by the Fermi $f(\xi)$ function (9.7), which is valid provided $2\Delta > \hbar\omega_D$ [27].

If an $f(\xi)$ function close to the Fermi function could be realized for ordinary superconductors ($2\Delta < \hbar\omega_D$), then the Meissner effect could disappear in this case also [28].

Also, the homogeneous state of a pumped superconductor with the Fermi or a near-Fermi $f(\xi)$ function may turn out to be unstable [29]. But this result is very sensitive to the form of the $f(\xi)$ function.

The determination of the form of the nonequilibrium quasiparticle distribution function and the dependence of the superconducting gap on the pump power is the principal problem in the investigation of the properties of nonequilibrium superconductors.

The quasiparticle distribution function for an optically pumped superconductor close to its phase-transition point is found in [30] on the basis of the kinetic equation. It is shown that $f(\xi)$ differs significantly from the equilibrium distribution; in particular, $f(\xi) < \frac{1}{2}$, a relation that has been experimentally confirmed [31-33]. This is explained by the fact that the quasiparticle scattering and recombination processes, which shape the energy distribution, occur at roughly the same rate, and the probabilities for scattering and recombination depend on the order parameter. The latter is due to the coherent nature of the interaction between the quasiparticles and the phonons and leads to a nonunique dependence of the gap $\Delta$ on the pump power $\beta$ [30]. To wit, starting from some critical power $\beta_c$, besides the normal solution corresponding to the decrease of the parameter $\Delta$ with increasing power, there appears a second solution, which describes the increase of the parameter $\Delta$ with increasing pump power and which merges with the first solution at the point $\Delta = \Delta_H$. This is a general result for the various methods of quasiparticle exictation (optical [30] and acoustic pumping, tunnel injection [34], etc.). The numerical solutions to the kinetic equation allow the determination of the dependence $\Delta(\beta)$ in the entire $\beta$ range [35].

It is shown in [36] that the state of a superconductor with the nonunique dependence $\Delta(\beta)$ is unstable for $\Delta < \Delta_H$. This instability is due to the decrease of the quasiparticle-recombination rate as the order parameter decreases on this branch of $\Delta(\beta)$. In this case the system can undergo either a first-order transition into the normal state as the pump power increases (with the presence of hysteresis) or a transition into the inhomogeneous state, consisting of regions of the normal and superconducting phases [36], the phase boundary in the case of optical pumping remaining stationary only at some power value $\beta = \beta_0 > \beta_c$. In the case when $\beta > \beta_0$ the normal-phase regions grow in time and fill the entire sample. These results allow us to explain the experimentally observed fact that the resistance increases smoothly with increasing pump power [37, 38]. The experimentally measured [39] temperature dependence of the critical power $\beta_c$ is in good agreement with the theoretical prediction [30].

The above-expounded ideas are applied in [34] to the case of tunnel injection. It is shown that the dependence of the parameter $\Delta$ on the voltage, $U$, across the tunnel junction has a nonunique character and that this leads to an $S$-shaped current-voltage characteristic of the tunnel junction and to the instability of the spatially homogeneous distribution of the parameter $\Delta$ and the quasiparticle concentration (current "filamentation"). The inhomogeneous state with tunnel injection differs significantly from the corresponding state in the case of optical pumping in that we can freeze the distribution of the superconducting and normal phases in the fixed-current regime. The results obtained in [34] allow a natural explanation of the experimental data given in [33, 40, 41] (the shape of the current-voltage characteristic, the discontinuous transition into the normal state, the hysteresis, the existence of "two" gaps.)†

In [42] Elesin develops a theory of threshold instability with a narrow quasiparticle source ($U \gtrsim 2\Delta_0$). He shows that a transition into an inhomogeneous state consisting of superconducting regions with different order parameters is possible.

Elesin [43] has studied the role of the effect of phonon reabsorption. Phonons emitted in the course of the energy relaxation of the quasiparticles can produce quasiparticle pairs, resulting in the "multiplication" and increase in the lifetime of the quasiparticles. It turns out that the dependence $\Delta(\beta)$ remains nonunique when allowance is made for the possible reabsorption, but the region of nonuniqueness with respect to the pump power $\beta$ shrinks.

In the following section we consider the possibility of the existence of a superconducting state in a system with a Fermi excitation distribution function $f(\xi)$ in the case when the system is not a superconductor in the absence of pumping because of the repulsive character of the interelectron interaction.

It is shown in [44] that, if the magnitude of the gap $\Delta$ is smaller than the Debye phonon energy, then the function $f(\xi)$ differs widely from the quasi-Fermi function (9.7). If the function $f(\xi)$ is a solution to the kinetic equation under these conditions, then the homogeneous state of the superconductor is a stable state (except for a narrow range of pump intensities [30, 45]).

In the opposite case of a wide gap a quasi-Fermi nonequilibrium excitation distribution with a nonzero chemical potential $\mu$ is possible. If in this case the gap $\Delta$ is a consequence of the interelectron attraction (e.g., in the case of the exciton mechanism of superconductivity) and, consequently, the pumping results in the suppression of $\Delta$, then the homogeneous state turns out to be unstable. The dependence of the period of the inhomogeneous state on the

†Let us note that these effects were first predicted by Elesin [26].

pump intensity is found in [44]. If the large magnitude of $\Delta$ is because of the interelectron Coulomb repulsion, then the homogeneous state is stable in the case of the quasi-Fermi function $f(\xi)$.

As done above, we shall neglect the effects of the warming up of the system, for which pulsed pumping is necessary under experimental conditions.

## 3. Superconductivity under Nonequilibrium Conditions in the Presence of Repulsive Interelectron Interaction

It can be seen from the expression, (9.1), for the superconducting gap $\Delta$ that the superconducting state is possible in the situation (which is the opposite of the normal situation) when the factor $[1 - 2f(\xi)]$ and the interaction potential $g$ change their signs concurrently, i.e., in the case of an inverted quasiparticle distribution $[f(\xi) > \frac{1}{2}]$ and repulsive interaction between the particles. It is pertinent here to recall Cooper's phenomenon [46], according to which at $T = 0$ a pair of electrons in the background of the Fermi step of the other electrons' distribution form a bound state when there is an arbitrarily weak attraction between them. Our assertion is that two electrons in the background of the Fermi step of the quasiparticle distribution $[f(\xi) = 1]$ form a bound state in the presence of an arbitrarily weak repulsion between them.

The main difficulty in the realization of such a state lies in the production and maintenance of an inverted population of excitations (quasiparticles). Below we shall show how to overcome this difficulty in the semiconductor model [47], in which the presence of the forbidden band makes the existence of an inverted distribution between the bands possible.

We shall also show that there exists in the conventional one-band model of a metal a "self-sustaining" superconducting state with $2\Delta > \hbar\omega_D$ and $f(\xi) > \frac{1}{2}$ in some energy range near the Fermi level when there is repulsion between the electrons.

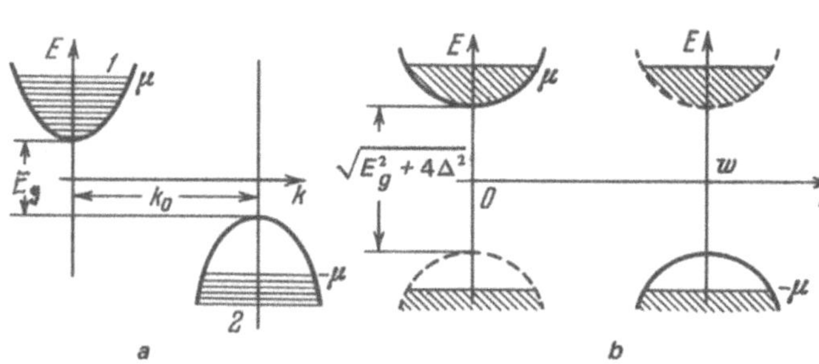


**Figure 9.1. (a) Dependence of the energy $E$ on the momentum $k$ for a pumped semiconductor: (1) conduction band, (2) valence band. (b) The dependence $E(k)$ for a pumped semiconductor in the superconducting state.**

Let us first consider a semiconductor (Fig. 9.1) with the Hamiltonian

$$H = \sum_{\alpha=1}^{2} \left\{ \sum_{\mathbf{p}} \xi_\alpha(\mathbf{p})\, a^+_{\alpha\mathbf{p}} a_{\alpha\mathbf{p}} + \frac{g_0}{2} \sum_{\mathbf{p}\mathbf{p}'} a^+_{\alpha\mathbf{p}} a^+_{\alpha-\mathbf{p}} a_{\alpha\mathbf{p}'} a_{\alpha-\mathbf{p}'} \right\} + \\ + \left[ \frac{g_1}{2} \sum_{\mathbf{p}\mathbf{p}'} a^+_{1\mathbf{p}} a^+_{1-\mathbf{p}} a_{2\mathbf{p}'} a_{2-\mathbf{p}'} + \text{H.C.} \right]. \qquad (9.8)$$

In the nonequilibrium state in the presence of pumping it is possible to have a band occupancy characterized by the Fermi quasilevels $\mu_1$ and $\mu_2$ for the electrons in the conduction band

$$\xi_1(\mathbf{p}) = \frac{p^2}{2m_1} + \frac{E_g}{2}$$

and in the valence band

$$\xi_2(\mathbf{p}) = -\frac{(\mathbf{p}-\mathbf{w})^2}{2m_2} - \frac{E_g}{2}.$$

It follows from the kinetic equation for the nonequilibrium electrons and holes of the semiconductor that, if the dielectric gap $E_g$ exceeds the highest phonon frequency in the system, i.e., if $E_g > \hbar\omega_D$, and the band extrema are the vector **w** apart in momentum space, then the recombination time, $\tau_R$, of the electrons excited by the external source can attain fairly large values—of the order of $10^{-5}$-$10^{-6}$ sec [48]. Since the energy relaxation time of the carriers (electrons in the conduction, and holes in the valence, band) is many orders of

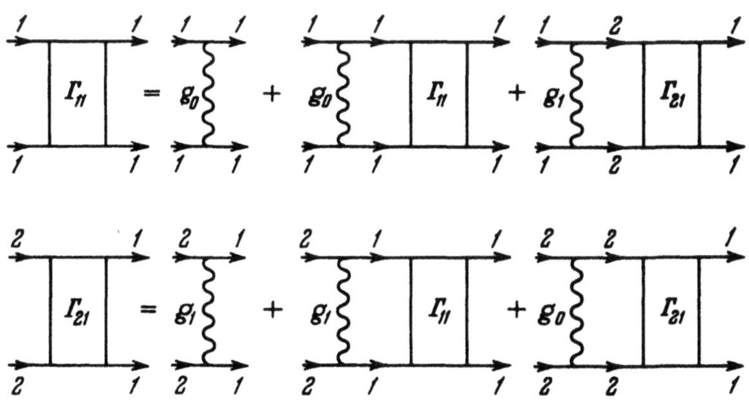

Figure 9.2. The equations for the vertex functions $\Gamma_{11}$ and $\Gamma_{21}$.

magnitude shorter than this value of $\tau_R$, the above-mentioned inverted population is entirely attainable: $\mu_1$ and $\mu_2$ then determine the Fermi quasilevels of the electrons and holes, as measured from the $\mu = 0$ level (the middle of the forbidden band).

An analysis of the stability of this nonequilibrium state against the superconducting pairing of the electrons within the boundaries of each band amounts to the solution of the system of equations for the vertex functions $\Gamma_{11}(q)$ and $\Gamma_{21}(q)$ depicted in Fig. 9.2, which in the case of a small four-momentum transfer $q = \{\omega, \mathbf{q}\}$ has the following form:

$$\Gamma_{11}(\omega_0) = [g_0 + (g_1^2 - g_0^2)\Pi_{22}]/\det, \qquad \Gamma_{21}(\omega_0) = g_1/\det, \tag{9.9}$$

where

$$\det = 1 - g_0(\Pi_{11} + \Pi_{22}) + (g_0^2 - g_1^2)\Pi_{11}\Pi_{22},$$

$$\Pi_{\alpha\alpha}(\omega_0, \mathbf{q}) = i(2\pi)^{-4}\int d\mathbf{p}\, d\omega\, G_\alpha(\omega, \mathbf{p})\, G_\alpha(\omega_0 - \omega, \mathbf{q} - \mathbf{p}).$$

In the limit $q = 0$ we easily find that

$$\Pi_{11} = \int \frac{d\mathbf{p}}{(2\pi)^3}\cdot\begin{cases} [2\xi_1(\mathbf{p}) - \omega_0 + i\delta]^{-1}, & \xi_1(\mathbf{p}) < \mu_1, \\ -[2\xi_1(\mathbf{p}) - \omega_0 - i\delta]^{-1}, & \xi_1(\mathbf{p}) > \mu_1, \end{cases} \tag{9.10}$$

$\Pi_{22}$ being obtainable from $\Pi_{11}$ by the substitution $1 \rightleftarrows 2$. In the subsequent computations of $\Pi_{\alpha\alpha}(\omega_0, \mathbf{q})$ we shall make the following assumption, which is justified for only quasi-two-dimensional structures:

$$\frac{d\mathbf{p}}{(2\pi)^3} = N(\xi)\, d\xi = d\xi\cdot\begin{cases} N(0), & |\xi| > E_g/2, \\ 0\ , & |\xi| < E_g/2. \end{cases} \tag{9.11}$$

Let us note that such semiconductors are being intensively studied at the present time (see, for example [9]).

Let us, moreover, assume that $m_1 = m_2$, i.e., $\mu_1 = -\mu_2$, and $N_1(0) = N_2(0)$. Let us further introduce the dimensionless constants, $\lambda_0 = N(0)g_0$ and $\lambda_1 = N(0)g_1$, characterizing the intraband and interband interactions of the electrons. The direct computation of the quantities $\Pi_{\alpha\alpha}(\omega_0, q = 0)$, using the formula (9.10) with the density of states (9.11), then leads to the following equation for the poles of the vertex functions:

$$1 - \lambda_0 \ln \frac{(2\mu)^2 - \omega_0^2}{2\tilde{\omega}\left(E_g^2 - \omega_0^2\right)^{1/2}} + \\ + \frac{\lambda_0^2 - \lambda_1^2}{4} \ln \frac{(2\mu - \omega_0)^2}{2\tilde{\omega}\,(E_g - \omega_0)} \ln \frac{(2\mu + \omega_0)^2}{2\tilde{\omega}\,(E_g + \omega_0)} = 0. \tag{9.12}$$

Let us, for simplicity, first set $|\lambda_0| = |\lambda_1|$ in (9.12); then

The vertex part will, depending on the sign of the logarithmic term, have a pole singularitv in the case of either attraction ($\lambda_0 < 0$) or repulsion ($\lambda_0 > 0$).

It can be seen that in the case of attraction the corresponding pole, $\omega_0$, is equal to $2\mu + i\Omega$ $[\Omega = 2\tilde{\omega}\exp(-1/|\lambda_0|)]$. The appearance of the real part $2\mu$ here is due to the chosen energy reference point and corresponds to the formation of an electron (hole) pair near the Fermi quasilevel.

Let us turn to the case of repulsion between the electrons (i.e., to the $\lambda_0 > 0$ case). Simple computations yeild the purely imaginary pole

$$\omega_0 = i\left(\Omega^2 - E_g^2\right)^{1/2},$$

where

$$\Omega = \frac{2\mu^2}{\tilde{\omega}} \exp\left(-\frac{1}{\lambda_0}\right). \tag{9.13}$$

The fact that the real part of the pole is now equal to zero indicates that the electron pairing occurs not at the Fermi quasilevel $\mu$, but near the extrema of the bands of the semiconductor. It can also be seen that instability occurs only in the case of a fairly small forbidden bandwidth $E_g < \Omega$. In the general case (when $|\lambda_0| \neq |\lambda_1|$) Eq. (9.12) is a second-order equation in the logarithmic term and therefore has two roots:

$$\frac{1}{2} \ln \frac{4\mu^2 + \Omega^2}{2\tilde{\omega}\left(E_g^2 + \Omega^2\right)^{1/2}} = \frac{\lambda_0}{\lambda_0^2 - \lambda_1^2} \pm \sqrt{\frac{\lambda_1^2}{(\lambda_0^2 - \lambda_1^2)^2} - \beta^2}, \\ \beta = \frac{1}{2}\left(\operatorname{arctg}\frac{\Omega}{E_g} - \operatorname{arctg}\frac{\Omega}{2\tilde{\omega}}\right). \tag{9.14}$$

Equation (9.14) determines the modulus of the purely imaginary pole of the vertex part. Of the two solutions to Eq. (9.14), we should select the one that corresponds to the greater $\Omega$ value. It can easily be verified that the two solutions correspond to phase differences of zero and $\pi$ between the order parameters for the two bands. The first type of solution is realized in the case of attraction between the electrons, while a phase difference of $\pi$ is energetically more advantageous in the case of repulsion. Allowing for the fact that $E_g \ll \tilde{\omega}$, we see that the modulus $\Omega$ is essentially determined by the logarithmic term only, and that $\beta \approx \pi/4$. Equation (9.14) can then easily be solved:

$$\Omega = \sqrt{\Omega_0^2 - E_g^2}, \tag{9.15}$$

$$\Omega_0 = \frac{2\mu^2}{\tilde{\omega}} \exp\left\{-2\left[\frac{\lambda_0}{\lambda_0^2-\lambda_1^2} - \sqrt{\frac{\lambda_1^2}{(\lambda_0^2-\lambda_1^2)^2} - \frac{\pi^2}{16}}\right]\right\}. \tag{9.16}$$

The expressions (9.15) and (9.16) show that the superconducting pairing occurs upon the simultaneous satisfaction of the two conditions (Fig. 9.3):

$$\text{a)}\ E_g < \Omega_0, \quad \text{b)}\ |\lambda_1| > \frac{\pi}{2}\lambda_0^2.$$

Of the systems that exhibit the superconducting instability when they are in the state in which the electron distribution in one band is inverted with respect to the distribution in the other band and the interelectron interaction is a repulsive one, materials with overlapping bands deserve special attention. For example, for a semimetal, when the electron distribution within each band is normal but the population of one band relative to the other is inverted (i.e., the Fermi quasilevels of the two overlapping bands do not coincide), the expression for the modulus of the imaginary pole exactly coincides with the expression (9.15), (9.16) obtained for a semiconductor, except that as $E_g$ we should now substitute the magnitude of the overlap between the conduction and valence bands of the semimetal (Fig. 9.4a). Similar properties are possessed by systems with the spectrum shown in Fig. 9.4b.

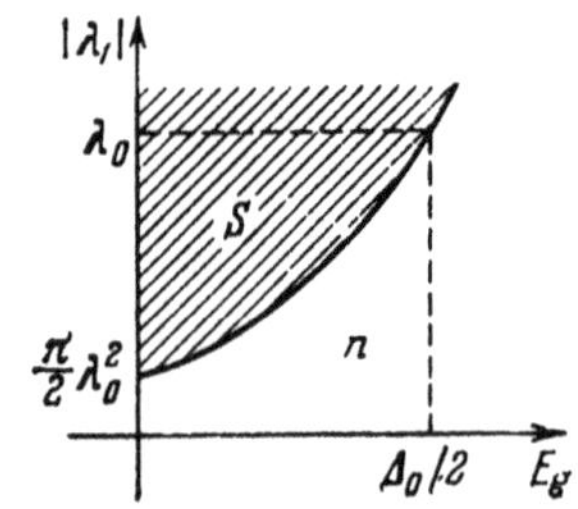


**Figure 9.3.** Relation between the magnitude of the forbidden bandwidth $E_g$ and the critical (for the onset of superconductivity) value of the parameter, $\lambda_1$, characterizing the interband matrix element $g_1$.

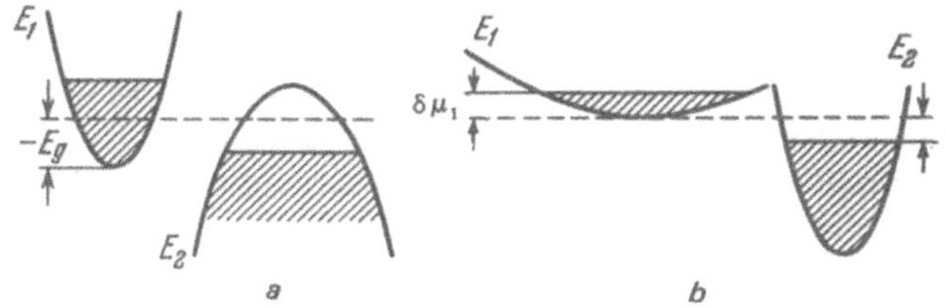


**Figure 9.4. The electron spectrum of two-band systems with overlapping bands in the inverted state: (a) semimetal, (b) two-band metal. The dashed line indicates the position of the Fermi level in the equilibrium state.**

The above-performed investigation of the singularities of the $\Gamma_{11}$ and $\Gamma_{21}$ functions shows that the instability occurs only in the particle-particle channel within each band. In the state with inverted population, the Coulomb interaction in the particle-hole channel corresponds to repulsion, and stationary interband electron-hole pairing [49] does not occur. Therefore, to describe the stable state of the system, it is sufficient to introduce, along with the ordinary Green function

$$G_1(\mathbf{p}, t) = -i\langle T a_{1\mathbf{p}}(t)\, a_{1\mathbf{p}}^{+}(0)\rangle, \tag{9.17}$$

the anomalous function

$$F_1^+(\mathbf{p}, t)\exp[i(E_{N+2} - E_N)t] = \langle N+2 | T a_{1-\mathbf{p}}^{+}(t)\, a_{1\mathbf{p}}^{+}(0) | N\rangle, \tag{9.18}$$

which characterizes only the intraband pairing of the electrons (the Green functions for the valence band are introduced in much the same way).

The form of the oscillating factor $\exp[i(E_{N+2} - E_N)t]$ is largely determined by the position of the pole singularity in the vertex functions $\Gamma_{11}(\omega_0)$ and $\Gamma_{21}(\omega_0)$. In particular, for the ordinary Cooper singularity, $E_{N+2} - E_N = 2\mu$ since in the case of attraction the instability develops for states close to the Fermi surface.

In our case the singularity is important when $\omega_0$ (as measured from the middle of the forbidden band) is small. We shall specially take this circumstance into consideration when deriving the equations of motion for the functions $G_1$ and $F_1^+$ by setting $E_{N+2} - E_N = 0$. Otherwise the derivation follows the steady-state technique for $T = 0$, and the equations for the functions (9.17) and (91.8) assume, after a Fourier expansion, the following form:

$$(\omega - \xi_1(\mathbf{p}))\, G_1(\omega, \mathbf{p}) - i\Delta_1 F_1^+(\omega, \mathbf{p}) = 1, \tag{9.19}$$

$$(\omega + \xi_1(\mathbf{p}))\, F_1^+(\omega, \mathbf{p}) + i\Delta_1^+ G_1(\omega, \mathbf{p}) = 0,$$

$$\Delta_1^+ = \lambda_0 F_1^+(0) + \lambda_1 F_2^+(0). \tag{9.20}$$

In the model with inverted population, the spectrum of the elementary excitations of the reconstructed phase is determined by the negative root, $-(\xi_1^2 + \Delta_1^2)^{1/2}$, of the determinant, $D$, of the system of equations (9.19) (see Fig. 9.1b). The smallest positive single-particle excitations in the system start from energies determined by the position of the chemical potential $\mu$. Therefore, the solution to the system (9.19) has the following form:

$$G_1(\omega, \mathbf{p}) = \frac{u_1^2}{\omega - \omega_1^+} + \frac{v_1^2}{\omega - \omega_1^-},$$

$$F_1^+(\omega, \mathbf{p}) = \frac{-i\Delta_1^+}{(\omega - \omega_1^+)(\omega - \omega_1^-)}, \tag{9.21}$$

$$\omega_1^{\pm} = \pm \{(\xi_1^2 + \Delta_1^2)^{1/2} - i\delta \operatorname{sgn}[(\xi_1^2 + \Delta_1^2)^{1/2} - \mu_1]\}, \tag{9.22}$$

$$u_1^2 = \frac{1}{2}\left[1 + \frac{\xi_1(\mathbf{p})}{(\xi_1^2(\mathbf{p}) + \Delta_1^2)^{1/2}}\right],$$

$$v_1^2 = \frac{1}{2}\left[1 - \frac{\xi_1(\mathbf{p})}{(\xi_1^2(\mathbf{p}) + \Delta_1^2)^{1/2}}\right].$$

The poles of the functions $G_1$ and $F_1^+$, (9.21), are bypassed in (9.22) in such a way as to ensure the correct population for the quasiequilibrium state at $T = 0$. A more general analysis based on the use of the kinetic equations is carried out below for the metallic model.

For a self-consistent determination of the order parameter $\Delta_1^+$ from the formula (9.20), we require the value of the anomalous average $F_2^+(0)$ for the valence-band electrons ($\alpha = 2$). The derivation of the corresponding equations is entirely similar to the derivation of (9.19), while the functions $G_2$ and $F_2^+$ can be formally obtained from (9.21) by means of the substitution $1 \rightleftarrows 2$, except that, by analogy with (9.21), we should take account of the singularities involved in the occupation of the states in the lower band (see Fig. 9.1). Then from the definition of the average for coincident arguments we find a system of self-consistent equations for the order parameters $\Delta_1^+$ and $\Delta_2^+$:

$$\Delta_1^+ = \lambda_0 F_1^+(0) + \lambda_1 F_2^+(0), \quad \Delta_2^+ = \lambda_0 F_2^+(0) + \lambda_1 F_1^+(0), \tag{9.23}$$

where

$$F_\alpha^+(0) = \int_{E_g/2}^{\tilde{\omega}} N(\xi)\, d\xi \frac{\Delta_\alpha^+}{2(\xi_\alpha^2 + \Delta_\alpha^2)^{1/2}} \operatorname{sgn}[(\mu_\alpha^2 - \Delta_\alpha^2)^{1/2} - \xi]. \tag{9.24}$$

For $m_1 = m_2$ the order parameters are also equal, i.e., $\Delta_1 = \Delta_2 \equiv \Delta$, and obey one self-consistent equation:

$$\frac{2}{\lambda_0 + \lambda_1} = \int_{E_g/2}^{\tilde{\omega}} N(\xi)\, d\xi \, \frac{\operatorname{sgn}[(\mu^2 - \Delta^2)^{1/2} - \xi]}{(\xi^2 + \Delta^2)^{1/2}}. \tag{9.25}$$

In the constant-density-of-states [i.e., $N(\xi) = N(0)$] approximation, Eq. (9.25) can be solved for $\Delta$ [cf. (9.16)]:

$$\Delta = \sqrt{\Delta_0 (\Delta_0 - E_g)}, \quad \Delta_0 = \frac{2\mu^2}{\tilde{\omega}} \exp\left(-\frac{2}{\lambda_0 + \lambda_1}\right).$$

Equation (9.25) also has a solution at arbitrarily small values of the constants $\lambda_0$ and $\lambda_1$ (for $E_g \longrightarrow 0$) if the function $N(\xi) \sim (\xi - E_g)^{-1/2}$, i.e., if it behaves as in the one-dimensional case. In this case the quantity $\Delta$ is determined by a power function of the constants $\lambda_0$ and $\lambda_1$. Thus, for the observation of superconductivity in systems with inverted quasiparticle distributions stemming from the interelectron repulsion we need a quasi-one-dimensional or layered semiconductor with a narrow forbidden band. Here the condition on the forbidden bandwidth is less rigid for quasi-one-dimensional semiconductors.

On the face of it, it may appear that an inverted distribution of the electrons within the same band (of a metal or a doped semiconductor) is impossible even in the presence of a constantly acting source that transfers electrons from states below the Fermi level $E_F$ into higher-lying states. On the other hand, in the absence of an inverted distribution $[f(\xi) < \frac{1}{2}]$ the equation for the gap $\Delta$ for the case of repulsive interaction ($\lambda > 0$) has only the trivial solution $\Delta = 0$.

But it follows from an analysis of the consistent system of equations for $\Delta$ and the kinetic equation for the excitation distribution function $f(\xi)$ [50] that a nontrivial "self-sustaining" solution $[\Delta \neq 0, f(\xi) > \frac{1}{2}]$ exists if

$$2\Delta < \hbar\omega_D. \tag{9.26}$$

The quasiparticle recombination across the gap $\Delta$ is a slow process in this case because of the impossibility of one-phonon processes.

The recombination time $\tau_R$ may turn out to be much longer than the energy-relaxation time for the quasiparticles above the gap, i.e., there will be a quasiparticle cutoff:

$$f(\xi) = \left[\exp\left(\frac{\sqrt{\xi^2 + \Delta^2} - \mu}{T}\right) + 1\right]^{-1}, \tag{9.27}$$

where $\mu$ is the Fermi quasilevel determined by the source intensity and the time

$\tau_R$. The solution to Eq. (9.1) for $T = 0$ then has the form

$$\Delta_0 = \frac{2\mu^2}{\tilde{\omega}} \exp\left(-\frac{1}{\lambda}\right). \tag{9.28}$$

This state is self-sustaining since the gap $\Delta$ (when $2\Delta > \hbar\omega_D$) guarantees an inverted excitation distribution, which in turn leads to the existence of a gap $\Delta$ for $\lambda > 0$.

The problem of inducing the system to go from the $\Delta = 0, f(\xi) < \frac{1}{2}$ state into the $\Delta \neq 0, f(\xi) > \frac{1}{2}$ superconducting state is an important one. One method is suggested in [27]. It consists of producing at the Fermi level a "start-up" dielectric gap $\Lambda > \hbar\omega_D$ by means of interband transitions occurring under the action of a strong electromagnetic field. If $\Delta_0 > \Lambda > \hbar\omega_D$, then superconductivity can occur as a result of the excitation of particles across the gap $\Lambda$ by another source. The magnitude of the superconducting gap $\Delta$ will be equal to $(\Delta_0^2 - \Lambda^2)^{1/2}$. After the establishment of this state, the strong electromagnetic field can be adiabatically switched off (so that $\Lambda \longrightarrow 0$). A steady self-sustaining state (with $\Delta = \Delta_0$) will then remain in the presence of the pump source, $I$, inducing the excitations across the gap $\Delta$.

If for such a one-band metallic model (with $\Delta = \Lambda = 0$) we prescribe in purely formal terms an inverted electron distribution in some energy band near the Fermi level, then we can use the results obtained above for the semiconductor model to analyze the stability properties of this state. Setting for this purpose $E_g = 0$ and $|\lambda_1| = |\lambda_0|$ in the expression (9.12), we can easily verify that such a state is unstable against electron–electron pairing.

Similarly, such an instability obtains in the presence of a field gap, $\Lambda$, that is smaller than the quantity $\Delta_0$.

The condition (9.26) was obtained under the assumption that the energy relaxation occurred as a result of electron–electron collisions. This is true when $(\mu/E_F) \ll 1$, i.e., when the electron-phonon Coulomb collisions are suppressed on account of the Pauli principle [27]:

$$\frac{1}{\tau_{ee}} = \frac{1}{\tau_{co}} \left(\frac{\mu}{E_F}\right)^2, \tag{9.29}$$

where $\tau_{Co}$ is the Coulomb collision time.

Thus, for $(\mu/E_F) \ll 1$ the time $\tau_{ee}$ can be much longer than the electron-phonon collision time $\tau_{ph}$, as was assumed above. When the condition (9.26) is satisfied at the same time, the characteristic time, $\tau_R$, of the excitation recombination across the gap $\Delta_0$ will be longer than the characteristic time, $\tau_{ph}$, of relaxation in energy terms, and an inverted population can therefore be realized.

The question, however, remains whether or not the finite excitation lifetime $\tau_R$ will suppress the Cooper pairing effect. Notice that the fulfillment of the condition (9.26) guarantees the suppression of the electron–phonon recombination processes, but that the recombination due to the electron–electron processes will not be suppressed therewith and will be determined by the expression (9.29). But it can be seen from a comparison of this expression with the expression, (9.28), for the magnitude of the gap $\Delta_0$ that the parameter $\mu/E_F$ ($E_F \approx \tilde{\omega}$), which ensures the condition $\tau_{\rm ph} \ll \tau_{ee}$, also leads to a small value of $\Delta_0$.

On the other hand, the quantity $1/\tau_{ee}$ is quadratic in the Coulomb-interaction constant $\lambda$, whereas the quantity $\Delta_0$, (9.28), depends exponentially on this same interaction constant. Therefore, even when the quantity $\lambda \approx 1$ (which, it is true, is typical for normal electron densities), the parameter $1/\tau_{ee}\Delta_0$ will be of the order of unity.

If $\tau_{ee}$ played the role of the lifetime of the Cooper pairs, then, as in the case of scattering by magnetic impurities [51], superconductivity would be impossible when $(1/\tau_{ee}\Delta_0) > 1$. But in the problem under consideration the time $\tau_R \approx \tau_{ee}$ governs the recombination of the quasiparticles with the quasiholes. As a result of the action of the pump source, the particles and holes get excited, becoming thermalized over a period of time equal to the energy relaxation time $\tau_{\rm ph}$ ($\ll \tau_{ee} \approx \tau_R$) and redistributed into coherent quasiparticle and quasihole states over a period $\tau_Q$ [52]. Since the formation of the gap $\Delta_0$ in the presence of repulsion is determined by an energy range of the order of $\mu \gg \Delta_0$ [see (9.28)] and the role of the coherence factors is minor at energies $\xi \gg \Delta_0$, the time $\tau_Q$ will be of the order of $\tau_{\rm ph}$ [52].

So, if $\tau_{\rm ph} \ll \tau_{ee}$ and the pump source acts constantly, then the recombination time will be longer than the time required for the establishment of the inverted coherent quasiparticle state, which will make the appearance of the superconducting state in the presence of repulsion possible. Thus, the recombination time $\tau_R$ (for the quasiparticles and not the pairs) should be compared not with the "pairing" time $\hbar/\Delta$ but with the time required for the establishment of the coherent quasiparticle state. Possibilities exist (at least in principle) of significantly increasing the time $\tau_R$ by means of the electron–electron collisions in comparison with (9.29). One of these possibilities lies in the use of quasi-one-dimensional systems. As shown in [53], in the one-dimensional case the electron–electron collisions do not lead to energy relaxation processes, i.e., $1/\tau_{ee}$ vanishes. The one-dimensional electron motion was brought about in those investigations by means of a strong magnetic field.

The assertion can be made that the electron–electron relaxation in quasi-one-dimensional superconductors is also largely suppressed. In other words, it can be shown that in this case

$$\frac{1}{\tau_{ee}} \approx \frac{1}{\tau_{\rm co}}\left(\frac{\Delta}{E_F}\right)^2,$$

i.e., that the relaxation in such superconductors differs from the relaxation in the three-dimensional case in that the quantity $\mu$ is replaced by $\Delta$ ($\mu >> \Delta$) in the expression (9.29).

Since in the case of systems with repulsion, superconductivity is possible only in the presence of an inverted population, i.e., under nonequilibrium conditions, the problem of the realization of any particular state cannot be solved by comparing the free energies. From this point of view, superconductivity is energetically disadvantageous in the presence of repulsion [49]. A method of investigating the stability properties of nonequilibrium superconducting systems is proposed in [54], where it is noted that, if the quasiparticle level $\mu$ is located in a region in which the density of states increases with increasing energy, then a system with an inverted population is stable against plasma oscillations [54].

## 4. Some Properties of Superconductors with an Inverted Population

Since the superconducting gap is not formed at the Fermi quasilevels $\pm\mu$, the minimum excitation energy of the quasiparticles is equal to zero, as in the nonsuperconducting state, i.e., as in the $\Delta = 0$ case. The question therefore arises whether an undamped current can exist in such a system in the presence of scattering. To investigate this question, we should find the current in an alternating electric field, $E(t) = E_0 e^{i\omega t}$, taking account of the elastic scattering by the impurities. We shall follow Mattis and Bardeen [55] in regards to the method of allowing for the impurity scattering.

Then, using the Keldysh technique [56], we obtain for the current an expression that coincides with the one given in [55], except that instead of the expression (9.2) for $f(\xi)$, we now have the nonequilibrium quasiparticle distribution functions, $f(\xi)$, obeying the kinetic equations [56]:

$$j(\mathbf{r}, t) = \frac{e^2 N(0)\, v_F}{2\pi^2 \hbar c} e^{i\omega t} \int \frac{\mathbf{R}\,[\mathbf{R} \cdot \mathbf{A}_\omega(\mathbf{r}')\, \mathcal{T}(\omega, R)\, e^{-R/l}\, d\mathbf{r}'}{R^4},$$

$$\mathcal{T}(\omega, R) = \int\limits_{-\infty}^{\infty} d\xi\, d\xi' \cos\alpha(\xi - \xi') \left[L(\omega, \xi, \xi') - \frac{f(\xi) - f(\xi')}{\xi' - \xi}\right],$$

$$L(\omega, \xi, \xi') = \frac{1}{4}\left[1 + \frac{\xi\xi' + \Delta^2}{\varepsilon\varepsilon'}\right]\left[\frac{f(\xi') - f(\xi)}{\varepsilon - \varepsilon' - \omega + i\delta} + \frac{f(\xi') - f(\xi)}{\varepsilon - \varepsilon' + \omega - i\delta}\right] +$$

$$+ \frac{1}{4}\left[1 - \frac{\xi\xi' + \Delta^2}{\varepsilon\varepsilon'}\right]\left[\frac{1 - f(\xi') - f(\xi)}{\varepsilon + \varepsilon' - \omega + i\delta} + \frac{1 - f(\xi) - f(\xi')}{\varepsilon + \varepsilon' + \omega - i\delta}\right],$$

$$\delta \to +0, \quad \mathbf{R} = \mathbf{r} - \mathbf{r}', \quad \alpha = \frac{R}{\hbar v_F},$$

$$\mathbf{E} = \frac{\partial \mathbf{A}}{\partial t}, \quad \varepsilon = \sqrt{\xi^2 + \Delta^2}.$$

Let us find the expression for the current in the $\omega \longrightarrow 0$ limiting case, which is the case of interest to us here, by substituting the expression (9.27) for $f(\xi)$:

$$\mathbf{j}(r, t) = \mathbf{E}(t)\left[2\sigma_N + \frac{i\pi\Delta}{\omega}\sigma_N\right], \quad \mu \gg \Delta, \tag{9.30}$$

where $\sigma_N = 2e^2 n_0 v_F l/3m$ is the normal conductivity, $l$ being the mean freepath.

It can be seen from the formula (9.30) that such a system can be described by the two-fluid model of superconductivity. The first term in (9.30) then corresponds to the normal (nonsuperfluid) component and is due to the scattering of the excitations close to the Fermi quasilevels $\pm\mu$.

The conductivity of the superfluid component (second term), on the other hand, becomes infinite as $\omega \longrightarrow 0$, i.e., it is insensitive to scattering. Notice that the sign of the second term is opposite to the sign of the corresponding term in the expression for the current of an equilibrium superconductor. The physical meaning of this result is clear from Fig. 9.5.

Figure 9.5b corresponds to the conducting state (i.e., to the pairing of electrons with a nonzero resultant momentum) without allowance for scattering. Allowance for scattering leads to the symmetrization of the momentum distribution of the electrons at the Fermi quasilevels $\pm\mu$ (Fig. 9.5c), while for $p \approx p_0$ the asymmetry remains on account of the suppression of the scattering as a result of the presence of the gap.

Notice that for $\Delta = 0$ we obtain for the current the usual expression:

$$\mathbf{j} = \sigma_N \mathbf{E}.$$

The corresponding calculation for the semiconductor model follows the above-presented scheme. In the constant-density-of-states approximation, the expression for the current is identical in form to (9.30), in which $\sigma_N$ must be taken to mean the conductivity, $\sigma_N = 2e^2 m\mu l/3\pi^2$, over a band with an electron concentration of $n = (2m\mu)^{3/2}/3\pi^2$ when the mean free path $l = v_F \tau_{tr}$.

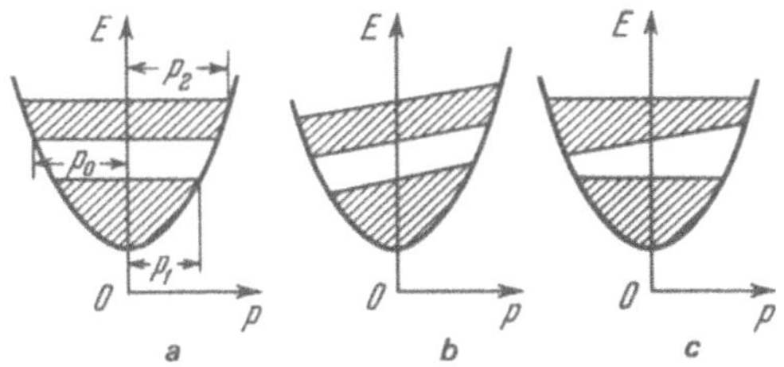


**Figure 9.5. The electron spectrum of a metal in a state with an inverted population: (a) nonconducting state; (b) conducting state without allowance for scattering; (c) conducting state with allowance for scattering.**

## 5. Anomalous Paramagnetism in a Nonequilibrium Superconductor

The paramagnetic component of the Fourier transform of the current in a nonequilibrium superconductor located in a weak constant magnetic field is given by the usual expression

$$\mathbf{j}_{\mathbf{q}}^{\Pi} = -\frac{2e^2}{m^2(2\pi)^4}\int d\omega\, d\mathbf{p}\, \mathbf{p}\,(\mathbf{pA})\,[G(\omega, \mathbf{p})\, G(\omega, \mathbf{p}+\mathbf{q}) + F^{+}(\omega, \mathbf{p})\, F(\omega, \mathbf{p}+\mathbf{q})], \tag{9.31}$$

where the functions $G$ and $F$ are given by the expressions (9.21).

Substituting (9.21) into (9.31) and integrating over $\omega$, we obtain for $\mathbf{j}_{\mathbf{q}}^{\Pi}$ an expression that coincides formally with the expression obtained in [2] if instead of the expression (9.2) for $f(\xi)$ we substitute the nonequilibrium quasiparticle distribution function:

$$\mathbf{j}_{\mathbf{q}}^{\Pi} = \frac{e^2\hbar^2}{m^2c\,(2\pi)^3}\int d\mathbf{p}\,(2\mathbf{p}+\mathbf{q})\,\mathbf{p}\mathbf{A}_{\mathbf{q}} L(\xi_{\mathbf{p}}, \xi_{\mathbf{p}+\mathbf{q}}),$$

$$L(\xi, \xi') = \frac{1}{2}\left(\frac{1-f(\xi)-f(\xi')}{\varepsilon+\varepsilon'}\right)\left(1-\frac{\xi\xi'+\Delta^2}{\varepsilon\varepsilon'}\right) + \frac{1}{2}\left(\frac{f(\xi')-f(\xi)}{\varepsilon-\varepsilon'}\right)\left(1+\frac{\xi\xi'+\Delta^2}{\varepsilon\varepsilon'}\right), \qquad \varepsilon = \sqrt{\xi^2+\Delta^2}.$$

As $q \longrightarrow 0$, the first term in $L(\xi, \xi')$ vanishes because of the vanishing of the coherent factor, while the second term goes over into $\partial f/\partial\varepsilon$.

Then the total current, including the diamagnetic term, assumes the form

$$\mathbf{j}_{\mathbf{q}\to 0} = -\frac{e^2 n_0}{mc}\left(1+\frac{2E_F}{p_F^5}\int_0^\infty p^4\,\frac{\partial f}{\partial\varepsilon}\,dp\right)\mathbf{A}, \qquad (\mathbf{A} \equiv \mathbf{A}_0). \tag{9.32}$$

Substituting into (9.32) the expression for $f(\xi)$ (for $T \ll \mu$), we finally find that

$$\mathbf{j}_0 = \frac{e^2 n_0}{mc}\left(\frac{2\mu}{\sqrt{\mu^2-\Delta^2}}-1\right)\mathbf{A} \approx +\frac{e^2 n_0}{mc}\mathbf{A}, \quad (\mu \gg \Delta), \tag{9.33}$$

i.e., that the current is equal in magnitude but opposite in sign to the current in an equilibrium superconductor. This means that the system under consideration possesses anomalous paramagnetism, which results in the penetration of the magnetic field into the sample, the magnetic field undergoing oscillations with a period of $(4\pi n_0 e^2/mc^2)^{-1/2}$ in the process [58].

The physical meaning of this result is clear from Fig. 9.5. The electrons with momenta close to $p = p_1$ and $p = p_2$ make a contribution to the paramagnetic current, whereas those with momenta equal to $p = p_0$ contribute nothing because of the presence of the gap. Thus, the paramagnetic current is doubled, so that the total current, including the diamagnetic term, is equal to (9.33).

The magnetic instability of such a state even in the absence of an external field can be inferred precisely from the anomalous paramagnetism of superconductors with repulsion on the basis of the Maxwell equations. In fact, to solve the problem of the stability of the homogeneous state of such superconductors, we should investigate the complete system of equations (including the self-consistency equations) describing the behavior of the nonequilibrium superconductor in a magnetic field. For equilibrium superconductors, the ordinary Ginzburg-Landau equations [59] with two constants–the effective charge, which, in this case, should be a multiple of the elementary charge, and the effective mass, which remains arbitrary–constitute such a system. Therefore, it appears, in accordance with (9.33), that is will be sufficient to assume the mass in the Ginzburg-Landau equations to be negative (which will correspond to going over from the quasiparticle to the quasihole description). But then the gradient term in the equation for the order parameter will also become negative. This will indicate that the inhomogeneous state will be favored in the system under consideration even in the absence of an external field. On the other hand, the application of various stability criteria to nonequilibrium superconductors with repulsion yields a definite result [44]: The homogeneous state is stable. This question can be resolved by deriving the Ginzburg-Landau equations for a nonequilibrium superconductor from the microscopic theory.

In the presence of an external magnetic field, we should replace in (9.19) $\mathbf{p}$ by $\mathbf{p} - \frac{e}{c}\mathbf{A}$ after writing the system of equations in the coordinate form. In the case of the semiconductor model, in the vicinity of the critical pump-intensity $\beta_c$, which is equal to

$$\beta_c = \left(\frac{\tilde{\omega} E q}{2}\right)^{1/2} \exp\left(\frac{1}{\lambda_0 + \lambda_1}\right)$$

the quantity $\Delta$ is small: $\Delta/\sqrt{2}Eq \equiv \psi = [(\beta - \beta_c)/\beta_c]^{1/2}$. Let us emphasize that the entire analysis is being carried out for the temperature $T = 0$. By expanding the system (9.19) in powers of $\Delta$ up to the third order and substituting the Green function thus found into the self-consistency equations (9.23) and the ordinary expression for the current density, we can derive the following equations [58]:

$$\left[\frac{1}{2m}\left(\frac{\partial}{\partial \mathbf{r}}-\frac{2ie}{c}\mathbf{A}\right)^2+3Eq\left(\frac{\beta-\beta_c}{\beta_c}\right)-|\psi|^2\right]\psi(\mathbf{r})=0$$

$$\mathbf{j}(\mathbf{r})=\frac{ie}{m}\left(\psi^*\frac{\partial\psi}{\partial \mathbf{r}}-\psi\frac{\partial\psi}{\partial \mathbf{r}}\right)+\frac{4e^2\mathbf{r}}{m}|\psi|^2\mathbf{A}(\mathbf{r}) \tag{9.34}$$

where $n = (2m\mu)^{3/2}/3\pi^2$ is the number of nonequilibrium electrons in the conduction band. It follows from (9.34) that the quantity $\psi(\mathbf{r})$ serves as the expansion parameter; therefore, analogous equations cannot be obtained for the metallic model. It can also be seen that the first equation is similar to the Ginzburg–Landau equation for an equilibrium superconductor, in which the role of temperature is played by the pump intensity $\beta$. The sign of the expression for the current density is opposite to that of the normal expression. This result is explained by the fact that the first equation describes the behavior of the "wave function" of the superconducting pairs, whereas the second equation contains the total current, which is the resultant of the diamagnetic pair current and the paramagnetic quasiparticle current. In this case the quasiparticle current turns out to be superconducting and, as indicated above, 2 times greater than the diagmagnetic pair current.

An analysis of (9.34) shows that the homogeneous state of such a superconductor is stable in the absence of a field. In the presence of an external magnetic field, this field penetrates into the sample, undergoing in the process oscillations with period $(mc^2E_g^2/8\pi ne^2\Delta^2)^{1/2}$, which in turn give rise to oscillations of the superconducting gap [58].

It should be noted that electron–hole pairing, corresponding to the appearance of dielectric gaps at the Fermi quasilevels, is possible in the presence of superconducting pairing. In this case the system will turn out to be diamagnetic, as in the case of ordinary superconductors.

## 6. Conclusion

Let us consider the effect of temperature on the superconducting state. As the temperature rises, a change of the order of unity occurs in the distribution function in a band of width $k_B T$ close to the Fermi quasilevel $\mu$. Since in the energy region close to the superconducting gap $f(\xi, T)$, (9.7), varies with temperature roughly like $\exp(-\mu/k_B T)$, it is clear that the transition temperature $T_c \sim \mu$.

Let us note that, under equilibrium conditions with attraction between the particles, the quantity $T_c$ is determined by the magnitude of the superconducting gap $\Delta$ for $T = 0$. In the case under consideration the quantity $T_c$ depends on the degree of inversion of the population and can be high even for a small gap.

Let us estimate the intensity, $I$, of the pump source at a frequency $\omega \gtrsim 2\mu$. The steady state sets in when

$$\varkappa I = \frac{n}{\tau_R},$$

where $\varkappa$ is the absorption coefficient. Then from the condition, (9.26), for the existence of a self-sustaining solution we obtain

$$I > \frac{1}{\varkappa\tau_R}\left[\frac{m^2\omega_D\tilde{\omega}\hbar}{2}\exp\left(\frac{1}{\lambda}\right)\right]^{3/4}. \tag{9.35}$$

For $\tau_R \approx 10^{-8}$ sec, $\tilde{\omega} \approx 1$ eV, $m \approx (0.1 - 1)m_0$, $\lambda = 0.5$, and $\varkappa = (10^3\text{-}10^4)$ cm$^{-1}$, we obtain for the intensity $I$ a value of the order of $10^3$-$10^5$ W/cm$^2$. The corresponding critical temperature, $T_c \approx \hbar/m(\varkappa\tau_R I)^{2/3}$, turns out to be of the order of $10^3$ K. We note that our estimate for the intensity is rather illustrative.

As noted above, to realize superconductivity in the $\lambda > 0$ case in the metallic model, we should create with the aid of a strong monochromatic field an inter-band-transition-induced dielectric gap $\Lambda$ extending over the entire Fermi surface. For this purpose, the shape of the second band's constant energy surface for the energy value at a distance $\hbar\Omega$ from the Fermi level should coincide with that of the Fermi surface. Since such a coincidence is possible only in the vicinity of the extremal points, we must have appropriate semiconductor in which the Fermi occupation is achieved by means of impurity doping.

As noted above, in semiconductors ($E_g > 0$) and semimetals ($E_g < 0$) superconductivity is possible in the presence of an inverted population if the condition $|E_g| < \Delta_0$ is fulfilled. Thus, it is necessary to investigate the narrow-band layered and quasi-one-dimensional semiconductors as well as semiconducting compounds of the $A_2B_6$ type and the alloys BiSb and $Bi_2Tl_3$; what is more, the most suitable materials for obtaining sufficiently long excitation lifetimes are fairly pure substances with band extrema that do not coincide in momentum space.

GaAs-type semiconductors with several conduction-band minima that do not coincide in momentum space may be suitable for the realization of the necessary conditions in two-band systems in which the Fermi level of the lower, partially filled band should touch the bottom of an upper band with significantly heavier effective mass (see Fig. 9.4b). In this case we can by means of donor-impurity doping get the Fermi level to be at the location necessary for the occurrence of the superconducting state.

On the face of it, it seems that the condition $|E_g| < \Delta_0$ cannot be fulfilled since, as a result of the electron–hole Coulomb attraction, the forbidden band-width in the stable dielectric state should be greater than the Coulomb binding energy (the exciton dielectric) [49]. But in the presence of an inverted population in the $\xi$ band extending from 0 to $\mu$, the Coulomb interaction will correspond to electron–hole repulsion. Therefore, steady-state electron–hole pairing is impossible.

# 10

# References

## Chapter 1


1. V. L. Ginzburg, *Annu. Rev. Mater. Sci.* **2**, 663 (1972).
2. O. Fischer, M. Decroux, S. Roth, R. Chevrel, and M. Sargent, *J. Phys. C* **8**, L474 (1975).
3. W. A. Little, *Phys. Rev. A* **134**, 1416 (1964).
4. V. L. Ginzburg, *Phys. Lett.* **13**, 101 (1964); *Zh. Éksp. Teor. Fiz.* **47**, 2318 (1964).
5. V. L. Ginzburg, *Usp. Fiz. Nauk* **95**, 91 (1968); **101**, 185 (1970); **118**, 315 (1976).
6. J. Bardeen and J. Schrieffer, *New Developments in the Study of Superconductivity* (Russian translation), Fizmatgiz (1962).
7. P. G. de Gennes, *Superconductivity of Metals and Alloys*, Benjamin, New York (1966).
8. J. R. Schrieffer, *Theory of Superconductivity*, Benjamin, New York, (1964).
9. V. M. Agranovich and V. L. Ginzburg, *Spatial Dispersion in Crystal Optics and the Theory of Excitons* John Wiley and Sons, New York (1966).
10. D. Pines, *Elementary Excitations in Solids*, Benjamin, New York (1963).
11. J. Bardeen, L. N. Cooper, and J. R. Schrieffer, *Phys. Rev.* **106**, 162; **108**, 1175 (1957).
12. M. L. McMillan, *Phys. Rev.* **167**, 331 (1968).
13. A. E. Karakozov, E. G. Maksimov, and S. A. Mashkov, *Zh. Éksp. Teor. Fiz.* **68**, 1937 (1975).
14. O. V. Dolgov and E. G. Maksimov, *Pis'ma Zh. Éksp. Teor. Fiz.* **28**, 3 (1978); *Rev. Mod. Phys.* **53**, 81 (1981).
15. B. T. Geilikman, *Usp. Fiz. Nauk* **109**, 65 (1973); L. Ley, *Phys. Rev. Lett.* **35**, 1796 (1975); W. E. Rudge and P. M. Grant, *Phys. Rev. Lett.* **35**, 1799 (1975); P. Mengel, P. M. Grant, W. E. Rudge, B. H. Schechtman, and D. W. Rice, *Phys. Rev. Lett.* **35**, 1803 (1975); *Physics Today* **28**, No. 6, 17 (1975); R. L. Civiak *et al.*, *Solid State Commun.* **18**, 1205 (1976); C. K. Chiang *et al.*, *Solid State Commun.* **18**, 1451 (1976).
16. Yu. V. Kopaev, *Zh. Éksp. Teor. Fiz.* **58**, 1012 (1970); A. I. Rusinov, Do Chan Kat, and Yu. V. Kopaev, *Zh. Éksp. Teor. Fiz.* **65**, 1984 (1973).
17. B. A. Volkov and Yu. V. Kopaev, *Pis'ma Zh. Eksp. Teor. Fiz.* **27**, 10 (1978).
18. B. A. Volkov, Yu. V. Kopaev, and V. V. Tugushev, *Pis'ma Zh. Éksp. Teor. Fiz.* **27**, 615 (1978); **30**, 317 (1979); *Zh. Éksp. Teor. Fiz.* **18**, 729 (1981).
19. B. A. Volkov, V. L. Ginzburg, and Yu. V. Kopaev, *Pis'ma Zh. Éksp. Teor. Fiz.* **27**, 221 (1978).
20. V. L. Ginzburg, *Solid State Commun.* **39**, 991 (1981).
21. N. B. Brandt, S. V. Kuvshinnikov, A. P. Rusakov, and M. V. Semenov, *Pis'ma Zh. Éksp. Teor. Fiz.* **27**, 37 (1978).
22. C. W. Chu, A. P. Rusakov, S. Huang, S. Early, T. H. Geballe, and C. Y. Huang, *Phys. Rev. B* **18**, 2116 (1978); see also: Comments, *Solide State Phys.* **9**, 115 (1979).

23. L. Landau, *Phys. Z. Sowjetunion* **4**, 43 (1933).
24. W. Hiesenberg, *Z. Naturforsch. Teil A* **2**, 185 (1947); *Ann. Phys. (Leipzig)* **3**, 289 (1948).
25. V. L. Ginzburg, *Usp. Fiz. Nauk* **48**, 25 (1952); *Fortschr. Phys.* **1**, 101 (1953).
26. V. L. Ginzburg and D. A. Kirzhnits, *Zh. Éksp. Teor. Fiz.* **46**, 397 (1964).
27. T. M. Rice, *Phys. Rev. A* **140**, 1189 (1965).
28. P. C. Hohenberg, *Phys. Rev.* **158**, 383 (1967).
29. W. A. Little, *Phys. Rev.* **156**, 396 (1967).
30. J. S. Langer and V. Ambegaokar, *Phys. Rev.* **164**, 498 (1967).
31. V. V. Shmidt, *Zh. Éksp. Teor. Fiz.* **54**, 263 (1968).
32. V. L. Ginzburg and D. A. Kirzhnits, *Dokl, Akad. Nauk SSSR* **176**, 553 (1967).
33. Y. Imry, *Ann. Phys. (N.Y.)* **51**, 1 (1969).
34. D. I. Scalapino, M. Sears, and R. A. Ferrel, *Phys. Rev. B* **6**, 3409 (1972).
35. G. Parisi, *Phys. Lett. A* **43**, 379 (1973).
36. V. L. Berezinskii, *Zh. Éksp. Teor. Fiz.* **59**, 907; **61**, 1144 (1971).
37. J. M. Kosterlitz and D. J. Thouless, *J. Phys. C* **6**, 1181 (1973).
38. V. L. Ginzburg, *Usp. Fiz. Nauk* **113**, 335 (1979).
39. *Proc. Intern. Conference on Organic Superconductors*, in: *J. Polymer Sci.*, part C, No. 29 (1970).
40. L. N. Bulaevskii, *Usp. Fiz. Nauk* **115**, 263 (1975).
41. D. Davis, H. Gutfreund, and W. A. Little, *Phys. Rev. B* **13**, 4766 (1976).
42. G. F. Zharkov and Yu. A. Uspenskii, *Zh. Éksp. Teor. Fiz.* **61**, 2123 (1971); **65**, 1460, 2511 (1973).
43. D. Allender, J. Bray, and J. Bardeen, *Phys. Rev. B* **7**, 1020 (1973); **8**, 4433 (1973).
44. V. M. Agranovich, A. G. Mal'shukov, and M. A. Mekhtiev, *Zh. Éksp. Teor. Fiz.* **63**, 2274 (1972); see also *Fiz. Tverd. Tela* **14**, 3684 (1972); **15**, 3392 (1973); J. Halbritter, *Solid State Commun.* **18**, 1447 (1976).
45. V. L. Ginzburg and D. A. Kirzhnits, *Phys. Rep. C* **4**, 345 (1972).
46. D. L. Miller, M. Strongin, O. F. Kammerer, and B. G. Streetman, *Phys. Rev. B* **13**, 4834 (1976); A. K. Ghosh and D. H. Douglass, *Phys. Rev. Lett.* **37**, 32 (1976).
47. A. D. Yoffe, *Annu. Rev. Mater. Sci.* **3**, 147 (1973).
48. L. N. Bulaevskii, *Usp. Fiz. Nauk* **116**, 449 (1975).
49. V. V. Lomonosov, Yu. A. Malov, and V. P. Smilga, *Fiz. Nizk. Temp.* **4**, 599 (1978).
50. Yu. E. Lozovik and V. I. Yudson, *Pis'ma Zh. Eksp. Teor. Fiz.* **22**, 566 (1975); **25**, 18 (1977); *Zh. Eksp. Teor. Fiz.* **76**, 670 (1979).
51. S. I. Shevchenko, *Fiz. Nizk. Temp.* **3**, 605 (1977).

## Chapter 2

1. O. V. Dolgov, Preprint FIAN, No. 92 (1977); Kratk. Soobshch. Fiz., No. 11, 8 (1977); O. V. Dolgov and I. I. Mazin, *Fiz. Tverd. Tela* **20**, No. 12 (1978); O. V. Dolgov and V. S. Karakhtanov, *Kratk. Soobshch. Fiz.*, No. 9 (1978); O. V. Dolgov, E. G. Maksimov, *Usp. Fiz. Nauk* (in press) (1982).
2. V. L. Ginzburg, *Usp. Fiz. Nauk* **97**, 601 (1969); V. L. Ginzburg and D. A. Kirzhnits, *Nature* **220**, 148 (1969); L. de Cezare, *Astrophys. Space Sci.* **25**, 133 (1973).
3. D. A. Kirzhnits, *Usp. Fiz. Nauk* **104**, 469 (1971).
4. O. V. Dolgov, thesis, Moscow Physics and Technology Institute (1975).
5. G. M. Éliashberg, *Zh. Éksp. Teor. Fiz.* **38**, 966 (1960); **39**, 1437 (1960); M. V. Buikov, *Dokl. Akad. Nauk SSSR* **132**, 561 (1960).
6. J. Schrieffer, *Theory of Superconductivity*, Benjamin, New York (1964).
7. D. A. Kirzhnits, E. G. Maksimov, and D. I. Khomskii, Preprint FIAN, No. 108 (1970);

Preprint IFM UNTsAN, Sverdlovsk (1973); D. A. Kirzhnits, E. G. Maksimov, and D. I. Khomskii, *J. Low Temp. Phys.* 10, 79 (1973).
8. V. L. Ginzburg, *Usp. Fiz. Nauk* 101, 185 (1970).
9. M. L. Cohen and P. W. Anderson, *Conference on d- and f-Band Superconductivity*, American Institute of Physics, Rochester, 1971, Proceedings, New York (1972), p. 17.
10. A. Fasolino, M. Parinello, and M. Tosi, *Phys. Lett. A* 66, 119 (1978).
11. O. V. Dolgov and E. G. Maksimov, *Pis'ma Zh. Éksp. Teor. Fiz.* 28, 3 (1978).
12. O. V. Dolgov, D. A. Kirzhnits, and E. G. Maksimov, *Rev. Mod. Phys.* 53 (1), 81 (1981).
13. A. A. Abrikosov, L. P. Gor'kov, and I. E. Dzyaloshinskii, *Methods of Quantum Field Theory in Statistical Physics* (in Russian), Fizmatgiz (1962).
14. D. A. Kirzhnits, *Usp. Fiz. Nauk* 119, No. 2, 357 (1976); *Sov. Phys.-Usp.* 19, 530 (1976).
15. D. Pines and F. Nozières, *Theory of Quantum Liquids*, Benjamin, New York (1966).
16. L. D. Landau and E. M. Lifshitz, *Electrodynamics of Continuous Media* Pergamon Press. New York (1957).
17. D. A. Kirzhnits and V. N. Sazonov, in: *The Einstein Collection*, 1973 (in Russian), Nauka (1974).
18. P. Martin, *Phys. Rev.* 161, 143 (1967).
19. D. A. Kirzhnits and Yu. A. Nepomnyashchii, *Zh. Éksp. Teor. Fiz.* 59, 2203 (1970); Yu. A. Nepomnyashchii, *Teor. Mat. Fiz.* 8, 413 (1971); Yu. A. Nepomnyashchii and A. A. Nepomnyashchii, *Teor. Mat. Fiz.* 9, 137 (1971).
20. S. Coleman and E. Weinberg, *Phys. Rev. D* 7, 1888 (1973); J. Illiopoulos, C. Itsykson, and A. Martin, *Rev. Mod. Phys.* 47, 165 (1975); D. A. Kirzhnits and A. D. Linde, *Zh. Éksp. Teor. Fiz.* 67, 1263 (1974).
21. M. A. Leontovich, *Introduction to Thermodynamics* (in Russian), Gostekhizdat, Moscow–Leningrad (1952); *Statistical Physics* (in Russian), Gostekhizdat, Moscow–Leningrad (1944).
22. D. A. Kirzhnits, Yu. E. Lozovik, and G. V. Shpatakovskaya, *Usp. Fiz. Nauk* 117, No. 1, 3 (1975).
23. I. O. Kulik, *Zh. Éksp. Teor. Fiz.* 47, 2159 (1964).
24. B. T. Geilikman, V. Z. Kresin, and N. F. Masharov, *Fiz. Nizk. Temp.* 1, 31, 154 (1975).
25. N. N. Bogolyubov, V. V. Tolmachev, and D. V. Shirkov, *A New Method in the Theory of Superconductivity* Consultants Bureau, New York (1959).
26. D. A. Kirzhnits, *Pis'ma Zh. Éksp. Teor. Fiz.* 9, 360 (1969).
27. V. P. Kopyshev, *Dokl. Akad. Nauk SSSR* 161, 1067 (1965); J. A. Reissland, *Physics of Phonons*, Wiley, New York (1973).
28. A. B. Migdal, *Zh. Éksp. Teor. Fiz.* 34, 1438 (1958).

## Chapter 3

1. I. R. Gavaler, *J. Appl. Phys. Lett.* 23, 480 (1973).
2. L. R. Testardi, I. H. Wernic, and W. A. Royer, *Solid State Commun.* 15, 1 (1974).
3. J. Bardeen, L. N. Cooper, and J. R. Schrieffer, *Phys. Rev.* 108, (1957).
4. W. L. McMillan, *Phys. Rev.* 167, 331 (1968).
5. A. E. Karakozov, E. G. Maksimov, and S. A. Mashkov, *Zh. Éksp. Teor. Fiz.* 68, 1937 (1975).
6. B. P. Allen and R. C. Dynes, *Phys. Rev. B* 12, 905 (1975).
7. L. B. Dubovskii and A. N. Kozlov, *Zh. Éksp. Teor. Fiz.* 68, 2224 (1975).
8. G. M. Éliashberg, *Zh. Éksp. Teor. Fiz.* 38, 366 (1968).
9. H. Fröhlich, *Phys. Rev.* 79, 845 (1950); *Proc. R. Soc. London Ser. A* 215, 291 (1952).
10. Q. Wentzel, *Phys. Rev.* 83, 168 (1951).

11. B. T. Matthias, in: *Superconductivity of d- and f-Band Metals*, American Institute of Physics Conf., Rochester, 1971, New York (1972), p. 367.
12. L. R. Testardi, L. H. Wernick, W. A. Royer, D. D. Bacon, and A. R. Storm, *J. Appl. Phys.* **45**, 446 (1974).
13. F. Bloch, *Z. Phys.* **59**, 208 (1930).
14. A. B. Migdal, *Zh. Éksp. Teor. Fiz.* **34**, 1438 (1958).
15. D. Mattis and W. Langer, *Phys. Rev. Lett.* **25**, 376 (1970); Yu. V. Kopaev and R. V. Timerov, *Zh. Éksp. Teor. Fiz.* **63**, 280 (1972).
16. A. M. Afanas'ev and Yu. Kagan, *Zh. Éksp. Teor. Fiz.* **43**, 1456 (1962).
17. L. N. Bulaevskii, *Usp. Fiz. Nauk* **115**, 271 (1975).
18. M. Born and R. Oppenheimer, *Ann. Phys. (Leipzig)* **84**, 457 (1927).
19. G. V. Chester, *Adv. Phys.* **10**, 357 (1961).
20. E. G. Brovman and Yu. M. Kagan, *Usp. Fiz. Nauk* **112**, 369 (1974).
21. B. T. Geilikman, *Usp. Fiz. Nauk* **115**, 403 (1975).
22. E. G. Maksimov, *Zh. Éksp. Teor. Fiz.* **69**, 2236 (1975).
23. W. Hanke and H. Bilz, in: *Neutron Inelastic Scattering*, Proc. Symposium, 1972, Vienna (1972), p. 3.
24. D. Pines and F. Nozières, *Theory of Quantum Liquids*, Benjamin, New York (1966).
25. J. Schrieffer, *Theory of Superconductivity*, Benjamin, New York (1964).
26. D. Bohm and D. Pines, *Phys. Rev.* **82**, 625 (1951).
27. J. Hubbard, *Proc. R. Soc. London Ser. A* **224**, 199 (1958).
28. S.-K Chan and V. Heine, *J. Phys. F* **3**, 795 (1973).
29. W. Harrison, *Pseudopotentials in the Theory of Metals*, Benjamin, New York (1966).
30. O. V. Dolgov and E. G. Maksimov, *Pis'ma Zh. Éksp. Teor. Fiz.* **28**, 3 (1978).
31. H. Beck and D. Straus, *Helv. Phys. Acta* **48**, 655 (1975).
32. É. G. Batyev and V. L. Pokrovskii, *Zh. Éksp. Teor. Fiz.* **46**, 262 (1964).
33. O. V. Dolgov, D. A. Kirzhnits, and E. G. Maksimov, Preprint, P. N. Lebedev Physical Institute, No. 278 (1978); *Rev. Mod. Phys.* **53**, 81 (1981).
34. I. T. Izuyama and M. Saitoch, *Solid State Commun.* **16**, 549 (1975).
35. I. O. Kulik, *Zh. Éksp. Teor. Fiz.* **66**, 2224 (1974).
36. J. W. Garland, *Phys. Rev.* **153**, 460 (1967).
37. G. Baum, *Ann. Phys. (N.Y.)* **14**, 1 (1961).
38. P. S. Keating, *Phys. Rev.* **175**, 1171 (1968).
39. N. S. Gillis, *Phys. Rev. B* **1**, 1872 (1970).
40. A. K. Rajagopal and M. H. Cohen, *Collect. Phen* **1**, 9 (1972).
41. E. S. Fradkin, *Tr. Fiz. Inst. Akad. Nauk SSSR* **29**, 7 (1965).
42. J. W. Geldart and R. Taylor, *Solid State Commun.* **9**, 7 (1971).
43. K. S. Sangwi, M. P. Tosi, A. Sjolander, and K. R. Land, *Phys. Rev.* **B1**, 1044 (1970).
44. F. Toigo and T. O. Woodruff, *Phys. Rev. B* **4**, 4312 (1971).
45. A. A. Kugler, *J. Stat. Phys.* **12**, 35 (1975).
46. V. Heine, P. Nozières, and J. W. Wilkins, *Philos. Mag.* **13**, 741 (1966).
47. L. Hedin and S. Lundquist, in: *Solid State Physics*, Vol. 23, edited by F. Seitz and D. Turnbull, Academic Press, New York (1969).
48. G. Arbman and V. von Barth, *J. Phys F* **5**, 1155 (1975).
49. M. Rasolt and S. H. Vosko, *Phys. Rev. Lett.* **32**, 297 (1974).
50. P. B. Allen, *Phys. Rev. B* **5**, 3857 (1972).
51. S. G. Das, *Phys. Rev. B* **7**, 2238 (1973).
52. I. R. Gomersall and B. L. Gyorffy, *J. Phys. F* **4**, 1204 (1974).
53. B. M. Klein and D. A. Papaconstantopoulos, *Phys. Rev. Lett.* **32**, 1193 (1974).
54. D. J. Scalapino, in: *Superconductivity*, Vol. 1, edited by R. D. Parks, Dekker, New York (1969).
55. J. M. Luttinger and P. Nozières, *Phys. Rev.* **127**, 1423, 1431 (1962).

56. T. M. Rice, *Ann. Phys. (N.Y.)* **31**, 100 (1965).
57. S. Adler, *Phys. Rev.* **126**, 413 (1962).
58. N. Wiser, *Phys. Rev.* **129**, 62 (1963).
59. S. K. Sinha, R. P. Gupta, and D. L. Price, *Phys. Rev. B* **9**, 2564, 2573 (1975).
60. S. K. Sinha and B. N. Harmon, *Phys. Rev. Lett.* **35**, 1515 (1975).
61. *Superconductivity in d- and f-Band Metals*, Proc. Second Rochester Conf., edited by D. H. Douglass, Plenum Press (1977).
62. *Proc. Puerto Rico Conf.*, San Juan, Puerto Rico, 1975, publ. in *Ferroelectrics* **16**, **17**, Nos. 1/2 (1977).
63. *Proc. Intern. Conf. on Lattice Dynamics*, edited by M. Balkanski Flammarion, Paris (1978).
64. P. B. Allen, in: *Dynamical Properties of Solids*, Vol. 3, edited by K. Horton and A. A. Maradudin, North Holland, Amsterdam and New York (1980).
65. W. Hanke, *Adv. Phys.* **27**, 287 (1978).
66. C. M. Varma and W. Weber, *Phys. Rev. Lett.* **39**, 1084 (1977).
67. S. J. Cale and D. G. Pettifor, *Solid State Commun.* **24**, 175 (1977).
68. K. Terakura, *J. Phys.* **C11**, 469 (1978).
69. J. F. Cooke, H. L. Davis, and T. M. Mostoller, *Phys. Rev. B* **9**, 2485 (1974).
70. R. P. Gupta and A. J. Freeman, *Phys. Rev. B* **13**, 4376 (1976).
71. E. G. Maksimov and I. I. Mazin, *Solid State Commun.* **27**, 527 (1974).

## Chapter 4

1. W. L. McMillan, *Phys. Rev.* **167**, 331 (1968).
2. R. C. Dynes, *Solid State Commun.* **10**, 615 (1973).
3. R. Benda, *Phys. Status Solidi* **38**, 451 (1969).
4. C. R. Leavens and I. P. Carbotte, *Can. J. Phys.* **49**, 724 (1971).
5. M. V. Medvedev, É. A. Pashitskii, and Yu. S. Pyatiletov, *Zh. Éksp. Teor. Fiz.* **65**, 1186 (1973).
6. A. E. Karakozov, E. G. Maksimov, and S. A. Mashkov, *Zh. Éksp. Teor. Fiz.* **68**, 1937 (1975).
7. G. M. Éliashberg, *Zh. Éksp. Teor. Fiz.* **38**, 366 (1960).
8. E. G. Batyev, *Fiz. Tverd. Tela* **6**, 3554 (1964).
9. D. I. Scalapino, in: *Superconductivity*, edited by R. D. Parks, Dekker, New York (1969).
10. J. Schrieffer, *Theory of Superconductivity*, Benjamin, New York (1964).
11. W. L. McMillan and I. M. Rowell, in: *Superconductivity*, edited by R. D. Parks, Dekker, New York (1969).
12. N. F. Berk and J. R. Schrieffer, *Phys. Rev. Lett.* **17**, 433 (1966).
13. G. Grimvall, *Phys. Kondens. Mater.* **9**, 283 (1969).
14. I. I. Sabo, *Phys. Rev. B* **1**, 1325 (1970).
15. D. A. Kirzhnits, E. G. Maksimov, and D. I. Khomskii, Preprint, Fiz. Inst. Akad Nauk SSSR, No. 108 (1970); *J. Low Temp. Phys.* **10**, 79 (1973).
16. P. B. Allen and R. C. Dynes, *Phys. Rev. B* **12**, 905 (1975).
17. W. Kessel, Z. Naturforsch. *Teil A* **29**, 445 (1974).
18. I. M. Rowell, W. L. McMillan, and W. L. Feldman, *Phys. Rev.* **178**, 837 (1969).
19. P. T. Truant and I. P. Carbotte, *Phys. Rev. B* **6**, 3642 (1972).
20. G. Bergmann, *Phys. Rev. B* **3**, 3757 (1971).
21. S. I. Vedeneev, A. I. Golovashkin, and G. P. Motulevich, *Fiz. Tverd. Tela* **16**, 668 (1974).
22. I. K. Yanson, *Zh. Éksp. Teor. Fiz.* **66**, 1035 (1974); *Fiz. Tverd. Tela* **16**, 3535 (1974).
23. E. G. Maksimov, *Zh. Éksp. Teor. Fiz.* **37**, 1562 (1969).

24. M. A. Coulthard, *J. Phys. F* **1**, 168 (1969).
25. J. P. Carbotte, P. T. Truant, and R. C. Dynes, *Can. J. Phys.* **48**, 1504 (1970).
26. B. L. Harman and S. K. Sinha, *Phys. Rev. B* **16**, 3519 (1977).
27. W. H. Butler, H. G. Smith, and N. Wakabayashi, *Phys. Rev. Lett.* **39**, 1004 (1977).
28. W. H. Butler, F. J. Pinski, and P. B. Allen, *Phys. Rev. B* **19**, 3708 (1979).
29. *Superconductivity in d- and f-Band Metals*, Proc. Second Rochester Conf., edited by D. H. Douglass, Plenum Press, New York (1977).
30. J. J. Hopfield, *Phys. Rev.* **86**, 443 (1969).
31. I. R. Gomersal and B. L. Gyorffy, *J. Phys. F* **4**, 1204 (1974).
32. I. F. Foulkes and I. R. Gomersall, *J. Phys. F* **5**, 153 (1975).
33. J. Appel and W. Kohn, *Phys. Rev. B* **4**, 2162 (1971).
34. O. V. Dolgov and I. I. Mazin, *Fiz. Tverd. Tela* **20**, 3717 (1978).
35. H. Rietschel, *Z. Phys. B* **30**, 271 (1978).
36. J. C. Phillips, *J. Appl. Phys.* **43**, 356 (1972).
37. D. G. Pettifor, *J. Phys. F* **7**, 1009 (1977).
38. D. A. Papaconstantopoulos, L. L. Boyer, B. M. Klein, A. R. Williams, V. L. Morruzzi, and J. F. Janak, *Phys. Rev. B* **15**, 4221 (1977).
39. W. H. Butler, J. J. Olson, J. S. Faulkner, and B. L. Gyorffy, *Phys. Rev. B* **14**, 3823 (1976).
40. W. H. Butler, *Phys. Rev. B* **15**, 5267 (1977).
41. P. B. Allen, in: *Dynamical Properties for Solids*, Vol. 3, edited by G. K. Horton and A. A. Maradudin, North Holland, Amsterdam (1980); G. Grimvall, Phys. Scripta, **14**, 63 (1976).
42. B. M. Klein and D. A. Papaconstantopoulos, *Phys. Rev. Lett.* **32**, 1193 (1974).
43. S. V. Vonsovskii, Yu. A. Izyumov, and É. Z. Kurmaev, *Superconductivity of the Transition Metals and their Alloys and Compounds* (in Russian), Nauka, Moscow (1977).
44. *Transition Metals*, 1977, edited by M. J. G. Lee, J. M. Perz, and E. Fawcett, The Institute of Physics, Bristol (1978).
45. B. T. Matthias, T. H. Geball, and V. R. Compton, *Rev. Mod. Phys.* **35**, 1 (1970).
46. B. T. Matthias, in: *Prog. Low Temp. Phys.*, Vol. VII, edited by C. J. Gorter, North Holland, Amsterdam (1957), p. 138.
47. E. G. Brovman and Yu. Kagan, *Usp. Fiz. Nauk* **112**, 363 (1974).
48. R. C. Dynes and J. M. Rowell, *Phys. Rev. B* **11**, 1884 (1975).
49. P. B. Allen and R. C. Dynes, *Phys. Rev. B* **11**, 1895 (1975).
50. R. I. Boughton, J. L. Olsen, and G. Palmy, in: *Prog. Low Temp. Phys.*, Vol. IV, edited by C. J. Gorter, North Holland, Amsterdam (1970), p. 163.
51. N. W. Ashcroft, *Phys. Rev. Lett.* **21**, 1748 (1968).
52. T. Schneider and E. Stoll, *Physica* **55**, 702 (1971).
53. L. G. Caron, *Phys. Rev. B* **9**, 5025 (1974).
54. R. P. Gupta and S. K. Sinha, in: *Superconductivity in d- and f-Band Metals*, Proc. Second Rochester Conf., edited by D. H. Douglass, Plenum Press, New York (1977), p. 593.
55. B. Adler, in: *Progress in Very High Pressure Research*, edited by T. Bundy, Wiley, New York (1960).
56. E. Ostgaard, *Phys. Lett. A* **45**, 371 (1973).
57. E. G. Brovman, Yu. Kagan, and A. Kholas, *Zh. Éksp. Teor. Fiz.* **61**, 2429 (1971).
58. E. G. Brovman, Yu. Kagan, and A. Kholas, *Zh. Éksp. Teor. Fiz.* **62**, 1492 (1972).
59. F. V. Grigor'ev, S. B. Kormer, O. L. Mikhailova, A. P. Tolochko, and V. D. Urlin, *Pis'ma Zh. Éksp. Teor. Fiz.* **16**, 286 (1972).
60. L. F. Bereshchagin, E. N. Yakovlev, and Yu. A. Timofeev, *Pis'ma Zh. Éksp. Teor. Fiz.* **21**, 190 (1975).
61. P. S. Hawke, T. J. Burgess, D. E. Duerre, J. G. Huebel, R. N. Keeler, H. Klapper, and W. C. Wallace, *Phys. Rev. Lett.* **41**, 994 (1978).

62. E. E. Salpeter, *Phys. Rev. Lett.* **28**, 566 (1972).
63. G. F. Chapline, Jr., *Phys. Rev. B* **6**, 2067 (1972).
64. S. V. Iordanskii, O. V. Lokushevskii, E. B. Vul, L. A. Sidorovich, and A. M. Finkel'shtein, *Pis'ma Zh. Éksp. Teor. Fiz.* **17**, 530 (1973).
65. G. Gladstone, M. A. Jensen, and J. R. Schrieffer, in: *Superconductivity*, edited by R. D. Parks, Dekker, New York (1969).
66. F. Y. Fradin and J. D. Williamson, *Phys. Rev. B* **10**, 2803 (1974).
67. I. R. Gomersall and B. L. Gyorffy, *Phys. Rev. Lett.* **33**, 1286 (1974).
68. S. T. Chui, *Phys. Rev. B* **9**, 2097 (1974).
69. S. T. Chui, *Ann. Phys. (N.Y.)* **80**, 361 (1973).
70. M. M. Collver and R. H. Hammond, *Phys. Rev. Lett.* **30**, 92 (1973).
71. W. Weber, *Phys. Rev. B* **8**, 5093 (1973).
72. M. Ishikawa and L. E. Toth, *Phys. Rev. B* **3**, 1856 (1971).
73. D. I. Khomskii, *Kratk. Soobshch. Fiz.* No. 6, 53 (1971).
74. B. T. Matthias, in: *Superconductivity in d- and f-Band Metals*, AIP Conf. Proc., No. 4, edited by D. H. Douglass, New York (1972).
75. L. R. Testardi, J. H. Wernick, W. A. Royer, D. D. Bacon, and A. R. Storn, *J. Appl. Phys.* **45**, 446 (1974).
76. J. D. Axe and G. Shirane, *Phys. Rev. B* **8**, 1965 (1973).
77. C. W. Chu, *Phys. Rev. Lett.* **33**, 1283 (1974).
78. C. W. Chu and L. R. Testardi, *Phys. Rev. Lett.* **32**, 766 (1974).
79. T. F. Smith, R. N. Shelton, and A. C. Lawson, *J. Phys. F* **3**, 2157 (1973).
80. C. W. Chu and T. F. Smith, *Phys. Rev. B* **11**, 1866 (1975).
81. M. L. Cohen and P. W. Anderson, in: *Superconductivity in d- and f-Band Metals*, AIP Conf. Proc., No. 4, edited by D. H. Douglass, New York (1972).
82. Yu. A. Bychkov, L. P. Gor'kov, and I. E. Dzyaloshinskii, *Zh. Éksp. Teor. Fiz.* **50**, 798 (1966).
83. I. E. Dzyaloshinskii and A. I. Larkin, *Zh. Éksp. Teor. Fiz.* **61**, 791 (1971).
84. Yu. V. Kopaev and R. V. Timerov, *Zh. Éksp. Teor. Fiz.* **63**, 290 (1972).
85. L. N. Bulaevskii, *Usp. Fiz. Nauk* **115**, 271 (1975).
86. J. Labbé and J. Friedel, *J. Phys. (Paris)* **27**, 153, 303 (1966).
87. L. P. Gor'kov, *Pis'ma Zh. Éksp. Teor. Fiz.* **21**, 656 (1975).
88. B. T. Geilikman, *Usp. Fiz. Nauk* **88**, 327 (1966).
89. É. A. Pashitskii, *Zh. Éksp. Teor. Fiz.* **64**, 2387 (1968).
90. H. Fröhlich, *Proc. Phys. Soc. London* **C1**, 544 (1968).
91. H. Gutfreund and Y. Unna, *J. Phys. Chem. Solids* **34**, 1523 (1973).
92. A. Rothwarf, *Phys. Rev. B* **2**, 3560 (1970).
93. V. M. Dubovik, *Fiz. Met. Metalloved.* **35**, 1193 (1973).
94. E. G. Maksimov and O. A. Pankratov, *Fiz. Tverd. Tela* **16**, 874 (1974).
95. B. N. Ganguly and E. F. Wood, *Phys. Rev. Lett.* **28**, 681 (1972).
96. H. Fröhlich, *Phys. Lett. A* **35**, 325 (1971).
97. V. Z. Kresin, *Dokl. Akad. Nauk SSSR* **165**, 1058 (1965).
98. V. M. Agranovich and M. A. Mekhtiev, *Fiz. Tverd. Tela* **13**, 2732 (1971).
99. B. T. Geilikman, *Usp. Fiz. Nauk* **109**, 65 (1973).
100. C. R. Leavens, *Solid State Commun.* **19**, 395 (1976).
101. M. L. Cohen and P. W. Anderson, *Proc. AIP Conf. on d- and f-Band Metals*, edited by D. H. Douglass, Am. Inst. Phys., New York (1972).
102. O. V. Dolgov, D. A. Kirkhnits, and E. G. Maksimov, Preprint, Fiz. Inst. Akad. Nauk SSSR, No. 278, Moscow (1978); *Rev. Mod. Phys.* **53**, 81 (1981).
103. J. Bardeen, in: *Superconductivity in d- and f-Band Metals*, Proc. Second Rochester Conf., edited by D. H. Douglass, Plenum, New York (1977), p. 1.
104. M. L. Cohen and S. G. Louie, in: *Superconductivity in d- and f-Band Metals*, Proc. Rochester Conf., edited by D. H. Douglass, Plenum, New York (1977), p. 7.

105. Y. Takada, *J. Phys. Soc. Jpn.* **45**, 786 (1978).
106. Yu. A. Uspenskii, *Zh. Éksp. Teor. Fiz.* 57, 143 (1979).

## Chapter 5

1. B. W. Batterman and C. B. Carret, *Phys. Rev. Lett.* **13**, 390 (1964); *Phys. Rev.* **145**, 296 (1966).
2. P. W. Anderson and E. J. Blount, *Phys. Rev. Lett.* **14**, 217 (1965).
3. W. L. McMillan, *Phys. Rev.* **167**, 333 (1968).
4. W. Chu, E. Bacher, A. S. Cooper, and I. P. Maita, *Phys. Rev. B* **4**, 320 (1971).
5. A. N. Kozlov and L. A. Maksimov, *Zh. Éksp. Teor. Fiz.* **48**, 1184 (1965).
6. Yu. V. Kopaev, *Trudy FIAN* **86**, 3 (1975).
7. B. A. Volkov, *Trudy FIAN*, **104**, 3 (1978).
8. B. A. Volkov and O. A. Pankratov, *Zh. Éksp. Teor. Fiz.* **75**, 1362 (1978).
9. N. I. Kulikov and V. V. Tugushev, *Fiz. Tverd. Tela* **23**, 625 (1981).
10. P. A. Fedders and P. C. Martin, *Phys. Rev.* **143**, 245 (1965).
11. T. M. Rice, *Phys. Rev.* **B2**, 3619 (1970).
12. B. A. Volkov, Yu. V. Kopaev, and A. I. Rusinov, *Zh. Éksp. Teor. Fiz.* **68**, 1899 (1975).
13. B. A. Volkov, Yu. V. Kopaev, and V. V. Tugushev, *Zh. Éksp. Teor. Fiz.* **75**, 735 (1978).
14. V. F. Elesin and Yu. V. Kopaev, *Pis'ma Zh. Éksp. Teor. Fiz.* **24**, 78 (1977).
15. M. Shaham, J. Berak, U. El-Hanany, and W. W. Warren, *Phys. Rev. Lett.* **39**, 570 (1977).
16. L. B. Coleman, M. J. Cohen, D. J. Sandman, F. G. Yamagishi, A. F. Garito, and A. J. Heeger, *Solid State Commun.* **12**, 1125 (1973).
17. E. M. Henley and L. Wilets, *Ann. Phys. (N.Y.)* **14**, 120 (1961).
18. Yu. A. Bychkov, L. P. Gor'kov, and I. E. Dzyaloshinskii, *Zh. Éksp. Teor. Fiz.* **50**, 738 (1966).
19. I. E. Dzyaloshinskii and A. I. Larkin, *Zh. Éksp. Teor. Fiz.* **61**, 791 (1971).
20. K. Levin, S. L. Cunningham, and D. L. Mills, *Phys. Rev. B* **10**, 3832 (1974).
21. L. P. Gor'kov and I. E. Dzyaloshinskii, *Zh. Éksp. Teor. Fiz.* **67**, 397 (1974).
22. R. E. Peierls, *Quantum Theory of Solids*, Oxford University Press, Oxford (1955).
23. L. V. Keldysh and Yu. V. Kopaev, *Fiz. Tverd. Tela* **6**, 2791 (1964).
24. B. I. Halperin and T. M. Rice, *Solid State Phys.* **21**, 115 (1968).
25. L. P. Gor'kov and T. T. Mnatsakanov, *Zh. Éksp. Teor. Fiz.* **63**, 684 (1972); A. A. Abrikosov, *Zh. Éksp. Teor. Fiz.* **65**, 1508 (1973).
26. B. A. Volkov and Yu. V. Kopaev, *Pis'ma Zh. Éksp. Teor. Fiz.* **27**, 10 (1978).
27. B. A. Volkov and V. L. Ginzburg, and Yu. V. Kopaev, *Pis'ma Zh. Éksp. Teor. Fiz.* **27**, 221 (1978).
28. B. A. Volkov, Yu. V. Kopaev, M. S. Nunuparov, and V. V. Tugushev, *Pis'ma Zh. Éksp. Teor. Fiz.* **30**, 317 (1979).
29. B. A. Volkov, V. G. Kantser, and Yu. V. Kopaev, *Zh. Éksp. Teor. Fiz.* **76**, 1856 (1979).
30. B. A. Volkov, Yu. V. Kopaev, and M. S. Nunuparov, *Fiz. Tverd. Tela* **21**, 2733 (1979).
31. Yu. V. Kopaev, *Fiz. Tverd. Fiz.* **8**, 2633 (1966); S. A. Gordyunin and L. P. Gor'kov, *Zh. Éksp. Teor. Fiz.* **63**, 1922 (1972).
32. B. A. Volkov and Yu. V. Kopaev, *Zh. Éksp. Teor. Fiz.* **64**, 2184 (1973).
33. Yu. V. Kopaev and R. Kh. Timerov, *Zh. Éksp. Teor. Fiz.* **63**, 290 (1972).
34. Yu. E. Lozovik and V. I. Yudson, *Fiz. Tverd. Tela* **17**, 1613 (1975).
35. W. Kohn, *Phys. Rev. Lett.* **19**, 789 (1967).

36. Yu. V. Kopaev, *Fiz. Tverd. Tela* **12**, 3 (1970).
37. Yu. V. Kopaev and A. I. Rusinov, *Fiz. Tverd. Tela* **11**, 1306 (1969).
38. A. A. Abrikosov, L. P. Gor'kov, and I. E. Dzyaloshinskii, *Methods of Quantum Field Theory in Statistical Physics* (in Russian), Fizmatgiz (1962).
39. Yu. V. Kopaev, *Zh. Éksp. Teor. Fiz.* **58**, 1012 (1970).
40. R. R. Guseinov and L. V. Keldysh, *Zh. Éksp. Teor. Fiz.* **63**, 2255 (1972).
41. A. I. Rusinov, Do Chan Kat, and Yu. V. Kopaev, *Zh. Éksp. Teor. Fiz.* **65**, 1984 (1973).
42. A. G. Aronov and É. B. Sonin, *Zh. Éksp. Teor. Fiz.* **63**, 1059 (1972).
43. S. C. Lo and K. W. Wong, *Nuovo Cimento B* **10**, 361, 383 (1972).
44. R. A. Bari, *Phys. Rev. B* **7**, 2128 (1973)
45. Y. Nagaoka, *Prog. Theor. Phys.* **52**, 1716 (1974).
46. I. G. Brankov and N. S. Tonchev, Preprint, Joint Institute for Nuclear Research, Dubna, P4-8150 (1974).
47. K. Leven, D. L. Mills, and S. L. Cunningham, *Phys. Rev. B* **10**, 3821 (1974).
48. A. M. Gabovich and E. A. Pashitskii, *Fiz. Met. Metalloved.* **39**, 37 (1975).
49. D. C. Mattis and W. D. Langer, *Phys. Rev. Lett.* **25**, 376 (1970).
50. I. G. Brankov and N. S. Tonchev, Preprint, Joint Institute for Nuclear Research, Dubna, PH-8940 (1975).
51. M. Cohen, in: *Superconductivity of Semiconductors and Transition Metals* (Russian Translation), Mir, Moscow (1972).
52. A. D. C. Grassie and A. Benyon, *Phys. Lett. A* **39**, 199 (1972).
53. D. M. Eagles, *Phys. Rev.* **186**, 456 (1969).
54. M. Schafroth, S. Butter, and J. Blatt, *Helv. Phys. Acta* **30**, 93 (1957).
55. V. L. Gurevich, A. I. Larkin, and Yu. A. Firsov, *Fiz. Tverd. Tela* **4**, 185 (1962).
56. L. D. Landau and E. M. Lifshitz, *Quantum Mechanics* (in Russian), Fizmatgiz (1963).
57. Yu. V. Kopaev and R. Kh. Timerov, *Zh. Éksp. Teor. Fiz.* **63**, 290 (1972).
58. A. N. Kozlov and L. A. Maksimov, *Zh. Éksp. Teor. Fiz.* **48**, 1184 (1965).
59. W. Kohn, *Phys. Rev. Lett.* **19**, 789 (1967).
60. R. Shanks, *Solid State Commun.* **15**, 753 (1974).
61. A. A. Abrikosov and L. A. Fal'kovskii, *Zh. Éksp. Teor. Fiz.* **43**, 1089 (1962); S. A. Gordyunin and L. P. Gor'kov, *Zh. Éksp. Teor. Fiz.* **63**, 1922 (1972); A. A. Abrikosov, *Zh. Éksp. Teor. Fiz.* **65**, 2063 (1973).
62. A. I. Golovashkin and G. P. Motulevich, *Fiz. Tverd. Tela* **13**, 1232 (1971).
63. E. E. Havinga and M. H. van Maaren, *Phys. Lett. A* **31**, 167 (1970).
64. B. G. Silbernagel, M. Weger, W. G. Clark, and I. W. Wernick, *Phys. Rev.* **153**, 535 (1967).
65. L. R. Testardi and T. B. Baternan, *Phys. Rev.* **154**, 402 (1967).
66. K. R. Keller and J. J. Hanak, *Phys. Rev.* **154**, 628 (1967).
67. L. Testardi, M. Veger, and I. Gol'dberg, *Superconductivity of Compounds with β-Tungsten Structure*, Mir, Moscow (1977).
68. Yu. V. Kopaev, V. V. Menyailenko, and S. N. Molotkov, *Zh. Eksp. Teor. Fiz.* **77**, 352 (1979).
69. A. A. Gorbatsevich, Yu. V. Kopaev, and S. N. Molotkov, *Fiz. Tverd. Tela* **21**, 2351 (1979).
70. Yu. V. Kopaev and S. N. Molotkov, *Fiz. Tverd. Tela* **21**, 1195 (1979).
71. P. Fulde and R. A. Ferrell, *Phys. Rev. A* **135**, 550 (1964); A. I. Larkin and Yu. N. Ovchinnikov, *Zh. Éksp. Teor. Fiz.* **47**, 1136 (1964).
72. W. Fogle and J. H. Perlstein, *Phys. Rev. B* **6**, 1402 (1972).
73. S. H. Liu, *Phys. Rev. B* **10**, 3619 (1974).
74. D. Tosatti and P. W. Anderson, *Solid State Commun.* **14**, 385 (1974).
75. H. P. Bronzel, C. R. Helms, and S. Kelemen, *Phys. Rev. Lett.* **35**, 1237 (1975).

## Chapter 6


1. W. A. Little, *Phys. Rev.* **134A**, 1416 (1961).
2. V. L. Ginzburg, *Phys. Lett.* **13**, 101 (1964); *Zh. Éksp. Teor. Fiz.* **47**, 2318 (1964); *Usp. Fiz. Nauk* **261**, 185 (1970).
3. I. F. Shchegolev, *Phys. Status Solidi A* **12**, 9 (1972).
4. H. R. Zeller, *Festkörperprobleme*, Vol. 13, Vieweg, Brownschweig (1973), p. 31.
5. V. Walatka, M. Labes, and J. Perlstein, *Phys. Rev. Lett.* **31**, 1139 (1973).
6. F. R. Gamble, R. A. Klemm, F. J. Di Salvo, and T. H. Geballe, *Science (N.Y.)* **168**, 568 (1976).
7. J. A. Wilson and A. D. Joffe, *Adv. Phys.* **18**, 193 (1969).
8. A. D. Yoffe, *Festkörperprobleme* **13**, 1 (1973); *Annu. Rev. Mater. Sci.* **3**, 147 (1973).
9. F. J. Di Salvo, B. G. Bagley, J. M. Voorhoeve, and J. V. Waszczak, *J. Phys. Chem. Solids* **34**, 1357 (1973).
10. A. H. Thompson, F. R. Gamble, and R. F. Koehler, *Phys. Rev.* **B5**, 2811 (1972).
11. R. C. Morris, R. V. Coleman, and R. Bhandari, *Phys. Rev. B* **5**, 895 (1972).
12. R. C. Morris and R. V. Coleman, *Phys. Rev. B* **7**, 991 (1973).
13. W. Y. Liang, *J. Phys.* **C4**, L378 (1971); **6**, 551 (1973).
14. L. N. Bulaevskii, *Usp. Fiz. Nauk* **116**, 449 (1975).
15. V. L. Kalikhman and Ya. S. Umanskii, *Usp. Fiz. Nauk* **108**, 503 (1972).
16. R. F. Frindt, *Phys. Rev. Lett.* **28**, 999 (1972).
17. F. Consadori, A. A. Fife, R. F. Frindt, and S. Gygax, *Appl. Phys. Lett.* **32**, 882 (1974).
18. F. R. Gamble, J. H. Osiecki, M. Cais, R. Pisharody, F. J. Di Salvo, and T. H. Geballe, *Science (N.Y.)* **174**, 493 (1971).
19. F. R. Gamble, J. H. Osiecki, and F. J. Di Salvo, *J. Chem. Phys.* **55**, 3525 (1971).
20. A. R. Beal and W. Y. Liang, *Philos. Mag.* **27**, 1397 (1973).
21. F. J. Di Salvo. G. W. Hull, L. H. Schwartz, J. M. Voorhoeve, and J. V. Waszczak, *J. Chem. Phys.* **59**, 1922 (1973).
22. G. S. Parry, C. B. Scruby, and P. M. Williams, *Philos. Mag.* **29**, 601 (1974).
23. J. V. Acrivos and J. R. Salem, *Philos. Mag.* **30**, 603 (1974).
24. E. Eherenfrend, A. C. Gossard, and F. R. Gamble, *Phys. Rev. B* **5**, 1708 (1972).
25. F. J. Di Salvo, *Proc. Conf. on Low Temp. Phys. (LT-13)*, Vol. 3, Boulder, Colorado (1972), p. 417.
26. R. B. Somoano and A. Rembaum, *Phys. Rev. Lett.* **27**, 402 (1971).
27. R. B. Somoano, V. Hadek, and A. Rembaum, *J. Chem. Phys.* **58**, 697 (1973).
28. J. V. Acrivos, Y. Liang, J. A. Wilson, and A. D. Yoffe, *J. Phys. C* **4**, 18 (1971).
29. G. V. Subba Rao, M. W. Shafer, S. Kawarazaki, and A. M. Toxen, *J. Solid State Chem.* **9**, 323 (1974).
30. C. B. Carter and P. W. Williams, *Philos. Mag.* **26**, 393 (1972).
31. E. I. Kats, *Zh. Éksp. Teor. Fiz.* **58**, 1471 (1970).
32. E. A. Antonova, S. A. Medvedev, and Yu. I. Shebalin, *Zh. Éksp. Teor. Fiz.* **57**, 329 (1969).
33. R. E. Jones, H. R. Shanks, D. K. Finnermore, and B. Morosin, *Phys. Rev. B* **6**, 835 (1972).
34. I. E. Dzyaloshinskii and E. I. Kats, *Zh. Éksp. Teor. Fiz.* **55**, 338 (1968).
35. T. Sambongi, *J. Low Temp. Phys.* **18**, No. 1/2, 139 (1975).
36. D. Jerome, A. J. Grant, and A. D. Yoffe, *Solid State Commun.* **9**, 2183 (1971).
37. T. F. Smith, L. E. De Long, A. R. Moodenbough, T. H. Geballe, and R. E. Schwall, *J. Phys. C* **5**, L230 (1972).
38. M. A. Il'ina, E. S. Itskevich, and G. A. Kalyuzhnaya, *Fiz. Tverd. Tela* **14**, 515 (1972).

39. A. H. Thompson, F. R. Gamble, and J. F. Revelli, *Solid State Commun.* **9**, 981 (1971).
40. J. P. Tidman, O. Singh, A. E. Curzon, and R. F. Frindt, *Philos. Mag.* **30**, 1191 (1975).
41. D. E. Moncton, J. D. Axe, and F. J. Di Salvo, *Phys. Rev. Lett.* **34**, 734 (1975).
42. L. N. Bulaevskii, *Usp. Fiz. Nauk* **115**, 263 (1975).
43. C. Berthier, P. Molinie, and D. Jerome, *Solid State Commun.* **18**, 1393 (1976).
44. J. C. Tsang, M. W. Shafer, and B. L. Crowder, *Phys. Rev. B* **11**, 155 (1975).
45. Yu. V. Kopaev, R. Kh. Timerov, *Zh. Éksp. Teor. Fiz.* **63**, 290 (1972).
46. N. N. Bogolyubov. V. V. Tolmachev, and B. V. Shirkov, *A New Method in the Theory of Superconductivity* (in Russian), Izd. AN SSSR (1958).
47. M. L. Cohen and P. W. Anderson, *Proc. AIP Conf. on d- and f-Band Superconductivity*, Rochester, 1971, Am. Inst. Phys., New York (1972), p. 17.
48. L. N. Bulaevskii and Yu. A. Kukharenko, *Zh. Éksp. Teor. Fiz.* **60**, 1518 (1971).
49. D. A. Kirzhnits, E. G. Maksimov, and D. I. Khomskii, *J. Low Temp. Phys.* **10**, No. 1/2, 79 (1973).
50. W. A. Little, *J. Low Temp. Phys.* **15**, No. 3/4, 365 (1973).
51. T. M. Rice, *Phys. Rev.* **140A**, 1889 (1965).
52. P. C. Hohenberg, *Phys. Rev.* **158**, 383 (1967).
53. V. L. Ginzburg and D. A. Kirzhnits, *Dokl. Akad. Nauk SSSR* **176**, 553 (1967).
54. H. T. Mikeshka and H. Schmidt, *J. Low Temp. Phys.* **2**, No. 3/4, 371 (1970).
55. V. L. Berezinskii, *Zh. Éksp. Teor. Fiz.* **61**, 1144 (1971).
56. W. A. Little, *Phys. Rev.* **156**, 396 (1967).
57. R. A. Ferrell, *Phys. Rev. Lett.* **13**, 330 (1964).
58. I. E. Dzyaloshinskii and E. I. Katz, *Zh. Éksp. Teor. Fiz.* **56**, 2373 (1968).
59. R. A. Klemm, M. R. Beasley, and A. Luther, *J. Low Temp. Phys.* **16**, No. 5/6, 607 (1974); *Phys. Rev. B* **12**, 877 (1975).
60. T. Tsuzuki, *J. Low Temp. Phys.* **9**, 525 (1973).
61. K. B. Efetov and A. I. Larkin, *Zh. Éksp. Teor. Fiz.* **66**, 2290 (1974).
62. W. E. Lawrence and S. Doniach, *Proc. Conf. on Low Temp. Phys.* (*LT-12*), Kyoto, 1970, p. 361.
63. L. N. Bulaevskii, *Zh. Éksp. Teor. Fiz.* **64**, 2241; **65**, 1278 (1973).
64. L. N. Bulaevskii and A. A. Guseinov, *Pis'ma Zh. Éksp. Teor. Fiz.* **19**, 742 (1974).
65. R. E. Schwall, G. R. Stewart, and T. H. Geballe, *J. Low Temp. Phys.* **22**, No. 5/6, 557 (1976).
66. J. A. Woollam and R. B. Somoano, *Phys. Rev. B* **13**, 3843 (1976).
67. D. A. Whitney, R. M. Fleming, and R. V. Coleman, *Phys. Rev. B* **15**, 3405 (1977).
68. N. B. Hannay, T. H. Geballe, B. T. Matthias, K. Ander, P. Schmidt, and D. Macnair, *Phys. Rev. Lett.* **14**, 225 (1965).
69. Y. Koike, H. Suematsu, K. Higuchi, and S. Tanuma, *Solid State Commun.* **27**, 623 (1978).
70. O. I. Kulik and I. K. Yanson, *The Josephson Effect in Superconducting Tunnel Structures* (in Russian), Nauka (1970).
71. A. A. Abrikosov, L. P. Gor'kov, and I. E. Dzyaloshinskii, *Methods of Quantum Field Theory in Statistical Physics* (in Russian), Fizmatgiz, Moscow (1962).
72. N. Boccara, J. P. Carton, and G. Sarma, *Phys. Lett.* **49A**, 165 (1974).
73. R. C. Morris and R. V. Coleman, *Phys. Rev. B* **7**, 991 (1973).
74. L. N. Bulaevskii and A. I. Rusinov, *Pis'ma Zh. Éksp. Teor. Fiz.* **21**, 147 (1975).
75. L. N. Bulaevskii, A. I. Rusinov, and A. A. Guseinov, *Zh. Éksp. Teor. Fiz.* **71**, 2356 (1976).
76. K. B. Efetov and A. I. Larkin, *Zh. Éksp. Teor. Fiz.* **69**, 764 (1975).
77. R. R. Gerhardts, *Phys. Rev. B* **9**, 2945 (1974).

78. L. H. Bulaevskii, *Zh. Éksp. Teor. Fiz.* **66**, 2212 (1974).
79. E. J. Di Salvo, R. Schwall, T. H. Geballe, F. R. Gamble, and J. H. Osiecki, *Phys. Rev. Lett.* **27**, 310 (1971).
80. D. E. Prober, M. R. Beasley, and R. E. Schwall, *Proc. Conf. on Low Temp. Phys.* (*LT-13*), Vol. 3, Boulder, Colorado (1972), p. 428; *Phys. Rev. B* **15**, 5245 (1977).
81. J. A. Wilson, F. J. Di Salvo, and S. Mahajan, *Adv. Phys.* **24**, 117 (1975).
82. D. D. Balla, A. A. Mamalui, M. A. Obolenskii, V. A. Starodub, and Kh. B. Chashka, *Proc. Twentieth All-Union Conf. on Low Temp. Phys.*, *LT-20*, pt. III, 1977, Moscow (1979).

## Chapter 7

1. N. F. Mott and E. A. Davis, *Electronic Properties in Non-Crystalline Materials*, Oxford University Press, Oxford (1971), Chap. 5.
2. E. H. Lieb and F. Y. Wu, Phys. *Rev. Lett.* **20**, 1445 (1968).
3. R. E. Borland, *Proc. R. Soc. London* 78, 926 (1961).
4. N. F. Mott and W. D. Twose, *Adv. Phys.* **10**, 107 (1961).
5. I. F. Shchegolev, *Phys. Status Solidi A* **12**, 9 (1972).
6. Z. G. Soos, *Annu. Rev. Phys. Chem.* **25**, 121 (1974).
7. V. V. Walatka, M. M. Labes, and J. H. Perlstein, *Phys. Rev. Lett.* **31**, 1139 (1973).
8. M. J. Cohen, L. B. Coleman, A. Garito, and A. J. Heeger, *Phys. Rev. B* **10**, 1298 (1974).
9. H. R. Zeller, *Festkörperprobleme* **13**, 31 (1973).
10. L. N. Bulaevskii, *Usp. Fiz. Nauk* **115**, 263 (1975).
11. Che-hsiung Hsu and M. M. Labes, *J. Chem. Phys.* **61**, 4640 (1974).
12. R. L. Greene, G. B. Street, and L. J. Suter, *Phys. Rev. Lett.* **34**, 577 (1975).
13. R. J. Warmach, T. A. Callcott, and C. R. Watson, *Phys. Rev. B* **12**, 3331 (1975).
14. R. B. Somoano, A. Gupta, V. Hadek, T. Dafta, M. Jones, R. Dech, and A. M. Hermann, *J. Chem. Phys.* **63**, 4970 (1975).
15. S. J. La Placa, P. W. R. Corfield, R. Thomas, and B. A. Scott, *Solid State Commun.* **17**, 635 (1975).
16. G. A. Toombs, *Phys. Rep.* **40C**, 182 (1978).
17. *Proc. Intern. Conf. on Organic Conductors and Semiconductors*, Siofok, Hungary, 1976, edited by L. Pal, G. Grüner, A. Janossy, and J. Solyom, Springer-Verlag, Berlin (1977).
18. *Proc. Intern. Conf. on Quasi-One-Dimensional Conductors*, Dubrovnik, Jugoslavia (1978).
19. S. K. Khanna, E. Ehrenfreund, A. F. Garito, and A. J. Heeger, *Phys. Rev. B* **10**, 2205 (1974).
20. F. F. Igoshin, A. N. Kir'yanov, V. N. Topnikov, and I. F. Shchegolev, *Zh. Éksp. Teor. Fiz.* **68**, 1203 (1975).
21. D. I. Fedutin, *Pis'ma Zh. Éksp. Teor. Fiz.* **17**, 299 (1973).
22. A. S. Berenblyum, L. I. Buravov, M. L. Khidekal', I. F. Shchegolev, and E. B. Yakimov, *Pis'ma Zh. Éksp. Teor. Fiz.* **13**, 619 (1971).
23. R. Comès, M. Lambert, H. Launois, and H. R. Zeller, *Phys. Rev. B* **8**, 571 (1973).
24. H. Fröhlich, *Proc. R. Soc. London Ser. A* **223**, 296 (1954).
25. P. A. Lee, T. M. Rice, and P. W. Anderson, *Solid State Commun.* **14**, 703 (1974).
26. H. G. Shuster, *Phys. Rev. B* **11**, 613 (1975).
27. (a) D. Allender, I. W. Bray, and J. Bardeen, *Phys. Rev.* **B9**, 119 (1974); **20**, 37 (1974); (b) M. J. Rice, *Solid State Commun* **16**, 1285 (1975).

28. L. Pietronero, S. Strässler, and G. A. Toombs, *Phys. Rev.* **B12**, 5213 (1975).
29. M. J. Rice, S. Strässler, and W. R. Schneider, *Proc. German Phys. Soc. Conf. on One-Dimensional Conductors*, University of Saarbrücken, 1974, Springer-Verlag (1975), p. 282.
30. M. J. Rice, A. R. Bishop, J. A. Krumhansl, and S. E. Trullinger, *Phys. Rev. Lett.* **36**, 432 (1976).
31. M. I. Rice, C. B. Duke, and N. O. Lipari, *Solid State Commun.* **17**, 1089 (1975).
32. M. J. Rice, *Phys. Rev. Lett.* **37**, 36 (1976).
33. L. N. Bulaevskii and D. M. Khomskii, *Zh. Éksp. Teor. Fiz.* **73**, 1148 (1977).
34. L. N. Bulaevskii, A. I. Buzdin, and D. I. Khomskii, *Proc. Intern. Conf. on Quasi-One-Dimensional Conductors*, Dubrovnik, Yugoslavia, 1978; A. I. Buzdin and L. N. Bulaevskii, *Zh. Éksp. Teor. Fiz.* **16**, No. 4 (1979).
35. P. A. Lee, T. M. Rice, and P. W. Anderson, *Phys. Rev. Lett.* **31**, 462 (1973).
36. D. J. Scalapino, M. Sears, and R. A. Ferrell, *Phys. Rev. B* **6**, 3409 (1972).
37. K. B. Efetov and A. I. Larkin, *Zh. Éksp. Teor. Fiz.* **66**, 2290 (1974).
38. D. J. Bergman, T. M. Rice, and P. A. Lee, *Phys. Rev. B* **15**, 1706 (1977).
39. B. Renker, H. Rietschel, L. Pintschovius, W. Gläser, P. Brüesch, D. Kuse, and M. J. Rice, *Phys. Rev. Lett.* **30**, 1144 (1973).
40. T. E. Phillips, T. J. Kisternmacher, J. P. Ferraris, and D. O. Cowan, *Chem. Commun.* **1973**, 471 (1973).
41. R. Comès, M. Lambert, H. Launois, and H. R. Zeller, *Phys. Status Solidi B* **58**, 587 (1973).
42. B. Renker, L. Pintschovius, W. Gläser, H. Rietschel, R. Comès, L. Liebert, and W. Drexel, *Phys. Rev. Lett.* **32**, 836 (1974).
43. I. W. Lynn, M. Iizumi, G. Shirane, S. A. Werner, and R. B. Saillant, *Phys. Rev. B* **12**, 1154 (1975).
44. C. F. Eagen, S. A. Werner, and K. B. Saillant, *Phys. Rev. B* **12**, 2036 (1975).
45. P. Brüesch, S. Strässler, and H. R. Zeller, *Phys. Rev. B* **12**, 219 (1975).
46. H. Niedoba, H. Launois, D. Brinkman, R. Bruger, and H. R. Zeller, *Phys. Status Solidi B* **58**, 309 (1973).
47. F. Mehran and B. A. Scott, *Phys. Rev. Lett.* **31**, 1347 (1973).
48. W. A. Reed, F. S. L. Hsu, R. J. Schutz, J. E. Graebner, and H. J. Guggenheim, *Phys. Rev. Lett.* **34**, 473 (1975).
49. L. N. Bulaevskii and M. V. Sadovskii, *Fiz. Tverd. Tela* **16**, 1159 (1974).
50. E. F. Steigmeier, R. London, and H. Auderset, *Solid State Commun.* **17**, 1447 (1975).
51. R. Comes, B. Renker, L. Pintschovius, R. Currat, M. Gläser, and G. Scheiber, *Phys. Status Solidi B* **71**, 171 (1975).
52. K. Carneiro, G. Shirane, S. A. Werner, and S. Kaiser, *Phys. Rev. B* **13**, 4258 (1976).
53. H. G. Schuster, *Solid State Commun.* **14**, 127 (1974).
54. P. N. Shen and C. M. Varma, *Solid State Commun.* **15**, 1905 (1974).
55. T. R. Kochler and P. A. Lee, *Phys. Rev. B* **16**, 5263 (1977).
56. K. B. Efetov and A. I. Larkin, *Zh. Éksp. Teor. Fiz.* **72**, 2350 (1977).
57. L. P. Gor'kov, *Pis'ma Zh. Éksp. Teor. Fiz.* **25**, 384 (1977).
58. H. Fukuyama and P. A. Lee, *Phys. Rev. B* **17**, 535, 542 (1978).
59. C. J. Fritchie, *Acta Crystallogr.* **20**, 892 (1966).
60. L. I. Buravov, I. N. Eremenko, R. B. Lyubovskii, L. N. Rozenberg, M. L. Khidekel', R. P. Shibaeva, I. F. Shchegolev, and É. B. Yagubskii, *Pis'ma Zh. Éksp. Teor. Fiz.* **20**, 457 (1974).
61. L. N. Bulaevskii, A. V. Zvorykina, Yu. S. Karimov, R. B. Lyubovskii, and I. F. Shchegolev, *Zh. Éksp. Teor. Fiz.* **62**, 725 (1972).

62. L. B. Coleman, M. I. Cohen, D. J. Sandman, F. G. Yamagashi, A. F. Garito, and A. J. Heeger, *Solid State Commun.* **12**, 1125 (1973).
63. R. P. Croff, A. Suna, and R. E. Merrifield, *Phys. Rev. Lett.* **33**, 418 (1974).
64. L. I. Buravov, D. N. Fedutin, and I. F. Shchegolev, *Zh. Éksp. Teor. Fiz.* **59**, 1125 (1970).
65. L. I. Buravov, M. L. Khidekel', I. F. Shchegolev, and É. B. Yagubskii, *Pis'ma Zh. Éksp. Teor. Fiz.* **12**, 142 (1970).
66. L. N. Bulaevskii and I. F. Shchegolev, *Proc. Intern. Conf. on Magnetism, ICM-73*, Vol. 1(2), Nauka (1974), p. 45.
67. L. N. Bulaevskii, A. A. Guseinov, O. N. Eremenko, V. N. Topnikov, and I. F. Shchegolev, *Fiz. Tverd. Tela* **17**, 781 (1975).
68. L. J. Azevedo and W. G. Clark, *Phys. Rev. B* **16**, 3252 (1977).
69. I. F. Shchegolev, M. K. Makova, D. N. Fedutin, I. S. Krainskii, M. L. Khidekel', and É. B. Yagubskii, *Zh. Éksp. Teor. Fiz.* **72**, 1140 (1977).
70. L. N. Bulaevskii, *Zh. Éksp. Teor. Fiz.* **75**, 748 (1978).
71. G. Teodorou and M. H. Cohen, *Phys. Rev. Lett.* **37**, 1014 (1976); G. Theodorou, *Phys. Rev. B* **16**, 2254, 2264, 2273 (1977).
72. J. P. Pouget, S. Megtert, and R. Comès, *Proc. Intern. Conf. on Quasi-One-Dimensional Conductors*, Dubrovnik, Jugoslavia, 1978.
73. L. N. Bulaevskii, O. B. Lyubovskii, and I. F. Shchegolev, *Pis'ma Zh. Eksp. Teor. Fiz.* **16**, 42 (1972).
74. A. N. Bloch, R. B. Weisman, and C. M. Varma, *Phys. Rev. Lett.* **28**, 753 (1972).
75. M. A. Butler, F. Wudl, and Z. G. Soos, *Phys. Rev. B* **12**, 4708 (1975).
76. A. A. Gogolin, V. I. Mel'nikov, and É. I. Rashba, *Zh. Éksp. Teor. Fiz.* **69**, 327 (1975).
77. A. A. Gogolin, S. N. Zolotukhin, V. I. Mel'nikov, É. I. Rashba, and I. F. Shchegolev, *Pis'ma Zh. Éksp. Teor. Fiz.* **22**, 564 (1975).
78. J. B. Torrance, *Phys. Rev. B* **17**, 3099 (1978).
79. V. Ya. Krivnov and A. A. Ovchinnikov, *Zh. Éksp. Teor. Fiz.* **73**, 2364 (1977).
80. D. J. Klein and W. A. Seitz, *Phys. Rev. B* **10**, 3217 (1974).
81. E. Conwell, *Phys. Rev. B* **18**, 1818 (1978).
82. W. D. Grobman, R. A. Pollak, D. E. Eastman, E. T. Mass, and B. A. Scott, *Phys. Rev. Lett.* **32**, 534 (1974).
83. M. A. Butler, I. P. Ferraris, A. N. Bloch, and D. O. Cowan, *Chem. Phys. Lett.* **24**, 600 (1974).
84. W. D. Grobman and B. D. Silverman, *Solid State Commun.* **19**, 319 (1976).
85. E. F. Skelton, A. N. Bloch, J. P. Ferraris, and D. O. Cowan, *Phys. Lett.* **32**, 1363 (1974).
86. R. H. Blessing and P. Coppens, *Solid State Commun.* **15**, 215 (1974).
87. I. C. Scott, A. F. Garito, and A. J. Heeger, *Phys. Rev. B* **10**, 3131 (1974).
88. D. E. Shafer, F. Wudl, G. A. Thomas, J. P. Ferraris, and D. O. Cowan, *Solid State Commun.* **14**, 347 (1974).
89. G. Soda, D. Jerome, M. Weger, J. M. Fabre, L. Giral, and K. Bechgaard, *Proc. Intern. Conf. on Organic Conductors and Semiconductors*, Siofok, Hungary, 1976, edited by L. Pal, G. Grüner, A. Janossy, and J. Solyom, Springer-Verlag, Berlin, (1977), p. 371.
90. E. Abrahams, J. Solyom, and F. Woynarovich, *Phys. Rev. B* **16**, 5238 (1977).
91. A. J. Berlinsky, J. F. Carolan, and L. Weiler, *Solid State Commun.* **15**, 795 (1974).
92. P. M. Grant, R. L. Greene, G. C. Weighton, and G. Castro, *Phys. Rev. Lett.* **31**, 1311 (1973).
93. A. A. Bright, A. F. Garito, and A. J. Heeger, *Solid State Commun.* **13**, 943 (1973).
94. G. A. Thomas et al., *Phys. Rev. B* **13**, 5105 (1976).
95. M. J. Cohen, L. B. Coleman, A. F. Garito, and A. J. Heeger, *Phys. Rev. B* **13**, 5111 (1976).

96. R. M. Herman, M. B. Salamon, G. D. Pasquali, and G. Stucky, *Solid State Commun.* 19, 137 (1976).
97. T. O. Poehler, A. N. Bloch, J. P. Ferraris, and D. O. Cowan, *Solid State Commun.* 15, 2337 (1974).
98. C. C. Jacobsen, D. B. Tanner, A. F. Garito, and A. J. Heeger, *Phys. Rev. Lett.* 32, 1301; 33, 1559 (1974).
99. M. B. Salamon, J. W. Bray, G. de Pasquali, R. A. Craven, G. Stucky, and A. Shchultz, *Phys. Rev. B* 11, 619 (1974).
100. J. F. Kwak, P. M. Chaikin, A. A. Russel, A. F. Garito, and A. J. Heeger, *Solid State Commun.* 16, 729 (1975).
101. R. A. Craven, M. B. Salamon, G. de Pasquali, R. M. Herman, G. Stucky, and A. Scultz, *Phys. Rev. Lett.* 31, 601 (1973).
102. F. Denoyer, R. Comès, A. F. Garito, and A. J. Heeger, *Phys. Rev. Let.* 35, 445 (1975).
103. S. Kagoshima, H. Anzai, K. Kajimura, and T. Ishiguro, *J. Phys. Soc. Jpn.* 39, 1143 (1975).
104. R. Comès, S. M. Shapiro, G. Shirane, A. F. Garito, and A. J. Heeger, *Phys. Rev. Lett.* 35, 1518 (1975).
105. R. Comes, S. M. Shapiro, G. Shirane, A. F. Garito, and A. J. Heeger, *Phys Rev. Lett.* 35, 1518 (1975).
106. Y. Tomkiewicz, B. A. Scott, L. J. Tao, and R. S. Title, *Phys. Rev. Lett.* 32, 1363 (1974).
107. J. P. Pouget, S. K. Khanna, F. Denoyer, R. Comes, A. F. Garito, and A. J. Heeger, *Phys. Rev. Lett.* 37, 437 (1976).
108. V. J. Emery, *Phys. Rev. Lett.* 37, 107 (1976).
109. H. Morawitz, *Phys. Rev. Lett.* 34, 1096 (1975).
110. M. J. Cohen, L. B. C. Coleman, A. F. Garito, and A. J. Heeger, *Phys. Rev. B* 10, 1298 (1974).
111. A. I. Buzdin and L. N. Bulaevskii, *Pis'ma Zh. Éksp. Teor. Fiz.* 26, 388 (1977).
112. J. E. Eldridge and F. E. Bates, *Proc. Intern. Conf. on Quasi-One-Dimensional Conductors*, Dubrovnik, Jugoslavia, 1978.
113. M. J. Cohen, Q. W. J. Gunning, and A. J. Heeger, *Proc. Intern. Conf. on Quasi-One-Dimensional Conductors*, Dubrovnik, Jugoslavia, 1978.
114. N. P. Ong and P. Monceau, *Phys. Rev. B* 16, 3443 (1977).
115. S. Uchida, K. Tanabe, and S. Tanaka, *Solid State Commun.* 27, 637 (1978).
116. A. I. Larkin and P. A. Lee, *Phys. Rev. B* 17, 1596 (1978).
117. K. Maki, *Phys. Rev. B* 18, 1641 (1978).
118. A. N. Bloch, D. O. Cowan, K. Bechgaard, R. E. Pyle, R. H. Banks, and T. O. Poehler, *Phys. Rev. Lett.* 34, 1561 (1975).
119. I. F. Shchegolev, R. B. Lyubovskii, *Proc. Intern. Conf. on Quasi-One-Dimensional Conductors*, Dubrovnik, Jugoslavia, 1978; S. P. Zolotukhin, V. F. Kalinskii, A. I. Kotov, R. B. Lyubovskii, M. L. Khidekel', R. P. Shibaeva, I. F. Shchegolev, and É. B. Yagubskii, *Pis'ma Zh. Éksp. Teor. Fiz.* 25, 480 (1977); V. N. Laukhin, A. I. Kotov, M. L. Khidekel', I. F. Shchegolev, and É. B. Yagubskii, *Pis'ma Zh. Éksp. Teor. Fiz.* 28, 284 (1978); *Solid State Commun.* 26, 725 (1978).
120. G. Soda, D. Jerome, M. Weger, J. Alizon, J. Gallice, H. Robert, J. M. Fabre, and L. Giral, *J. Phys. (Paris)* 38, 931 (1977).
121. P. Haen, F. Lapierre, P. Monceau, M. Nunez Reguerio, and J. Richard, *Solid State Commun.* 26, 725 (1978).
122. M. Bondeulle and P. Michel, *Acta Crystallogr. Sect. A* 28, Suppl., 199 (1972).
123. V. Walatka, M. Labes, and J. Perlstein, *Phys. Lett.* 31, 1139 (1973).
124. D. E. Parry and J. H. Thomas, *J. Phys. C* 8, L45 (1975).

125. W. D. Gill, R. L. Greene, G. B. Street, and W. A. Little, *Phys. Rev. Lett.* **35**, 1732 (1975).
126. A. A. Bright, M. J. Cohen, A. F. Garito, A. J. Heeger, C. M. Mikulski, P. J. Russo, and A. G. MacDiarmid, *Phys. Rev. Lett.* **34**, 206 (1975).
127. R. L. Greene, P. M. Grant, and G. B. Street, *Phys. Rev. Lett.* **34**, 89 (1975).
128. L. J. Azevedo, W. G. Clark, G. Deutscher, R. L. Greene, G. B. Street, and L. J. Suter, *Solid State Commun.* **19**, 197 (1976).
129. R. L. Civiak, C. Elbaum, W. Junker, G. Gough, H. I. Kao, L. F. Nichols, and M. M. Labes, *Solid State Commun.* **18**, 1205 (1976).
130. W. J. Friesen, A. J. Berlinsky, B. Bergensen, C. Weiler, and T. M. Rice, *J. Phys. C* **8**, 3549 (1975).
131. W. E. Rudge and P. M. Grant, *Phys. Rev. Lett.* **35**, 1799 (1975).
132. A. A. Bright and P. Soven, *Solid State Commun.* **17**, 2003 (1975).
133. P. M. Grant, R. L. Greene, and G. B. Street, *Phys. Rev. Lett.* **35**, 1741 (1975).
134. A. A. Bright, M. J. Cohen, A. F. Garito, A. J. Heeger, C. M. Mikulski, and A. G. MacDiarmid, *Appl. Phys. Lett.* **26**, 612 (1975).
135. Yu. V. Kopaev and R. Kh. Timerov, *Zh. Éksp. Teor. Fiz.* **63**, 290 (1972).
136. L. A. Turkevich and R. A. Klemm, Preprint, 1978; *Phys. Rev. B* **19**.
137. G. Binnig and H. E. Hoenig, *J. Phys. (Paris)* **C6** No. 8 (Proc. LT-15), 441 (1978).
138. K. Levin, D. L. Kills, and S. L. Cunningham, *Phys. Rev. B* **10**, 3832 (1974).
139. Yu. A. Bychkov, L. P. Gor'kov, and I. E. Dzyaloshinskii, *Zh. Éksp. Teor. Fiz.* **50**, 738 (1973).
140. I. E. Dzyaloshinskii and A. I. Larkin, *Zh. Éksp. Teor. Fiz.* **61**, 791 (1971).
141. L. P. Gor'kov and I. E. Dzyaloshinskii, *Zh. Éksp. Teor. Fiz.* **67**, 397 (1974).
142. A. I. Larkin and V. I. Mel'nikov, *Zh. Éksp. Teor. Fiz.* **71**, 2199 (1976).
143. C. Weyl, W. M. Englert, K. Beckgaard, G. Jekanno, and S. Etemad, *Solid State Commun.* **19**, 925 (1976).
144. D. Davis, H. Gutfreund, and W. A. Little, *Phys. Rev. B* **13**, 4766 (1976).
145. V. L. Kresin, *Phys. Lett.* **49A**, 117 (1974).
146. S. S. Parkin, M. Riboult, D. Jerome, and K. Bechgaard, *J. Phys. C Solid St. Phys.* **14**, 5305 (1981).
147. L. N. Bulaevskii and A. I. Buzdin, *Chemica Scripta* **17**, 67 (1981).

## Chapter 8

1. V. L. Ginzburg, *Zh. Éksp. Teor. Fiz.* **47**, 2318 (1964); *Phys. Lett.* **13**, 101 (1964).
2. V. L. Ginzburg, *Usp. Fiz. Nauk* **95**, 91 (1968); *Contemp. Phys.* **9**, 355 (1968).
3. V. L. Ginzburg, *Usp. Fiz. Nauk* **101**, 185 (1970).
4. V. L. Ginzburg, *Annu. Rev. Mater. Sci.* **2**, 663 (1972).
5. L. Cooper, *Phys. Rev. Lett.* **6**, 689 (1961).
6. D. A. Kirzhnits and E. G. Maksimov, *Pis'ma Zh. Éksp. Teor. Fiz.* **2**, 442 (1965); *Fiz. Met. Metalloved.* **22**, 520 (1956).
7. D. A. Kirzhnits, E. G. Maksimov, and D. I. Khomskii, *J. Low Temp. Phys.* **10**, 79 (1978).
8. G. F. Kharkov and Yu. A. Uspenskii, *Zh. Éksp. Teor. Fiz.* **61**, 2123 (1971); **65**, 1460 (1973).
9. Yu. A. Uspenskii and G. F. Zharkov, *Zh. Éksp. Teor. Fiz.* **65**, 2511 (1973).
10. M. L. McMillan, *Phys. Rev.* **167**, 331 (1968).
11. L. D. Landau and E. M. Lifshitz, *Electrodynamics of Continuous Media* (in Russian), Gostekhizdat (1957).
12. V. P. Silin and A. A. Rukhadze, *Electromagnetic Properties of Plasmas and Plasma-like Media* (in Russian), Fizmatgiz (1961).

13. V. M. Agranovich and V. L. Ginzburg, *Spatial Dispersion in Crystal Optics and the Theory of Excitons* (in Russian), Nauka (1968).
14. G. E. H. Reuter and E. H. Sondheimer, *Proc. R. Soc. London Ser. A* 195 (1948).
15. J. C. Inkson, *Surf. Sci.* 88, 69 (1971).
16. S. Rangarajan, *Phys. Rev. B* 10, 872 (1974).
17. D. Allender, J. Bray, and J. Bardeen, *Phys. Rev. B* 7, 1020 (1973); 8, 4433 (1973).
18. J. E. Rowe and H. Ibach, *Phys. Rev. Lett.* 32, 421 (1974).
19. J. E. Rowe, M. Traum, and N. V. Smith, *Phys. Rev. Lett.* 33, 1335 (1974).
20. D. Pines, *Elementary Excitations in Solids*, Benjamin, New York (1963).
21. J. C. Inkson, *J. Phys. C* 5, 2599 (1972); 6, 1181 (1973).
22. S.-L. Drechsler, *Electron-Electron Interaction and Superconductivity in Sandwiches*, Ergänizungen zum Manuskript fur einen Vortrag in Kurtschatow-Institut (24.10-29.10.1977).
23. S. Rangarajan and Jha S. Sundhashu, *Pramana* 6, No. 3, 161 (1976).
24. S. Srinivasan and Jha S. Sundhashu, "Exciton-Enhancement of Superconducting Transition Temperature in a Metal-Semiconductor System with Spherical Interface," Preprint, Tata Institute of Fundamental Research, (TH) 77-10.
25. É. A. Pashitskii and P. M. Tomchuk, *Pis'ma Zh. Éksp. Teor. Fiz.* 9, 367 (1969).
26. É. A. Pashitskii, *Zh. Éksp. Teor. Fiz.* 56, 662 (1971).
27. V. A. Borisyuk, *Fiz. Tverd. Tela* 12, 485 (1970).
28. É. A. Pashitskii and Yu. A. Romanov, *Ukr. Fiz. Zh.* 15, 1594 (1970).
29. A. M. Gabovich and D. P. Moiseev, *Fiz. Tverd. Tela* 17, 269 (1975).
30. A. M. Gabovich, É. A. Pashitskii, and S. K. Uvarova, *Fiz. Nizk. Temp.* 1, 984 (1975).
31. É. A. Pashitskii, *Zh. Éksp. Teor. Fiz.* 55, 2387 (1968).
32. É. A. Pashitskii and V. M. Chernousenko, *Zh. Éksp. Teor. Fiz.* 60, 1483 (1971).
33. É. A. Pashitskii, V. L. Makarov, and S. D. Tereshchenko, *Fiz. Tverd. Tela* 16, 427 (1974).
34. H. Fröhlich, *Phys. Lett.* 26A, 169 (1968); *J. Phys.* C1, 544 (1968).
35. B. T. Geilikman, *Fiz. Tverd. Tela* 12, 1881 (1970).
36. B. T. Geilikman, *Usp. Fiz. Nauk* 108, 65 (1973).
37. V. Heine, *Phys. Rev.* 138A, 1689 (1965).
38. I. E. Tamm, *Principles of the Theory of Electricity* (in Russian), Nauka (1970).
39. J. Walter and M. L. Cohen, *Phys. Rev. B* 2, 1821 (1970).
40. G. Srinivasan, *Phys. Rev.* 178, 1244 (1969).
41. H. Ehrenreich and H. R. Phillipp, *Phys. Rev.* 129, 1550 (1963).
42. J. C. Inkson, *J. Phys. C* 8, 164 (1975).
43. J. Vannimenus and H. F. Budd, *Solid State Commun.* 15, 1739 (1974).
44. D. C. Langreth, *Phys. Rev. B* 11, 2155 (1975).
45. J. C. Inkson and P. W. Anderson, *Phys. Rev. B* 8, 4429 (1973).
46. M. Weger and I. B. O. Goldberg, *Solid State Phys.*, Vol. 28, edited by H. Ehrenreich, F. Seitz, and D. Turnbull, Academic Press, New York (1973).
47. M. Schlüter, J. R. Chelikowsky, S. G. Lonie, and M. L. Cohen, *Phys. Rev. B* 10, 4200 (1975).
48. S. Ciraci, I. P. Batra, and W. A. Tiller, *Phys. Rev.* **B12**, No. 12, 5811 (1975).
49. G. Chiarotti, S. Nannarone, R. Pastore, and P. Chiaradia, *Phys. Rev. B* 4, 3398 (1971).
50. J. Halbritter, *Proc. LT-14*, Vol. 2 (1975), p. 227.
51. M. L. Cohen and P. W. Anderson, in: *Superconductivity in d- and f-Band Metals*, edited by D. H. Douglass, Am. Inst. Phys. New York (1972).
52. S. L. Anler, *Phys. Rev.* 126, 413 (1962).
53. N. Wiser, *Phys. Rev.* 129, 62 (1963).
54. J. Ziman, *Principles of the Theory of Solids*, Cambridge University Press, Cambridge (1972).

55. W. Harrison, *Solid State Theory*, McGraw Hill, New York (1970).
56. P. Hohenberg and W. Kohn, *Phys. Rev.* **136**, 864 (1964).
57. W. Kohn and L. J. Sham, *Phys. Rev.* **140**, 1133 (1965).
58. N. D. Lang and W. Kohn, *Phys. Rev. B* **1**, 4555 (1970); **B3**, 1215 (1971).
59. G. D. Maham, *Phys. Rev. B* **12**, 5585 (1975).
60. N. D. Lang, *Solid State Commun.* 7, 1057 (1969); *Phys. Rev.* **B4**, 4234 (1971).
61. N. D. Lang, *Density Functional Formalism and the Structure of Metal Surfaces in Solid State Physics*, Academic Press, New York (1973).
62. R. Gomer and C. S. Smith (editors), *Structure and Properties of Solid Surfaces*, University of Chicago Press, Chicago (1953).
63. J. R. Smith, *Phys. Rev.* **181**, 522 (1969).
64. A. Eguiluz, S. C. Ying, and J. J. Quinn, *Phys. Rev. B* **11**, 2118 (1975).
65. R. Evans, *J. Phys. C* 7, 2808 (1974).
66. A. J. Bennet and C. B. Duke, *Phys. Rev.* **188**, 1050 (1969).
67. D. S. Bourdeau, *Phys. Rev. B* **1**, 4551 (1970).
68. I. M. Lifshits and A. M. Kosevich, *Izv. Akad. Nauk SSSR Ser. Fiz.* **19**, 395 (1955).
69. V. B. Sandomirskii, *Zh. Éksp. Teor. Fiz.* **52**, 158 (1967).
70. S. S. Nedorezov, *Zh. Éksp. Teor. Fiz.* **56**, 299 (1969).
71. A. P. Van Gelder, *Physica* **59**, 510 (1972).
72. W. L. McCubbin and W. H. Young, *Phys. Rev. B* **11**, 2421 (1975).
73. P. J. Feibelman, *Phys. Rev. B* **3**, 220 (1971); **3**, 2974 (1971); **9**, 5077 (1974).
74. P. J. Feibelman, C. B. Duke, and A. Baghci, *Phys. Rev.* **5**, 2436 (1972).
75. J. Harris and A. Griffin, *Phys. Lett.* **34A**, 51 (1971).
76. A. J. Bennett, *Phys. Rev. B* **1**, 203 (1970).
77. D. E. Beck and V. Celli, *Phys. Rev. Lett.* **28**, 1124 (1972).
78. A. J. Bennett and C. B. Duke, in: *Structure and Chemistry of Solid Surfaces*, edited by G. A. Somorjai, Wiley, New York (1969).
79. A. D. Boardman and E. D. Issac, *Phys. Lett. A* **53**, 339 (1975).
80. A. D. Boardman, B. V. Paranjape, and R. Teshima, *Surf. Sci.* **49**, 275 (1975).
81. A. Eguiluz and J. J. Quinn, *Phys. Lett. A* **53**, 151 (1975).
82. Yu. Kagan and L. B. Dubovskii, *Zh. Éksp. Teor. Fiz.* **72**, No. 2, 646 (1977).
83. V. M. Golyakov, M. N. Mikheeva, M. B. Tsetlin, *Zh. Éksp. Teor. Fiz.* **68**, 736 (1975).
84. M. N. Mikheeva, M. B. Tsetlin, A. A. Teplov, V. M. Golyakov, A. P. Demidov, *Letters to Zh. Éksp. Teor. Fiz.* **15**, 303 (1972).
85. V. M. Agranovich, A. G. Mal'shukov, and M. A. Mekhtiev, *Fiz. Tverd. Tela* **14**, 459 (1972); *Zh. Éksp. Teor. Fiz.* **63**, 2274 (1972); *Izv. Akad. Nauk SSSR Ser. Fiz.* **37**, 324 (1973).
86. V. M. Agranovich and V. I. Yudson, *Opt. Commun.* **5**, 422 (1972).
87. A. G. Mal'shukov, *Fiz. Tverd. Tela* **16**, 2274 (1974).
88. P. H. Ritchie, *Phys. Rev.* **106**, 874 (1957).
89. H. Morawitz and M. R. Philpott, *Phys. Rev. B* **10**, 4863 (1974).
90. V. M. Agranovich, *Fiz. Tverd. Tela* **14**, 3684 (1972).
91. J. Heinriks, *Phys. Rev.* 8, 1708 (1973).
92. V. M. Agranovich, *Theory of Exictons* (in Russian), Nauka (1968).
93. A. Many, Y. Goldstein, and N. B. Grover, *Semiconductor Surfaces*, North-Holland Publishing Co., Amsterdam (1972).
94. F. Evangelisty, A. Frova, and J. V. Fishbach, *Phys. Rev. Lett.* **29**, 1001 (1972); *Surf. Sci.* 841 (1973).
95. A. Frova, F. Evangelisty, and M. Zanini, *Phys. Status Solidi* **24**, 315 (1974).
96. J. R. Schrieffer, in: *Semiconductor Surface Physics*, edited by R. H. Kingston, University of Pennsylvania Press, Philadelphia (1957), p. 55.

97. J. A. Pals, *Phys. Lett. A* **39**, 101 (1972).
98. J. A. Appelbaum and D. R. Hamann, *Phys. Rev. B* **10**, 4973 (1974).
99. E. Ni Foo and H. S. Wong, *Phys. Rev. B* **10**, 4819 (1974).
100. H. Sixl, *Phys. Lett. A* **53**, 333 (1975).
101. W. Mönch, *Surf. Sci.* **63**, 79 (1977).
102. G. Margaritondo, J. E. Rowe, S. B. Christman, *Phys. Rev. B* **14**, 5396 (1976).
103. S. G. Louie, M. L. Cohen, *Phys. Rev. B* **13**, 2461 (1976).
104. J. R. Chelikowski, *Phys. Rev. B* **16**, 3618 (1977).
105. L. W. Gurney, *Phys. Rev.* **47**, 479 (1935).
106. A. J. Bennett and L. M. Falicov, *Phys. Rev.* **151**, 512 (1966).
107. T. B. Grimley, *Proc. Phys. Soc. London* **90**, 751 (1967); **92**, 776 (1967).
108. A. Madhuker and B. Bell, *Phys. Rev. Lett.* **34**, 1631 (1975).
109. O. V. Konstantinov and A. Ya. Shik, *Zh. Éksp. Teor. Fiz.* **58**, 1661 (1970).
110. R. Fuchs and K. L. Kliewer, *Phys. Rev. A* **140**, 2076 (1965).
111. K. L. Kliewer and R. Fuchs, *Phys. Rev.* **144**, 495 (1966); **150**, 573 (1966).
112. R. Fuchs and K. L. Kliewer, *Phys. Rev. B* **3**, 2270 (1971).
113. R. Ruppin and R. Engelman, *Rep. Prog. Phys.* **33**, 149 (1970).
114. A. A. Maradudin and D. L. Mills, *Phys. Rev. B* **7**, 2787 (1973).
115. B. Fisher, N. Marshall, and H. J. Quesser, *Surf. Sci.* **34**, 50 (1973).
116. B. Fisher and H. J. Quesser, *Solid State Commun.* **16**, 1125 (1975).
117. F. Abeles and T. Lopez-Rios, *Solid State Commun.* **16**, 843 (1975).
118. H. Morawitz and M. R. Philpott, *Phys. Rev. B* **10**, 4863 (1974).
119. R. Engelman and R. Ruppin, *J. Phys. C* **1**, 614, 630, 1515 (1968).
120. R. Brako, J. Hrnčević, and M. Šunjič, *Z. Phys. B* **21**, 193 (1975).
121. M. Šunjič and A. A. Lucas, *Phys. Rev. B.* **3**, 719 (1971).
122. A. A. Lucas, *Phys. Rev. B* **4**, 2939 (1971).
123. M. Natta, *J. Phys. (Paris)*, *C1* **31**, 53 (1970).
124. R. Ruppin and R. Engelman, *Rep. Prog. Phys.* **33**, 149 (1970).
125. R. Ruppin, *Surf. Sci.* **34**, 20 (1973).
126. F. Stern, *Phys. Rev. Lett.* **18**, 546 (1967).
127. A. Fetter, *Ann. Phys. (N.Y.)* **81**, 367 (1973).
128. K. W. Chiu and J. J. Quinn, *Phys. Rev. B* **9**, 4724 (1974).
129. A. Equiluz, T. K. Lee, J. J. Quinn, and K. W. Chiu, *Phys. Rev.* **B11**, 4989 (1975).
130. A. V. Chaplik, *Zh. Éksp. Teor. Fiz.* **62**, 746 (1972).
131. D. Pugh, *Phys. Rev. Lett.* **12**, 531 (1964).
132. R. O. Jones, *Phys. Rev. Lett.* **20**, 992 (1968).
133. K. Hirabajashi, *J. Phys. Soc. Jpn.* **27**, 1475 (1969).
134. D. Levin and S. Freeman, *Phys. Rev. B* **2**, 3255 (1970).
135. C. M. Chaves, N. Majlis, and M. Cardona, *Solid State Commun.* **4**, 271, 631 (1966).
136. R. O. Jones, *J. Phys. C* **5**, 1615 (1972).
137. F. Yndurain and M. Elices, *Surf. Sci.* **29**, 540 (1972).
138. J. A. Appelbaum and D. R. Hamann, *Phys. Rev. B* **6**, 2116 (1972); *Phys. Rev. Lett.* **31**, 106 (1973); **32**, 225 (1974).
139. G. Iadonisi and B. Preziosi, *Nuovo Cimento B* **27**, 193 (1975).
140. S. G. Louie and M. L. Cohen, *Phys. Rev. Lett.* **35**, 866 (1975).
141. J. A. Appelbaum, G. A. Baraff, and D. R. Hamann, *Phys. Rev. Lett.* **35**, 729 (1975).
142. G. P. Alldredge and L. Kleinman, *Phys. Rev. Lett.* **28**, 1264 (1972).
143. D. S. Boudreau, *Surf. Sci.* **28**, 344 (1971).
144. E. Caruthers, L. Kleinman, and G. P. Alldredge, *Phys. Rev. B* **8**, 4570 (1973); *Phys. Rev. Lett.* **35**, 738 (1975).
145. J. J. Hopfield, *Phys. Rev.* **172**, 1555 (1968).

146. J. J. Hopfield and D. G. Thomas, *Phys. Rev.* **132**, 563 (1963).
147. D. J. Chadi and R. M. White, *Phys. Rev. B* **11**, 5077 (1975).
148. N. V. Zavaritskii, *Dokl. Akad. Nauk SSSR* **82**, 229 (1952); **86**, 501 (1952).
149. W. Buckel and R. Hilsch, *Z. Phys.* **138**, 109 (1954).
150. I. S. Kukhareva, *Zh. Éksp. Teor. Fiz.* **43**, 1173 (1962).
151. M. Strongin, O. K. Kammerer, and A. Paskin, *Phys. Rev. Lett.* **14**, 949 (1965).
152. B. Abeles, R. W. Cohen, and G. W. Cullen, *Phys. Rev. Lett.* **17**, 631 (1966).
153. J. W. Garland, K. H. Bennemann, and F. M. Mueller, *Phys. Rev. Lett.* **21**, 1315 (1968).
154. M. Strongin, R. S. Thompson, O. F. Kammerer, and J. E. Crow, *Phys. Rev. B* **1**, 1078 (1970).
155. J. J. Hauser, *Phys. Rev. B* **3**, 1611 (1971).
156. P. Townsend, S. Gregory, and R. G. Taylor, *Phys. Rev. B* **5**, 54 (1972).
157. W. Felsh and R. E. Glover, *Solid State Commun.* **10**, 1033 (1972).
158. W. Rühl, *Z. Phys.* **186**, 190 (1965).
159. D. C. Naugle, J. W. Baker, and R. E. Allen, *Phys. Rev. B* **7**, 3028 (1973).
160. P. Hilsh and D. G. Naugle, *Z. Phys.* **201**, 1 (1967).
161. D. G. Naugle, *Phys. Lett. A* **25**, 688 (1967).
162. M. Strongin and O. F. Kammerer, *J. Appl. Phys.* **39**, 2509 (1968).
163. R. E. Glover, S. K. Ghosh, and W. E. Daniels, in: *Basic Problems in Thin Film Physics*, Göttingen (1966), p. 536.
164. A. G. Aslamazov and I. A. Larkin, *Fiz. Tverd. Tela* **10**, 1104 (1968); *Phys. Lett. A* **26**, 238 (1968).
165. K. Maki, *Prog. Theor. Phys.* **40**, 193 (1968).
166. R. S. Thompson, *Phys. Rev. B* **1**, 327 (1970).
167. C. G. Granquist and T. Glaeson, *Solid State Commun.* **16**, 123 (1975).
168. M. Strongin, O. F. Kammerer, H. H. Farrell, and D. L. Miller, *Phys. Rev. Lett.* **30**, 129 (1973).
169. V. I. Zaporozhchenko, and V. I. Rakhovskii, *Dokl. Akad. Nauk SSSR* **217**, 544 (1974).
170. *Proc. Intern. Conf. on Organic Superconductors*, ed. Little, Honolulu (1969).
171. *Proc. Low Temp. Physics*, LT-13 (1973).
172. *Proc. Low Temp. Physics*, LT-14 (1975).
173. H. Kallman, G. Vaubel, and H. Baessler, *Phys. Status Solidi B* **44**, 813 (1971).
174. H. Kallman, Z. Naturforsch. *Teil A* **26**, 799 (1971).
175. H. Baessler, in: *Sixth Molecular Cryst. Symposium*, Shloss Elman, Germany (1974).
176. V. M. Agranovich, Yu. V. Konobeev, and M. A. Mekhtiev, *Fiz. Tverd. Tela* **10**, 1754 (1968).
177. J. Singh and H. Baessler, *Phys. Status Solidi B* **62**, 147 (1974).

## Chapter 9

1. L. P. Gor'kov and G. M. Éliashberg, *Zh. Éksp. Teor. Fiz.* **54**, 612 (1968); G. M. Éliashberg, *Zh. Éksp. Teor. Fiz.* **61**, 1254 (1971).
2. J. Bardeen, L. Cooper, and J. Schrieffer, *Phys. Rev.* **108**, 1175 (1957).
3. G. M. Éliashberg, *Zh. Éksp. Teor. Fiz.* **38**, 966 (1960).
4. B. D. Josephson, *Adv. Phys.* **14**, 419 (1965).
5. J. Clarke, *Phys. Rev. Lett.* **28**, 1363 (1972); M. Tinkham, *Phys. Rev.* **B6**, 1747 (1972).
6. V. L. Ginzburg and G. F. Zharkov, *Pis'ma Zh. Éksp. Teor. Fiz.* **20**, 223 (1974).
7. V. M. Buimistrov, *Pis'ma Zh. Éksp. Teor. Fiz.* **8**, 274 (1968); N. Kumar and K. P. Sinha, *Phys. Rev.* **174**, 482 (1968); V. D. Blazhin and A. S. Selivanenko, *Fiz. Tekh. Poluprovodn.* **4**, 239 (1970).

8. L. N. Bulaevskii, Yu. V. Kopaev, and Yu. A. Kukharenko, *Fiz. Tverd. Tela* **14**, 114 (1972).
9. J. A. Wilson and A. D. Joffe, *Adv. Phys.* **18**, 193 (1969).
10. N. Kumar, *Phys. Rev. B* **9**, 4993 (1974).
11. V. F. Elesin and Yu. V. Kopaev, *Fiz. Tverd. Tela* **14**, 669 (1972).
12. V. M. Galitskii, S. P. Goreslavskii, and V. F. Elesin, *Zh. Éksp. Teor. Fiz.* **57**, 207 (1969).
13. V. F. Elesin and Yu. V. Kopaev, *Fiz. Tverd. Tela* **16**, 840 (1974).
14. L. V. Keldysh, *Usp. Fiz. Nauk* **100**, 514 (1970).
15. V. V. Kapaev and Yu. V. Kopaev, *Zh. Éksp. Teor. Fiz.* **69**, 2171 (1975).
16. G. M. Éliashberg, *Pis'ma Zh. Éksp. Teor. Fiz.* **11**, 186 (1970); B. I. Ivlev and G. M. Éliashberg, *Pis'ma Zh. Éksp. Teor. Fiz.* **13**, 464 (1971).
17. B. I. Ivlev, *Zh. Éksp. Teor. Fiz.* **59**, 1036 (1970).
18. V. M. Dmitriev, E. V. Khristenko, A. F. Trubitsin, and F. F. Mende, *Usp. Fiz. Nauk* **15**, 1611 (1970); M. T. Levinsen, *Phys. Rev. A* **9**, 145 (1974); F. E. Gzegers-Hensen and M. T. Levinsen, *Phys. Rev. Lett.* **27**, 947 (1971); M. T. Jahn and Y. H. Kao, *J. Low. Temp. Phys.* **13**, 175 (1973); Yu. I. Latyshev and F. Ya. Nad', Abstracts of Papers Presented at the Eighteenth All-Union Conf. on Low Temperature Physics, Kiev (1974), p. 155.
19. T. I. Tredwell and E. H. Jacobsen, *Phys. Rev. Lett.* **35**, 244 (1975).
20. K. E. Gray, *Solid State Commun.* **26**, 633 (1978).
21. A. G. Aronov and V. L. Gurevich, *Pis'ma Zh. Éksp. Teor. Fiz.* **15**, 564 (1972); *Fiz. Tverd. Tela* **14**, 2675 (1972).
22. A. G. Aronov and V. L. Gurevich, *Zh. Éksp. Teor. Fiz.* **63**, 1809 (1972).
23. L. R. Testardi, *Phys. Rev. B* **4**, 2189 (1971).
24. C. Owen and P. Scalapino, *Phys. Rev. Lett.* **28**, 1559 (1972).
25. W. Parker and W. Williams, *Phys. Rev. Lett.* **20**, 924 (1972).
26. R. A. Vardanyan and B. I. Ivlev, *Zh. Éksp. Teor. Fiz.* **65**, 231 (1973); V. F. Elesin, *Zh. Éksp. Teor. Fiz.* **66**, 1755 (1974).
27. V. M. Galitskii, V. F. Elesin, and Yu. V. Kopaev, *Pis'ma Zh. Éksp. Teor. Fiz.* **18**, 50 (1973).
28. A. G. Aronov, *Pis'ma Zh. Éksp. Teor. Fiz.* **18**, 387 (1973).
29. J. J. Chang and P. J. Scalapino, *Phys. Rev. B* **10**, 404 (1974); V. G. Baru and A. A. Sukhanov, *Pis'ma Zh. Éksp. Teor. Fiz.* **21**, 209 (1975).
30. V. F. Elesin, *Zh. Éksp. Teor. Fiz.* **71**, 1490 (1976).
31. J. Fuchs, P. W. Epperlein, M. Welte, and J. Eisenmeng, *Phys. Rev. Lett.* **38**, 16, 919 (1977).
32. B. Lambert, D. Salin, and J. Joffrin, *J. Phys. (Paris)* **38**, L-343 (1977).
33. H. W. Willemsen and K. E. Gray, *Phys. Rev. Lett.* **41**, 12, 812 (1978).
34. V. F. Elesin and E. B. Levchenko, *Fiz. Tverd. Tela* **20**, 2459 (1978).
35. V. F. Elesin, V. E. Kondrashov, B. N. Shamraev, and A. P. Sukhikh, *Fiz. Tverd. Tela* **19** (1977).
36. V. F. Elesin, *Zh. Éksp. Teor. Fiz.* **73**, 355 (1977).
37. R. C. Dynes, V. Narayanamurti, and J. P. Carno, *Phys. Rev. Lett.* **39**, 229 (1977).
38. A. I. Golovashkin, K. V. Mittsin, and G. P. Motulevich, *Zh. Éksp. Teor. Fiz.* **68**, 1408 (1975).
39. A. I. Golovashkin, O. M. Ivanenko, K. V. Mittsin, G. P. Motulevich, and A. L. Shubin, *Zh. Éksp. Teor. Fiz.* **75**, 1515 (1978).
40. P. Hu, R. C. Dynes, and V. Narayanamurti, *Phys. Rev. B* **10**, 2786 (1974).
41. I. Iguchi, *Phys. Lett. A* **64**, 415 (1978).
42. V. F. Elesin, *Zh. Éksp. Teor. Fiz.* **77**, No. 6 (1979).

43. V. F. Elesin, *Fiz. Tverd. Tela* 19, 2977 (1977).
44. V. F. Elesin, V. V. Kapaev, and Yu. V. Kopaev, *Zh. Éksp. Teor. Fiz.* 71, 714 (1976).
45. V. F. Elesin, V. E. Kondrashov, and Yu. M. Sklyarov, *Fiz. Tverd. Tela* 18, 3393 (1976).
46. L. N. Cooper, *Phys. Rev.* 104, 1189 (1957).
47. D. A. Kirzhnits and Yu. V. Kopaev, *Pis'ma Zh. Éksp. Teor. Fiz.* 17, 379 (1973).
48. B. M. Vul (editor), *Excitons in Semiconductors* (in Russian), Nauka, (1971).
49. A. N. Kozlov and L. A. Maksimov, *Zh. Éksp. Teor. Fiz.* 48, 1184 (1965).
50. V. F. Elesin, Yu. V. Kopaev, and R. Kh. Timerov, *Zh. Éksp. Teor. Fiz.* 65, 2343 (1973).
51. A. A. Abrikosov and L. P. Gor'kov, *Zh. Éksp. Teor. Fiz.* 35, 1158 (1958).
52. I. Clarke, *Phys. Rev. Lett.* 28, 1363 (1972); M. Tinkham, *Phys. Rev. B* 6, 1747 (1972).
53. A. S. Aleksandrov and V. F. Elesin, *Zh. Éksp. Teor. Fiz.* 58, 1062 (1970).
54. A. G. Aronov and V. L. Gurevich, *Zh. Éksp. Teor. Fiz.* 65, 1111 (1973).
55. D. C. Mattis and J. Bardeen, *Phys. Rev.* 111, 412 (1958).
56. L. V. Keldysh, *Zh. Éksp. Teor. Fiz.* 47, 1515 (1964).
57. V. L. Ginzburg, *Usp. Fiz. Nauk* 69, 537 (1959).
58. B. G. Idlis and Yu. V. Kopaev, *Zh. Éksp. Teor. Fiz.* 71, 2305 (1976).
59. V. L. Ginzburg and L. D. Landau, *Zh. Éksp. Teor. Fiz.* 20, 1064 (1950).

GPSR Compliance

*The European Union's (EU) General Product Safety Regulation (GPSR) is a set of rules that requires consumer products to be safe and our obligations to ensure this.*

*If you have any concerns about our products, you can contact us on ProductSafety@springernature.com*

In case Publisher is established outside the EU, the EU authorized representative is:

Springer Nature Customer Service Center GmbH
Europaplatz 3
69115 Heidelberg, Germany

**Batch number: 10682411**

Printed by Printforce, the Netherlands